LONDON MATHEMATICAL SOCIETY LECTURE NOTE SERIES

Managing Editor: Professor M. Reid, Mathematics Institute,
University of Warwick, Coventry CV4 7AL, United Kingdom

The titles below are available from booksellers, or from Caml
http://www.cambridge.org/mathematics

325 Lectures on the Ricci flow, P. TOPPING
326 Modular representations of finite groups of Lie type,
327 Surveys in combinatorics 2005, B.S. WEBB (ed)
328 Fundamentals of hyperbolic manifolds, R. CANAR
329 Spaces of Kleinian groups, Y. MINSKY, M. SAKUM
330 Noncommutative localization in algebra and topology, A.
331 Foundations of computational mathematics, Santander 2005, L.M PARDO, A. PINKUS,
M.J. TODD (eds)
332 Handbook of tilting theory, L. ANGELERI HÜGEL, D. HAPPEL & H. KRAUSE (eds)
333 Synthetic differential geometry (2nd Edition), A. KOCK
334 The Navier-Stokes equations, N. RILEY & P. DRAZIN
335 Lectures on the combinatorics of free probability, A. NICA & R. SPEICHER
336 Integral closure of ideals, rings, and modules, I. SWANSON & C. HUNEKE
337 Methods in Banach space theory, J.M.F. CASTILLO & W.B. JOHNSON (eds)
338 Surveys in geometry and number theory, N. YOUNG (ed)
339 Groups St Andrews 2005 &I, C.M. CAMPBELL, M.R. QUICK, E.F. ROBERTSON & G.C. SMITH (eds)
340 Groups St Andrews 2005 &II, C.M. CAMPBELL, M.R. QUICK, E.F. ROBERTSON & G.C. SMITH (eds)
341 Ranks of elliptic curves and random matrix theory, J.B. CONREY, D.W. FARMER, F. MEZZADRI &
N.C. SNAITH (eds)
342 Elliptic cohomology, H.R. MILLER & D.C. RAVENEL (eds)
343 Algebraic cycles and motives I, J. NAGEL & C. PETERS (eds)
344 Algebraic cycles and motives II, J. NAGEL & C. PETERS (eds)
345 Algebraic and analytic geometry, A. NEEMAN
346 Surveys in combinatorics 2007, A. HILTON & J. TALBOT (eds)
347 Surveys in contemporary mathematics, N. YOUNG & Y. CHOI (eds)
348 Transcendental dynamics and complex analysis, P.J. RIPPON & G.M. STALLARD (eds)
349 Model theory with applications to algebra and analysis I, Z. CHATZIDAKIS, D. MACPHERSON,
A. PILLAY & A. WILKIE (eds)
350 Model theory with applications to algebra and analysis II, Z. CHATZIDAKIS, D. MACPHERSON,
A. PILLAY & A. WILKIE (eds)
351 Finite von Neumann algebras and masas, A.M. SINCLAIR & R.R. SMITH
352 Number theory and polynomials, J. MCKEE & C. SMYTH (eds)
353 Trends in stochastic analysis, J. BLATH, P. MÖRTERS & M. SCHEUTZOW (eds)
354 Groups and analysis, K. TENT (ed)
355 Non-equilibrium statistical mechanics and turbulence, J. CARDY, G. FALKOVICH & K. GAWEDZKI
356 Elliptic curves and big Galois representations, D. DELBOURGO
357 Algebraic theory of differential equations, M.A.H. MACCALLUM & A.V. MIKHAILOV (eds)
358 Geometric and cohomological methods in group theory, M.R. BRIDSON, P.H. KROPHOLLER &
I.J. LEARY (eds)
359 Moduli spaces and vector bundles, L. BRAMBILA-PAZ, S.B. BRADLOW, O. GARCÍA-PRADA &
S. RAMANAN (eds)
360 Zariski geometries, B. ZILBER
361 Words: Notes on verbal width in groups, D. SEGAL
362 Differential tensor algebras and their module categories, R. BAUTISTA, L. SALMERÓN & R. ZUAZUA
363 Foundations of computational mathematics, Hong Kong 2008, F. CUCKER, A. PINKUS & M.J. TODD (eds)
364 Partial differential equations and fluid mechanics, J.C. ROBINSON & J.L. RODRIGO (eds)
365 Surveys in combinatorics 2009, S. HUCZYNSKA, J.D. MITCHELL & C.M. RONEY-DOUGAL (eds)
366 Highly oscillatory problems, B. ENGQUIST, A. FOKAS, E. HAIRER & A. ISERLES (eds)
367 Random matrices: High dimensional phenomena, G. BLOWER
368 Geometry of Riemann surfaces, F.P. GARDINER, G. GONZÁLEZ-DIEZ & C. KOUROUNIOTIS (eds)
369 Epidemics and rumours in complex networks, M. DRAIEF & L. MASSOULIÉ
370 Theory of p-adic distributions, S. ALBEVERIO, A.YU. KHRENNIKOV & V.M. SHELKOVICH
371 Conformal fractals, F. PRZYTYCKI & M. URBAŃSKI
372 Moonshine: The first quarter century and beyond, J. LEPOWSKY, J. MCKAY & M.P. TUITE (eds)
373 Smoothness, regularity and complete intersection, J. MAJADAS & A. G. RODICIO
374 Geometric analysis of hyperbolic differential equations: An introduction, S. ALINHAC
375 Triangulated categories, T. HOLM, P. JØRGENSEN & R. ROUQUIER (eds)
376 Permutation patterns, S. LINTON, N. RUŠKUC & V. VATTER (eds)
377 An introduction to Galois cohomology and its applications, G. BERHUY
378 Probability and mathematical genetics, N. H. BINGHAM & C. M. GOLDIE (eds)
379 Finite and algorithmic model theory, J. ESPARZA, C. MICHAUX & C. STEINHORN (eds)
380 Real and complex singularities, M. MANOEL, M.C. ROMERO FUSTER & C.T.C WALL (eds)
381 Symmetries and integrability of difference equations, D. LEVI, P. OLVER, Z. THOMOVA &
P. WINTERNITZ (eds)

382 Forcing with random variables and proof complexity, J. KRAJÍČEK
383 Motivic integration and its interactions with model theory and non-Archimedean geometry I, R. CLUCKERS, J. NICAISE & J. SEBAG (eds)
384 Motivic integration and its interactions with model theory and non-Archimedean geometry II, R. CLUCKERS, J. NICAISE & J. SEBAG (eds)
385 Entropy of hidden Markov processes and connections to dynamical systems, B. MARCUS, K. PETERSEN & T. WEISSMAN (eds)
386 Independence-friendly logic, A.L. MANN, G. SANDU & M. SEVENSTER
387 Groups St Andrews 2009 &in Bath I, C.M. CAMPBELL *et al* (eds)
388 Groups St Andrews 2009 &in Bath II, C.M. CAMPBELL *et al* (eds)
389 Random fields on the sphere, D. MARINUCCI & G. PECCATI
390 Localization in periodic potentials, D.E. PELINOVSKY
391 Fusion systems in algebra and topology, M. ASCHBACHER, R. KESSAR & B. OLIVER
392 Surveys in combinatorics 2011, R. CHAPMAN (ed)
393 Non-abelian fundamental groups and Iwasawa theory, J. COATES *et al* (eds)
394 Variational problems in differential geometry, R. BIELAWSKI, K. HOUSTON & M. SPEIGHT (eds)
395 How groups grow, A. MANN
396 Arithmetic differential operators over the *p*-adic integers, C.C. RALPH & S.R. SIMANCA
397 Hyperbolic geometry and applications in quantum chaos and cosmology, J. BOLTE & F. STEINER (eds)
398 Mathematical models in contact mechanics, M. SOFONEA & A. MATEI
399 Circuit double cover of graphs, C.-Q. ZHANG
400 Dense sphere packings: a blueprint for formal proofs, T. HALES
401 A double Hall algebra approach to affine quantum Schur-Weyl theory, B. DENG, J. DU & Q. FU
402 Mathematical aspects of fluid mechanics, J.C. ROBINSON, J.L. RODRIGO & W. SADOWSKI (eds)
403 Foundations of computational mathematics, Budapest 2011, F. CUCKER, T. KRICK, A. PINKUS & A. SZANTO (eds)
404 Operator methods for boundary value problems, S. HASSI, H.S.V. DE SNOO & F.H. SZAFRANIEC (eds)
405 Torsors, étale homotopy and applications to rational points, A.N. SKOROBOGATOV (ed)
406 Appalachian set theory, J. CUMMINGS & E. SCHIMMERLING (eds)
407 The maximal subgroups of the low-dimensional finite classical groups, J.N. BRAY, D.F. HOLT & C.M. RONEY-DOUGAL
408 Complexity science: the Warwick master's course, R. BALL, V. KOLOKOLTSOV & R.S. MACKAY (eds)
409 Surveys in combinatorics 2013, S.R. BLACKBURN, S. GERKE & M. WILDON (eds)
410 Representation theory and harmonic analysis of wreath products of finite groups, T. CECCHERINI-SILBERSTEIN, F. SCARABOTTI & F. TOLLI
411 Moduli spaces, L. BRAMBILA-PAZ, O. GARCÍA-PRADA, P. NEWSTEAD & R.P. THOMAS (eds)
412 Automorphisms and equivalence relations in topological dynamics, D.B. ELLIS & R. ELLIS
413 Optimal transportation, Y. OLLIVIER, H. PAJOT & C. VILLANI (eds)
414 Automorphic forms and Galois representations I, F. DIAMOND, P.L. KASSAEI & M. KIM (eds)
415 Automorphic forms and Galois representations II, F. DIAMOND, P.L. KASSAEI & M. KIM (eds)
416 Reversibility in dynamics and group theory, A.G. O'FARRELL & I. SHORT
417 Recent advances in algebraic geometry, C.D. HACON, M. MUSTAŢĂ & M. POPA (eds)
418 The Bloch-Kato conjecture for the Riemann zeta function, J. COATES, A. RAGHURAM, A. SAIKIA & R. SUJATHA (eds)
419 The Cauchy problem for non-Lipschitz semi-linear parabolic partial differential equations, J.C. MEYER & D.J. NEEDHAM
420 Arithmetic and geometry, L. DIEULEFAIT *et al* (eds)
421 O-minimality and Diophantine geometry, G.O. JONES & A.J. WILKIE (eds)
422 Groups St Andrews 2013, C.M. CAMPBELL *et al* (eds)
423 Inequalities for graph eigenvalues, Z. STANIĆ
424 Surveys in combinatorics 2015, A. CZUMAJ *et al* (eds)
425 Geometry, topology and dynamics in negative curvature, C.S. ARAVINDA, F.T. FARRELL & J.-F. LAFONT (eds)
426 Lectures on the theory of water waves, T. BRIDGES, M. GROVES & D. NICHOLLS (eds)
427 Recent advances in Hodge theory, M. KERR & G. PEARLSTEIN (eds)
428 Geometry in a Fréchet context, C. T. J. DODSON, G. GALANIS & E. VASSILIOU
429 Sheaves and functions modulo *p*, L. TAELMAN
430 Recent progress in the theory of the Euler and Navier-Stokes equations, J.C. ROBINSON, J.L. RODRIGO, W. SADOWSKI & A. VIDAL-LÓPEZ (eds)
431 Harmonic and subharmonic function theory on the real hyperbolic ball, M. STOLL
432 Topics in graph automorphisms and reconstruction (2nd Edition), J. LAURI & R. SCAPELLATO
433 Regular and irregular holonomic D-modules, M. KASHIWARA & P. SCHAPIRA
434 Analytic semigroups and semilinear initial boundary value problems (2nd Edition), K. TAIRA
435 Graded rings and graded Grothendieck groups, R. HAZRAT
436 Groups, graphs and random walks, T. CECCHERINI-SILBERSTEIN, M. SALVATORI & E. SAVA-HUSS (eds)
437 Dynamics and analytic number theory, D. BADZIAHIN, A. GORODNIK & N. PEYERIMHOFF (eds)
438 Random walks and heat kernels on graphs, M.T. BARLOW
439 Evolution equations, K. AMMARI & S. GERBI (eds)
440 Surveys in combinatorics 2017, A. CLAESSON *et al* (eds)
441 Polynomials and the mod 2 Steenrod algebra I, G. WALKER & R.M.W. WOOD
442 Polynomials and the mod 2 Steenrod algebra II, G. WALKER & R.M.W. WOOD
443 Asymptotic analysis in general relativity, T. DAUDÉ, D. HÄFNER & J.-P. NICOLAS (eds)
444 Geometric and cohomological group theory, P.H. KROPHOLLER, I.J. LEARY, C. MARTÍNEZ-PÉREZ & B.E.A. NUCINKIS (eds)

London Mathematical Society Lecture Note Series: 442

Polynomials and the mod 2 Steenrod Algebra

Volume 2: Representations of $\mathsf{GL}(n, \mathbb{F}_2)$

GRANT WALKER
University of Manchester

REGINALD M. W. WOOD
University of Manchester

CAMBRIDGE
UNIVERSITY PRESS

University Printing House, Cambridge CB2 8BS, United Kingdom

One Liberty Plaza, 20th Floor, New York, NY 10006, USA

477 Williamstown Road, Port Melbourne, VIC 3207, Australia

4843/24, 2nd Floor, Ansari Road, Daryaganj, Delhi – 110002, India

79 Anson Road, #06-04/06, Singapore 079906

Cambridge University Press is part of the University of Cambridge.

It furthers the University's mission by disseminating knowledge in the pursuit of education, learning, and research at the highest international levels of excellence.

www.cambridge.org
Information on this title: www.cambridge.org/9781108414456
DOI: 10.1017/9781108304092

First published 2018

Printed in the United Kingdom by Clays, St Ives plc

A catalogue record for this publication is available from the British Library.

ISBN – 2 Volume Set 978-1-108-41406-7 Paperback
ISBN – Volume 1 978-1-108-41448-7 Paperback
ISBN – Volume 2 978-1-108-41445-6 Paperback

Contents of Volume 2

Contents of Volume 1

Preface

This book is about the mod 2 Steenrod algebra A_2 and its action on the polynomial algebra $\mathsf{P}(n) = \mathbb{F}_2[x_1,\ldots,x_n]$ in n variables, where $\mathbb{F}_2$ is the field of two elements. Polynomials are graded by degree, so that $\mathsf{P}^d(n)$ is the set of homogeneous polynomials of degree d. Although our subject has its origin in the work of Norman E. Steenrod in algebraic topology, we have taken an algebraic point of view. We have tried as far as possible to provide a self-contained treatment based on linear algebra and representations of finite matrix groups. In other words, the reader does not require knowledge of algebraic topology, although the subject has been developed by topologists and is motivated by problems in topology.

There are many bonuses for working with the prime $p = 2$. There are no coefficients to worry about, so that every polynomial can be written simply as a sum of monomials. We use a matrix-like array of 0s and 1s, which we call a 'block', to represent a monomial in $\mathsf{P}(n)$, where the rows of the block are formed by the reverse binary expansions of its exponents. Thus a polynomial is a set of blocks, and the sum of two polynomials is the symmetric difference of the corresponding sets. Using block notation, the action of A_2 on $\mathsf{P}(n)$ can be encoded in computer algebra programs using standard routines on sets, lists and arrays. In addition, much of the literature on the Steenrod algebra and its applications in topology concentrates on the case $p=2$. Often a result for $p=2$ has later been extended to all primes, but there are some results where no odd prime analogue is known.

We begin in Chapter 1 with the algebra map $\mathsf{Sq} : \mathsf{P}(n) \to \mathsf{P}(n)$ defined on the generators by $\mathsf{Sq}(x_i) = x_i + x_i^2$. The map Sq is the total Steenrod squaring operation, and the Steenrod squares $Sq^k : \mathsf{P}^d(n) \to \mathsf{P}^{d+k}(n)$ are its graded parts. The linear operations Sq^k can be calculated using induction on degree and the Cartan formula $Sq^k(fg) = \sum_{i+j=k} Sq^i(f)Sq^j(g)$, which is equivalent to the multiplicative property of Sq. A general Steenrod operation is a sum of compositions of Steenrod squares.

The multiplicative monoid $\mathsf{M}(n)$ of $n \times n$ matrices over $\mathbb{F}_2$ acts on the right of $\mathsf{P}(n)$ by linear substitution of the variables. Thus $\mathsf{P}^d(n)$ gives a representation over $\mathbb{F}_2$ of $\mathsf{M}(n)$ and of $\mathsf{GL}(n)$, the general linear group of invertible matrices. This matrix action commutes with the action of the Steenrod squares, and the interplay between the two gives rise to a host of interesting algebraic problems.

One of these, the 'hit' problem, is a constant theme here. A polynomial f is 'hit' if there are polynomials f_k such that $f = \sum_{k>0} Sq^k(f_k)$. The hit polynomials form a graded subspace $\mathsf{H}(n)$ of $\mathsf{P}(n)$, and the basic problem is to find the dimension of the quotient space $\mathsf{Q}^d(n) = \mathsf{P}^d(n)/\mathsf{H}^d(n)$. We call $\mathsf{Q}(n)$ the space of 'cohits'. Since $\mathsf{P}(n)$ is spanned by monomials, $\mathsf{Q}(n)$ is spanned by their equivalence classes, which we refer to simply as 'monomials in $\mathsf{Q}(n)$'. A monomial whose exponents are integers of the form $2^j - 1$ is called a 'spike', and cannot appear as a term in a hit polynomial. It follows that a monomial basis for $\mathsf{Q}(n)$ must include all the spikes.

At a deeper level, the hit problem concerns the structure of $\mathsf{Q}(n)$ as a representation of $\mathsf{GL}(n)$ or $\mathsf{M}(n)$. We develop the tools needed to answer this in the 1- and 2-variable cases in Chapter 1. These include the 2-variable version of the maps introduced by Masaki Kameko to solve the 3-variable case, and a map which we call the duplication map. We hope that this opening chapter is accessible to graduate students and mathematicians with little or no background in algebraic topology, and that it will serve as an appetizer for the rest of the book.

Chapter 2 introduces a second family of Steenrod operations, the conjugate Steenrod squares $Xq^k : \mathsf{P}^d(n) \to \mathsf{P}^{d+k}(n)$. These are useful in the hit problem because of a device known as the 'χ-trick'. This states that the product of f and $Xq^k(g)$ is hit if and only if the product of $Xq^k(f)$ and g is hit. We use the χ-trick to prove that $\mathsf{Q}^d(n) = 0$ if and only if $\mu(d) > n$, where $\mu(d)$ is the smallest number of integers of the form $2^j - 1$ (with repetitions allowed) whose sum is d. This establishes the 1986 conjecture of Franklin P. Peterson which first stimulated interest in the hit problem.

Here is a rough guide to the structure of the rest of the book, in terms of three main themes: the Steenrod algebra A_2, the Peterson hit problem, and matrix representations. Volume 1 contains Chapters 1 to 15, and Volume 2 contains Chapters 16 to 30.

Chapters 3 to 5 develop A_2 from an algebraic viewpoint.

Chapters 6 to 10 provide general results on the hit problem, together with a detailed solution for the 3-variable case.

Chapters 11 to 14 introduce the Hopf algebra structure of A_2 and study its structure in greater depth.

Chapter 15 introduces the theme of modular representations by relating the hit problem to invariants and the Dickson algebra.

Chapters 16 to 20 develop the representation theory of $\mathsf{GL}(n,\mathbb{F}_2)$ via its action on 'flags', or increasing sequences of subspaces, in an n-dimensional vector space $\mathsf{V}(n)$ over $\mathbb{F}_2$.

Chapter 21 explores a fundamental relation between linear maps $\mathsf{V}(m) \to \mathsf{V}(n)$ and Steenrod operations, leading to a maximal splitting of $\mathsf{P}(n)$ as a direct sum of A_2-modules.

Chapter 22 studies the A_2-summands of $\mathsf{P}(n)$ corresponding to the Steinberg representation of $\mathsf{GL}(n,\mathbb{F}_2)$.

Chapters 23 and 24 develop the relation between flag modules and the dual hit problem.

Chapters 25 and 26 study the hit problem for symmetric polynomials over $\mathbb{F}_2$.

Chapters 27 and 28 study the splitting of $\mathsf{P}(n)$ as an A_2-module obtained using a cyclic subgroup of order $2^n - 1$ in $\mathsf{GL}(n,\mathbb{F}_2)$.

Chapters 29 and 30 return to Peterson's original problem, with a partial solution of the 4-variable case.

The contents of Chapters 3 to 30 are summarized below in more detail.

In Chapter 3 we interpret the operations Sq^k as generators of a graded algebra A_2, subject to a set of relations called the Adem relations. The algebra A_2 is the mod 2 Steenrod algebra, and the operations Sq^k of Chapter 1 provide $\mathsf{P}(n)$ with the structure of a left A_2-module. If $f = \sum_{k>0} Sq^k(f_k)$, then f can be reduced to a set of polynomials of lower degree modulo the action of the positively graded part A_2^+ of A_2. A monomial basis for $\mathsf{Q}(n)$ gives a minimal generating set for $\mathsf{P}(n)$ as an A_2-module. Thus the hit problem is an example of the general question of finding a minimal generating set for a module over a ring. The structure of A_2 itself is completely determined by its action on polynomials, in the sense that two expressions in the generators Sq^k are equal in A_2 if and only if the corresponding operations on $\mathsf{P}(n)$ are equal for all n. For example, the results $Sq^1 Sq^{2k}(f) = Sq^{2k+1}(f)$ and $Sq^1 Sq^{2k+1}(f) = 0$ of Chapter 1 imply the Adem relations $Sq^1 Sq^{2k} = Sq^{2k+1}$ and $Sq^1 Sq^{2k+1} = 0$.

In Chapter 3 we also establish the two most important bases for A_2 as a vector space over $\mathbb{F}_2$. These are the admissible monomials in the generators Sq^k, due to Henri Cartan and Jean-Pierre Serre, and the basis introduced by John W. Milnor by treating A_2 as a Hopf algebra. As mentioned above, we represent a monomial by a 'block' whose rows are the reversed binary expansions of its exponents, and whose entries are integers 0 or 1. We use blocks to keep track of Steenrod operations on monomials. This 'block technology' and 'digital engineering' works well for the prime 2, and greatly facilitates our understanding of techniques which can appear opaque when expressed in more standard notation.

Chapter 4 begins with the multiplication formula for elements of the Milnor basis. This combinatorial formula helps to explain the ubiquity of the Milnor basis in the literature, as a product formula is not available for other bases of A_2. We also discuss the compact formulation of the Adem relations due to Shaun R. Bullett and Ian G. Macdonald. We use this to construct the conjugation χ of A_2, which interchanges Sq^k and Xq^k.

Chapter 5 provides combinatorial background for the algebra A_2, the hit problem and the representation theory of $\mathsf{GL}(n)$ over $\mathbb{F}_2$. Sequences of non-negative integers appear in various forms, and we distinguish 'finite sequences' from 'sequences'. A 'finite sequence' has a fixed number of entries, called its 'size', while a 'sequence' is an infinite sequence $R = (r_1, r_2, \ldots)$ with only a finite number of nonzero terms, whose 'length' is the largest ℓ for which $r_\ell > 0$. However, a sequence R is usually written as a finite sequence $(r_1, \ldots, r_n)$, where $n \geq \ell$, by suppressing some or all of the trailing 0s. The modulus of a sequence or a finite sequence is the sum of its terms. For example, the degree of a monomial is the modulus of its sequence of exponents. The set of all sequences indexes the Milnor basis of A_2.

For brevity, we call a sequence R 'decreasing' if $r_i \geq r_{i+1}$ for all i, i.e. if it is non-increasing or weakly decreasing. Thus a decreasing sequence of modulus d is a partition of d. Such a partition can alternatively be regarded as a multiset of positive integers with sum d. We discuss two special types of partition; 'binary' partitions, whose parts are integers of the form 2^j, and 'spike' partitions, whose parts are integers of the form $2^j - 1$.

We introduce two total order relations on sequences, the left (lexicographic) order and the right (reversed lexicographic) order, and two partial order relations, dominance and 2-dominance. The ω-sequence $\omega(f) = (\omega_1, \omega_2, \ldots, \omega_k)$ of a monomial f is defined by writing f as a product $f_1 f_2^2 \cdots f_k^{2^{k-1}}$, where f_i is a product of ω_i distinct variables. In terms of blocks, $\omega(f)$ is the sequence of column sums of the block representing f, and the degree of f is $\omega_1 + 2\omega_2 + 4\omega_3 + \cdots + 2^{k-1}\omega_k$, the '2-degree' of $\omega(f)$. The set of decreasing sequences of 2-degree d has a minimum element $\omega^{\min}(d)$, which plays an important part in the hit problem, and is the same for the left, right and 2-dominance orders. We end Chapter 5 by relating this combinatorial material to the admissible and Milnor bases of A_2.

In Chapter 6 we return to the hit problem and introduce 'local' cohit spaces $\mathsf{Q}^\omega(n)$. A total order relation on ω-sequences of monomials gives a filtration on $\mathsf{P}^d(n)$ with quotients $\mathsf{P}^\omega(n)$. For the left and right orders, this passes to a filtration on $\mathsf{Q}^d(n)$ with quotients $\mathsf{Q}^\omega(n)$. A polynomial f in $\mathsf{P}^\omega(n)$ is 'left reducible' if it is the sum of a hit polynomial and monomials with ω-sequences $< \omega$ in the left order, and similarly for the right order. We prove the theorem of William M. Singer that $\mathsf{Q}^\omega(n) = 0$ if $\omega < \omega^{\min}(d)$ in the left order. We introduce the 'splicing' technique for manufacturing hit equations, extend the Kameko and duplication maps of Chapter 1 to the n-variable case, and determine $\mathsf{Q}^\omega(n)$ for 'head' sequences $\omega = (n-1,\dots,n-1)$ and 'tail' sequences $\omega = (1,\dots,1)$.

We begin Chapter 7 by proving that $\dim \mathsf{Q}^d(n)$ is bounded by a function of n independent of d. Thus only finitely many isomorphism classes of $\mathbb{F}_2\mathsf{GL}(n)$-modules can be realized as cohit modules $\mathsf{Q}^d(n)$. We extend splicing techniques and show that $\mathsf{Q}^\omega(n) = 0$ if ω is greater than every decreasing ω-sequence in the left order. A correspondence between blocks with decreasing ω-sequences and Young tableaux is used to define 'semi-standard' blocks (or monomials), and we show that $\mathsf{Q}^\omega(n)$ is spanned by such blocks when $\omega = \omega^{\min}(d)$.

In Chapter 8, we obtain reduction theorems for $\mathsf{Q}^\omega(n)$ when the sequence ω has a 'head' of length $\geq n-1$ or a 'tail' of length $\geq n$. It follows that $\dim \mathsf{Q}^d(n) = \prod_{i=1}^n (2^i - 1)$ for degrees $d = \sum_{i=1}^n (2^{a_i} - 1)$, when $a_i - a_{i+1} \geq i+1$ for $i < n$ and when $a_i - a_{i+1} \geq n-i+1$ for $i < n$. We complete a solution of the 3-variable hit problem by giving bases for $\mathsf{Q}^\omega(3)$ in the remaining cases.

The techniques introduced so far are useful for obtaining upper bounds for $\dim \mathsf{Q}^d(n)$, but are less efficient for obtaining lower bounds, where we may wish to prove that no linear combination of a certain set of monomials is hit. Chapter 9 introduces the dual problem of finding $\mathsf{K}^d(n)$, the simultaneous kernel of the linear operations $Sq_k : \mathsf{DP}^d(n) \to \mathsf{DP}^{d-k}(n)$ dual to Sq^k for $k > 0$. Here $\mathsf{DP}(n)$ is a 'divided power algebra' over $\mathbb{F}_2$, whose elements are sums of dual or 'd-monomials' $v_1^{(d_1)} \cdots v_n^{(d_n)}$. As a $\mathbb{F}_2\mathsf{GL}(n)$-module, $\mathsf{K}^d(n)$ is the dual of $\mathsf{Q}^d(n)$ defined by matrix transposition, and so $\dim \mathsf{K}^d(n) = \dim \mathsf{Q}^d(n)$. Thus we aim to find upper bounds for $\dim \mathsf{Q}^d(n)$ by using spanning sets in $\mathsf{Q}^d(n)$, and lower bounds by using linearly independent elements in $\mathsf{K}^d(n)$.

An advantage of working in the dual situation is that $\mathsf{K}(n)$ is a subalgebra of $\mathsf{DP}(n)$. Since the dual spikes are in $\mathsf{K}(n)$, they generate a subalgebra $\mathsf{J}(n)$ of $\mathsf{K}(n)$ which is amenable to calculation. In the cases $n = 1$ and 2, $\mathsf{J}(n) = \mathsf{K}(n)$, and when $n = 3$, $\mathsf{K}^d(n)/\mathsf{J}^d(n)$ has dimension 0 or 1. We explain how to construct the dual $\mathsf{K}^\omega(n)$ of $\mathsf{Q}^\omega(n)$ with respect to an order relation. We study the duals of the Kameko and duplication maps, and solve the dual hit problem for $n \leq 3$. In Chapter 10 we extend these results by determining $\mathsf{K}^d(3)$ and

$\mathsf{Q}^d(3)$ as modules over $\mathbb{F}_2\mathsf{GL}(3)$. Here the flag module $\mathsf{FL}(3)$, given by the permutation action of $\mathsf{GL}(3)$ on subspaces of the defining module $\mathsf{V}(3)$, plays an important part. We describe tail and head modules in terms of the exterior powers of $\mathsf{V}(3)$.

Hopf algebras are introduced in Chapter 11. A Hopf algebra A has a 'coproduct' $A \to A \otimes A$ compatible with the product $A \otimes A \to A$, and an 'antipode' $A \to A$. We show that the coproduct $Sq^k \mapsto \sum_{i+j=k} Sq^i \otimes Sq^j$ and the conjugation χ provide the mod 2 Steenrod algebra A_2 with the structure of a Hopf algebra. For a graded Hopf algebra A of finite dimension in each degree, the graded dual A^* is also a Hopf algebra. In this sense, the divided power algebra $\mathsf{DP}(n)$ is dual to the polynomial algebra $\mathsf{P}(n)$. We show that the graded dual A_2^* of A_2 is a polynomial algebra on generators ξ_j of degree $2^j - 1$ for $j \geq 1$, and determine its structure maps. We conclude this chapter with the formula of Zaiqing Li for conjugation in A_2, which complements Milnor's product formula of Chapter 4.

Chapters 12 and 13 give more detail on the internal structure of A_2. In Chapter 12 we focus on two important families of Hopf subalgebras of A_2, namely the subalgebras of 'Steenrod qth powers' A_q, where q is a power of 2, and the finite subalgebras $\mathsf{A}_2(n)$ generated by Sq^k for $k < 2^{n+1}$. We also introduce some more additive bases of A_2. We continue in Chapter 13 by introducing a 'cap product' action of the dual algebra A_2^* on A_2, which can be used to obtain relations in A_2 by a process which we call 'stripping'. We use the 'halving' map (or Verschiebung) of A_2 to explain why its action on $\mathsf{P}(n)$ reproduces itself by doubling exponents of monomials Sq^A and squaring polynomials. This map sends Sq^k to 0 if k is odd and to $Sq^{k/2}$ if k is even. Since it is the union of the finite subalgebras $\mathsf{A}_2(n)$, the algebra A_2 is nilpotent. We apply the stripping technique to obtain the nilpotence order of certain elements of A_2.

Chapter 14 is devoted to a proof of the 2-dominance theorem of Judith H. Silverman and Dagmar M. Meyer. This deep result states that a monomial f in $\mathsf{P}^d(n)$ is hit if $\omega(f)$ is not greater than $\omega^{\min}(d)$ in the 2-dominance order. This strengthens the Peterson conjecture of Chapter 2 and the theorem of Singer from Chapter 6. One consequence is the Silverman–Singer criterion, which states that if g and h are homogeneous polynomials such that $\deg g < (2^k - 1)\mu(\deg h)$, where μ is the numerical function of Chapter 2, then $f = gh^{2^k}$ is hit.

In Chapter 15, we consider the Dickson algebra $\mathsf{D}(n)$ of $\mathsf{GL}(n)$-invariants in $\mathsf{P}(n)$. Following Nguyen H. V. Hung and Tran Ngoc Nam, we show that all Dickson invariants of positive degree are hit in $\mathsf{P}(n)$ when $n \geq 3$. There is a large class of similar problems: given a subgroup G of $\mathsf{GL}(n)$, the subalgebra of G-invariant polynomials $\mathsf{P}(n)^G$ is an A_2-module, and the 'relative' hit problem asks for the elements of $\mathsf{P}(n)^G$ which are hit in $\mathsf{P}(n)$. The corresponding 'absolute' hit problem asks for a minimal generating set for $\mathsf{P}(n)^G$. We consider the absolute problem for the Weyl subgroup $G = \mathsf{W}(n)$ of permutation matrices in $\mathsf{GL}(n)$ in Chapter 25.

In the chapters which follow, we shift attention to the representation theory of $\mathsf{GL}(n)$ over $\mathbb{F}_2$. We begin in Chapter 16 by studying the flag module $\mathsf{FL}(n)$, which is defined by the permutation action of $\mathsf{GL}(n)$ on the right cosets of the Borel subgroup $\mathsf{B}(n)$ of lower triangular matrices. This module is isomorphic to $\mathsf{Q}^d(n)$ when the degree d is 'generic' in the sense of Chapter 8. The Bruhat decomposition $A = BWB'$ of a matrix A in $\mathsf{GL}(n)$ is used to define certain subspaces of $\mathsf{FL}(n)$ which we call 'Schubert cells'. Here $B, B' \in \mathsf{B}(n)$ are lower triangular matrices and $W \in \mathsf{W}(n)$ is a permutation matrix. We show that $\mathsf{FL}(n)$ is the direct sum of 2^{n-1} submodules $\mathsf{FL}_I(n)$, where $I \subseteq \{1, 2, \ldots, n-1\}$ is the set of dimensions of the subspaces in the 'partial' flags given by right cosets of parabolic subgroups of $\mathsf{GL}(n)$.

The main aim of Chapter 17 is to construct a full set of 2^{n-1} irreducible $\mathbb{F}_2\mathsf{GL}(n)$-modules $L(\lambda)$. Following C. W. Curtis, $L(\lambda)$ is defined as the head of the summand $\mathsf{FL}_I(n)$ of $\mathsf{FL}(n)$, where λ is the 'column 2-regular' partition corresponding to I, i.e. $\lambda_i - \lambda_{i+1} = 1$ if $i \in I$, 0 if $i \notin I$. The summand of $\mathsf{FL}(n)$ corresponding to complete flags is the Steinberg module $\mathsf{St}(n)$. We use the Hecke algebra $\mathsf{H}_0(n)$ of endomorphisms of $\mathsf{FL}(n)$ which commute with the action of $\mathsf{GL}(n)$, and follow the methods of R. W. Carter and G. Lusztig.

In Chapter 18, we review the background from modular representation theory that we use to study $\mathsf{P}(n)$ and $\mathsf{DP}(n)$ as $\mathbb{F}_2\mathsf{GL}(n)$-modules. We explain the role of idempotents in obtaining direct sum decompositions, and introduce the Steinberg idempotent $e(n) = \overline{\mathsf{B}}(n)\overline{\mathsf{W}}(n) \in \mathbb{F}_2\mathsf{GL}(n)$, the sum of all products BW, where $B \in \mathsf{B}(n)$ and $W \in \mathsf{W}(n)$. We study $e(n)$ and the conjugate idempotent $e'(n) = \overline{\mathsf{W}}(n)\overline{\mathsf{B}}(n)$ by means of an embedding of $\mathsf{H}_0(n)$ in the group algebra $\mathbb{F}_2\mathsf{GL}(n)$ due to N. J. Kuhn. We also discuss Brauer characters and the representation ring $R_2(\mathsf{GL}(n))$.

In Chapter 19 we use idempotents in $\mathbb{F}_2\mathsf{GL}(n)$ to split $\mathsf{P}(n)$ as a direct sum of A_2-submodules $\mathsf{P}(n, \lambda)$, each occurring $\dim L(\lambda)$ times. We discuss the problem of determining the number of factors isomorphic to $L(\lambda)$ in a composition series for $\mathsf{P}^d(n)$. Following Ton That Tri, we use the Mui algebra of $\mathsf{B}(n)$-invariants in $\mathsf{P}(n)$ to determine the minimum degree d in which $\mathsf{P}^d(n)$ has a submodule isomorphic to $L(\lambda)$.

As Weyl modules and their duals are central topics of modular representation theory, it is no surprise that they appear here also. As these modules are defined over infinite coefficient fields, we begin Chapter 20 by reviewing some results on modular representations of the algebraic group $\overline{\mathsf{G}}(n)$ of nonsingular $n \times n$ matrices over $\overline{\mathbb{F}}_2$, the algebraic closure of $\mathbb{F}_2$. We then introduce the 'restricted' Weyl module $\Delta(\lambda, n)$ over $\mathbb{F}_2$ and its transpose dual $\nabla(\lambda, n)$, and show that if λ is column 2-regular and if the ordering on ω-sequences is suitably chosen, then $\Delta(\lambda, n) \cong \mathsf{K}^{\omega}(n)$ and $\nabla(\lambda, n) \cong \mathsf{Q}^{\omega}(n)$, where ω is the partition conjugate to λ. We use the theory of polynomial $\overline{\mathsf{G}}(n)$-modules to determine the minimum degree d in which $L(\lambda)$ occurs as a composition factor of $\mathsf{P}^d(n)$.

Chapter 21 gives a self-contained proof of an important result of J. F. Adams, J. Gunawardena and H. R. Miller. This states that every degree-preserving A_2-module map $\mathsf{P}(m) \to \mathsf{P}(n)$ is given by a sum of linear substitutions given by the action of $m \times n$ matrices over $\mathbb{F}_2$. It follows that the A_2-summands in a maximal splitting of $\mathsf{P}(n)$ obtained using idempotents in $\mathbb{F}_2\mathsf{M}(n)$ rather than $\mathbb{F}_2\mathsf{GL}(n)$ are indecomposable. Hence such a splitting is a maximal direct sum decomposition of $\mathsf{P}(n)$ as an A_2-module.

Chapter 22 is concerned with the A_2-summands of $\mathsf{P}(n)$ corresponding to the Steinberg representation $\mathsf{St}(n)$ of $\mathsf{GL}(n)$. We discuss the 'internal' model $\mathsf{MP}(n)$ of the A_2-module $\mathsf{P}(\mathsf{St}(n))$ defined by Stephen A. Mitchell and Stewart B. Priddy using admissible monomials of length n in A_2 itself. Although the hit problem for $\mathsf{P}(n)$ can be split into a corresponding problem for $\mathsf{P}(n, \lambda)$ for each λ, the Steinberg summand is the only case where this problem has been solved for all n, and we give minimal generating sets for the summands given by the idempotents $e(n)$ and $e'(n)$ of Chapter 18.

In Chapter 23, we identify the module $\mathsf{J}^d(n)$ of $\mathsf{K}^d(n)$ generated by the dual spikes in degrees $d = \sum_{i=1}^{n}(2^{a_i} - 1)$, where $a_1 > a_2 > \cdots > a_n$, in terms of the flag module $\mathsf{FL}(n)$. We show that $\mathsf{Q}^d(n) \cong \mathsf{FL}(n)$ in 'generic' degrees d, and extend the method in Chapter 24 to obtain results of Tran Ngoc Nam on $\mathsf{J}(n)$ relating cohit modules to partial flag modules. Following Nguyen Sum, we give counterexamples for $n \geq 5$ to Kameko's conjecture that $\dim \mathsf{Q}^d(n) \leq \dim \mathsf{FL}(n)$ for all d.

In Chapters 25 and 26 we discuss the hit problem for the action of A_2 on the algebra of symmetric polynomials $\S(n)$, the invariants in $\mathsf{P}(n)$ of the group $\mathsf{W}(n)$ of permutations of the variables. More generally, we discuss the 'absolute' hit problem for any subgroup G of $\mathsf{W}(n)$, and show that the Peterson conjecture, the Kameko map and Singer's minimal spike theorem have analogues for $\mathsf{P}(n)^G$. We solve the symmetric hit problem for $n \leq 3$, using the dual problem to obtain the lower bound in the case $n = 3$. Following Singer, we introduce the 'bigraded Steenrod algebra' $\widetilde{\mathsf{A}}_2$, which is obtained by omitting the relation $Sq^0 = 1$ in the definition of A_2, and apply $\widetilde{\mathsf{A}}_2$ to the dual problem.

In Chapters 27 and 28 we consider the cyclic subgroup $\mathsf{C}(n)$ of order $2^n - 1$ in $\mathsf{GL}(n)$. This is obtained by regarding $\mathsf{P}^1(n)$ as the underlying vector space of the Galois field $\mathbb{F}_{2^n}$. The action of $\mathsf{C}(n)$ on $\mathsf{P}(n)$ is diagonalized over $\mathbb{F}_{2^n}$ by a change of variables which 'twists' the action of A_2, in the sense that $Sq^1(t_i) = t_{i-1}$ in the new variables, which are indexed mod n. Following H. E. A. Campbell and P. S. Selick, we show that the polynomial algebra $\widetilde{\mathsf{P}}(n) = \mathbb{F}_2[t_1,\ldots,t_n]$ splits as the direct sum of $2^n - 1$ A_2-modules $\widetilde{\mathsf{P}}(n,j)$ corresponding to the 1-dimensional representations of $\mathsf{C}(n)$. In particular, $\widetilde{\mathsf{P}}(n,0)$ can be identified with the ring of $\mathsf{C}(n)$-invariants of $\mathsf{P}(n)$. In Chapter 28, we solve the dual cyclic hit problem for $n=3$ by using the twisted analogue $\widetilde{\mathsf{J}}(n)$ of the d-spike module $\mathsf{J}(n)$.

In Chapters 29 and 30, we collect some results on the hit problem in the 4-variable case as further illustration of our methods. Nguyen Sum has extended the method introduced by Kameko to find a monomial basis for $\mathsf{Q}^d(4)$ for all d. We include without proof some of the results of Sum, and also some results which we have verified only by computer using MAPLE. Thus there remain some challenging aspects of the hit problem even in the case $n=4$.

The Steenrod algebra was originally defined for all primes p, but we have restricted attention to the case $p=2$. All the problems have analogues for odd primes, but in general much less is known, and a number of difficulties arise in trying to extend our techniques to the odd prime case. The 2-variable hit problem for the action of A_p on the polynomial algebra $\mathbb{F}_p[x,y]$ has been solved by Martin D. Crossley, but little appears to be known even for the 3-variable case. In common with many authors on the Steenrod algebra, we have therefore confined ourselves to the prime 2.

There are several good textbooks on topology which include material on the Steenrod algebra and its applications, such as those by Brayton I. Gray [70] and by Robert E. Mosher and Martin C. Tangora [147], in addition to the classic Annals of Mathematics Study [196], based on lectures by Steenrod himself, and the Cartan seminars [33]. A treatment of the Steenrod algebra from an algebraic viewpoint, including Steenrod operations over an arbitrary finite field, is given in Larry Smith's book [190] on invariant theory. The book of Harvey R. Margolis [129] treats the general theory of modules over the Steenrod algebra. Still other approaches to the Steenrod algebra are possible. The survey article [233] treats Steenrod operations as linear differential operators with polynomial coefficients, and is a precursor for this book.

We sometimes introduce definitions and constructions for a small number of variables and extend them to the general case in later chapters. Although this can involve a certain amount of repetition, it has the advantage of leading to interesting results at an early stage by elementary methods. We hope that our approach will appeal to readers whose main interests are in algebra, especially in the modular representation theory of linear groups, or in the combinatorics related to symmetric polynomials and to the invariant theory of finite groups.

In order to avoid interruptions to the text, citations and background material are collected in the 'Remarks' sections at the end of each chapter. The occasional reference to topology may occur in these, but we have not tried to explain the topology. We have also omitted important topics such as the Singer transfer map and its applications to the homotopy groups of spheres through the Adams spectral sequence. These would require another volume, which we are not qualified to write. For similar reasons, we do not treat the theory of analytic functors and the category $\mathcal{U}/\mathcal{N}il$ of unstable A_2-modules modulo nilpotent objects due to Hans-Werner Henn, Jean Lannes and Lionel Schwartz. Finally, in a subject which crosses several disciplines, notation presents a problem because the traditional symbols of one area may be in conflict with those of another. A list of symbols for the main ingredients of our subject appears at the end of the book, together with an index of the main terms defined in the text.

We should like to offer our sincere thanks to the School of Mathematics of the University of Manchester for providing us with office space and computing facilities during work on this text. We should also like to thank several colleagues at Manchester for mathematical help and support, and in particular Peter Eccles, Nige Ray, Peter Rowley and Bob Sandling. Of colleagues farther afield, we should like to thank Nguyen H. V. Hung, Ali S. Janfada, Bill Singer and Larry Smith for their interest in our project. The first author would also like to thank Stephen R. Doty for teaching him something about modular representation theory. Finally we should like to thank Roger Astley and his colleagues at Cambridge University Press for their encouragement, support and patience. We set out with the modest aim of providing a beginning graduate student in topology and algebra with a basic primer on the Steenrod algebra, illustrated by our favourite application to a problem proposed by Frank Peterson, but the project has expanded substantially in scope over the past eight years.

16
The action of $\mathsf{GL}(n)$ on flags

16.0 Introduction

This chapter begins a more systematic approach to the study of the polynomial algebra $\mathsf{P}(n)$ as a $\mathbb{F}_2\mathsf{GL}(n)$-module, based on the flag module $\mathsf{FL}(n)$. We have seen in Chapter 10 that a close approximation to the $\mathbb{F}_2\mathsf{GL}(3)$-modules $\mathsf{Q}(3)$ and $\mathsf{K}(3)$ is given by quotients of $\mathsf{FL}(3)$, the permutation module of dimension 21 given by the action of $\mathsf{GL}(3)$ on right cosets of the lower triangular subgroup $\mathsf{L}(3)$.

We identify elements of the defining module $\mathsf{V}(n)$ for $\mathbb{F}_2\mathsf{GL}(n)$ with row vectors v, so that $A \in \mathsf{GL}(n)$ acts by $v \to v \cdot A = vA$. A (complete) flag X in $\mathsf{V}(n)$ is a nested sequence of subspaces, one of each dimension. There is a natural correspondence between flags and right cosets $\mathsf{L}(n)A$ of the lower triangular subgroup $\mathsf{L}(n)$. The module $\mathsf{FL}(n)$ is defined by the permutation action $(\mathsf{L}(n)A) \cdot B = \mathsf{L}(n)AB$, where $A, B \in \mathsf{GL}(n)$, on these cosets. It can be alternatively described as the representation of $\mathsf{GL}(n)$ induced from the trivial 1-dimensional representation of $\mathsf{L}(n)$. If the base field $\mathbb{F}_2$ is replaced by $\mathbb{F}_p$ for a general prime p, then this remains true if we take $\mathsf{L}(n)$ to be the Borel subgroup $\mathsf{B}(n, \mathbb{F}_p)$ of all lower triangular matrices, rather than $\mathsf{L}(n, \mathbb{F}_p)$, the subgroup of lower triangular matrices with 1s on the diagonal. For this reason we shall use the alternative notation $\mathsf{B}(n)$ for $\mathsf{L}(n)$ in the case $p = 2$ also.

Much of the combinatorial structure of the group $\mathsf{GL}(n)$ is determined by the Weyl subgroup $\mathsf{W}(n)$ of permutation matrices. We distinguish $\mathsf{W}(n)$ from the symmetric group $\Sigma(n)$ of permutations of the set $Z[n] = \{1, 2, \ldots, n\}$. Because we work with the right action of $\mathsf{GL}(n)$ on $\mathsf{V}(n)$, we associate to a permutation $\rho \in \Sigma(n)$ the matrix W obtained by applying ρ to the *columns* of the identity matrix I_n. This fixes an anti-isomorphism between $\Sigma(n)$ and $\mathsf{W}(n)$. In Section 16.1 we treat the length and descent set of a permutation ρ, and the Bruhat order on $\Sigma(n)$, from the point of view of permutation matrices.

In Section 16.2 we discuss the Bruhat decomposition $A = BWB'$ of $A \in \mathsf{GL}(n)$, where $B, B' \in \mathsf{B}(n)$ and $W \in \mathsf{W}(n)$. Since W is uniquely determined by A, we obtain a decomposition of $\mathsf{GL}(n)$ as the disjoint union of the double cosets $\mathsf{B}(n)W\mathsf{B}(n)$, or 'Bruhat cells'. Although the matrices B and B' are not unique in general, we obtain a unique decomposition by restricting B' to be in a subgroup $\mathsf{B}(W) \subseteq \mathsf{B}(n)$ determined by W, which we call a 'Bruhat subgroup'. In Section 16.3 we define a natural correspondence between cosets $\mathsf{B}(n)A$ and complete flags in $\mathsf{V}(n)$, and introduce the flag module $\mathsf{FL}(n)$. The Bruhat decomposition of $\mathsf{GL}(n)$ corresponds to a decomposition of $\mathsf{FL}(n)$ into $\mathbb{F}_2$-subspaces $\mathsf{Sch}(W)$, called Schubert cells.

In the case $n = 2$, the module $\mathsf{FL}(2)$ gives the representation of $\mathsf{GL}(2) \cong \Sigma(3)$ which permutes the three nonzero elements $u = v_1, v = v_2$ and $w = v_1 + v_2$ of $\mathsf{V}(2)$. The elements of $\mathsf{FL}(2)$ correspond to formal sums of u, v, w, and $\mathsf{FL}(2)$ is isomorphic to the direct sum of the 1-dimensional module generated by $u + v + w$ and the 2-dimensional module generated by $u + v$. In the general case, $\mathsf{FL}(n)$ is the direct sum of 2^{n-1} indecomposable submodules indexed by subsets $I \subseteq Z[n-1] = \{1, \ldots, n-1\}$, which correspond to 'partial' flags of type I, i.e. nested sequences of subspaces with dimensions in I.

The partial flag module $\mathsf{FL}^I(n)$ is introduced in Section 16.4. It is given by the permutation action of $\mathsf{GL}(n)$ on cosets $\mathsf{P}^I(n)A$, where $\mathsf{P}^I(n)$ is the parabolic subgroup of type I containing $\mathsf{B}(n)$. The elements of $\mathsf{P}^I(n)$ are block lower triangular matrices. For $J \subseteq I \subseteq Z[n-1]$, $\mathsf{FL}^J(n)$ can be embedded as a direct summand in $\mathsf{FL}^I(n)$ by associating to a partial flag of type J the sum of all partial flags of type I which contain it, and so we obtain a quotient module $\mathsf{FL}_I(n)$ of $\mathsf{FL}^I(n)$ by factoring out all such summands for $J \subset I$. In Section 16.5 we show that $\mathsf{FL}(n)$ is the direct sum of the modules $\mathsf{FL}_I(n)$. We shall prove in Chapter 17 that the modules $\mathsf{FL}_I(n)$ are indecomposable, so that this decomposition of $\mathsf{FL}(n)$ is maximal. In Section 16.6, we give a more detailed treatment of the module $\mathsf{FL}(3)$.

16.1 Permutation matrices

Let $\mathsf{W}(n) \subset \mathsf{GL}(n)$ be the subgroup of **permutation matrices**, i.e. matrices with one entry 1 in each row and column and all other entries 0. The defining module $\mathsf{V}(n)$ is the right $\mathbb{F}_2\mathsf{GL}(n)$-module obtained by identifying elements of $\mathbb{F}_2^n$ with row vectors, so that $\mathsf{GL}(n)$ acts by matrix multiplication. In particular, a permutation matrix $W = (w_{i,j})$ acts on $\mathsf{V}(n)$ by $x_i \cdot W = x_j$ if $w_{i,j} = 1$. We denote the symmetric group of permutations of $Z[n] = \{1, 2, \ldots, n\}$ by $\Sigma(n)$, and we associate to W the permutation ρ defined by $w_{i,j} = 1$ if and only if $\rho(i) = j$.

We use one-row notation $(\rho(1),\ldots,\rho(n))$ for permutations. For example,

$$W = \begin{pmatrix} 0 & 1 & 0 & 0 \\ 0 & 0 & 0 & 1 \\ 0 & 0 & 1 & 0 \\ 1 & 0 & 0 & 0 \end{pmatrix} \longleftrightarrow (2,4,3,1) = \rho. \tag{16.1}$$

Thus the matrix W is obtained by applying the permutation ρ to the columns, or the inverse permutation ρ^{-1} to the rows, of the identity matrix I_n.

This choice of notation is awkward in one respect. Since $(\rho_1 \circ \rho_2)(i) = \rho_1(\rho_2(i))$, our notation for permutations gives a left action of the symmetric group $\Sigma(n)$ on $Z[n]$, whereas the action of $\mathsf{GL}(n)$ on $\mathsf{V}(n)$ is on the right. Hence the bijection $\rho \leftrightarrow W$ is an anti-isomorphism: if $\rho_1 \leftrightarrow W_1$ and $\rho_2 \leftrightarrow W_2$, then $\rho_1 \circ \rho_2 \leftrightarrow W_2 W_1$. The definitions which follow are usually introduced for $\Sigma(n)$, but we approach them from the point of view of $\mathsf{W}(n)$.

Definition 16.1.1 The **length** $\mathrm{len}(W)$ of $W \in \mathsf{W}(n)$ is the number of submatrices of W of the form

$$J = \begin{pmatrix} 0 & 1 \\ 1 & 0 \end{pmatrix}.$$

The submatrices are given by the entries in any two rows and any two columns: for example, in (16.1) $\mathrm{len}(W) = 4$. The identity matrix I_n is the unique element of length 0 in $\mathsf{W}(n)$, and the anti-diagonal matrix W_0 corresponding to the reversal $\rho_0 = (n, n-1, \ldots, 1)$ is the unique element of maximal length $n(n-1)/2$. The elements of length 1 are the switch matrices S_i obtained by exchanging rows i and $i+1$ of I_n for $1 \le i \le n-1$. More generally, the switch matrix $S_{i,j}$ obtained by exchanging rows i and j of I_n has length $2|i-j|-1$.

Definition 16.1.2 Let $W \in \mathsf{W}(n)$. For $1 \le i \le n-1$, i is a **descent** of W if rows i and $i+1$ of W contain a submatrix J. More generally, for $1 \le i < j \le n$, (i,j) is an **inversion** of W if rows i and j of W contain a submatrix J. We write $\mathrm{des}(W)$ for the set of descents of W, and $\mathrm{inv}(W)$ for the set of inversions.

Thus the length of W is the number of its inversions. For the matrix W in (16.1), $\mathrm{des}(W) = \{2,3\}$ and $\mathrm{inv}(W) = \{(1,4),(2,3),(2,4),(3,4)\}$. The identity matrix I_n is the unique matrix with no descents, and the reversal matrix W_0 is the unique matrix with descent set $Z[n-1]$. The switch matrix S_i has descent set $\{i\}$ and inversion set $\{(i,i+1)\}$. For $j > i+1$, the switch matrix $S_{i,j}$ has descent set $\{i, j-1\}$ and inversion set consisting of (i,j) and all (i,k) and (k,j)

for $i < k < j$. The set $\operatorname{inv}(W)$ is transitive, in the sense that $(i,j) \in \operatorname{inv}(W)$ if $(i,k),(k,j) \in \operatorname{inv}(W)$ for some k such that $i < k < j$.

If $\rho \in \Sigma(n)$ is the permutation associated to W, we write the length and the descent and inversion sets of W alternatively as $\operatorname{len}(\rho)$, $\operatorname{des}(\rho)$ and $\operatorname{inv}(\rho)$. Thus $i \in \operatorname{des}(\rho)$ if $\rho(i) > \rho(i+1)$, and $(i,j) \in \operatorname{inv}(\rho)$ if $i < j$ and $\rho(i) > \rho(j)$.

Proposition 16.1.3 *For $W \in \mathsf{W}(n)$ and the reversal matrix W_0*

(i) $len(W^{-1}) = len(W)$,
(ii) $i \in des(WW_0)$ *if and only if* $i \notin des(W)$,
(iii) $i \in des(W_0W)$ *if and only if* $n-i \notin des(W)$,
(iv) $len(WW_0) = len(W_0W) = n(n-1)/2 - len(W)$.

Proof For (i) we observe that $W^{-1} = W^{\mathrm{tr}}$ and transposition of a matrix preserves submatrices of the form J and I_2. For (ii) and (iii) we observe that reversal of the rows or columns exchanges these submatrices, and, in the case of column reversal, inversions and non-inversions are exchanged. Hence $(i,j) \in \operatorname{inv}(WW_0)$ if and only if $(i,j) \notin \operatorname{inv}(W)$, and $(i,j) \in \operatorname{inv}(W_0W)$ if and only if $(n+1-j, n+1-i) \notin \operatorname{inv}(W)$. In particular (iv) follows. □

Proposition 16.1.4 *For $W \in \mathsf{W}(n)$*

(i) $$len(S_iW) = \begin{cases} len(W)+1, & \text{if } i \notin des(W), \\ len(W)-1, & \text{if } i \in des(W); \end{cases}$$

(ii) $$len(WS_i) = \begin{cases} len(W)+1, & \text{if } i \notin des(W^{-1}), \\ len(W)-1, & \text{if } i \in des(W^{-1}). \end{cases}$$

Proof For (i), the inversion $(i,i+1)$ is added or removed from $\operatorname{des}(W)$, while inversions not involving row i or $i+1$ are unchanged, and inversions involving only one of these rows are replaced by inversions involving only the other. Since $\operatorname{len}(WS_i) = \operatorname{len}((WS_i)^{-1}) = \operatorname{len}(S_iW^{-1})$, (ii) follows by applying (i) to W^{-1}. □

Proposition 16.1.5 *The switch matrices S_i, $1 \le i \le n-1$, generate the group $\mathsf{W}(n)$. For $W \in \mathsf{W}(n)$, the minimum number r of factors in a product $W = S_{i_1} \cdots S_{i_r}$ is $len(W)$, where this product is I_n in the case $r = 0$.*

Proof This follows from Proposition 16.1.4(i) by iteration. By choosing $i_1 \in \operatorname{des}(W)$, $i_2 \in \operatorname{des}(S_{i_1}W)$ and so on, we can reduce the length of the product matrix by 1 at each step. Since I_n is the only matrix with no descents, we obtain $S_{i_r} \cdots S_{i_1}W = I_n$ after $r = \operatorname{len}(W)$ steps. Hence $W = S_{i_1} \cdots S_{i_r}$. □

Definition 16.1.6 A product $W = S_{i_1} \cdots S_{i_r}$ of length $r = \mathrm{len}(W)$ is a **reduced word** for W.

Example 16.1.7 By iterative switching of rows i and $i+1$ where i is a descent, we find that the matrix W of (16.1) has 3 reduced words, namely $S_2S_3S_2S_1$, $S_3S_2S_1S_3$ and $S_3S_2S_3S_1$.

Remark 16.1.8 The generators $S_1, \ldots, S_{n-1}$ of $\mathsf{W}(n)$ satisfy the relations $S_i^2 = I_n$, $S_iS_j = S_jS_i$ if $|i-j| > 1$, and $S_iS_{i+1}S_i = S_{i+1}S_iS_{i+1}$. It is a standard result of group theory that these are a set of defining relations. Thus it is possible to use these relations to convert any word in $S_1, \ldots, S_{n-1}$ to a reduced word, or to interchange two reduced words for the same element of $\mathsf{W}(n)$.

Proposition 16.1.9 *There is a reduced word of the form S_iW' for a permutation matrix $W \in \mathsf{W}(n)$ if and only if $i \in des(W)$, and there is a reduced word of the form $W'S_j$ for W if and only if $j \in des(W^{-1})$.*

Proof If S_iW' is a reduced word for W, then W' is a reduced word for S_iW, so the first statement follows from Proposition 16.1.4(i). The second statement follows similarly from Proposition 16.1.4(ii), or by replacing W by W^{-1}. □

Proposition 16.1.10 *Let ρ be the permutation associated to W, and let $(i,j) \in inv(W)$, so that the $(i,\rho(i))$th and $(j,\rho(j))$th entries of W are the entries 1 in a submatrix*

$$\begin{pmatrix} 0 & 1 \\ 1 & 0 \end{pmatrix}$$

of W. Let W' be the matrix obtained by exchanging rows i and j of W, or columns $\rho(j)$ and $\rho(i)$. If the (k,ℓ)th entries of W are 0 for all other values of k and ℓ such that $i \le k \le j$ and $\rho(j) \le \ell \le \rho(i)$, then $len(W') = len(W) - 1$, and otherwise $len(W') < len(W) - 1$.

Proof Consider W as a partitioned matrix of the form

$$W = \left(\begin{array}{c|c|c} * & * & * \\ \hline * & * & * \\ \hline * & * & * \end{array}\right)$$

where the rows are divided into rows $k < i$, $i \le k \le j$ and $k > j$, and the columns into columns $\ell < \rho(j)$, $\rho(j) \le \ell \le \rho(i)$ and $\ell > \rho(i)$. By hypothesis, the only entries 1 in the central submatrix are the $(i,\rho(i))$ and $(j,\rho(j))$ entries. The result follows by case by case consideration of the relative positions of the two 1s in the central submatrix with respect to a 1 in one of the other submatrices. □

Example 16.1.11 For W as in (16.1), $\mathrm{inv}(W) = \{(1,4),(2,3),(2,4),(3,4)\}$. The condition of Proposition 16.1.10 is satisfied for $(i,j) = (1,4)$, $(2,3)$ and $(3,4)$ but not for $(i,j) = (2,4)$. Exchanging these rows gives a matrix W' of length 3 in the first three cases, but of length 1 in the fourth case.

Definition 16.1.12 Let $W, W' \in \mathsf{W}(n)$ with $\mathrm{len}(W) = \mathrm{len}(W') + 1$. Then W covers W' in the **Bruhat order** if $W = S_{i,j}W'$ where $1 \leq j \leq n$, or equivalently if $W = W'S_{\rho(j),\rho(i)}$, where ρ is the permutation associated to W. Given W_1 and W_2 in $\mathsf{W}(n)$, we write $W_1 \geq W_2$ if W_1 is connected to W_2 by a chain of such coverings.

Proposition 16.1.10 gives an equivalent formulation of this definition. Note that the condition of Proposition 16.1.10 is always satisfied when adjacent rows or columns of W are switched, i.e. in the cases $j = i+1$ and $\rho(i) = \rho(j)+1$.

Example 16.1.13 The permutation matrix

$$W = \begin{pmatrix} 0 & 0 & 1 & 0 \\ 0 & 0 & 0 & 1 \\ 1 & 0 & 0 & 0 \\ 0 & 1 & 0 & 0 \end{pmatrix} \longleftrightarrow (3,4,1,2) \tag{16.2}$$

has inversion set $\{(1,3),(1,4),(2,3),(2,4)\}$, and W covers $S_{2,3}W \leftrightarrow (3,1,4,2)$ (adjacent rows), $WS_{2,3} \leftrightarrow (2,4,1,3)$ (adjacent columns), $S_{1,3}W = WS_{1,3} \leftrightarrow (1,4,3,2)$ and $S_{2,4}W = WS_{2,4} \leftrightarrow (3,2,1,4)$ in the Bruhat order.

For $W_1, W_2 \in \mathsf{W}(n)$, we can use the following criterion to determine whether $W_1 \geq W_2$ in the Bruhat order. We first define this as a second partial order $\geq_s$ on $\mathsf{W}(n)$, and then we shall prove that $\geq_s$ is the same as the Bruhat order.

Definition 16.1.14 For $W \in \mathsf{W}(n)$ and $1 \leq i,j \leq n$, let $s_{i,j}(W)$ be the number of entries 1 in the north-east corner submatrix of W given by elements in rows $1,\ldots,i$ and columns $j,\ldots,n$ of W. Then $W_1 \geq_s W_2$ for $W_1, W_2 \in \mathsf{W}(n)$ if and only if $s_{i,j}(W_1) \geq s_{i,j}(W_2)$ for $1 \leq i,j \leq n$.

In terms of the permutation $\rho \in \Sigma(n)$ associated to W, $s_{i,j}$ is the number of integers $a \in \mathsf{Z}[n]$ such that $a \leq i$ and $\rho(a) \geq j$.

Proposition 16.1.15 *Given $W_1, W_2 \in \mathsf{W}(n)$, $W_1 \geq W_2$ in Bruhat order if and only if $W_1 \geq_s W_2$.*

Proof To prove necessity of the sum criterion in Definition 16.1.14, it suffices to consider the case where W_1 covers W_2. Let $\rho_1, \rho_2 \in \Sigma(n)$ be the permutations associated to $W_1, W_2 \in \mathsf{W}(n)$. If W_1 covers W_2 and $S_{k,\ell}W_1 = W_2$, then it is straightforward to check that

$$s_{i,j}(W_2) = \begin{cases} s_{i,j}(W_1) - 1, & \text{if } k \leq i \leq \ell \text{ and } \rho_1(\ell) < j \leq \rho_1(k), \\ s_{i,j}(W_1), & \text{otherwise,} \end{cases} \tag{16.3}$$

and hence $W_1 \geq_s W_2$.

To prove sufficiency, assume that $W_1 \geq_s W_2$. We shall construct W' such that $W' = S_{k,\ell}W_1$ where $(k,\ell) \in \mathrm{inv}(W_1)$, so that W_1 covers W' in the Bruhat order, and show that $W' \geq_s W_2$. The result then follows by induction on $r = \mathrm{len}(W_1) - \mathrm{len}(W_2)$, and iteration of the construction produces a chain of length r from W_1 to W_2.

We choose k and ℓ as follows. Let the kth row be the first where W_1 and W_2 differ. Then by the sum criterion the entry 1 in row k of W_2 precedes the entry 1 in W_1, i.e. $\rho_2(k) < \rho_1(k)$. Since W_1 and W_2 agree in rows $< k$, the entry 1 in column $\rho_2(k)$ is in a row $> k$. Let ℓ be minimal such that $\ell > k$ and W_1 has an entry 1 in a column $\rho_1(\ell)$ such that $\rho_2(k) \leq \rho_1(\ell) < \rho_1(k)$. Then the submatrix of W_1 given by rows $k, \ldots, \ell$ and columns $\rho_1(\ell), \ldots, \rho_1(k)$ has only two entries 1, at its north-east corner $(k, \rho_1(k))$ and at its south-west corner $(\ell, \rho_1(\ell))$. Thus the matrix $W' = S_{k,\ell}W_1$ obtained by exchanging rows k and ℓ of W_1 covers W' in the Bruhat order.

It remains to prove that $W' \geq_s W_2$. By (16.3) it suffices to prove that $s_{i,j}(W') \geq s_{i,j}(W_2)$ for $k \leq i < \ell$ and $\rho_1(\ell) < j \leq \rho_1(k)$, or equivalently that $s_{i,j}(W_1) > s_{i,j}(W_2)$ for the corresponding submatrices A_1 and A_2 of W_1 and W_2. Consider the submatrices $B_1 \supset A_1$ and $B_2 \supset A_2$ of W_1 and W_2 given by rows $k, \ldots, \ell - 1$ and columns $\rho_2(k) \leq j \leq \rho_1(k)$. By our choice of ℓ, all entries of the first column of B_1 are 0s, but the $(k, \rho_2(k))$ entry of W_2 is 1. Since the sum criterion holds for B_1 and B_2, strict inequality must hold when the first column is removed, and in particular for A_1 and A_2. □

Example 16.1.16 Let $W_1, W_2 \in \mathsf{W}(6)$ be the matrices associated to the permutations $(3,6,5,1,4,2)$ and $(2,4,6,1,3,5)$ respectively. The proof of Proposition 16.1.15 gives an algorithm for constructing a chain in decreasing Bruhat order showing that $W_1 > W_2$, with corresponding permutations $(3,6,5,1,4,2) \to (2,6,5,1,4,3) \to (2,5,6,1,4,3) \to (2,4,6,1,5,3) \to (2,4,6,1,3,5)$.

16.2 The Bruhat decomposition of $\mathsf{GL}(n)$

Let $\mathsf{B}(n)$ be the lower triangular subgroup of $\mathsf{GL}(n)$. For $W \in \mathsf{W}(n)$, the right coset $\mathsf{B}(n)W$ contains all matrices obtained by applying the associated permutation ρ to the columns of a lower triangular matrix, while the left coset $W\mathsf{B}(n)$ contains all matrices obtained by applying ρ^{-1} to the rows. For example, with W as in (16.1),

$$\mathsf{B}(n)W = \begin{pmatrix} 0 & 1 & 0 & 0 \\ 0 & * & 0 & 1 \\ 0 & * & 1 & * \\ 1 & * & * & * \end{pmatrix}, \quad W\mathsf{B}(n) = \begin{pmatrix} * & 1 & 0 & 0 \\ * & * & * & 1 \\ * & * & 1 & 0 \\ 1 & 0 & 0 & 0 \end{pmatrix}, \tag{16.4}$$

where the stars represent elements 0 or 1 in $\mathbb{F}_2$. Thus matrices in $\mathsf{B}(n)W$ are obtained by replacing the 0s in W below the 1s by stars, while matrices in $W\mathsf{B}(n)$ are obtained by replacing the 0s in W to the left of the 1s by stars.

We wish to express a matrix $A \in \mathsf{GL}(n)$ as a product BWB', where B and B' are lower triangular and W is a permutation matrix. We begin by using lower triangular transvections $T_{i,j}$ which map row i to row i + row j, where $i > j$, so as to reduce A to a matrix A' in which every entry below the last 1 in each row is 0.

Example 16.2.1 In the case $n = 4$, a typical reduction is

$$A = \begin{pmatrix} 1 & 1 & 0 & 0 \\ 0 & 1 & 1 & 1 \\ 1 & 0 & 1 & 0 \\ 1 & 1 & 1 & 1 \end{pmatrix} \quad \mapsto \quad A' = \begin{pmatrix} 1 & 1 & 0 & 0 \\ 1 & 0 & 1 & 1 \\ 1 & 0 & 1 & 0 \\ 1 & 0 & 0 & 0 \end{pmatrix},$$

obtained by adding row 1 to row 2, row 1 to row 4 and row 2 to row 4.

Since these row operations are equivalent to premultiplying A by lower triangular matrices, $A' = BA$ where $B \in \mathsf{B}(n)$ is the result of applying the same sequence of transvections to the identity matrix I_n. Further, by (16.4) $A' = WB'$ where $W \in \mathsf{W}(n)$ and $B' \in \mathsf{B}(n)$. Thus $A = B^{-1}WB'$. In Example 16.2.1

$$B = \begin{pmatrix} 1 & 0 & 0 & 0 \\ 1 & 1 & 0 & 0 \\ 0 & 0 & 1 & 0 \\ 0 & 1 & 0 & 1 \end{pmatrix}, \; W = \begin{pmatrix} 0 & 1 & 0 & 0 \\ 0 & 0 & 0 & 1 \\ 0 & 0 & 1 & 0 \\ 1 & 0 & 0 & 0 \end{pmatrix}, \; B' = \begin{pmatrix} 1 & 0 & 0 & 0 \\ 1 & 1 & 0 & 0 \\ 1 & 0 & 1 & 0 \\ 1 & 0 & 1 & 1 \end{pmatrix}.$$

The matrix W is given by the last 1 in each row of A'. Note that for $i > j$ the (i,j)th entry of B' is 1 if columns i and j contain a submatrix J, and is 0 if columns i and j contain a submatrix I_2.

Theorem 16.2.2 (Bruhat decomposition) *Every matrix $A \in \mathsf{GL}(n)$ is a product $A = BWB'$, where W is a permutation matrix uniquely determined by A, and B and B' are lower triangular.*

Proof The existence of the decomposition follows from the procedure illustrated above. To prove that W is unique, we show that if $B_1W_1B_1' = B_2W_2B_2'$ then $W_1 = W_2$. Let $A = W_1B_1'(B_2')^{-1} = B_1^{-1}B_2W_2$, so that $A \in W_1\mathsf{B}(n) \cap \mathsf{B}(n)W_2$. Let ρ_1 and ρ_2 be the permutations associated to W_1 and W_2 respectively, let $1 \leq i \leq n$, and let $j_1 = \rho_1(i)$, $j_2 = \rho_2(i)$. Then by (16.4) $a_{i,j_1} = a_{i,j_2} = 1$ and $a_{i,j} = 0$ for all $j > j_1$, where $A = (a_{i,j})$. Hence $j_2 \leq j_1$. Since this is true for all i, $\rho_2 = \rho_1$ and so $W_2 = W_1$. □

Theorem 16.2.2 shows that $\mathsf{GL}(n)$ is the disjoint union of the double cosets $\mathsf{B}(n)W\mathsf{B}(n)$, where $W \in \mathsf{W}(n)$.

Definition 16.2.3 Given $W \in \mathsf{W}(n)$, the corresponding **Bruhat cell** of $\mathsf{GL}(n)$ is the double coset $\mathsf{B}(n)W\mathsf{B}(n)$ consisting of all matrices $A = BWB'$ where $B, B' \in \mathsf{B}(n)$.

The Bruhat cells do not all have the same size. There is a unique largest Bruhat cell $\mathsf{B}(n)W_0\mathsf{B}(n)$ with $|\mathsf{B}(n)|^2 = 2^{n(n-1)}$ elements, and a unique smallest Bruhat cell $\mathsf{B}(n) = \mathsf{B}(n)I_n\mathsf{B}(n)$. It follows from Proposition 16.2.7 below that $|\mathsf{B}(n)W\mathsf{B}(n)| = 2^m$, where $m = n(n-1)/2 + \mathrm{len}(W)$.

To prove this, we shall refine the Bruhat decomposition so as to obtain a unique factorization of each matrix $A \in \mathsf{GL}(n)$. Consider the left coset $W\mathsf{B}(n)$ of (16.4). By using lower triangular transvections we can reduce A to a matrix A' in which every entry below the last 1 in each row is 0. As in Example 16.2.1, A' has the form

$$\begin{pmatrix} * & 1 & 0 & 0 \\ * & 0 & * & 1 \\ * & 0 & 1 & 0 \\ 1 & 0 & 0 & 0 \end{pmatrix}, \tag{16.5}$$

and conversely every element of $W\mathsf{B}(n)$ can be recovered by applying the same set of transvections to such matrices A'. Since the stars in the reduced matrix correspond to the stars in 2×2 submatrices of the form $\left(\begin{smallmatrix} * & 1 \\ 1 & 0 \end{smallmatrix}\right)$, the number of stars is $\mathrm{len}(W)$. In terms of the permutation ρ associated to W, the (i,j)th entry of the reduced matrix is a star if and only if $j < \rho(i)$ and $i < \rho^{-1}(j)$. The reduced matrices form the left coset $W\mathsf{B}(W)$, where $\mathsf{B}(W)$ is the subset of $\mathsf{B}(n)$ given by matrices

$$\begin{pmatrix} 1 & 0 & 0 & 0 \\ * & 1 & 0 & 0 \\ * & 0 & 1 & 0 \\ * & 0 & * & 1 \end{pmatrix}. \tag{16.6}$$

This argument shows that the Bruhat cell $\mathsf{B}(n)W\mathsf{B}(n) = \mathsf{B}(n)W\mathsf{B}(W)$, and we note that $\mathsf{B}(W)$ is a subgroup of $\mathsf{B}(n)$. For $1 \le j < i \le n$, the (i,j)th entry of (16.6) is a star if $(j,i) \in \mathrm{inv}(W^{-1})$, and is 0 otherwise. In terms of the permutation $\rho = (2,4,3,1)$ corresponding to W, the stars correspond to pairs (i,j) such that $i > j$ but i appears before j in one-line notation.

Definition 16.2.4 For $W \in \mathsf{W}(n)$, the **Bruhat subgroup** $\mathsf{B}(W) \subseteq \mathsf{B}(n)$ is the group of all matrices $B \in \mathsf{B}(n)$ such that for $1 \le j < i \le n$ the (i,j)th entry $b_{i,j}$ of B is 0 if $(j,i) \notin \mathrm{inv}(W^{-1})$. Equivalently $b_{i,j} = 0$ if columns i and j of W contain a submatrix I_2. In terms of the permutation ρ associated to W, $b_{i,j} = 0$ if j precedes i in one-line notation for ρ. We shall also write $\mathsf{B}(\rho)$ for $\mathsf{B}(W)$.

Proposition 16.2.5 *Let* $W \in \mathsf{W}(n)$. *Then*

(i) $\mathsf{B}(W)$ *is a subgroup of* $\mathsf{B}(n)$ *of order* $2^{len(W)}$,
(ii) $\mathsf{B}(W) \cap \mathsf{B}(W_0W) = I_n$ *and* $\mathsf{B}(n) = \mathsf{B}(W)\mathsf{B}(W_0W)$,
(iii) $W\mathsf{B}(W)W^{-1} \subseteq \mathsf{U}(n)$, *the upper triangular subgroup of* $\mathsf{GL}(n)$.

Proof (i) Let $C = AB$ where $A, B \in \mathsf{B}(W)$, and let $1 \le j < i \le n$. Since $c_{i,j} = \sum_{j \le k \le i} a_{i,k} b_{k,j}$, if $c_{i,j} = 1$ then $a_{i,k} = 1$ and $b_{k,j} = 1$ for some k. Hence columns k and i of W contain a submatrix J, as do columns j and k, and so columns j and i also contain a submatrix J. Hence $C \in \mathsf{B}(W)$. Since $|\mathrm{inv}(W^{-1})| = \mathrm{len}(W^{-1}) = \mathrm{len}(W)$, $\mathsf{B}(W)$ has order $2^{\mathrm{len}(W)}$.

(ii) follows by observing that W_0W is the row-reversal of W, and this exchanges submatrices I_2 and J in a given pair of columns i and j.

(iii) Let $B' = WBW^{-1}$ where $B \in \mathsf{B}(W)$. Then $b_{i,j} = b'_{\rho^{-1}(i),\rho^{-1}(j)}$, where ρ is the permutation corresponding to W. In 'star' notation, $b_{i,j} = *$ for $1 \le j < i \le n$ if and only if columns j and i of W contain a submatrix J, so that $\rho^{-1}(j) > \rho^{-1}(i)$. Hence $B' \in \mathsf{U}(n)$.

□

Example 16.2.6 Let W correspond to $\rho = (2,4,3,1)$, as in (16.1). Then applying W^{-1} to (16.5) gives the subgroup of $\mathsf{U}(4)$ whose elements have the form

$$\begin{pmatrix} 1 & 0 & 0 & * \\ 0 & 1 & * & * \\ 0 & 0 & 1 & * \\ 0 & 0 & 0 & 1 \end{pmatrix}.$$

The next result is the strong form of the Bruhat decomposition.

Proposition 16.2.7 *Every matrix $A \in \mathsf{GL}(n)$ can be uniquely expressed in the form $A = BWB'$ where $B \in \mathsf{B}(n)$, $W \in \mathsf{W}(n)$ and $B' \in \mathsf{B}(W)$.*

Proof Using lower triangular transvections we can factor A as $A = BA'$, where $B \in \mathsf{B}(n)$ and A' is row-reduced, i.e. all entries below the last 1 in each column are 0. These 1s define a matrix $W \in \mathsf{W}(n)$, and we can permute the rows of A' to give a lower triangular matrix B'. The matrix $B' = W^{-1}A' \in \mathsf{B}(W)$ by definition of $\mathsf{B}(W)$. Hence $A = BA' = BWB'$. Hence the required factorization exists.

To prove uniqueness, A lies in a unique Bruhat cell $\mathsf{B}(n)W\mathsf{B}(n)$ by Theorem 16.2.2. Suppose that $B_1WB_1' = A = B_2WB_2'$, where $B_1, B_2 \in \mathsf{B}(n)$ and $B_1', B_2' \in \mathsf{B}(W)$. Then $BW = WB'$ where $B = B_2^{-1}B_1 \in \mathsf{B}(n)$ and $B' = B_2'(B_1')^{-1} \in \mathsf{B}(W)$. Then since $B = WB'W^{-1}$, B is upper triangular by Proposition 16.2.5(iii). Hence $B = I_n$, and it follows that $B_1 = B_2$ and $B_1' = B_2'$. □

The next result relates the Bruhat decomposition to multiplication in $\mathsf{GL}(n)$. If $A_1 \in \mathsf{B}(n)W_1\mathsf{B}(n)$ and $A_2 \in \mathsf{B}(n)W_2\mathsf{B}(n)$, then, in order to assign the product A_1A_2 to a Bruhat cell, it suffices to do this for products of the form W_1BW_2, where $B \in \mathsf{B}(n)$. Recall that for $1 \leq i \leq n-1$, S_i is the switch matrix obtained by exchanging rows i and $i+1$ of the identity matrix I_n.

Proposition 16.2.8 *For $W \in \mathsf{W}(n)$ and $B \in \mathsf{B}(n)$,*

(i) *$S_iBW \in \mathsf{B}(n)S_iW\mathsf{B}(n)$ if $i \notin des(W)$, and*
(ii) *$S_iBW \in \mathsf{B}(n)W\mathsf{B}(n)$ or $\mathsf{B}(n)S_iW\mathsf{B}(n)$ if $i \in des(W)$,*

and both cases occur.

Proof Recall that in all cases $S_iBW \in \mathsf{B}(n)W'\mathsf{B}(n)$ where $W' \in \mathsf{W}(n)$ is obtained by row reducing $A = S_iBW$, as in Example 16.2.1, to obtain a matrix A' in which every entry below the last 1 in each row is 0. These are the entries 1 of W'. Since row reduction is carried out from top to bottom, the reduction of the ith row depends only on the first i rows of A. Since $A = S_iBW$ is obtained by exchanging rows i and $i+1$ of the matrix BW, and row reduction of BW gives W, the row reduction A' of A is in either $S_iW\mathsf{B}(W)$ or $W\mathsf{B}(W)$.

To decide which case occurs, it is sufficient to consider the 2×2 submatrix of BW given by the entries in rows i, $i+1$ and columns $\rho(i)$, $\rho(i+1)$, where ρ is the permutation associated to W. In case (i), $\rho(i) < \rho(i+1)$ by Proposition 16.1.4, and so (see (16.4)) this submatrix is $\left(\begin{smallmatrix}1&0\\0&1\end{smallmatrix}\right)$ or $\left(\begin{smallmatrix}1&0\\1&1\end{smallmatrix}\right)$. Exchanging the rows, we obtain $W' = S_iW$. In case (ii), $\rho(i) > \rho(i+1)$ by Proposition 16.1.4, and so (see (16.4)) this submatrix is $\left(\begin{smallmatrix}0&1\\1&0\end{smallmatrix}\right)$ or $\left(\begin{smallmatrix}0&1\\1&1\end{smallmatrix}\right)$. After exchanging the rows, the case $\left(\begin{smallmatrix}0&1\\1&0\end{smallmatrix}\right)$ leads to $W' = S_iW$, but the case $\left(\begin{smallmatrix}0&1\\1&1\end{smallmatrix}\right)$ gives $\left(\begin{smallmatrix}1&1\\1&0\end{smallmatrix}\right)$ after row reduction, and so $W' = W$. □

In particular, case (ii) arises when $W = S_i$, and when $B = L_i$ is the transvection obtained by adding row i of I_n to row $i+1$, $S_iL_iS_i = U_i \notin \mathsf{B}(n)$. By replacing W and B in Proposition 16.2.8 by their inverses and noting that $\mathrm{len}(W^{-1}) = \mathrm{len}(W)$, we obtain a corresponding result for WBS_i.

Proposition 16.2.9 *For* $W \in \mathsf{W}(n)$ *and* $B \in \mathsf{B}(n)$,

(i) $WBS_i \in \mathsf{B}(n)WS_i\mathsf{B}(n)$ *if* $i \notin des(W^{-1})$,
(ii) $WBS_i \in \mathsf{B}(n)W\mathsf{B}(n)$ *or* $\mathsf{B}(n)WS_i\mathsf{B}(n)$ *if* $i \in des(W^{-1})$,

and both cases occur. □

16.3 The flag module FL(n)

In this section we construct the permutation module $\mathsf{FL}(n)$ given by the action of $\mathsf{GL}(n)$ on complete flags in $\mathsf{V}(n)$, and express $\mathsf{FL}(n)$ as the direct sum of subspaces called Schubert cells, indexed by permutations $\rho \in \Sigma(n)$, We identify $\mathsf{V}(n)$ with $\mathsf{DP}^1(n)$, so that the action of $A = (a_{i,j}) \in \mathsf{GL}(n)$ on $\mathsf{V}(n)$ is given by $v_i \cdot A = \sum_{j=1}^n a_{i,j}v_j$ for $1 \le i \le n$, as in Definition 9.2.1. For $1 \le i \le n$, we identify the subspace of $\mathsf{V}(n)$ spanned by $v_1, \ldots, v_i$ with $\mathsf{V}(i)$, and define $\mathsf{V}(0) = \{0\}$. Although we work over $\mathbb{F}_2$, we shall use geometric language to study the action of $\mathsf{GL}(n)$ on $\mathsf{V}(n)$.

Definition 16.3.1 A **(complete) flag** X in $\mathsf{V}(n)$ is a sequence of subspaces $\{0\} = X^0 \subset X^1 \subset X^2 \subset \cdots \subset X^{n-1} \subset X^n = \mathsf{V}(n)$, where $\dim X^i = i$ for $0 \le i \le n$. The flag R such that $R^i = \mathsf{V}(i)$ for all i is called the **reference flag**. The vector space over $\mathbb{F}_2$ spanned by the set of flags in $\mathsf{V}(n)$ is denoted by $\mathsf{FL}(n)$.

Since the action of $\mathsf{GL}(n)$ on $\mathsf{V}(n)$ preserves inclusions of subspaces, it induces a permutation action of $\mathsf{GL}(n)$ on the set of flags. Thus for $A \in \mathsf{GL}(n)$ and a flag X, the flag $X \cdot A$ is given by $(X \cdot A)^i = X^iA$ for $0 \le i \le n$. Let $u_i = v_iA$ be the ith row of A, for $1 \le i \le n$, and let $X(A)$ be the flag such that $X(A)^i$ is spanned by $u_1, \ldots, u_i$. Since $u_i = v_iA$, $X(A)^i = \mathsf{V}(i)A$ and $X(A) = R \cdot A$. In particular, since $\mathsf{V}(i)A = \mathsf{V}(i)$ for all i if and only if A is lower triangular, $\mathsf{B}(n)$ is the stabilizer of the reference flag R.

Proposition 16.3.2 *The function* $\mathsf{GL}(n) \to \mathsf{FL}(n)$ *which associates to* $A \in \mathsf{GL}(n)$ *the flag* $X(A) = R \cdot A$ *induces a bijection between right cosets* $\mathsf{B}(n)A$ *of* $\mathsf{B}(n)$ *and flags in* $\mathsf{V}(n)$. *Thus* $\dim \mathsf{FL}(n) = 1 \cdot 3 \cdot 7 \cdots (2^n - 1)$.

Proof If A_1 and A_2 are in the same right coset of $\mathsf{B}(n)$, then $A_2 = BA_1$, where $B \in \mathsf{B}(n)$. Hence $\mathsf{V}(i)A_2 = \mathsf{V}(i)(BA_1) = (\mathsf{V}(i)B)A_1 = \mathsf{V}(i)A_1$. Conversely, suppose that $\mathsf{V}(i)A_1 = \mathsf{V}(i)A_2$ for $1 \le i \le n$, and let $A_2 = CA_1$ where $C \in \mathsf{GL}(n)$. Then $\mathsf{V}(i)A_2 = \mathsf{V}(i)(CA_1) = (\mathsf{V}(i)C)A_1$, and since A_1 is nonsingular it follows that $\mathsf{V}(i) = \mathsf{V}(i)C$ for $1 \le i \le n$. Hence $C \in \mathsf{B}(n)$. The second statement follows since $|\mathsf{GL}(n)| = (2^n - 1)(2^n - 2) \cdots (2^n - 2^{n-1})$, $|\mathsf{B}(n)| = 2^{n(n-1)/2}$ and the index $[\mathsf{GL}(n) : \mathsf{B}(n)] = |\mathsf{GL}(n)|/|\mathsf{B}(n)| = 1 \cdot 3 \cdot 7 \cdots (2^n - 1)$. □

Proposition 16.3.3 *The flag module* $\mathsf{FL}(n)$ *is isomorphic as a right* $\mathbb{F}_2\mathsf{GL}(n)$-*module to the permutation module* $\mathbb{F}_2(\mathsf{GL}(n)/\mathsf{B}(n))$ *spanned by the right cosets* $\mathsf{B}(n)A$, *with the action defined by* $(\mathsf{B}(n)A_1) \cdot A_2 = \mathsf{B}(n)(A_1A_2)$ *for* $A_1, A_2 \in \mathsf{GL}(n)$.

Proof It follows from associativity of matrix multiplication that the bijection $\mathsf{B}(n)A \leftrightarrow X(A) = R \cdot A$ commutes with the right action of $\mathsf{GL}(n)$. □

The permutation module $\mathbb{F}_2(\mathsf{GL}(n)/\mathsf{B}(n))$ can be regarded as the $\mathbb{F}_2\mathsf{GL}(n)$-module induced from the trivial 1-dimensional $\mathbb{F}_2\mathsf{B}(n)$-module. The stabilizer of the flag corresponding to $\mathsf{B}(n)A$ is the subgroup $A^{-1}\mathsf{B}(n)A$ conjugate to $\mathsf{B}(n)$. The Bruhat decomposition of $\mathsf{GL}(n)$ gives a corresponding decomposition of the set of right cosets.

Definition 16.3.4 Let $W \in \mathsf{W}(n)$ correspond to $\rho \in \Sigma(n)$ as in (16.1). The **Schubert cell** $\mathsf{Sch}(W)$ or $\mathsf{Sch}(\rho)$ is the $\mathbb{F}_2$-subspace of $\mathsf{FL}(n)$ spanned by the set of right cosets $\mathsf{B}(n)A$ such that $\mathsf{B}(n)A \subseteq \mathsf{B}(n)W\mathsf{B}(n)$, or the corresponding set of flags.

We have seen in Section 16.2 that each coset $\mathsf{B}(n)A$ contains a matrix $A' = BA$ such that every entry below the last 1 in each row is 0, and that these row-reduced matrices have the form WB where $W \in \mathsf{W}(n)$ and $B \in \mathsf{B}(W)$. It follows from Proposition 16.2.7 that each coset $\mathsf{B}(n)A$ contains a unique row-reduced matrix. Thus we have a bijection between complete flags X and row-reduced matrices $A' = A_X$ in the corresponding cosets, and it is convenient to use these matrices in calculations with flags.

The proof of Proposition 16.2.7 yields the following description of the row-reduced matrices for cosets in $\mathsf{Sch}(W)$. The corresponding flags are given by the sequences of subspaces of $\mathsf{V}(n)$ obtained by adding the rows of A_X successively to the spanning set. Thus the Schubert cell $\mathsf{Sch}(W)$ and the Bruhat subgroup $\mathsf{B}(W)$ are related as follows.

Proposition 16.3.5 *For $W \in \mathsf{W}(n)$, the row-reduced matrices A_X which correspond to flags X in $\mathsf{Sch}(W)$ are the elements of the left coset $W\mathsf{B}(W)$. The Schubert cell $\mathsf{Sch}(W)$ is a vector space of dimension $2^{len(W)}$ over $\mathbb{F}_2$. The flags X for which A_X is obtained by replacing the entries 0 in W which are not below or to the right of a 1 by stars.* □

Example 16.3.6 In (16.1), flags $X \in \mathsf{Sch}(\rho)$ correspond to the matrices

$$A_X = \begin{pmatrix} * & 1 & 0 & 0 \\ * & 0 & * & 1 \\ * & 0 & 1 & 0 \\ 1 & 0 & 0 & 0 \end{pmatrix},$$

where the stars stand for either 0 or 1 in $\mathbb{F}_2$.

The next result gives an alternative view of the Schubert cell decomposition of $\mathsf{FL}(n)$, which is useful in describing its structure as a $\mathbb{F}_2\mathsf{GL}(n)$-module.

Proposition 16.3.7 *The Schubert cells are the orbits of the action of $\mathsf{B}(n)$ on the set of flags in $\mathsf{V}(n)$. Hence if $\sigma_W \in \mathsf{FL}(n)$ is the sum of the flags in $\mathsf{Sch}(W)$, then the set $\{\sigma_W : W \in \mathsf{W}(n)\}$ is a $\mathbb{F}_2$-basis for the $\mathsf{B}(n)$-invariant elements of $\mathsf{FL}(n)$.*

Proof Writing flags as right cosets of $\mathsf{B}(n)$, we recall that $A \in \mathsf{GL}(n)$ acts on $\mathsf{FL}(n)$ by $(\sum_X \mathsf{B}(n)X)\cdot A = \sum_X \mathsf{B}(n)(XA)$. For all $B \in \mathsf{B}(n)$, $\mathsf{B}(n)X$ and $\mathsf{B}(n)XB$ lie in the same Bruhat cell $\mathsf{B}(n)X\mathsf{B}(n)$. Conversely, if $\mathsf{B}(n)Y \subseteq \mathsf{B}(n)X\mathsf{B}(n)$ then there is an element $B \in \mathsf{B}(n)$ such that $\mathsf{B}(n)XB = \mathsf{B}(n)Y$. Hence the orbits of the action of $\mathsf{B}(n)$ are the sets of cosets which lie in the same Bruhat cell $\mathsf{B}(n)W\mathsf{B}(n)$, $W \in \mathsf{W}$, or equivalently the sets of flags which lie in the same Schubert cell. This proves the first statement, and the second follows. □

16.4 Partial flags and parabolic subgroups

We generalize the results of the preceding section by considering permutation modules $\mathsf{FL}^I(n)$, where $I \subseteq Z[n-1] = \{1,\dots,n-1\}$, given by the action of $\mathsf{GL}(n)$ on partial flags in $\mathsf{V}(n)$. These are nested sequences of subspaces of $\mathsf{V}(n)$ ordered by inclusion, in which the dimensions of the subspaces are in I. For example, when $I = \{k\}$, the partial flag module $\mathsf{FL}^I(n)$ is given by the permutation action of $\mathsf{GL}(n)$ on the 'Grassmannian', i.e. the set of k-dimensional subspaces of $\mathsf{V}(n)$. Thus $\mathsf{FL}^I(n)$ is a permutation module for $\mathbb{F}_2\mathsf{GL}(n)$, and $\mathsf{FL}^I(n) = \mathsf{FL}(n)$ when $I = Z[n-1]$.

Definition 16.4.1 Given $I = \{i_1, i_2, \ldots, i_r\} \subseteq Z[n-1]$, a **partial flag of type** I in $\mathsf{V}(n)$ is a sequence X^I of subspaces $\{0\} = X^0 \subset X^{i_1} \subset X^{i_2} \subset \cdots \subset X^{i_r} \subset X^n = \mathsf{V}(n)$, where $\dim X^i = i$. The **reference** partial flag R^I of type I is defined by $R^i = \mathsf{V}(i)$ for all $i \in I$. The **lower parabolic subgroup** $\mathsf{P}^I(n)$ is the stabilizer of R^I in $\mathsf{GL}(n)$. The **partial flag module** $\mathsf{FL}^I(n)$ is the right $\mathbb{F}_2\mathsf{GL}(n)$-module given by the permutation action of $\mathsf{GL}(n)$ on the partial flags of type I.

We shall show that partial flags of type I correspond to right cosets of $\mathsf{P}^I(n)$ in $\mathsf{GL}(n)$. If $A \in \mathsf{P}^I(n)$, then the first i_1 rows of A form a basis for $\mathsf{V}(i_1)$, and so the submatrix A_{i_1} of A formed by the first i_1 rows and columns is nonsingular, i.e. it is in $\mathsf{GL}(i_1)$, and all other entries of A in the first i_1 rows are 0s. As the first i_2 rows of A form a basis for $\mathsf{V}(i_2)$, the submatrix A_{i_2} formed by the first i_2 rows and columns is in $\mathsf{GL}(i_2)$, and all other entries of A in the first i_2 rows are 0s. As the nonzero entries of A_{i_2} in the last $i_2 - i_1$ columns all lie in the last $i_2 - i_1$ rows, the submatrix formed by these rows and columns is in $\mathsf{GL}(i_2 - i_1)$.

Continuing in this way, we see that $\mathsf{P}^I(n)$ is a 'staircase group' consisting of matrices of the form illustrated below in the case $n = 6$, $I = \{2,3\}$. The diagonal submatrices are nonsingular and those below them are arbitrary, so this subgroup of $\mathsf{GL}(6)$ has order $2^{11} \cdot 6 \cdot 1 \cdot 168$.

$$\left(\begin{array}{cc|c|ccc} * & * & 0 & 0 & 0 & 0 \\ * & * & 0 & 0 & 0 & 0 \\ \hline * & * & 1 & 0 & 0 & 0 \\ \hline * & * & * & * & * & * \\ * & * & * & * & * & * \\ * & * & * & * & * & * \end{array}\right) \tag{16.7}$$

The staircase group $\mathsf{P}^I(n)$ has a subgroup $\mathsf{B}^I(n) \subseteq \mathsf{B}(n)$ consisting of the matrices whose diagonal blocks are identity matrices, and a subgroup $\mathsf{GL}^I(n)$ given by the matrices whose subdiagonal blocks are zero. Every matrix $A \in \mathsf{P}^I(n)$ has unique factorizations of the form $A = B_1A'$ and $A = A'B_2$, where $A' \in \mathsf{GL}^I(n)$ is obtained from A by replacing all 1s in subdiagonal blocks by 0s, and $B_1, B_2 \in \mathsf{B}^I(n)$. For example

$$\left(\begin{array}{cc|c} 1 & 1 & 0 \\ 1 & 0 & 0 \\ \hline 1 & 0 & 1 \end{array}\right) = \left(\begin{array}{cc|c} 1 & 0 & 0 \\ 0 & 1 & 0 \\ \hline 0 & 1 & 1 \end{array}\right)\left(\begin{array}{cc|c} 1 & 1 & 0 \\ 1 & 0 & 0 \\ \hline 0 & 0 & 1 \end{array}\right) = \left(\begin{array}{cc|c} 1 & 1 & 0 \\ 1 & 0 & 0 \\ \hline 0 & 0 & 1 \end{array}\right)\left(\begin{array}{cc|c} 1 & 0 & 0 \\ 0 & 1 & 0 \\ \hline 1 & 0 & 1 \end{array}\right).$$

The first factorization is achieved by blockwise row reduction and the second by blockwise column reduction, with the aim in both cases of reducing subdiagonal blocks to zero. As a group, $\mathsf{P}^I(n)$ is the semidirect product $\mathsf{B}^I(n) \rtimes \mathsf{GL}^I(n)$. In the general case, $\mathsf{B}^I(n)$ coincides with the Bruhat subgroup $\mathsf{B}(W)$, where $W \in \mathsf{W}(n)$ corresponds to the permutation which reverses successive subsequences of $(1,2,\ldots,n)$ of lengths $c_1,\ldots,c_{r+1}$, and $\mathsf{GL}^I(n) \cong \mathsf{GL}(c_1) \times \cdots \times \mathsf{GL}(c_{r+1})$.

Definition 16.4.2 A **composition** of $n \geq 1$ is a finite sequence C with sum $|C| = n$. For $I = \{i_1, i_2, \ldots, i_r\}$ where $0 = i_0 < i_1 < \cdots < i_r < i_{r+1} = n$, the associated composition of n is $C = (c_1, \cdots, c_{r+1})$, where $c_j = i_j - i_{j-1}$, so that $i_j = c_1 + \cdots + c_j$ for $1 \leq j \leq r+1$.

Proposition 16.4.3 *The partial flag module* $\mathsf{FL}^I(n)$ *is isomorphic as a right* $\mathbb{F}_2\mathsf{GL}(n)$*-module to the permutation module* $\mathbb{F}_2(\mathsf{GL}(n)/\mathsf{P}^I(n))$ *on the right cosets of* $\mathsf{P}^I(n)$ *in* $\mathsf{GL}(n)$*, with action given by* $(\mathsf{P}^I(n)A)A' = \mathsf{P}^I(n)(AA')$ *for* $A, A' \in \mathsf{GL}(n)$.

Proof We generalize the bijection of Proposition 16.3.2 to give a bijection between partial flags of type I in $\mathsf{V}(n)$ and right cosets $\mathsf{P}^I(n)A$ of $\mathsf{P}^I(n)$. As before, we associate to $A \in \mathsf{GL}(n)$ the partial flag $X(A) = R^I \cdot A$ of type I, so that $(X(A))^i = \mathsf{V}(i)A$ is the subspace of $\mathsf{V}(n)$ spanned by the first i rows of A. Clearly $X(A)$ is the reference partial flag R^I if and only if $A \in \mathsf{P}^I(n)$. The proof is completed as for Propositions 16.3.2 and 16.3.3, with $\mathsf{P}^I(n)$ replacing $\mathsf{B}(n)$ and with i restricted to elements of I. □

Proposition 16.4.4 *The number of partial flags of type I is the 2-multinomial coefficient*

$$\dim \mathsf{FL}^I(n) = |\mathsf{GL}(n)|/|\mathsf{P}^I(n)| = \binom{n}{c_1,\ldots,c_{r+1}}_2 = \frac{(n!)_2}{(c_1!)_2 \cdots (c_{r+1}!)_2},$$

where $(k!)_2 = 1 \cdot 3 \cdot 7 \cdots (2^k - 1)$ *for* $k \geq 0$.

Proof Since $|\mathsf{GL}(n)|/|\mathsf{B}(n)| = (n!)_2$ by Proposition 16.3.3, it is equivalent to prove that $|\mathsf{P}^I(n)|/|\mathsf{B}(n)| = (c_1!)_2 \cdots (c_{r+1}!)_2$. A matrix $P \in \mathsf{P}^I(n)$ can be written uniquely in the form $P = BA$ where $B \in \mathsf{B}^I(n)$ and $A \in \mathsf{GL}^I(n)$. Now A

is the direct sum of diagonal blocks A_i. We may identify A_i with an element of $\mathsf{GL}(c_i)$ for each $i \in I$, and write $A_i = B_i A_i'$ where $B_i \in \mathsf{B}(c_i)$ is lower triangular and A_i' is row-reduced, so that the elements A_i are a set of coset representatives for $\mathsf{B}(c_i)$ in $\mathsf{GL}(c_i)$. Combining the lower triangular factors, we have a unique factorization $P = B'A'$ where $B' \in \mathsf{B}(n)$ and A' is a block diagonal matrix with blocks A_i' for $i \in I$. Hence $|\mathsf{P}^I| = |\mathsf{B}(n)| \cdot \prod_{i \in I} |\mathsf{GL}(c_i)|/|\mathsf{B}(c_i)| = |\mathsf{B}(n)| \cdot (c_1!)_2 \cdots (c_{r+1}!)_2$. □

We next obtain a Schubert cell decomposition of the partial flag module $\mathsf{FL}^I(n)$. We begin with the Bruhat decomposition of $\mathsf{P}^I(n)$.

Proposition 16.4.5 *For $I \subseteq Z[n-1]$, let $\mathsf{W}^I(n) = \mathsf{W}(n) \cap \mathsf{GL}^I(n)$. Then every matrix $P \in \mathsf{P}^I(n)$ can be expressed in the form $P = BWB'$ where $B, B' \in \mathsf{B}(n)$ and $W \in \mathsf{W}^I(n)$. The permutation matrix W is uniquely determined by P, and there is a unique choice of B and B' such that $B' \in \mathsf{B}(W)$. Thus $\mathsf{P}^I(n)$ is the union of the Bruhat cells $\mathsf{B}(n)W\mathsf{B}(n)$ where $W \in \mathsf{W}^I(n)$.*

Proof As in the preceding proof, we can factor $P \in \mathsf{P}^I(n)$ uniquely in the form $P = BA$ where $B \in \mathsf{B}^I(n)$ and $A \in \mathsf{GL}^I(n)$. The element A is a block diagonal matrix with nonsingular diagonal blocks A_i, $1 \leq i \leq r+1$. The results follow by applying Theorem 16.2.2 and Proposition 16.2.7 to each submatrix A_i. □

Thus the group $\mathsf{W}^I(n)$ consists of permutation matrices with diagonal blocks in $\mathsf{W}(c_1) \times \cdots \times \mathsf{W}(c_{r+1})$, and is generated by the switch matrices S_i for $i \notin I$. When $n = 6$ and $I = \{2,3\}$, for example, $\mathsf{W}^I(n)$ is generated by S_1, S_4 and S_5, as we see from (16.7). In particular, $W^I(n) = \mathsf{W}(n)$ when $I = \emptyset$ and $\mathsf{W}^I(n) = I_n$ when $I = Z[n-1]$.

In order to define Schubert cells in $\mathsf{FL}^I(n)$, we require a convenient set of representatives for the right cosets $\mathsf{P}^I(n)A$, $A \in \mathsf{GL}(n)$. Since $\mathsf{W}^I(n) \subset \mathsf{P}^I(n)$, we begin by selecting a special set of representatives for the right cosets $\mathsf{W}^I(n)W$. These will parameterize the Schubert cell decomposition of $\mathsf{GL}(n)/\mathsf{P}^I(n)$.

Example 16.4.6 When $I = \{2\}$, $\mathsf{W}^I(4)$ is generated by S_1 and S_3, and its right cosets in $\mathsf{W}(4)$ are represented by the identity matrix I_4 and the five matrices

$$\begin{pmatrix} 1&0&0&0\\ 0&0&1&0\\ 0&1&0&0\\ 0&0&0&1 \end{pmatrix}, \begin{pmatrix} 1&0&0&0\\ 0&0&0&1\\ 0&1&0&0\\ 0&0&1&0 \end{pmatrix}, \begin{pmatrix} 0&1&0&0\\ 0&0&1&0\\ 1&0&0&0\\ 0&0&0&1 \end{pmatrix}, \begin{pmatrix} 0&1&0&0\\ 0&0&0&1\\ 1&0&0&0\\ 0&0&1&0 \end{pmatrix}, \begin{pmatrix} 0&0&1&0\\ 0&0&0&1\\ 1&0&0&0\\ 0&1&0&0 \end{pmatrix} \tag{16.8}$$

with descent set $\{2\}$.

Proposition 16.4.7 *For* $I \subseteq Z[n-1]$*, the permutation matrices* W *such that* $des(W) \subseteq I$ *are a set of representatives for the right cosets of* $\mathsf{W}^I(n)$ *in* $\mathsf{W}(n)$*.*

Proof Let $\rho \in \Sigma(n)$ be the permutation corresponding to $W \in \mathsf{W}(n)$. With notation as in Definition 16.4.2, $\mathrm{des}(\rho) \subseteq I$ if and only if ρ is increasing on each subinterval $\{i_{j-1}+1,\ldots,i_j\}$, $1 \leq j \leq r+1$. Thus successive blocks of $c_1,\ldots,c_r$ rows of W are in the same order as in the identity I_n, and pairs of consecutive rows not in this order can occur only between blocks. Premultiplication by elements $W' \in \mathsf{W}^I(n)$ permutes the rows within each block, and if $W' \neq I_n$, then the order of some pair of consecutive rows in one of the blocks must be exchanged, so that the $\mathrm{des}(W'W) \not\subseteq I$. Hence W is the unique element of the coset $\mathsf{W}^I(n)W$ with descent set in I. $\square$

When $\mathrm{des}(\rho) \subseteq I$, the one-line notation for ρ is the concatenation of an r-tuple of increasing sequences of lengths $c_1,\ldots,c_{r+1}$. Hence

Proposition 16.4.8 *The number of Schubert cells in* $\mathsf{FL}^I(n)$ *is*

$$[\mathsf{W}(n) : \mathsf{W}^I(n)] = \binom{n}{c_1,\ldots,c_{r+1}} = \frac{n!}{c_1!\cdots c_{r+1}!},$$

where $C = (c_1,\ldots,c_{r+1})$ *is the composition of* n *associated to* I*.* $\square$

This also follows from $\mathsf{W}^I(n) \cong \mathsf{W}(c_1) \times \cdots \times \mathsf{W}(c_{r+1})$. The following result will be useful in relating the parabolic subgroups $\mathsf{P}^I(n)$ to the Bruhat decomposition of $\mathsf{GL}(n)$.

Proposition 16.4.9 *Let* $V \in \mathsf{W}^I(n)$*,* $B \in \mathsf{B}(n)$ *and* $W \in \mathsf{W}(n)$*. Then* $VBW \in \mathsf{B}(n)W'\mathsf{B}(n)$*, where* $W' = V'W$ *for some* $V' \in \mathsf{W}^I(n)$*.*

Proof Since the subgroup $\mathsf{W}^I(n)$ of $\mathsf{W}(n)$ is generated by the elements S_i, $i \notin I$, V can be written as a word $V_1 \cdots V_m$ in these generators. By Proposition 16.2.8, V_jBW is either in $\mathsf{B}(n)V_jW\mathsf{B}(n)$ or in $\mathsf{B}(n)W\mathsf{B}(n)$. Hence, by induction on m, $VBW \in \mathsf{B}(n)V'W\mathsf{B}(n)$, where V' is a word obtained by omitting certain of the factors V_j from the word $V_1 \cdots V_m$. In particular, $V' \in \mathsf{W}^I(n)$. $\square$

We next obtain a set of representatives for the right cosets $\mathsf{P}^I(n)A$.

Proposition 16.4.10 *Let* $I \subseteq Z[n-1]$ *and* $A \in \mathsf{GL}(n)$*. Then the coset* $\mathsf{P}^I(n)A$ *contains a unique matrix* A' *such that* $A' = WB$*, where* $des(W) \subseteq I$ *and* $B \in \mathsf{B}(W)$*.*

Proof We first prove existence of such a matrix A'. Let $A \in \mathsf{GL}(n)$ lie in the Bruhat cell $\mathsf{B}(n)W\mathsf{B}(n)$, and let $W = W_1W_2$ where $W_1 \in \mathsf{W}^I(n)$ and $\mathrm{des}(W_2) \subseteq I$. Then $A = B_1W_1W_2B_2$ where $B_1, B_2 \in \mathsf{B}(n)$. Using Proposition 16.2.7, let

$W_2B_2 = B'_1W_2B'_2$ where $B'_1 \in \mathsf{B}(n)$ and $B'_2 \in \mathsf{B}(W_2)$. Then $B_1W_1B'_1 \in \mathsf{P}^I(n)$ since $W_1 \in \mathsf{W}^I(n)$, and so $A' = W_2B'_2$ is an element of the coset $\mathsf{P}^I(n)A$ of the required form.

We prove uniqueness in two steps. First we prove that W is unique. Suppose that $\mathsf{P}^I WB = \mathsf{P}^I W'B'$ where $\mathrm{des}(W)$ and $\mathrm{des}(W')$ are subsets of I and $B \in \mathsf{B}(W)$, $B' \in \mathsf{B}(W')$. Then $W'B' = (B_1VB_2)WB = B_1(VB_2W)B$, where $B_1, B_2 \in \mathsf{B}(n)$ and $V \in \mathsf{W}^I(n)$. By Proposition 16.4.9, $VB_2W = B_3V'WB_4$ where $B_3, B_4 \in \mathsf{B}(n)$ and $V' \in \mathsf{W}^I(n)$. By Proposition 16.2.7, it follows that $W' = V'W$ and since $\mathrm{des}(W)$ and $\mathrm{des}(W')$ are subsets of I, $W = W'$ and $V' = I_n$, the identity matrix.

Next we prove that $B \in \mathsf{B}(W)$ is unique. Let $PWB = WB'$ where $P \in \mathsf{P}^I(n)$, $\mathrm{des}(W) \subseteq I$ and $B, B' \in \mathsf{B}(W)$. Let $B'' = B'B^{-1} \in \mathsf{B}(W)$. Then $P = WB''W^{-1}$ is upper triangular by Proposition 16.2.5(iii). Thus in order to show that $P = I_n$ we need only show that the diagonal blocks in P are lower triangular, i.e. the (i,j)th entry of P is 0 if $i_{s-1} < i < j \leq i_s$ for $s = 1, \ldots, r+1$. Since $\mathrm{des}(W) \subseteq I$, $\rho(i) < \rho(j)$ where ρ is the permutation associated to W. Since $B'' \in \mathsf{B}(W)$, the last 1 in row i of WB'' is in column $\rho(i)$, and so the $(i, \rho(j))$th entry of WB'' is 0. But this is the (i,j)th entry of $P = WB''W^{-1}$. □

We next define a Schubert cell decomposition of the partial flag module $\mathsf{FL}^I(n)$.

Definition 16.4.11 Let $W \in \mathsf{W}(n)$ with $\mathrm{des}(W) \subseteq I$ correspond to $\rho \in \Sigma(n)$. Then the **Schubert cell** $\mathsf{Sch}(W)$ (or $\mathsf{Sch}(\rho)$) is the $\mathbb{F}_2$-subspace of $\mathsf{FL}^I(n)$ spanned by the set of right cosets $\mathsf{P}^I(n)A$ such that $\mathsf{P}^I(n)A \subseteq \mathsf{P}^I(n)W\mathsf{B}(n)$, or the corresponding set of partial flags of type I.

Example 16.4.12 For $n = 4$ and $I = \{2\}$, the diagrams below show the Bruhat order on permutations $\rho \in \Sigma(4)$ with $\mathrm{des}(\rho) \subseteq I$, and a geometric description of the flags in the corresponding Schubert cells of $\mathsf{FL}^2(4)$.

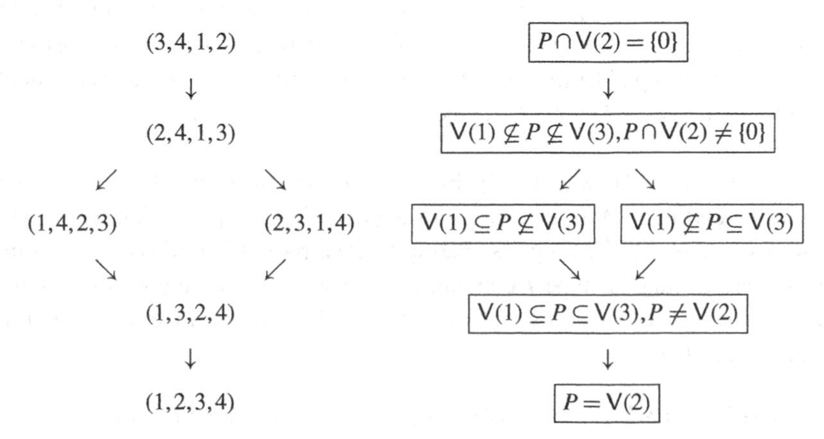

It follows from Proposition 16.4.10 that the set of partial flags in each Schubert cell is indexed in the same way as for complete flags, but only permutation matrices W with descent set in I are used, and we then have a bijection between partial flags $X \in \mathsf{Sch}(W)$ and row-reduced matrices in $W\mathsf{B}(n)$. We use these matrices in calculations with partial flags. By Proposition 16.3.5, $\dim \mathsf{Sch}(W) = 2^{\mathrm{len}(W)}$. This leads to the following formula, which relates the number of partial flags of type I, given by Proposition 16.4.4, to permutation matrices.

Proposition 16.4.13 *For* $I \subseteq Z[n-1]$, $\dim \mathsf{FL}^I(n) = \sum_{des(W) \subseteq I} 2^{len(W)}$. □

16.5 The direct sum decomposition of FL(n)

In this section we show that the $\mathbb{F}_2\mathsf{GL}(n)$-module $\mathsf{FL}(n)$ has a direct sum decomposition $\mathsf{FL}(n) \cong \bigoplus_I \mathsf{FL}_I(n)$, with 2^{n-1} summands $\mathsf{FL}_I(n)$ indexed by subsets $I \subseteq Z[n-1]$. The module $\mathsf{FL}_I(n)$ is a direct summand of the partial flag module $\mathsf{FL}^I(n)$. We shall prove in Chapter 17 that the modules $\mathsf{FL}_I(n)$ are indecomposable, so that this direct sum decomposition of $\mathsf{FL}(n)$ is maximal.

Proposition 16.5.1 *For each* $I \subseteq Z[n-1]$, *the partial flag module* $\mathsf{FL}^I(n)$ *can be embedded as a direct summand in* $\mathsf{FL}(n)$. *More generally, if* $J \subseteq I \subseteq Z[n-1]$, *then* $\mathsf{FL}^J(n)$ *can be embedded as a direct summand in* $\mathsf{FL}^I(n)$.

Proof Consider the $\mathbb{F}_2\mathsf{GL}(n)$-maps $\mathsf{FL}^J(n) \xrightarrow{i} \mathsf{FL}(n) \xrightarrow{j} \mathsf{FL}^J(n)$ which associate to a partial flag X of type J the sum of the complete flags containing X, and to a complete flag X' its restriction to a partial flag of type J. The composition $j \circ i$ maps a partial flag X to $X + \cdots + X = kX$, where k is the number of complete flags containing X. But $k = |\mathsf{P}^J(n)|/|\mathsf{B}(n)|$ is odd, since $\mathsf{B}(n)$ is a Sylow 2-subgroup of $\mathsf{GL}(n)$. Hence $j \circ i$ is the identity map. The same argument applies to the general case $J \subseteq I$, since $|\mathsf{P}^J(n)|/|\mathsf{P}^I(n)|$ is odd. □

For $I \subseteq Z[n-1]$, we identify $\mathsf{FL}^I(n)$ with its image in $\mathsf{FL}(n)$ under the above embedding. More generally, for $J \subseteq I \subseteq Z[n-1]$, we identify $\mathsf{FL}^J(n)$ with its image in $\mathsf{FL}^I(n)$, by associating to each partial flag of type J the sum of all partial flags of type I containing it. It is clear that these embeddings are consistent when $K \subseteq J \subseteq I$. Next we consider how the modules $\mathsf{FL}^I(n)$ fit together in $\mathsf{FL}(n)$.

Proposition 16.5.2 *For* $I, J \subseteq Z[n-1]$, $\mathsf{FL}^I(n) \cap \mathsf{FL}^J(n) = \mathsf{FL}^{I \cap J}(n)$.

Proof The inclusion $\mathsf{FL}^{I\cap J}(n) \subseteq \mathsf{FL}^{I}(n) \cap \mathsf{FL}^{J}(n)$ follows from Proposition 16.5.1. By replacing $\mathsf{FL}(n)$ by $\mathsf{FL}^{I\cup J}(n)$, we may assume that $I \cup J = Z[n-1]$.

Consider first the case $I \cap J = \emptyset$, $I \cup J = Z[n-1]$. Let $\mathcal{S}$ be a set of complete flags in $\mathsf{V}(n)$ which is closed with respect to the operation of replacing a flag X by a flag Y if $X^i = Y^i$ for all $i \in I$ or for all $j \in J$, so that the sum of the flags in $\mathcal{S}$ is an element of $\mathsf{FL}^{I}(n) \cap \mathsf{FL}^{J}(n)$. Since the only nonzero element of $\mathsf{FL}^{\emptyset}(n)$ is the sum of all complete flags, we wish to show that if $\mathcal{S}$ is non-empty then $\mathcal{S}$ contains all complete flags. Since $I \cup J = Z[n-1]$, a complete flag obtained by changing any one subspace in a flag in $\mathcal{S}$ is again in $\mathcal{S}$.

Let $X \in \mathcal{S}$, let Y be any complete flag and let $L = Y^1$. If $L = X^1$, we may work in the quotient space $\mathsf{V}(n)/L$ and the result will follow by induction on n. Hence we may assume that $L \neq X^1$. Define s such that $L \subseteq X^s$, $L \not\subseteq X^{s-1}$, so that s is uniquely defined and $2 \leq s \leq n$. For $1 \leq j \leq s$, define $Z^j = X^{j-1} \oplus L$. Thus $Z^1 = L$, $Z^s = X^s$ and the sequence Z of subspaces $\{0\} \subset Z^1 \subset Z^2 \subset \cdots \subset Z^{s-1} \subset X^s \subset \cdots \subset X^{n-1} \subset \mathsf{V}(n)$ is a complete flag. Using the inclusions

$$\begin{array}{ccc} X^{j-1} & \subset & X^j \\ \cap & & \cap \\ Z^j & \subset & Z^{j+1} \end{array}$$

we may successively replace X^j by Z^j for $j = s-1, s-2, \ldots, 1$, obtaining at each stage a complete flag in $\mathcal{S}$. Thus $Z \in \mathcal{S}$. Since $L = Z^1$, the result follows as before by induction on n.

For the general case, let X and Y be complete flags in $\mathsf{V}(n)$ which restrict to the same partial flag in $\mathsf{FL}^{I\cap J}(n)$. Since we are assuming that $I \cup J = Z[n-1]$, we may divide the interval from 1 to $n-1$ into subintervals whose end points are in $I \cap J$ and whose interior points are in I or J, but not in both. The general case follows by applying the argument for the case $I \cap J = \emptyset$ to each subinterval. □

Theorem 16.5.3 *For* $I \subseteq Z[n-1]$ *let* $\mathsf{FL}_I(n) = \mathsf{FL}^I(n)/\sum_{J\subset I} \mathsf{FL}^J(n)$. *Then* (i) $\mathsf{FL}^I(n) \cong \bigoplus_{J\subseteq I} \mathsf{FL}_J(n)$ *and* (ii) $\dim \mathsf{FL}_I(n) = \sum_{des(\rho)=I} 2^{len(\rho)}$.

Proof We fix n and argue by induction on $|I|$. For $I = \emptyset$, $\mathsf{FL}_{\emptyset}(n) = \mathsf{FL}^{\emptyset}(n)$, the 1-dimensional submodule generated by the sum of all the complete flags. Let $I = \{i_1, \ldots, i_r\}$, where $r \geq 1$. Then by Proposition 16.5.1 $\mathsf{FL}^I(n)$ has r direct summands $\mathsf{FL}^{I_k}(n)$, where $I_k = I \setminus \{i_k\}$. By the induction hypothesis, each of these is the direct sum of 2^{r-1} submodules of the form $\mathsf{FL}_J(n)$ with $J \subseteq I$. We collect a maximal subset of $\mathsf{FL}_J(n)$'s such that any two have zero intersection, as follows.

We select all the 2^{r-1} summands $\mathsf{FL}_J(n)$ of $\mathsf{FL}^{I_1}(n)$. By Proposition 16.5.2, $\mathsf{FL}^{I_1}(n) \cap \mathsf{FL}^{I_2}(n) = \mathsf{FL}^{I_1 \cap I_2}(n)$ contains 2^{r-2} of these summands $\mathsf{FL}_J(n)$, and so we select the remaining 2^{r-2} summands. Continuing in this way, we can select 2^{r-3} summands of $\mathsf{FL}^{I_3}(n)$ which are not in $\mathsf{FL}^{I_1}(n) + \mathsf{FL}^{I_2}(n)$, and so on until we have $2^{r-1} + 2^{r-2} + \cdots + 1 = 2^r - 1$ summands of $\mathsf{FL}^I(n)$, and these are isomorphic to the modules $\mathsf{FL}_J(n)$ for $J \subset I$. The complementary summand of $\sum_{J \subset I} \mathsf{FL}^J(n)$ in $\mathsf{FL}^I(n)$ is then isomorphic to $\mathsf{FL}_I(n)$. This proves (i), and (ii) follows by induction using Proposition 16.4.13. □

In particular, the anti-diagonal matrix W_0 is the unique element of $\mathsf{W}(n)$ with descent set $Z[n-1]$. Since $\mathrm{len}(W_0) = n(n-1)/2$, $\mathsf{FL}_I(n)$ has dimension $2^{n(n-1)/2}$ when $I = Z[n-1]$. We prove in Proposition 17.6.8 that it is an irreducible $\mathbb{F}_2\mathsf{GL}(n)$-module, the Steinberg module $\mathsf{St}(n)$.

Example 16.5.4 The cases $n = 2$ and $n = 3$ are discussed in Sections 16.0 and 16.6 respectively. By Proposition 16.3.2, $\dim \mathsf{FL}(4) = 315$ and $\dim \mathsf{FL}(5) = 9765$. We use Theorem 16.5.3 to tabulate I and $\dim \mathsf{FL}_I(n)$ for $n = 4$ and 5 as follows.

I	Ø	{1}	{2}	{1,2}	{1,3}	{1,2,3}
		{3}		{2,3}		
	1	14	34	56	76	64

I	Ø	{1}	{2}	{1,2}	{1,3}	{1,4}	{2,3}	{1,2,3}	{1,2,4}	{1,2,3,4}
		{4}	{3}	{3,4}	{2,4}			{2,3,4}	{1,3,4}	
	1	30	154	280	900	404	776	960	1456	1024

For $I \subseteq Z[n-1]$, let $j \in J$ if and only if $n - j \in I$. Then the dimensions of the subspaces in a partial flag of type J are the codimensions in $\mathsf{V}(n)$ of the subspaces in a partial flag of type I. If $(c_1, \ldots, c_{r+1})$ is the composition of n associated to I, its reversal $(c_{r+1}, \ldots, c_1)$ is associated to J. Partial flag modules of types I and J have the same dimension, and are paired in the tables above.

Definition 16.5.5 For $I \subseteq Z[n-1]$, its **transpose** $\mathrm{tr}(I) = \{j : n - j \in I\}$.

Proposition 16.5.6 *For $I \subseteq Z[n-1]$, let $J = tr(I)$. Then transpose duality maps $\mathsf{FL}^I(n)$ to $\mathsf{FL}^J(n)$ and $\mathsf{FL}_I(n)$ to $\mathsf{FL}_J(n)$. In particular, $\mathsf{FL}(n)^{tr} \cong \mathsf{FL}(n)$.*

Proof Recall from Section 9.1 that the transpose dual M^{tr} of a $\mathbb{F}_2\mathsf{GL}(n)$-module M is defined by using the action $\langle f \cdot A, x\rangle = \langle f, x \cdot A^{\text{tr}}\rangle$ of $\mathsf{GL}(n)$ on the dual space $M^* = \text{Hom}(M, \mathbb{F}_2)$. The annihilator of a partial flag of type I in $\mathsf{V}(n)$ is a partial flag of transpose type J in $\mathsf{V}(n)^*$. By Proposition 9.2.3, the modules $\mathsf{V}(n)$ and $\mathsf{V}(n)^{\text{tr}}$ are isomorphic, and so it suffices to construct an isomorphism $\theta : \mathsf{FL}^I(\mathsf{V}(n))^{\text{tr}} \to \mathsf{FL}^J(\mathsf{V}(n)^{\text{tr}})$, where $\mathsf{FL}^I(\mathsf{V}(n)) = \mathsf{FL}^I(n)$ and $\mathsf{FL}^J(\mathsf{V}(n)^{\text{tr}})$ is the partial flag module of type J in $\mathsf{V}(n)^{\text{tr}}$.

We consider the relation between subspaces of a finite-dimensional $\mathbb{F}_2\mathsf{GL}(n)$-module M and their annihilators in M^{tr}. Given a subspace $X \subseteq M$, its annihilator $\text{Ann}(X) \subseteq M^*$ is the set of linear maps $f : M \to \mathbb{F}_2$ such that $\langle f, x\rangle = 0$ for all $x \in X$, i.e. $X \subseteq \text{Ker}(f)$. Then $\dim \text{Ann}(X) = \dim M - \dim X$. Since $\langle f, x \cdot A\rangle = \langle f \cdot A^{\text{tr}}, x\rangle$, $f \in \text{Ann}(X \cdot A)$ if and only if $X \subseteq \text{Ker}(f \cdot A^{\text{tr}})$. Equivalently, $f \cdot A^{\text{tr}} \in \text{Ann}(X)$, or $f \in \text{Ann}(X) \cdot (A^{\text{tr}})^{-1}$. Hence $\text{Ann}(X \cdot A) = \text{Ann}(X) \cdot (A^{\text{tr}})^{-1}$.

To construct θ, we first consider the Grassmannian case $I = \{k\}$, $J = \{n-k\}$ where $0 \le k \le n$. Then $\mathsf{FL}^I(\mathsf{V}(n))^{\text{tr}}$ has a basis given by the projections $p_X : \mathsf{FL}^I(\mathsf{V}(n)) \to \mathbb{F}_2$ defined by $p_X(X) = 1$ and $p_X(Y) = 0$ if X and Y are subspaces of $\mathsf{V}(n)$ of dimension k and $X \neq Y$. We define $\theta : \mathsf{FL}^I(\mathsf{V}(n))^{\text{tr}} \to \mathsf{FL}^J(\mathsf{V}(n)^{\text{tr}})$ by $\theta(p_X) = \text{Ann}(X)$. Clearly θ is bijective. To prove that it is a $\mathbb{F}_2\mathsf{GL}(n)$-module isomorphism, we must show that $\theta(p_X \cdot A) = \text{Ann}(X) \cdot A = \text{Ann}(X) \cdot (A^{\text{tr}})^{-1}$ for $A \in \mathsf{GL}(n)$. Now $\langle p_X \cdot A, Y\rangle = \langle p_X, Y \cdot A^{\text{tr}}\rangle = 1$ if and only if $Y \cdot A^{\text{tr}} = X$. Hence $\theta(p_X \cdot A) = \text{Ann}(Y)$ where $Y = X \cdot (A^{\text{tr}})^{-1}$, as required.

The argument generalizes to all $I \subseteq Z[n-1]$, since the annihilator of a partial flag of type I in $\mathsf{V}(n)$ is a partial flag of transpose type J in $\mathsf{V}(n)^*$. □

Remark 16.5.7 As it is a permutation representation, the partial flag module $\mathsf{FL}^I(n)$ is self-dual for *contragredient* duality for all $I \subseteq Z[n-1]$. The contragredient dual M^* of M is defined by letting $\mathsf{GL}(n)$ act on $\text{Hom}(M, \mathbb{F}_2)$ by $\langle f \cdot A, x\rangle = \langle f, x \cdot A^{-1}\rangle$, where $x \in M$, $f \in \text{Hom}(M, \mathbb{F}_2)$ and $A \in \mathsf{GL}(n)$. It follows that all the modules $\mathsf{FL}_I(n)$ are self-dual for contragredient duality.

16.6 The structure of $\mathsf{FL}(3)$

The structure of the flag module $\mathsf{FL}(3)$ has been determined in Section 10.5 by means of its isomorphisms with $\mathsf{K}^d(3)$ and $\mathsf{Q}^d(3)$ for 'generic' degrees d. As in Chapter 10, we write $\mathsf{FL}^{\{1\}}(n)$ as $\mathsf{FL}^1(n)$, etc. Partial flag modules can be identified with submodules of $\mathsf{FL}(3)$ shown in diagram (10.1) as follows. The image of the embedding $\mathsf{FL}^{\emptyset}(3) \to \mathsf{FL}(3)$ of Proposition 16.5.1 is the summand $\mathsf{FL}_{\emptyset}(3)$ generated by the sum of all 21 flags. The 7 lines in $\mathsf{V}(3)$ give a basis for $\mathsf{FL}^1(3)$, which is identified under the embedding $\mathsf{FL}^1(3) \to \mathsf{FL}(3)$ with $\mathsf{FL}_1(3) \oplus \mathsf{FL}_{\emptyset}(3)$, and similarly the 7 planes give a basis for $\mathsf{FL}^2(3)$ which is identified with $\mathsf{FL}_2(3) \oplus \mathsf{FL}_{\emptyset}(3)$.

The structure of the summands $\mathsf{FL}_1(3)$ and $\mathsf{FL}_2(3)$ can be seen as follows. The map $\mathsf{FL}^1(3) \to \mathsf{V}(3)$ which associates to a line L in $\mathsf{V}(3)$ the nonzero element of L is a $\mathbb{F}_2\mathsf{GL}(3)$-map, and so $\mathsf{FL}^1(3)$ has a quotient module $\cong \mathsf{V}(3)$. Similarly $\mathsf{FL}^2(3)$ has a quotient module $\cong \mathsf{V}(3)^*$. Since $\mathsf{FL}_1(3)$ and $\mathsf{FL}_2(3)$ are transpose duals by Proposition 16.5.6, $\mathsf{FL}_1(3)$ has a submodule $\cong \mathsf{V}(3)^*$ and $\mathsf{FL}_2(3)$ has a submodule $\cong V(3)$. The sum of the 3 lines in any plane P is a generator of a submodule of $\mathsf{FL}^1(3)$ of dimension 4, isomorphic to $\mathsf{I}(3) \oplus \mathsf{V}(3)^*$. The sum of the complementary set of 4 lines generates the submodule $\mathsf{V}(3)^*$ of $\mathsf{FL}_1(3)$. Generators of submodules of $\mathsf{FL}^2(3)$ and $\mathsf{FL}_2(3)$ can be described in a similar way.

The diagram below shows a geometric description of the Schubert cells and the Bruhat order. We shall show that a generator for the summand $\mathsf{FL}_{1,2}(3)$ is given by the sum of the flags in the top Schubert cell $\mathsf{Sch}(3,2,1)$.

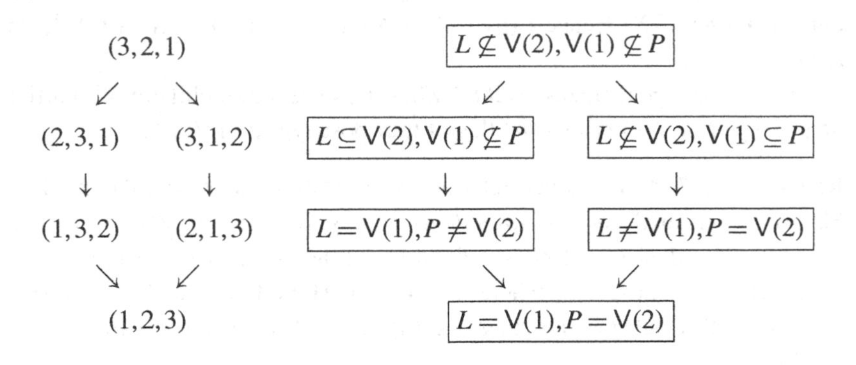

The diagrams below show the Schubert cell decomposition of $\mathsf{FL}(3)$, where a flag (L,P) is denoted by the row-reduced matrix A in the corresponding coset $\mathsf{B}(3)A$. The rows of the left hand diagram give the decomposition in terms of flags with the same line L, spanned by row 1 of A, and the rows of the right hand diagram give the decomposition in terms of flags with the same plane P, spanned by rows 1 and 2 of A.

$\begin{pmatrix}100\\010\\001\end{pmatrix}$	$\begin{pmatrix}100\\001\\010\end{pmatrix}$	$\begin{pmatrix}100\\011\\010\end{pmatrix}$	$\begin{pmatrix}100\\010\\001\end{pmatrix}$	$\begin{pmatrix}010\\100\\001\end{pmatrix}$	$\begin{pmatrix}110\\100\\001\end{pmatrix}$
$\begin{pmatrix}010\\100\\001\end{pmatrix}$	$\begin{pmatrix}010\\001\\100\end{pmatrix}$	$\begin{pmatrix}010\\101\\100\end{pmatrix}$	$\begin{pmatrix}100\\001\\010\end{pmatrix}$	$\begin{pmatrix}001\\100\\010\end{pmatrix}$	$\begin{pmatrix}101\\100\\010\end{pmatrix}$
$\begin{pmatrix}110\\100\\001\end{pmatrix}$	$\begin{pmatrix}110\\001\\100\end{pmatrix}$	$\begin{pmatrix}110\\101\\100\end{pmatrix}$	$\begin{pmatrix}100\\011\\010\end{pmatrix}$	$\begin{pmatrix}011\\100\\010\end{pmatrix}$	$\begin{pmatrix}111\\100\\010\end{pmatrix}$
$\begin{pmatrix}001\\100\\010\end{pmatrix}$	$\begin{pmatrix}001\\010\\100\end{pmatrix}$	$\begin{pmatrix}001\\110\\100\end{pmatrix}$	$\begin{pmatrix}010\\001\\100\end{pmatrix}$	$\begin{pmatrix}001\\010\\100\end{pmatrix}$	$\begin{pmatrix}011\\010\\100\end{pmatrix}$
$\begin{pmatrix}101\\100\\010\end{pmatrix}$	$\begin{pmatrix}101\\010\\100\end{pmatrix}$	$\begin{pmatrix}101\\110\\100\end{pmatrix}$	$\begin{pmatrix}010\\101\\100\end{pmatrix}$	$\begin{pmatrix}101\\010\\100\end{pmatrix}$	$\begin{pmatrix}111\\010\\100\end{pmatrix}$
$\begin{pmatrix}011\\100\\010\end{pmatrix}$	$\begin{pmatrix}011\\010\\100\end{pmatrix}$	$\begin{pmatrix}011\\110\\100\end{pmatrix}$	$\begin{pmatrix}110\\001\\100\end{pmatrix}$	$\begin{pmatrix}001\\110\\100\end{pmatrix}$	$\begin{pmatrix}111\\110\\100\end{pmatrix}$
$\begin{pmatrix}111\\100\\010\end{pmatrix}$	$\begin{pmatrix}111\\010\\100\end{pmatrix}$	$\begin{pmatrix}111\\110\\100\end{pmatrix}$	$\begin{pmatrix}110\\101\\100\end{pmatrix}$	$\begin{pmatrix}101\\110\\100\end{pmatrix}$	$\begin{pmatrix}011\\110\\100\end{pmatrix}$

Proposition 16.6.1 *Let* $\overline{\mathsf{B}}(3) \in \mathbb{F}_2\mathsf{GL}(3)$ *be the sum of the lower triangular matrices, let* $W_0 = \begin{pmatrix} 0&0&1 \\ 0&1&0 \\ 1&0&0 \end{pmatrix}$ *and let* $S = W_0\overline{\mathsf{B}}(3) = \sum_{i=1}^{8} A_i$*, the sum of the row-reduced representatives for the flags in* $\mathsf{Sch}(3,2,1)$*. Then* $\mathsf{B}(3)S = \sum_{i=1}^{8} \mathsf{B}(3)A_i$ *generates a* $\mathbb{F}_2\mathsf{GL}(3)$*-submodule* St *of* $\mathsf{FL}(3)$ *which has basis* $\mathsf{B}(3)S \cdot A_j, 1 \leq j \leq 8$.

Proof Since $\mathsf{B}(3)A = \mathsf{B}(3)$ for $A \in \mathsf{GL}(3)$ if and only if $A \in \mathsf{B}(3)$, the stabilizer of S in $\mathsf{GL}(3)$ is $\mathsf{B}(3)$, and so the $\mathsf{GL}(3)$-orbit of $S \in \mathbb{F}_2\mathsf{GL}(3)$ consists of 21 elements $S \cdot A$, where A runs through a set of coset representatives for $\mathsf{B}(3)$ in $\mathsf{GL}(3)$. We may take the row-reduced matrices A shown above as coset representatives, where the stars represent 0 or 1.

The element $\mathsf{B}(3)S \in \mathsf{FL}(3)$ is the sum of the 8 flags which are in general position with respect to the reference flag $R = (\mathsf{V}(1), \mathsf{V}(2))$. More generally, $\mathsf{B}(3)S \cdot A$ is the sum of the 8 flags which are in general position when the flag corresponding to the coset $\mathsf{B}(3)A$ is taken as the reference flag.

The 21 elements $\mathsf{B}(3)S \cdot A$ span St, and satisfy 14 linear relations, 7 each of two kinds, i.e. $\mathsf{B}(3)S \cdot A + \mathsf{B}(3)S \cdot A' + \mathsf{B}(3)S \cdot A'' = 0$, where the cosets $\mathsf{B}(3)A, \mathsf{B}(3)A', \mathsf{B}(3)A''$ correspond to three flags containing the same line or the same plane. These relations are illustrated by the diagram on page 25.

Thus the sums of the two sets of 7 relations are equal, as both are given by the sum of all 21 flags. It is clear from the diagram that the set $\{\mathsf{B}(3)S \cdot A_j, 1 \leq j \leq 8\}$ given by the row-reduced representatives A_i for the flags in $\mathsf{Sch}(3,2,1)$ form a spanning set for St. We label the matrices A_j as follows:

$$A_1 = \begin{pmatrix} 0&0&1 \\ 0&1&0 \\ 1&0&0 \end{pmatrix}, A_2 = \begin{pmatrix} 0&1&1 \\ 0&1&0 \\ 1&0&0 \end{pmatrix}, A_3 = \begin{pmatrix} 0&0&1 \\ 1&1&0 \\ 1&0&0 \end{pmatrix}, A_4 = \begin{pmatrix} 0&1&1 \\ 1&1&0 \\ 1&0&0 \end{pmatrix},$$

$$A_5 = \begin{pmatrix} 1&1&1 \\ 1&1&0 \\ 1&0&0 \end{pmatrix}, A_6 = \begin{pmatrix} 1&0&1 \\ 0&1&0 \\ 1&0&0 \end{pmatrix}, A_7 = \begin{pmatrix} 1&0&1 \\ 1&1&0 \\ 1&0&0 \end{pmatrix}, A_8 = \begin{pmatrix} 1&1&1 \\ 0&1&0 \\ 1&0&0 \end{pmatrix}.$$

In order to show that the elements $\mathsf{B}(3)S \cdot A_j$, $1 \leq j \leq 8$ are linearly independent in $\mathsf{FL}(3)$, we consider the coset $\mathsf{B}(3)A_i \cdot A_j = \mathsf{B}(3)A_iA_j$ for $1 \leq i,j \leq 8$.

When $i = 1$, the product matrix A_iA_j is lower triangular for all j, and so the coset $\mathsf{B}(3)A_i \cdot A_j = \mathsf{B}(3)$ is represented by the identity matrix I_3. When $i = 2$ or 3, we obtain the cosets represented by the two matrices of the form $\begin{pmatrix} * & 1 & 0 \\ 1&0&0 \\ 0&0&1 \end{pmatrix}$

or $\begin{pmatrix} 1&0&0 \\ 0&*&1 \\ 0&1&0 \end{pmatrix}$ respectively four times each. When $i = 4$ or 5, we obtain the cosets represented by the four matrices of the form $\begin{pmatrix} *&1&0 \\ *&0&1 \\ 1&0&0 \end{pmatrix}$ or $\begin{pmatrix} *&*&1 \\ 1&0&0 \\ 0&1&0 \end{pmatrix}$ respectively twice each, and when $i = 6, 7$ or 8, we obtain the cosets represented by the eight matrices of the form $\begin{pmatrix} *&*&0 \\ *&1&0 \\ 1&0&0 \end{pmatrix}$ once each.

Assume that $\sum_{j=1}^{8} c_j \mathsf{B}(3)S \cdot A_j = 0$ in $\mathsf{FL}(3)$, where $c_j \in \mathbb{F}_2$. By considering the occurrences of flags in $\mathsf{Sch}(2,3,1)$ and $\mathsf{Sch}(3,1,2)$ in the sum, we obtain 8 relations of the form $c_i = c_j$ which imply that all the coefficients c_j are equal, and finally by considering the occurrences of any flag in $\mathsf{Sch}(3,2,1)$ in the sum we obtain $c_j = 0$ for all j. Hence the elements $\mathsf{B}(3)S \cdot A_j \in \mathsf{St}$ are linearly independent, and so they form a basis of St. □

We have seen that St is a cyclic submodule of $\mathsf{FL}(3)$ generated by the $\mathsf{B}(3)$-invariant element $\mathsf{B}(3)S$. To prove that St is an irreducible $\mathbb{F}_2\mathsf{GL}(3)$-module, we consider the $\mathsf{B}(3)$-invariant elements of St.

Proposition 16.6.2 (i) $\mathsf{B}(3)S$ *is the unique nonzero* $\mathsf{B}(3)$*-invariant element of* St, *and* (ii) St *is an irreducible* $\mathbb{F}_2\mathsf{GL}(3)$*-module.*

Proof By Proposition 10.1.3, (i) implies (ii). An element of $\mathsf{FL}(3)$ is a sum of flags $\sum_A \mathsf{B}(3)A$ where the sum is over a subset Z of the set of 21 row-reduced matrices A shown above, and it is $\mathsf{B}(3)$-invariant if and only if Z is $\mathsf{B}(3)$-invariant. By Proposition 16.3.7, the set of cosets $\mathsf{B}(3)A$ for $A \in Z$ is a union of Schubert cells, the case $\mathsf{B}(3)S$ corresponding to $\mathsf{Sch}(1,2,3)$. Thus the $\mathbb{F}_2$-space of $\mathsf{B}(3)$-invariant elements of $\mathsf{FL}(3)$ has dimension 6, with basis corresponding to the Schubert cells.

It remains to show that $\mathsf{B}(3)S$ is the only nonzero $\mathsf{B}(3)$-invariant in St. Again we observe that an element $\sum_{A\in Z} \mathsf{B}(3)SA$ is $\mathsf{B}(3)$-invariant if and only if the set of cosets $\mathsf{B}(3)SA$ for $A \in Z$ is a union of Schubert cells. However, the relations $\mathsf{B}(3)S \cdot A + \mathsf{B}(3)S \cdot A' + \mathsf{B}(3)S \cdot A'' = 0$ corresponding to triples (A, A', A'') representing flags with common lines and planes in $\mathsf{V}(3)$ imply that the sum $\sum_{A\in Z} \mathsf{B}(3)SA$ when Z is the set of row-reduced matrices A in a fixed Schubert cell is the same for all the Schubert cells, and so it is equal to $\mathsf{B}(3)S$. □

16.7 Remarks

An excellent treatment of the Bruhat order from a combinatorial viewpoint is given in [20, Chapter 2]. Our treatment of the Bruhat decomposition owes much to [6, Chapter 2], where it is shown that the parabolic subgroups $\mathsf{P}^I(n)$ are the only subgroups of $\mathsf{GL}(n)$ which contain $\mathsf{B}(n)$.

Proposition 16.2.9 asserts that the subgroups B(n) and W(n), with generating set $S = \{S_1, \ldots, S_{n-1}\}$, form a *BN*-pair, or Tits system, in GL(n). The general situation is much simplified in our case, since the 'maximal torus' T reduces to the identity matrix I_n. The axioms for a *BN*-pair for a group G are as follows [79, Section 1.7]: (i) G is generated by B and N; (ii) $T = B \cap N$ is a normal subgroup of N; (iii) $W = N/T$ is generated by a set S of involutions; (iv) for $s \in S$ and $w \in W$, $WBS \subseteq BWB \cup BWSB$; and (v) for $s \in S$, $SBS \neq B$. An extended treatment of the work of J. Tits can be found in [22].

17

Irreducible $\mathbb{F}_2\mathsf{GL}(n)$-modules

17.0 Introduction

In this chapter we use the flag module $\mathsf{FL}(n)$ to construct a complete set of inequivalent irreducible $\mathbb{F}_2\mathsf{GL}(n)$-modules. We have seen in Chapter 16 that $\mathsf{FL}(n)$ is the direct sum of 2^{n-1} submodules $\mathsf{FL}_I(n)$ which correspond to partial flags given by subspaces of $\mathsf{V}(n)$ with dimensions in $I \subseteq Z[n-1]$. We show that each module $\mathsf{FL}_I(n)$ has a unique irreducible submodule and a unique irreducible quotient. In particular, $\mathsf{FL}_I(n)$ is indecomposable. By a general result in modular representation theory, the number of inequivalent irreducible $\mathbb{F}_2\mathsf{GL}(n)$-modules is the same as the number of conjugacy classes of elements of odd order in $\mathsf{GL}(n)$. There is one such conjugacy class of matrices for each choice of characteristic polynomial, and 2^{n-1} in all.

The usual notation for the irreducible $\mathbb{F}_2\mathsf{GL}(n)$-modules is derived from the representation theory of the algebraic group $\overline{\mathsf{G}}(n) = \mathsf{GL}(n, \overline{\mathbb{F}}_2)$, where $\overline{\mathbb{F}}_2$ is the algebraic closure of $\mathbb{F}_2$. We shall see in Chapter 20 that $\overline{\mathsf{G}}(n)$ has a family of irreducible representations $\overline{L}(\lambda)$ indexed by partitions λ of length $\leq n$. On restriction of the action to $\mathsf{GL}(n)$, the representations $\overline{L}(\lambda)$ such that λ is column 2-regular of length $\leq n-1$ give a complete set of 2^{n-1} inequivalent irreducible $\mathbb{F}_2\mathsf{GL}(n)$-modules $L(\lambda)$. Each irreducible $\mathbb{F}_2\mathsf{GL}(n)$-module $L(\lambda)$ occurs once as a submodule and once as a quotient module of $\mathsf{FL}(n)$. More precisely, $\mathsf{FL}_I(n)$ has a quotient module $L(\lambda)$, where I is the set of parts of the conjugate partition λ^{tr}, and its transpose dual $\mathsf{FL}_J(n)$, where $J = \mathrm{tr}(I)$, has a submodule $L(\lambda)$.

Our method is based on the **Hecke algebra** $\mathsf{H}_0(n) = \mathrm{End}_{\mathsf{GL}(n)}(\mathsf{FL}(n))$. The elements of $\mathsf{H}_0(n)$ are $\mathbb{F}_2\mathsf{GL}(n)$-module maps $f : \mathsf{FL}(n) \to \mathsf{FL}(n)$, which form an algebra over $\mathbb{F}_2$ under addition and composition. For example, as in Section 16.0 let u, v, w in $\mathsf{V}(2)$ correspond to the cosets $\mathsf{B}(2)$, $\mathsf{B}(2)\left(\begin{smallmatrix}0&1\\1&0\end{smallmatrix}\right)$ and $\mathsf{B}(2)\left(\begin{smallmatrix}1&1\\1&0\end{smallmatrix}\right)$ respectively. Then $\mathsf{Sch}(I_2) = \{u\}$ is the reference flag and

$\mathsf{Sch}(S) = \{v, w\}$ where $S = \left(\begin{smallmatrix} 0 & 1 \\ 1 & 0 \end{smallmatrix}\right)$. The Hecke algebra $\mathsf{H}_0(2)$ has basis elements f_{I_2} and f_S defined by $f_{I_2}(u) = u$ and $f_S(u) = v + w$. Hence f_{I_2} is the identity map of $\mathsf{FL}(2)$ and f_S is the projection of $\mathsf{FL}(2)$ on the summand $\mathsf{FL}_1(2)$. Writing $f_{I_2} = 1$ and $f_S = h$, $\mathsf{H}_0(2) = \{0, 1, h, 1 + h\}$ where $h^2 = h$ and $1 + h$ is the projection of $\mathsf{FL}(2)$ on the summand $\mathsf{FL}_\emptyset(2)$ generated by $u + v + w$.

The algebra $\mathsf{H}_0(n)$ has dimension $n!$ as a vector space over $\mathbb{F}_2$, and is a close relative of the group algebra $\mathbb{F}_2\Sigma(n)$ of the symmetric group. More precisely, $\mathsf{H}_0(n)$ is the monoid algebra over $\mathbb{F}_2$ of the 'Schubert monoid' $\mathsf{Sch}(n)$, whose elements f_W correspond to permutation matrices $W \in \mathsf{W}(n)$. However, the relations $S_i^2 = I_n$, $1 \leq i \leq n-1$ for the generating switch matrices $S_1, \ldots, S_{n-1}$ in $\mathsf{W}(n)$ are replaced by the relations $h_i^2 = h_i$ for the corresponding generators $h_i = f_{S_i}$ of $\mathsf{Sch}(n)$. The generators $h_1, \ldots, h_{n-1}$ also satisfy the commutativity relations $h_ih_j = h_jh_i$ if $|i - j| > 1$ and the braid relations $h_ih_jh_i = h_jh_ih_j$ if $|i - j| = 1$.

Together with the relations $S_i^2 = I_n$, the corresponding relations $S_iS_j = S_jS_i$ if $|i - j| > 1$ and $S_iS_jS_i = S_jS_iS_j$ if $|i - j| = 1$ form a set of defining relations between the generators S_i for $\mathsf{W}(n)$. These relations allow us to manipulate words $a_1a_2 \cdots a_r$ in $h_1, \ldots, h_{n-1}$ in much the same way as words in the generators S_i of $\mathsf{W}(n)$. However, a word with adjacent letters equal is shortened using the relation $h_i^2 = h_i$ instead of $S_i^2 = 1$. It follows that $W \in \mathsf{W}(n)$ and the corresponding basis element $f_W \in \mathsf{Sch}(n)$ are given by corresponding reduced words in the generators S_i and h_i. In particular, the identity matrix I_n corresponds to the identity element $f_{I_n} = 1 \in \mathsf{H}_0(n)$.

The reason for the notation H_0 is that the corresponding algebra over the real numbers can be regarded as a member of a family of deformations of the group algebra of $\mathsf{W}(n)$ obtained by replacing the relations $S_i^2 = I_n$ in $\mathsf{W}(n)$ by the relations $h_i^2 = q + (1 - q)h_i$, where $0 \leq q \leq 1$. Thus $\mathbb{F}_2\mathsf{Sch}(n)$ is the singular Hecke algebra obtained by taking $q = 0$.

Throughout this chapter, we treat complete flags as right cosets of the lower triangular subgroup $\mathsf{B}(n)$ of $\mathsf{GL}(n)$. In Section 17.1 we obtain the $\mathbb{F}_2$-basis $\{f_W, W \in \mathsf{W}(n)\}$ for the algebra $\mathsf{H}_0(n)$ and a product formula for basis elements. In Section 17.2 we view $\mathsf{H}_0(n)$ as the monoid algebra of the Schubert monoid $\mathsf{Sch}(n)$. We begin to study the structure of $\mathsf{H}_0(n)$ in Section 17.3 using the permutation subgroups $\mathsf{W}^I(n)$ of the parabolic subgroups $\mathsf{P}^I(n)$, where $I \subseteq Z[n-1]$. In Sections 17.4 and 17.5 we find elements g_I and s_I in $\mathsf{H}_0(n)$ which map the reference flag $\mathsf{B}(n)$ to generators of $\mathsf{FL}_I(n)$ and its unique irreducible submodule, or *socle* respectively. In Section 17.6 we show that these submodules give a complete set of irreducible $\mathbb{F}_2\mathsf{GL}(n)$-modules, and that the summands $\mathsf{FL}_I(n)$ of $\mathsf{FL}(n)$ are indecomposable.

17.1 The Hecke algebra $H_0(n)$

We begin with some general observations about induced modules. Given a finite group G, a subgroup B of G and an arbitrary coefficient field F, we consider the right FG-module M given by the permutation action of G on the set of right cosets of B, i.e. $Bx \cdot g = Bxg$ for $x, g \in G$. We have in mind the case $G = \mathsf{GL}(n)$, $B = \mathsf{B}(n)$ and $F = \mathbb{F}_2$, when M is the flag module $\mathsf{FL}(n)$. As the action is transitive, M is the cyclic module generated by any right coset of B, and in particular by B itself. We choose coset representatives by first expressing G as the disjoint union of double cosets Bz_iB, $1 \leq i \leq r$, and then expressing each double coset as the disjoint union of right cosets $By_{i,j}$, $1 \leq j \leq s(i)$. Thus $y_{i,j} = z_ib_{i,j}$ where $b_{i,j} \in B$, and elements of M can be written uniquely in the form $\sum_{i,j} c_{i,j}By_{i,j}$, where $c_{i,j} \in F$.

Linear endomorphisms $f : M \to M$ which commute with the action of G form an algebra $\mathrm{End}_{FG}(M)$ over F under addition and composition of maps, with identity element id_M. Since B generates M, f is determined by $f(B) = \sum_{i,j} c_{i,j}By_{i,j}$, where $c_{i,j} \in F$, and for all $g \in G, f(Bg) = f(B)g = \sum_{i,j} c_{i,j}By_{i,j}g$. In particular, $Bg = B$ for $g \in B$ and the cosets $By_{i,j}g$ are a permutation of the cosets $By_{i,j}$ for each i. Since B acts transitively on the cosets, the coefficient $c_{i,j} \in F$ is independent of j. Hence $f = \sum_i c_if_i$, where $f_i(B) = \sum_j By_{i,j}$ and $c_i \in F$.

For $1 \leq i \leq r$, the map $f_i : M \to M$ given by $f_i(Bg) = \sum_j By_{i,j}g$ is well defined, since $Bg = Bg'$ implies that $g' = bg$ where $b \in B$, and so $f_i(Bg') = \sum_j By_{i,j}g' = \sum_j (By_{i,j}b)g$. Since the cosets $By_{i,j}b$ are a permutation of the cosets $By_{i,j}$, $f_i(Bg') = f_i(Bg)$. Further f_i commutes with the action of G on M, since for $g_1, g_2 \in G$ we have $f_i(Bg_1 \cdot g_2) = f_i(Bg_1g_2) = \sum_j By_{i,j}(g_1g_2) = (\sum_j By_{i,j}g_1)g_2 = f_i(Bg_1) \cdot g_2$.

We summarize these observations as follows, where the notation $\sqcup$ denotes the disjoint union of sets.

Proposition 17.1.1 *Let G be a finite group, B a subgroup of G, F a field and M the right FG-module given by the permutation action of G on right cosets of B. Let $G = \sqcup_{i=1}^r Bz_iB$, and for each i let $Bz_iB = \sqcup_{j=1}^{s(i)} By_{i,j}$. Then the maps f_i defined by $f_i(B) = \sum_{j=1}^{s(i)} By_{i,j}$, $1 \leq i \leq r$, form a F-basis for $\mathrm{End}_{FG}(M)$.* □

We next consider the product in $\mathrm{End}_{FG}(M)$. Let $f_i(Bg) = \sum_x Bxg$, where $g \in G$ and each x represents a coset $Bx \subseteq Bz_iB$, and similarly let $f_j(Bg) = \sum_y Byg$, where each y represents a coset $By \subseteq Bz_jB$. Then $f_i \circ f_j(B) = f_i(\sum_y By) = \sum_{x,y} Bxy$. By counting the number $a_{i,j}^k$ of products xy in each double coset Bz_kB, we obtain structural equations $f_i \circ f_j = \sum_{i=1}^r a_{i,j}^k f_k$ for $\mathrm{End}_{FG}(M)$, where the coefficients $a_{i,j}^k$ are in F.

As in Proposition 16.3.7 we denote the sum of the flags in the Schubert cell $\mathsf{Sch}(W)$ by σ_W. Then the next result follows by applying Proposition 17.1.1 to the Bruhat decomposition of $\mathsf{GL}(n)$.

Proposition 17.1.2 *Given $W \in \mathsf{W}(n)$, let $f_W : \mathsf{FL}(n) \to \mathsf{FL}(n)$ be the $\mathbb{F}_2\mathsf{GL}(n)$-module map defined by $f_W(\mathsf{B}(n)) = \sigma_W$. Then $\{f_W : W \in \mathsf{W}(n)\}$ is a $\mathbb{F}_2$-basis for the* **Hecke algebra** $\mathsf{H}_0(n) = End_{\mathbb{F}_2\mathsf{GL}(n)}(\mathsf{FL}(n))$ □

Using Proposition 16.1.4, we can calculate products in $\mathsf{H}_0(n)$ which involve basis elements corresponding to the generators $S_1,\ldots,S_{n-1}$ of $\mathsf{W}(n)$.

Definition 17.1.3 For $1 \le i \le n-1$, let $h_i = f_{S_i}$ be the element of $\mathsf{H}_0(n)$ corresponding to the switch matrix $S_i \in \mathsf{W}(n)$.

Thus $h_i(\mathsf{B}(n)) = \sigma_{S_i} = \mathsf{B}(n)S_i + \mathsf{B}(n)S_iL_i$, the sum of the two flags in $\mathsf{Sch}(S_i)$, where $L_i \in \mathsf{B}(n)$ is the transvection obtained by replacing the $(i+1,i)$th entry of the identity matrix I_n by 1.

Proposition 17.1.4 *The elements h_i, $1 \le i \le n-1$, are idempotents in $\mathsf{H}_0(n)$, i.e. $h_i^2 = h_i$.*

Proof In effect, this is a calculation in $\mathsf{GL}(2)$. We have $h_i^2(\mathsf{B}(n)) = h_i(\mathsf{B}(n)S_i) + h_i(\mathsf{B}(n)S_iL_i) = h_i(\mathsf{B}(n))S_i + h_i(\mathsf{B}(n))S_iL_i = \mathsf{B}(n)S_iS_i + \mathsf{B}(n)S_iL_iS_i + \mathsf{B}(n)S_iS_iL_i + \mathsf{B}(n)S_iL_iS_iL_i$. Since $S_i^2 = I_n$, $L_i \in \mathsf{B}(n)$ and $S_iL_iS_i = L_iS_iL_i$, this reduces to $\mathsf{B}(n) + \mathsf{B}(n)S_iL_i + \mathsf{B}(n) + \mathsf{B}(n)S_i = h_i(\mathsf{B}(n))$. □

The next result generalizes this calculation.

Proposition 17.1.5 *Let $W \in \mathsf{W}(n)$ and let $S = S_i$, $1 \le i \le n-1$. Then*

$$\text{(i) } h_if_W = \begin{cases} f_{SW}, & \textit{if } len(SW) = len(W)+1, \textit{ i.e. } i \notin des(W), \\ f_W, & \textit{if } len(SW) = len(W)-1, \textit{ i.e. } i \in des(W), \end{cases}$$

$$\text{(ii) } f_Wh_i = \begin{cases} f_{WS}, & \textit{if } len(WS) = len(W)+1, \textit{ i.e. } i \notin des(W^{-1}), \\ f_W, & \textit{if } len(WS) = len(W)-1, \textit{ i.e. } i \in des(W^{-1}), \end{cases}$$

where the descent set $des(W)$ is defined in 16.1.2. Further, $h_if_W = h_if_{SW}$ and $f_Wh_i = f_{WS}h_i$.

Proof Let $h_i(\mathsf{B}(n)) = \sum_X \mathsf{B}(n)X$ and $f_W(\mathsf{B}(n)) = \sum_Y \mathsf{B}(n)Y$, with sums over representatives X, Y for the cosets $\mathsf{B}(n)X \subseteq \mathsf{B}(n)S\mathsf{B}(n)$, $\mathsf{B}(n)Y \subseteq \mathsf{B}(n)W\mathsf{B}(n)$. Then $h_if_W = \sum_{V\in\mathsf{W}(n)} c_Vf_V$ in $\mathsf{H}_0(n)$, where c_V is the number of pairs (X,Y) such that $XY \in \mathsf{B}(n)V\mathsf{B}(n)$, reduced mod 2. By Proposition 16.2.7, we may choose $X = SB$, $Y = WB'$ where B, B' are elements of the Bruhat subgroups $\mathsf{B}(S)$, $\mathsf{B}(W)$ respectively. Since $|\mathsf{B}(W)| = 2^{\mathrm{len}(W)}$, there are $2^{\mathrm{len}(W)+1}$ products XY to be assigned to Bruhat cells $\mathsf{B}(n)V\mathsf{B}(n)$.

We separate the cases $\mathrm{len}(SW) = \mathrm{len}(W) + 1$ and $\mathrm{len}(SW) = \mathrm{len}(W) - 1$, and use Proposition 16.2.8. In the first case, $XY = (SB)(WB') = (SBW)B' \in \mathsf{B}(n)SW\mathsf{B}(n)$, so $V = SW$ and since $\mathsf{B}(n)SW\mathsf{B}(n)$ contains $2^{\mathrm{len}(W)+1}$ cosets, the coefficient $c_V = 1$. Thus $h_i f_W = f_{SW}$.

In the second case, $XY = (SB)(WB') = (SBW)B' \in \mathsf{B}(n)SW\mathsf{B}(n) \cup \mathsf{B}(n)W\mathsf{B}(n)$, so $h_i f_W = r f_{SW} + s f_W$ for some integers r and s. In this case, $\mathsf{B}(n)SW\mathsf{B}(n)$ contains $2^{\mathrm{len}(W)-1}$ cosets $\mathsf{B}(n)Y$.

Counting the products XY, we find that $2^{\mathrm{len}(W)-1}r + 2^{\mathrm{len}(W)}s = 2^{\mathrm{len}(W)+1}$, i.e. $r + 2s = 4$. Again by case (ii) of Proposition 16.2.8, both cases $XY \in \mathsf{B}(n)SW\mathsf{B}(n)$ and $XY \in \mathsf{B}(n)W\mathsf{B}(n)$ occur. Hence $r > 0$ and $s > 0$, and so $r = 2$ and $s = 1$. Reducing mod 2, we obtain $h_i f_W = f_W$ in $\mathsf{H}_0(n)$. Using Proposition 16.2.9, the product $f_W h_i$ is evaluated in the same way. Finally, the statements $h_i f_W = h_i f_{SW}$ and $f_W h_i = f_{WS} h_i$ follow by exchanging W with WS or SW. □

The next result follows from Proposition 17.1.5 by induction on the length $\mathrm{len}(W)$, $W \in \mathsf{W}(n)$.

Proposition 17.1.6 *Let $W_1, W_2 \in \mathsf{W}(n)$ be expressed as products of the generating switch matrices S_i, $1 \le i \le n-1$. Then $f_{W_1} f_{W_2} = f_W$ for some $W \in \mathsf{W}(n)$, where $W = W_1' W_2 = W_1 W_2'$ for elements $W_1', W_2' \in \mathsf{W}(n)$ which are products obtained by omitting suitable factors in the products for W_1 and W_2 respectively. In particular, $f_{W_1} f_{W_2} = f_{W_1 W_2}$ if and only if $\mathrm{len}(W_1) + \mathrm{len}(W_2) = \mathrm{len}(W_1 W_2)$.* □

17.2 The Schubert monoid $\mathsf{Sch}(n)$

By Proposition 17.1.5, the set of basis elements f_W, $W \in \mathsf{W}(n)$, is closed under the product in $\mathsf{H}_0(n)$.

Definition 17.2.1 The **Schubert monoid** $\mathsf{Sch}(n)$ is the set of $\mathbb{F}_2\mathsf{GL}(n)$-module maps $f_W : \mathsf{FL}(n) \to \mathsf{FL}(n)$, $W \in \mathsf{W}(n)$, with the product given by composition.

Thus $\mathsf{H}_0(n)$ is the monoid algebra $\mathbb{F}_2\mathsf{Sch}(n)$.

Example 17.2.2 The multiplication table for $\mathsf{Sch}(3)$ is shown below, where $a = h_1 = f_{S_1}$, $b = h_2 = f_{S_2}$ and $\rho \in \Sigma(3)$ corresponds to $W \in \mathsf{W}(3)$ by (16.1).

$\rho \leftrightarrow W$	f_W	1	a	b	ab	ba	aba
$(1,2,3)$	1	1	a	b	ab	ba	aba
$(2,1,3)$	a	a	a	ab	ab	aba	aba
$(1,3,2)$	b	b	ba	b	aba	ba	aba
$(3,1,2)$	ab	ab	aba	ab	aba	aba	aba
$(2,3,1)$	ba	ba	ba	aba	aba	aba	aba
$(3,2,1)$	aba	aba	aba	aba	aba	aba	aba

The elements $g_\emptyset = 1+a+b+ab+ba+aba$, $g_1 = a+ba$, $g_2 = b+ab$ and $g_{1,2} = aba$ are idempotents in $\mathsf{H}_0(3) = \mathbb{F}_2\mathsf{Sch}(3)$ with sum 1 and pairwise products 0, and hence the maps $g_I : \mathsf{FL}(3) \to \mathsf{FL}(3)$ are projections on to direct summands. In Example 17.4.2, we identify each summand $g_I(\mathsf{FL}(3))$ with the corresponding module $\mathsf{FL}_I(3)$.

We next obtain generators and defining relations for $\mathsf{H}_0(n)$.

Theorem 17.2.3 *The Hecke algebra* $\mathsf{H}_0(n)$ *is generated by the idempotents* h_i, $1 \le i \le n-1$. *The elements* h_i *and* h_j *commute if* $|i-j| > 1$, *and satisfy the braid relation* $h_ih_jh_i = h_jh_ih_j$ *if* $|i-j| = 1$.

Proof The first statement follows from Proposition 17.1.6. The relations follow from Proposition 17.1.5 by using the relations $S_iS_j = S_jS_i$ if $|i-j| > 1$ and $S_iS_jS_i = S_jS_iS_j$ if $|i-j| = 1$ in $\mathsf{W}(n)$. □

Consider left multiplication by h_i as a linear map $\mathsf{H}_0(n) \to \mathsf{H}_0(n)$. Since h_i is idempotent, the only possible eigenvalues of h_i are 0 and 1. By Proposition 17.1.5(i), the elements f_W such that $i \in \mathrm{des}(W)$ form a basis of the 1-eigenspace of h_i, while the elements $f_W + f_{S_iW}$ such that $i \notin \mathrm{des}(W)$ form a basis of the 0-eigenspace. Since $i \in \mathrm{des}(W)$ if and only if $i \notin \mathrm{des}(S_iW)$, both eigenspaces have dimension $n!/2$. We observe that the 1-eigenspace of left multiplication by h_i is the left ideal $h_i\mathsf{H}_0(n)$, while the 0-eigenspace is the left ideal $(1+h_i)\mathsf{H}_0(n)$. We can similarly describe the corresponding eigenspaces for right multiplication by h_i by using Proposition 17.1.5(ii).

Every element $f_W \in \mathsf{Sch}(n)$ can be expressed as a word in $h_1, \dots, h_{n-1}$. As for $\mathsf{W}(n)$, a word is **reduced** if it has minimal length among all words representing the same element f_W, and we define the **length** $\mathrm{len}(f_W)$ of f_W to be this minimal length. Comparing the defining relations for the generators S_i of $\mathsf{W}(n)$ with those of Theorem 17.2.3, we see that the relations $S_i^2 = I_n$ are replaced by $h_i^2 = h_i$ for each i, but the commutativity relations are the same. This leads to the following observation.

Proposition 17.2.4 (i) *A word $h_{i_1} \cdots h_{i_r}$ in* $\mathsf{Sch}(n)$ *is reduced if and only if the corresponding word $S_{i_1} \cdots S_{i_r}$ in* $\mathsf{W}(n)$ *is reduced. The elements $W \in \mathsf{W}(n)$ and $f_W \in \mathsf{H}_0(n)$ are given by corresponding reduced words, and len$(f_W) =$ len(W).*

(ii) *There is a reduced word $h_{i_1} \cdots h_{i_r}$ for f_W if and only if $i_1 \in des(W)$, and there is a reduced word $h_{j_1} \cdots h_{j_r}$ for f_W if and only if $j_r \in des(W^{-1})$.*

(iii) *The elements $1 + h_i$, $1 \le i \le n-1$, are idempotent and generate $\mathsf{H}_0(n)$, and the map $\iota : \mathsf{H}_0(n) \to \mathsf{H}_0(n)$ defined by $\iota(h_i) = 1 + h_i$ for $1 \le i \le n-1$ is an involution.*

Proof (i) is clear from the preceding discussion, and (ii) follows from (i) and Proposition 16.1.9. Finally (iii) follows by checking that ι preserves the defining relations of Theorem 17.2.3. □

Proposition 17.2.5 *len$(f_{W_1} f_{W_2}) \ge \max(len(f_{W_1})$, len$(f_{W_2}))$ for all $W_1, W_2 \in$* $\mathsf{W}(n)$.

Proof By Proposition 17.1.5(i), either len$(h_i f_W) =$ len$(f_W) + 1$ or len$(h_i f_W) =$ len(f_W). It follows by induction on len(f_{W_1}) that len$(f_{W_1} f_{W_2}) \ge$ len(f_{W_2}). Similarly, we obtain len$(f_{W_1} f_{W_2}) \ge$ len(f_{W_1}) by induction on len(f_{W_2}) using Proposition 17.1.5(ii). □

Recall that the reversal matrix $W_0 \in \mathsf{W}(n)$ is the matrix with 1s on the anti-diagonal from top right to bottom left, corresponding to the permutation $\rho_0 = (n, n-1, \ldots, 1) \in \Sigma(n)$.

Proposition 17.2.6 (i) *For $1 \le i \le n-1$, $h_i f_{W_0} = f_{W_0} = f_{W_0} h_i$.*

(ii) *The elements f_{W_0} and $\iota(f_{W_0})$ are idempotent.*

Proof (i) follows from Proposition 17.1.5, since W_0 is the unique element of maximal length in $\mathsf{W}(n)$. Since f_{W_0} is a word in $h_1, \ldots, h_{n-1}$, it follows from (i) that f_{W_0} is idempotent. By Proposition 17.2.4(iii), $\iota(f_{W_0})$ is also idempotent. □

Proposition 17.2.7 *The idempotent $\iota(f_{W_0}) = \sum_{W \in \mathsf{W}(n)} f_W$.*

Proof The reversal matrix $W_0 \in \mathsf{W}(n)$ is given by the reduced word $W_0 = S_1(S_2S_1)(S_3S_2S_1) \cdots (S_{n-1}S_{n-2} \cdots S_1)$ in the generators $S_1, \ldots, S_{n-1}$. By Proposition 17.2.4(i), f_{W_0} is given by the corresponding product of the generators $h_1, \ldots, h_{n-1}$ of $\mathsf{H}_0(n)$. We argue by induction on n, the base case $n = 2$ being true by definition of ι. The inductive step is then the statement

$$E_n = E_{n-1}(1+h_{n-1})(1+h_{n-2})\cdots(1+h_1), \tag{17.1}$$

where $E_n = \sum_{W\in\mathsf{W}(n)} f_W$ and $\mathsf{W}(n-1)$ is identified with the subgroup of $\mathsf{W}(n)$ generated by $S_1,\ldots,S_{n-2}$.

By the induction hypothesis, for $i \le n-2$ we have $E_{n-1}(1+h_i) = \iota(f_{W'_0}h_i)$, where W'_0 is the reversal matrix in $\mathsf{W}(n-1)$. Using Proposition 17.2.6(i), it follows that $E_{n-1}(1+h_i) = E_{n-1}$. Hence the right hand sum in (17.1) reduces to $E_{n-1} + E_{n-1}h_{n-1}(1+h_{n-2})\cdots(1+h_1)$.

Since h_j commutes with h_{n-1} for $j \le n-3$, we can repeat the argument to obtain $E_{n-1}h_{n-1}(1+h_{n-3})\cdots(1+h_1) = E_{n-1}(1+h_{n-3})\cdots(1+h_1)h_{n-1} = E_{n-1}h_{n-1}$. Continuing in this way, we can reduce the right hand sum in (17.1) to $E_{n-1}+E_{n-1}h_{n-1}+E_{n-1}h_{n-1}h_{n-2}+\cdots+E_{n-1}h_{n-1}h_{n-2}\cdots h_1$, where each of the n terms is the sum of $(n-1)!$ reduced words. Since the corresponding reduced words in $\mathsf{W}(n)$ give all $n!$ elements, it follows from Proposition 17.2.4(i) that (17.1) holds in $\mathsf{H}_0(n)$. This completes the induction. □

17.3 The permutation matrices W_0^I

Recall from Section 16.4 that for $I = \{i_1, i_2, \ldots, i_r\} \subseteq Z[n-1]$, the subgroup $\mathsf{W}^I(n)$ of $\mathsf{W}(n)$ is generated by the switch matrices S_j where $j \notin I$, so that the elements of $\mathsf{W}^I(n)$ are block diagonal matrices with blocks of size c_j, $1 \le j \le r+1$, where $c_j = i_j - i_{j-1}$ and $i_0 = 0$, $i_{r+1} = n$. By restricting to matrices S_j where $j \notin I$, it follows from Proposition 17.1.5 that $\mathsf{Sch}^I(n) = \{f_W : W \in \mathsf{W}^I(n)\}$ is closed under multiplication, and so its monoid algebra $\mathsf{H}_0^I(n) = \mathbb{F}_2\mathsf{Sch}^I(n)$ is a subalgebra of $\mathsf{H}_0(n)$.

There is a unique matrix W_0^I of maximum length in $\mathsf{W}^I(n)$, corresponding to the permutation which reverses successive subsets of length $c_1,\ldots,c_{r+1}$ in $Z[n]$. Thus W_0^I has order 2, $\mathrm{len}(W_0^I) = \sum_{i=1}^{r+1} c_i(c_i-1)/2$ and the descent set $\mathrm{des}(W_0^I) = Z[n-1]\setminus I$. The right coset $\mathsf{W}^I(n)W_0$ consists of block anti-diagonal matrices with blocks of size $c_1,\ldots,c_{r+1}$. For example, when $n=4$ and $I=\{2\}$, $\mathsf{W}^I(n)W_0$ is the set of four matrices

$$\left(\begin{array}{cc|cc} 0&0&1&0\\ 0&0&0&1\\ \hline 1&0&0&0\\ 0&1&0&0 \end{array}\right), \left(\begin{array}{cc|cc} 0&0&1&0\\ 0&0&0&1\\ \hline 0&1&0&0\\ 1&0&0&0 \end{array}\right), \left(\begin{array}{cc|cc} 0&0&0&1\\ 0&0&1&0\\ \hline 1&0&0&0\\ 0&1&0&0 \end{array}\right), \left(\begin{array}{cc|cc} 0&0&0&1\\ 0&0&1&0\\ \hline 0&1&0&0\\ 1&0&0&0 \end{array}\right). \tag{17.2}$$

The corresponding permutations $\rho \in \Sigma(n)$ map the interval $[i_{j-1}+1, i_j]$ to the interval $[n-i_j+1, n-i_{j-1}]$ for $1 \le j \le r+1$. Thus $\rho(i_j) > n-i_j$ and $\rho(i_j+1) \le n-i_j$, and so $I \subseteq \mathrm{des}(\rho)$. There is a unique such $\rho = \rho^I$ which preserves the order within each subinterval. The corresponding matrix $V^I = W_0^I W_0$ is the unique element of minimum length in $\mathrm{W}^I(n)W_0$, and is also the unique element of maximum length such that $\mathrm{des}(V^I) = I$. For example, in the case $n=4$, $I=\{2\}$, V^I is the first matrix listed above.

Definition 17.3.1 Let $C=(c_1,\dots,c_{r+1})$ be the composition of n associated to $I \subseteq Z[n-1]$. Then $V^I = W_0^I W_0$ is the block anti-diagonal matrix with blocks are the identity matrices of size $c_1,\dots,c_{r+1}$.

By Proposition 16.1.3(iv), $\mathrm{len}(V^I) = n(n-1)/2 - \mathrm{len}(W_0^I)$. Conjugation with W_0 maps W_0^I to W_0^J where $J = \mathrm{tr}(I)$. Hence $(V^I)^{-1} = W_0 W_0^I = W_0^J W_0 = V^J$.

Example 17.3.2 For $n=4$ and $I \neq \emptyset$, the table below lists the permutations ρ corresponding to the elements $W \in \mathrm{W}^I(4)W_0$, with $\rho^I \leftrightarrow V^I$ listed first.

I	C	ρ
$\{1,2,3\}$	$(1,1,1,1)$	$(4,3,2,1)$
$\{1,2\}$	$(1,1,2)$	$(4,3,1,2),(4,3,2,1)$
$\{1,3\}$	$(1,2,1)$	$(4,2,3,1),(4,3,2,1)$
$\{2,3\}$	$(2,1,1)$	$(3,4,2,1),(4,3,2,1),$
$\{1\}$	$(1,3)$	$(4,1,2,3),(4,1,3,2),(4,2,1,3),$ $(4,2,3,1),(4,3,1,2),(4,3,2,1)$
$\{2\}$	$(2,2)$	$(3,4,1,2),(3,4,2,1),$ $(4,3,1,2),(4,3,2,1)$
$\{3\}$	$(3,1)$	$(2,3,4,1),(2,4,3,1),(3,2,4,1),$ $(3,4,2,1),(4,2,3,1),(4,3,2,1)$

Recall from Proposition 16.1.3(ii) that the descent sets of W and WW_0 are complementary subsets of $Z[n-1]$. We introduce a short notation for the complement of a subset I of $Z[n-1]$.

Definition 17.3.3 For $I \subseteq Z[n-1]$, $-I = Z[n-1] \setminus I$.

For $I \subseteq Z[n-1]$, $j \in \mathbb{Z}[n-1]$ is in the transpose subset $J = \mathrm{tr}(I)$ if and only if $n-j \in I$ (Definition 16.5.5). Thus $-(\mathrm{tr}(I)) = \mathrm{tr}(-I)$, and we write this as

$-\mathrm{tr}(I)$. Then $\mathrm{des}(WW_0) = -I$ and $\mathrm{des}(W_0W) = -\mathrm{tr}(I)$ by Proposition 16.1.3. The next result generalizes Proposition 16.1.9.

Proposition 17.3.4 *Let* $I \subseteq Z[n-1]$ *and let* S_I *be a reduced word for* W_0^{-I}. *Then* (i) *there is a reduced word of the form* S_IW' *for a permutation matrix* $W \in \mathsf{W}(n)$ *if and only if* $I \subseteq des(W)$, *and* (ii) *there is a reduced word of the form* $W'S_J$ *for* W *if and only if* $J \subseteq des(W^{-1})$.

Proof We need only prove (i), since (ii) follows from (i) by replacing W by W^{-1} and observing that W_0^{-I} has order 2. Note that W_0^{-I} is the element of minimum length in $\mathsf{W}(n)$ with descent set I. If S_IW' is a reduced word for W, then all inversions of S_I are inversions of W, and so $I \subseteq \mathrm{des}(W)$. To prove the converse, we give a procedure which reduces W to W_0^{-I} by exchanging pairs $(j_1, j_1+1), \ldots, (j_k, j_k+1)$ of consecutive columns, so that exactly one inversion is removed at each step and $WS_{j_1}\cdots S_{j_k} = W_0^{-I}$. Thus $\mathrm{len}(W) = k + \mathrm{len}(W_0^{-I})$ and S_IW' is reduced word for W, where $W' = S_{j_k}\cdots S_{j_1}$.

The reduction procedure matches the columns of the current matrix with the columns of W_0^{-I}, starting from the right. For example with $n=4$ and $I=\{1,3\}$,

$$W = \begin{pmatrix} 0001 \\ 0100 \\ 0010 \\ 1000 \end{pmatrix} \to \begin{pmatrix} 0010 \\ 0100 \\ 0001 \\ 1000 \end{pmatrix} \to \begin{pmatrix} 0010 \\ 1000 \\ 0001 \\ 0100 \end{pmatrix} \to \begin{pmatrix} 0100 \\ 1000 \\ 0001 \\ 0010 \end{pmatrix} = W_0^{-I},$$

by switching columns 3 and 4 to make column 4 agree with W_0^{-I}, and then switching columns 1 and 2, then 2 and 3 to make column 3 agree with W_0^{-I}. By inspection of the matrix W_0^{-I}, the entries 1 moved to the left are always in higher rows than those moved to the right, and so the number of inversions is reduced by exactly 1 at each step, while the descent set contains I. □

17.4 Generators of the summands of $\mathsf{FL}(n)$

In this section we shall find an element $g_I \in \mathsf{H}_0(n)$ such that the flag $g_I(\mathsf{B}(n))$ generates the direct summand $\mathsf{FL}_I(n)$ of $\mathsf{FL}(n)$. By Proposition 16.5.1, the partial flag module $\mathsf{FL}^I(n)$ is embedded in $\mathsf{FL}(n)$ as a direct summand by associating to a partial flag X^I of type I the sum of all complete flags X which contain it.

Recall that the partial flags of type I can be regarded as the right cosets of the parabolic subgroup $\mathsf{P}^I(n)$, so that the reference flag R^I of type I corresponds to $\mathsf{P}^I(n)$ itself. Thus the complete flags X which contain a given partial flag X^I correspond to the right coset of $\mathsf{B}(n)$ in $\mathsf{P}^I(n)$. By Proposition 16.4.4, the number of such complete flags is $|\mathsf{P}^I(n)|/|\mathsf{B}(n)| = (c_1!)_2 \cdots (c_{r+1}!)_2$. In terms of the Schubert cell decomposition of $\mathsf{FL}(n)$, these are the flags in $\mathsf{Sch}(W)$ for $W \in \mathsf{W}^I(n)$. Hence the self-map of $\mathsf{FL}(n)$ given by the idempotent $\iota(f_{W_0^I})$ maps the reference flag R to the sum of all the complete flags which contain R^I, and so $\iota(f_{W_0^I})$ is the projection of $\mathsf{FL}(n)$ on the direct summand $\mathsf{FL}^I(n)$.

In the case $I = \{i\}$, $W_0^{-I} = S_i$ and so $f_{W_0^{-I}} = h_i$. This suggests the following notation for the idempotents $f_{W_0^I}$ and $\iota(f_{W_0^I})$.

Definition 17.4.1 For $I \subseteq Z[n-1]$, $h_I = f_{W_0^{-I}}$, $e_I = \iota(h_I)$ and $g_I = e_{-I}h_I$.

This indexing is chosen so that $h_{\{i\}} = h_i$, and we omit the brackets in this and similar notation. Thus by Propositions 17.2.4 and 17.2.7, h_I is the longest reduced word in the generators h_i, $i \in I$, and e_I is the sum of all reduced words in these generators. Since $\mathsf{W}^{-I}(n)$ is the subgroup of $\mathsf{W}(n)$ generated by the matrices S_i with $i \in I$, by the remarks above $e_{-I} = \iota(f_{W_0^I}) = \sum_{W \in \mathsf{W}^I(n)} f_W$ is the projection of $\mathsf{FL}(n)$ on to $\mathsf{FL}^I(n)$. Thus $e_{-I}(\mathsf{B}(n))$ generates $\mathsf{FL}^I(n)$.

Example 17.4.2 We tabulate the elements e_I, h_I and g_I in $\mathsf{H}_0(3)$.

I	e_I	h_I	g_I
{1,2}	$(1+a)(1+b)(1+a)$	aba	aba
{1}	$1+a$	a	$(1+b)a$
{2}	$1+b$	b	$(1+a)b$
$\emptyset$	1	1	$(1+a)(1+b)(1+a)$

As we saw in Section 16.6, the indecomposable summands $\mathsf{FL}_1(3)$ and $\mathsf{FL}_2(3)$ of $\mathsf{FL}(3)$ have dimension 6, with submodules $\mathsf{V}(3)^*$ and $\mathsf{V}(3)$ and quotients $\mathsf{V}(3)$ and $\mathsf{V}(3)^*$ respectively. As a submodule of $\mathsf{FL}^1(3)$, a generator of this copy of $\mathsf{V}(3)^*$ is obtained by choosing a plane P in $\mathsf{V}(3)$ and taking the sum of the 4 lines which do not lie in P. Under the standard embedding of $\mathsf{FL}^1(3)$ in $\mathsf{FL}(3)$, this translates into the sum of 12 flags. If $P = \mathsf{V}(2)$, then the row-reduced matrices representing these 12 flags have first row of the form $(*\ *\ 1)$, and so they are the flags in $\mathsf{Sch}(3,2,1)$ and $\mathsf{Sch}(3,1,2)$.

Similarly $\mathsf{FL}^2(3)$ has a submodule isomorphic to $\mathsf{V}(3)$ generated by the sum of the 4 planes which do not contain a fixed line L in $\mathsf{V}(3)$, and again this translates into the sum of 12 flags. If $L = \mathsf{V}(1)$, then the row-reduced matrices representing these 12 flags have last row (1 0 0), and so they are the flags in $\mathsf{Sch}(3,2,1)$ and $\mathsf{Sch}(2,3,1)$.

By Example 17.2.2, the four elements $g_{1,2} = aba$, $g_1 = a + ba$, $g_2 = b + ab$ and $g_\emptyset = 1 + a + b + ab + ba + aba$ of $\mathsf{H}_0(3)$ are idempotents with sum 1 and pairwise products 0. It follows (see Section 18.1) that they give decompositions $\mathsf{H}_0(3) = \bigoplus_I \mathsf{H}_0(3)g_I$ and $\mathsf{H}_0(3) = \bigoplus_I g_I\mathsf{H}_0(3)$ of $\mathsf{H}_0(3)$ as a direct sum of left or right ideals. The corresponding elements $g_I(\mathsf{B}(3))$ of $\mathsf{FL}(3)$ generate the summands in a direct sum decomposition, and arguing as above (e.g. $a + ba = (1+b)a$ is in the image of $\mathsf{FL}^1(3)$) we see that $g_I(\mathsf{B}(3))$ generates $\mathsf{FL}_I(3)$. The diagram below shows the composition factors of $\mathsf{FL}(3)$, labelled by elements of $\mathsf{H}_0(3)$ which map $\mathsf{B}(3)$ to a generator of the corresponding module.

The action of $\mathsf{H}_0(3)$ on $\mathsf{FL}(3)$ can be read off from this diagram. For example, the nilpotent element $s_2 = g_2 a$ maps $\mathsf{FL}_1(3)$ to the submodule $\mathsf{V}(3)$ of $\mathsf{FL}_2(3)$ since $s_2 g_1 = s_2$ and maps the other three summands to 0 since $s_1 g_I = 0$ for $I \neq \{1\}$.

$$\mathsf{St}(3) = \mathsf{FL}_{1,2}(3)$$

$$g_{1,2}$$

$$\begin{array}{llcclll} g_1 & \mathsf{V}(3) & & & & \mathsf{V}(3)^* & g_2 \\ & \downarrow & = \mathsf{FL}_1(3) & & \mathsf{FL}_2(3) = & \downarrow & \\ s_1 = g_1 b & \mathsf{V}(3)^* & & & & \mathsf{V}(3) & g_2 a = s_2 \end{array}$$

$$\mathsf{I}(3) = \mathsf{FL}_\emptyset(3)$$

$$g_\emptyset$$

Proposition 17.4.3 *Let* $I \subseteq Z[n-1]$ *and* $j \notin I$. *Then*

(i) $h_j f_{W_0^I} = f_{W_0^I} = f_{W_0^I} h_j$,
(ii) *the elements* $f_{W_0^I}$ *and* $\iota(f_{W_0^I})$ *are idempotent,*
(iii) $\iota(f_{W_0^I}) = \sum_W f_W$, *where the sum is over all* W *in* $\mathsf{W}^I(n)$.

Proof In the case $I = \{1,2,\ldots,k\}$, $1 \leq k \leq n-1$, these statements follow immediately from Proposition 17.2.6 by identifying $\mathsf{W}^I(n)$ with $\mathsf{W}(k+1)$. By reindexing, the same results hold for any sequence I of consecutive integers in $Z[n-1]$. In the general case, $W^I(n) \cong \mathsf{W}(c_1) \times \cdots \times \mathsf{W}(c_{r+1})$, where the generating switch matrices S_j of different factors commute. By Theorem 17.2.3, the corresponding generators h_j of $\mathsf{H}_0^I(n)$ commute. Hence $f_{W_0^I}$ is the product of commuting idempotents, and so it is itself idempotent. Since ι is an involution, $\iota(f_{W_0^I})$ is also idempotent. The sum formula for $\iota(f_{W_0^I})$ follows in a similar way from Proposition 17.2.7. □

Proposition 17.4.4 *Let $I \subseteq Z[n-1]$. Then*

(i) *e_I and h_I are idempotent,*
(ii) *$e_I = \sum_W f_W$, where the sum is over all $W \in \mathsf{W}^{-I}(n)$,*
(iii) *for all $i \in I$, $h_i h_I = h_I = h_I h_i$ and $h_i e_I = 0 = e_I h_i$,*
(iv) *for $J \subseteq I$, $h_J h_I = h_I = h_I h_J$ and $e_J e_I = e_I = e_I e_J$.*

Proof (i), (ii) and the first part of (iii) restate Proposition 17.4.3. Applying the involution ι to the first part of (iii) gives the second part. Since h_J is a word in the elements h_i, $i \in I$, the first part of (iv) follows from the first part of (iii), and the second part then follows by applying ι. □

Proposition 17.4.5 *For $I \subseteq Z[n-1]$ and $W \in \mathsf{W}(n)$,*

(i) *$f_W = h_I f_{W'}$ for some $W' \in \mathsf{W}(n)$ if and only if $I \subseteq des(W)$,*
(ii) *$f_W = f_{W'} h_J$ for some $W' \in \mathsf{W}(n)$ if and only if $J \subseteq des(W^{-1})$.*

Proof Both statements follow from Proposition 17.3.4 using the correspondence between reduced words in $\mathsf{W}(n)$ and $\mathsf{Sch}(n)$, Proposition 17.2.4(i). □

Proposition 17.4.6 *Given $I \subseteq Z[n-1]$ and $W \in \mathsf{W}(n)$ with $des(W) = I$,*

(i) *$e_{-I} f_W = f_W +$ elements $f_{W'}$ such that $len(W') > len(W)$,*
(ii) *$g_I = h_I +$ elements $f_{W'}$ such that $len(W') > len(W_0^{-I})$.*

Proof We prove (i), since (ii) is the special case of (i) obtained by taking $W = W_0^{-I}$, so that $f_W = h_I$ and $\mathrm{len}(W_0^{-I}) = \mathrm{len}(h_I)$. Since $e_{-I} = \sum_{V \in \mathsf{W}^I(n)} f_V$, $e_{-I} f_W = \sum_{V \in \mathsf{W}^I(n)} f_V f_W$. If $V \neq I_n$ then V is a product of switch matrices S_i with $i \notin I$. Since $\mathrm{des}(W) = I$, it follows from Proposition 17.1.5 that $\mathrm{len}(f_V f_W) > \mathrm{len}(f_W)$. □

Theorem 17.4.7 (Norton basis) *The elements $e_{-I} f_W$, where $I \subseteq Z[n-1]$ and $des(W) = I$, form a $\mathbb{F}_2$-basis for $\mathsf{H}_0(n)$. Thus $\mathsf{H}_0(n) = \oplus_I g_I \mathsf{H}_0(n)$, where $\dim g_I \mathsf{H}_0(n)$ is the number of elements $W \in \mathsf{W}(n)$ with $des(W) = I$.*

Proof By Proposition 17.4.6, the set of $n!$ elements $e_{-I} f_W$ of $\mathsf{H}_0(n)$ is triangularly related to the basis given by the elements f_W, $W \in \mathsf{W}(n)$ of $\mathsf{Sch}(n)$, and so it is a basis of $\mathsf{H}_0(n)$. □

We note that by Proposition 17.4.5(i), we may write $f_W = h_I f_{W'}$ for some $W' \in \mathsf{W}(n)$, and since $g_I = e_{-I} h_I$ each element of the Norton basis can be written in the form $g_I f_{W'}$ for some $W' \in \mathsf{W}(n)$. The Norton basis for $\mathsf{H}_0(4)$ is shown in the last column of the table at the end of the chapter.

We shall use this basis to find generators for the indecomposable summands $\mathsf{FL}_I(n)$ of $\mathsf{FL}(n)$.

Proposition 17.4.8 *The elements $g_I(\mathsf{B}(n))$ for $I \subseteq Z[n-1]$ generate $\mathsf{FL}(n)$ as a $\mathbb{F}_2\mathsf{GL}(n)$-module.*

Proof Every flag X in $\mathsf{V}(n)$ corresponds to a right coset $\mathsf{B}(n)A$, where $A \in \mathsf{GL}(n)$. By Theorem 17.4.7, the identity element of $\mathsf{H}_0(n)$ can be written as $1 = \sum_I g_I f_I'$, where $f_I' \in \mathsf{H}_0(n)$ and the sum is over all $I \subseteq Z[n-1]$. Hence $\mathsf{B}(n)A = \sum_I g_I f_I'(\mathsf{B}(n)A) = \sum_I g_I(f_I'(\mathsf{B}(n)A))$. Since $\mathsf{FL}(n)$ is the cyclic $\mathbb{F}_2\mathsf{GL}(n)$-module generated by $\mathsf{B}(n)$, for each I there is an element $c_I \in \mathbb{F}_2\mathsf{GL}(n)$ such that $f_I'(\mathsf{B}(n)A) = \mathsf{B}(n) \cdot c_I$. Since the left action of $\mathsf{H}_0(n)$ and the right action of $\mathbb{F}_2\mathsf{GL}(n)$ on $\mathsf{FL}(n)$ commute, $X = \sum_I g_I(\mathsf{B}(n) \cdot c_I) = \sum_I g_I(\mathsf{B}(n)) \cdot c_I$, as required. □

If $\mathrm{des}(W) = I$, then by Proposition 17.4.5 $f_W = h_I f_{W'}$ for some $W' \in \mathsf{W}(n)$. Since h_I is idempotent, $h_I f_W = f_W$. Since $g_I = e_{-I} h_I$, $g_I f_W = e_{-I} f_W = g_I f_{W'}$.

Proposition 17.4.9 *For $I \subseteq Z[n-1]$, $g_I(\mathsf{B}(n))$ generates $\mathsf{FL}_I(n)$.*

Proof By the remarks at the beginning of this section, $e_{-I} = \iota(f_{\mathsf{W}_0^I}) \in \mathsf{H}_0(n)$ maps the reference flag $\mathsf{B}(n)$ to a generator of the partial flag module $\mathsf{FL}^I(n)$, so that e_{-I} is the projection of $\mathsf{FL}(n)$ on a direct summand isomorphic to $\mathsf{FL}^I(n)$. Since $g_I(\mathsf{B}(n)) = e_{-I} h_I(\mathsf{B}(n))$, $g_I(\mathsf{B}(n))$ is in the image of the projection e_{-I}. By Theorem 16.5.3, $\mathsf{FL}_I(n) = \mathsf{FL}^I(n)/\sum_{J \subset I} \mathsf{FL}^J(n)$. We shall prove by contradiction that $g_I(\mathsf{B}(n)) \notin \sum_{J \subset I} \mathsf{FL}^J(n)$. The result follows, since $e_{-I}(\mathsf{B}(n))$ generates $\mathsf{FL}^I(n)$.

By Proposition 17.4.4(i), $h_I^2 = h_I$. By Proposition 17.4.6(ii), $g_I = h_I +$ elements $f_{W'}$ such that $\mathrm{len}(W') > \mathrm{len}(h_I)$. Hence $h_I g_I = h_I +$ elements $h_I f_{W'}$ such that $\mathrm{len}(f_{W'}) > \mathrm{len}(h_I)$. By Proposition 17.2.5 $\mathrm{len}(h_I f_{W'}) \geq \mathrm{len}(f_{W'}) > \mathrm{len}(h_I)$. Hence $h_I g_I \neq 0$, and it follows that $h_I g_I(\mathsf{B}(n)) \neq 0$.

Suppose that $g_I(\mathsf{B}(n)) \in \sum_{J \subset I} \mathsf{FL}^J(n)$. Since e_{-J} is the projection of $\mathsf{FL}(n)$ on to $\mathsf{FL}^J(n)$, $g_I(\mathsf{B}(n)) = \sum_{J \subset I} e_{-J}(X_J)$ where $X_J \in \mathsf{FL}(n)$ is a sum of flags. Hence $h_I g_I(\mathsf{B}(n)) = \sum_{J \subset I} h_I e_{-J}(X_J)$. But by Proposition 17.4.4(iii) $h_I e_{-J} = 0$ if $J \subset I$, since $h_i e_{-J} = 0$ for $i \in I \setminus J$ and there is a reduced word for h_I ending in h_i. Hence $h_I g_I(\mathsf{B}(n)) = 0$. This is a contradiction, and it follows that $g_I(\mathsf{B}(n)) \notin \sum_{J \subset I} \mathsf{FL}^J(n)$. □

Remark 17.4.10 Since $\mathsf{FL}(n)$ is the cyclic $\mathbb{F}_2\mathsf{GL}(n)$-module generated by $\mathsf{B}(n)$, it follows from Proposition 17.4.9 that the image of the $\mathbb{F}_2\mathsf{GL}(n)$-module map $g_I : \mathsf{FL}(n) \to \mathsf{FL}(n)$ is $\mathsf{FL}_I(n)$. However, g_I is not idempotent in general. The elements g_I for $n = 4$ are tabulated at the end of this chapter. In this case, $g_{1,3}$ is not idempotent since $g_{1,3}^2 = g_{1,3} + g_{1,3}bac$ in $\mathsf{H}_0(4)$, but $g_{1,3}^2$ is idempotent.

17.5 The submodule $s_I(\mathsf{FL}(n))$ of $\mathsf{FL}_I(n)$

In this section we fix subsets I and J of $Z[n-1] = \{1,\dots,n-1\}$ such that $J = \mathrm{tr}(I)$, so that $j \in J$ if and only if $n-j \in I$. We introduce a cyclic submodule $s_I(\mathsf{FL}(n))$ of the summand $\mathsf{FL}_I(n)$ of $\mathsf{FL}(n)$. In Section 17.6 we shall see that $s_I(\mathsf{FL}(n))$ is the unique irreducible submodule of $\mathsf{FL}_I(n)$.

The coset $\mathsf{W}^I(n)W_0$ consists of block anti-diagonal matrices W, with blocks of size $c_1,\dots,c_{r+1}$ where $(c_1,\dots,c_{r+1})$ is the composition of n associated to I. Hence $I \subset \mathrm{des}(W)$, and since $W^{-1} = W^{\mathrm{tr}}$, $W^{-1} \in W_0\mathsf{W}^I(n)$ is a block anti-diagonal matrix with blocks of size $c_{r+1},\dots,c_1$, so that $J \subseteq \mathrm{des}(W^{-1})$. In the case $n=4$, $I=\{2\}$, the coset $\mathsf{W}^I(n)W_0$ is illustrated in (17.2).

Definition 17.5.1 For $I \subseteq Z[n-1]$, let $s_I \in \mathsf{H}_0(n)$ be the sum of the elements f_W, where $W \in \mathsf{W}^I(n)W_0$.

Thus $s_I(\mathsf{B}(n))$ is the sum of the flags in the Schubert cells $\mathsf{Sch}(W)$, $W \in \mathsf{W}^I(n)W_0$. In particular, $s_\emptyset$ is the sum of all $n!$ basis elements f_W, so that $s_\emptyset(\mathsf{B}(n))$ is the sum of all the flags in $\mathsf{FL}(n)$, and is $\mathsf{GL}(n)$-invariant. When $I = Z[n-1]$, $s_I = f_{W_0}$ and $s_I(\mathsf{B}(n))$ is the sum of the flags in $\mathsf{Sch}(W_0)$, which are in 'general position' with respect to the reference flag. For $n=3$ and $I=\{1\}$, the elements of $\mathsf{W}^I(n)W_0$ correspond to the permutations $(3,1,2)$ and $(3,2,1)$ in $\Sigma(3)$, so that $s_1 = ab + aba$ where $a = h_1$ and $b = h_2$, and similarly $s_2 = ba + aba$. For $n=4$, s_I is the sum of the 6 elements f_W where W corresponds to a permutation $\rho \in \Sigma(4)$ such that $\rho(1) = 4$. It can be written as a sum of 6 reduced words in $a = h_1$, $b = h_2$ and $c = h_3$ by using the table at the end of this chapter.

Proposition 17.5.2 *The stabilizer of* $s_I(\mathsf{B}(n))$ *in* $\mathsf{GL}(n)$ *is the parabolic subgroup* $\mathsf{P}^J(n)$, *where* $J = tr(I)$.

Proof Since $s_I(\mathsf{B}(n))$ is the sum of all the flags in certain Schubert cells and the right action of elements of $\mathsf{B}(n)$ permutes the flags in each Schubert cell, the stabilizer of $s_I(\mathsf{B}(n))$ contains $\mathsf{B}(n)$, and so it is a parabolic subgroup of $\mathsf{GL}(n)$. Since $W_0S_jW_0 = S_{n-j}$ for $1 \le j \le n-1$, $\mathsf{W}^I(n)W_0S_j = \mathsf{W}^I(n)W_0$ if and only if $S_{n-j} \in \mathsf{W}^I$, so that $n-j \notin I$. Thus the Schubert cells in $s_I(\mathsf{B}(n))$ are permuted by S_j if and only if $j \notin J$, and the result follows. □

Conjugation with W_0 maps $\mathsf{W}^I(n)$ to $\mathsf{W}^J(n)$, so that we can alternatively write s_I in the form $s_I = \sum_{W \in W_0\mathsf{W}^J(n)} f_W$. In particular, conjugation with W_0 maps W_0^I to W_0^J. Recall from Definition 17.3.1 that V^I is the element of minimum length in $\mathsf{W}^I(n)W_0$, so that its anti-diagonal blocks are identity matrices. Thus $V^I = W_0^IW_0 = W_0W_0^J$, and since $W_0^I \in \mathsf{W}^I$, $\mathsf{W}^I(n)V^I = \mathsf{W}^I(n)W_0 = W_0W_0^J = V^I\mathsf{W}^J(n)$, so that we can replace W_0 by V^I in Definition 17.5.1.

Proposition 17.5.3 *For* $I \subseteq Z[n-1]$ *and* $J = tr(I)$, $s_I = e_{-I}f_{V^I} = f_{V^I}e_{-J}$.

Proof By the preceding remarks, s_I is the sum of the basis elements f_W where $W = W^I V^I$ and $W^I \in \mathsf{W}^I(n)$. We shall prove that $\text{len}(W) = \text{len}(W^I) + \text{len}(V^I)$, from which it follows that $f_W = f_{W^I} f_{V^I}$ by Proposition 17.1.6. Since $e_{-I} = \sum_{W \in \mathsf{W}^I(n)} f_W$, this gives the factorization $s_I = e_{-I} f_{V^I}$.

Recall that $\text{len}(W_1) + \text{len}(W_2) = \text{len}(W_0)$ if $W_1 W_2 = W_0$ in $\mathsf{W}(n)$. Thus $\text{len}(V^I) = \text{len}(W_0) - \text{len}(W_0^I)$, since $V^I = W_0^I W_0$. More generally, $\text{len}(W_1) + \text{len}(W_2) = \text{len}(W_0^I)$ if $W_1 W_2 = W_0^I$ where $W_1, W_2 \in \mathsf{W}^I(n)$. Thus $\text{len}(W) = \text{len}(W^I V^I) = \text{len}(W^I W_0^I W_0) = \text{len}(W_0) - \text{len}(W^I W_0^I) = \text{len}(W_0) - (\text{len}(W_0^I) - \text{len}(W^I)) = \text{len}(V^I) + \text{len}(W^I)$, as required.

To prove the second factorization $s_I = f_{V^I} e_{-J}$, we write s_I as the sum of the basis elements f_W where $W = V^I W^J$ and $W^J \in \mathsf{W}^J(n)$, and use a similar argument to show that $\text{len}(W) = \text{len}(V^I) + \text{len}(W^J)$. □

Example 17.5.4 Let $n = 4$, and let $h_1 = a$, $h_2 = b$ and $h_3 = c$. When $I = \{1\}$, $J = \{3\}$ and $\mathsf{W}^I(n)$ is the subgroup of $\mathsf{W}(4)$ generated by S_2 and S_3. Hence $W_0^I = S_2 S_3 S_2$ and $h_{-I} = bcb$. Then $e_{-I} = \iota(bcb) = (1+b)(1+c)(1+b) = 1 + b + c + bc + cb + bcb$ is the sum of the elements $f_W \in \mathsf{H}_0(4)$ such that $W \in \mathsf{W}^I(n)$. The element $V^I \in \mathsf{W}(4)$ corresponds to the permutation $(4,1,2,3)$, so that $f_{V^I} = abc$. Thus Proposition 17.5.3 states that $s_1 = (1+b)(1+c)(1+b)abc = abc(1+a)(1+b)(1+a)$.

Proposition 17.5.5 *For* $I \subseteq Z[n-1]$,

$$\text{(i) } h_i s_I = \begin{cases} s_I, & \text{if } i \in I, \\ 0, & \text{if } i \notin I, \end{cases} \qquad \text{(ii) } s_I h_j = \begin{cases} s_I, & \text{if } j \in tr(I), \\ 0, & \text{if } j \notin tr(I). \end{cases}$$

Thus $\mathsf{H}_0(n)s_I = s_I\mathsf{H}_0(n)$ *is the* 1*-dimensional vector space over* $\mathbb{F}_2$ *generated by* s_I.

Proof (i) If $W \in \mathsf{W}^I(n)W_0$, then $I \subseteq \text{des}(W)$, and so by Proposition 17.2.4(ii) there is a reduced word for f_W beginning with h_i. Since h_i is idempotent, it follows that $h_i f_W = f_W$. Summing over $W \in \mathsf{W}^I(n)W_0$, $h_i s_I = s_I$. On the other hand, if $i \notin I$ then $h_i e_{-I} = 0$ by Proposition 17.4.4(iii), and so $h_i s_I = 0$ by Proposition 17.5.3.

(ii) If $W \in \mathsf{W}^I(n)W_0$, then $W^{-1} \in W_0\mathsf{W}^I(n)$ and $J \subseteq \text{des}(W^{-1})$. By Proposition 17.2.4(ii), there is a reduced word for f_W ending in h_j if $j \in J$. Hence $f_W h_j = f_W$ when $j \in J$. Summing over $W \in \mathsf{W}^I(n)W_0$, $s_I h_j = s_I$. On the other hand, if $j \notin J$ then $e_{-J}h_j = 0$ by Proposition 17.4.4(iii), and so $s_I h_j = 0$ by Proposition 17.5.3. □

Since $h_i f_W = f_W$ for all $i \in I$ when $I \subseteq \text{des}(W)$, and h_I is a word in the elements $h_i, i \in I$, we have $h_I f_W = f_W$ for all $W \in \mathsf{W}^I(n)W_0$. In particular, $h_I f_{V^I} = f_{V^I}$. Hence $g_I f_{V^I} = e_{-I} h_I f_{V^I} = e_{-I} f_{V^I} = s_I$.

Proposition 17.5.6

$$s_I g_K = \begin{cases} s_I, & \text{if } K = J = tr(I), \\ 0, & \text{otherwise.} \end{cases}$$

Proof Consider first $s_I g_J$, where $J = \text{tr}(I)$. Recall that $g_J = e_{-J} h_J$. Since $e_{-J} = \sum_{W \in \mathsf{W}^J(n)} f_W$ and, by Proposition 17.5.5, $\mathsf{W}^J(n)$ is generated by the switch matrices $S_j, j \notin J$, $s_I h_j = 0$, and hence $s_I f_W = 0$ for all $f_W \neq 1$ in the sum e_{-J}. Hence $s_I e_{-J} = s_I$, and so $s_I g_J = s_I h_J$. Now $h_J = f_{W_0^{-J}}$ is a product of generators h_j with $j \in J$, and for such j we have $s_I h_j = s_I$ by Proposition 17.5.5. Hence $s_I g_J = s_I$.

Now assume that $K \neq J$ and consider $s_I g_K$. Assume first that $K \not\subseteq J$. Since $g_K = e_{-K} h_K$ and $s_I e_{-K} = s_I$ or 0 by Proposition 17.5.5, it suffices to prove that $s_I h_K = 0$. Let $j \in K, j \notin J$. Then there is a reduced word for W_0^{-K} beginning with S_j, and $s_I h_j = 0$ by Proposition 17.5.5. Hence $s_I h_K = 0$.

Now suppose $K \subset J$. Since $g_K = e_{-K} h_K$, it suffices to prove that $s_I e_{-K} = 0$. As above, $s_I e_{-K} = \sum_{W \in \mathsf{W}^K(n)} s_I f_W$. Each W in the sum is a product of switch matrices S_j with $j \notin K$. Now $s_I f_W = 0$ if W is not a product of switch matrices S_j with $j \in J$, so $s_I e_{-K}$ is the sum of products $s_I f_W = s_I$, where W is a product of S_j's for $j \in J \setminus K$. Since the subgroup $\mathsf{W}^{-(J \setminus K)}$ has even order, $s_I e_{-K} = 0$. □

Proposition 17.5.7 *For $I \subseteq \mathsf{Z}[n-1]$, $s_I(\mathsf{FL}(n))$ is a submodule of $\mathsf{FL}_I(n)$.*

Proof We show that $s_I = g_I t_I$ for some $t_I \in \mathsf{H}_0(n)$, so that the result follows from Proposition 17.4.9. Since $s_I = e_{-I} f_{V^I}$ and $g_I = e_{-I} h_I$, it suffices to prove that $f_{V^I} = h_I t_I$. Since $\text{des}(V^I) = I$, it follows from Proposition 17.3.4 that there is a reduced word in $S_1, \ldots, S_{n-1}$ of the form $S_I W'$ for V^I, where S_I is a reduced word for W_0^{-I}. By Proposition 17.2.4 there is a corresponding reduced word for f_{V^I} in $h_1, \ldots, h_{n-1}$. Since $h_I = f_{W_0^{-I}}$, this reduced word expresses f_{V^I} in the form $h_I t_I$ for some $t_I \in \mathsf{H}_0(n)$. □

17.6 Irreducible submodules of FL(*n*)

In this section we show that $s_I(\mathsf{FL}(n))$ is the unique irreducible submodule of $\mathsf{FL}_I(n)$ for all $I \subseteq Z[n-1]$. Thus $\mathsf{FL}(n)$ has 2^{n-1} irreducible submodules $s_I(\mathsf{FL}(n))$, and we show that they form a complete set of irreducible $\mathbb{F}_2\mathsf{GL}(n)$ modules, i.e. every irreducible $\mathbb{F}_2\mathsf{GL}(n)$ module is isomorphic to $s_I(\mathsf{FL}(n))$ for exactly one I. The following consequence of Proposition 16.3.7 is useful in analyzing the submodule structure of $\mathsf{FL}(n)$.

Proposition 17.6.1 *The map $f_W \mapsto f_W(\mathsf{B}(n))$ gives an additive isomorphism between the Hecke algebra $\mathsf{H}_0(n)$ and the $\mathbb{F}_2$-space of $\mathsf{B}(n)$-invariant elements of $\mathsf{FL}(n)$.* □

Theorem 17.6.2 *The element $s_I(\mathsf{B}(n))$ is the unique nonzero $\mathsf{B}(n)$-invariant in $s_I(\mathsf{FL}(n))$.*

Proof By Proposition 16.3.7, the $\mathsf{B}(n)$-invariants in $\mathsf{FL}(n)$ are the elements which are sums of all the flags in some set of Schubert cells, i.e. they are of the form $F = \sum_{W \in S} f_W(\mathsf{B}(n))$ where $S \subseteq \mathsf{W}(n)$. We aim to prove that if $F \in s_I(\mathsf{FL}(n))$ then $S = W^I(n)W_0$ and $F = s_I(\mathsf{B}(n))$. If $F \in s_I(\mathsf{FL}(n))$, then $F = s_I(X)$ for some $X \in \mathsf{FL}(n)$. It follows from Proposition 17.5.5 that $h_iF = F$ if $i \in I$ and $h_iF = 0$ if $i \notin I$.

For $1 \leq i \leq n-1$, a general $\mathsf{B}(n)$-invariant $F \in \mathsf{FL}(n)$ can be written as $F = F_1(i) + F_2(i)$, where $F_1(i)$ and $F_2(i)$ are the sums of terms $f_W(\mathsf{B}(n))$ with $i \in \mathrm{des}(W)$ and $i \notin \mathrm{des}(W)$ respectively. Recall from Proposition 17.1.5 that $h_if_W = f_W$ if $i \in \mathrm{des}(W)$ and $h_if_W = f_{S_iW}$ if $i \notin \mathrm{des}(W)$. Hence $h_iF = F_1(i) + F_2'(i)$, where $F_2'(i) = \sum_{i \notin \mathrm{des}(W), W \in S} f_{S_iW}(\mathsf{B}(n))$. Thus $F_1(i) + F_2'(i) = F$ if $i \in I$ and $F_1(i) + F_2'(i) = 0$ if $i \notin I$.

Since an element $h \in \mathsf{H}_0(n)$ is determined uniquely by $h(\mathsf{B}(n))$, we obtain the following equations in $\mathsf{H}_0(n)$, where all sums are over elements of S with descent sets as stated.

$$\text{(i)} \sum_{i \notin \mathrm{des}(W)} f_W = \sum_{i \notin \mathrm{des}(W)} f_{S_iW},\ i \in I, \quad \text{(ii)} \sum_{j \in \mathrm{des}(W)} f_W = \sum_{j \notin \mathrm{des}(W)} f_{S_jW},\ j \notin I.$$

If $i \notin \mathrm{des}(W)$, then $i \in \mathrm{des}(W')$ where $W' = S_iW$. Hence the left side of (i) is a sum of elements f_W with $i \notin \mathrm{des}(W)$, while the right side is a sum of elements $f_{W'}$ with $i \in \mathrm{des}(W')$. Thus both sums are 0, and it follows that $i \in \mathrm{des}(W)$ for all $W \in S$. Hence $I \subseteq \mathrm{des}(W)$ for all $W \in S$.

By the same argument, both sides of (ii) are sums of elements f_W with $j \in \mathrm{des}(W)$, and (ii) states that the set of elements $W \in S$ such that $j \in \mathrm{des}(W)$ is the same as the set of elements $W' = S_jW$ where $W \in S$ and $j \notin \mathrm{des}(W)$. It follows that if $W \in S$ then $S_jW \in S$, whether $j \in \mathrm{des}(W)$ or not. Hence S is closed under premultiplication by S_j for $j \notin I$, and so it is a union of cosets $\mathsf{W}^I(n)A$ in $\mathsf{W}(n)$.

To complete the proof, we observe that all cosets $\mathsf{W}^I(n)A$ of $\mathsf{W}^I(n)$ in $\mathsf{W}(n)$ other than $\mathsf{W}^I(n)W_0$ contain elements W such that $I \not\subseteq \mathrm{des}(W)$. We may assume that $I \neq \emptyset$ since $\mathsf{W}^{\emptyset}(n) = \mathsf{W}(n)$. Recall that $\mathsf{W}^I(n)$ is the subgroup of block permutation matrices with blocks of size $c_1,\ldots,c_{r+1}$ where $C=(c_1,\ldots,c_{r+1})$ is the composition of n associated to I.

In particular, $c_1=i_1$ is the smallest element of I, and the elements of a coset $\mathsf{W}^I(n)A$ consist of permutation matrices such that the entries 1 in the first c_1 rows appear in a fixed set of columns. Unless these columns are the last c_1 columns, we can find such a matrix which does not have i_1 in its descent set. Hence if S is nonempty then $S=\mathsf{W}^I(n)W_0$. □

By Proposition 17.6.1, the Hecke algebra $\mathsf{H}_0(n)$ is faithfully represented by its image under the map $f\mapsto f(\mathsf{B}(n))$, which consists of the $\mathsf{B}(n)$-invariant elements of $\mathsf{FL}(n)$. As observed in Section 17.1, Proposition 17.1.5 is equivalent to the statement that the 1-eigenspace of left multiplication by h_i in $\mathsf{H}_0(n)$ is $h_i\mathsf{H}_0(n)$, and the 0-eigenspace is $(1+h_i)\mathsf{H}_0(n)$. Hence Theorem 17.6.2 can be interpreted as a statement about $\mathsf{H}_0(n)$, as follows.

Proposition 17.6.3 *Given $I\subseteq Z[n-1]$, the intersection of the left ideals $h_i\mathsf{H}_0(n)$ for $i\in I$ and $(1+h_i)\mathsf{H}_0(n)$ for $i\notin I$ is the 1-dimensional subspace generated by s_I.* □

Proposition 17.6.4 *For $I,J\subseteq Z[n-1]$,*

(i) *$s_I(\mathsf{FL}(n))$ is an irreducible $\mathbb{F}_2\mathsf{GL}(n)$ module,*
(ii) *$s_I(\mathsf{FL}(n))\cong SF_J(n)$ if and only if $I=J$.*

Proof (i) follows from Definition 17.5.1 and Theorem 17.6.2 using the irreducibility criterion of Proposition 10.1.3.

(ii) An isomorphism $s_I(\mathsf{FL}(n))\to SF_J(n)$ must map the unique $\mathsf{B}(n)$-invariant element $s_I(\mathsf{B}(n))$ of $s_I(\mathsf{FL}(n))$ to the unique $\mathsf{B}(n)$-invariant element $s_J(\mathsf{B}(n))$ of $SF_J(n)$, so $s_I(\mathsf{B}(n))$ and $s_J(\mathsf{B}(n))$ must have the same stabilizer in $\mathsf{GL}(n)$. By Proposition 17.5.2 the stabilizer of $s_I(\mathsf{B}(n))$ is the parabolic subgroup $\mathsf{P}^{tr(I)}(n)$. Since $\mathsf{P}^{tr(I)}(n)$ and $\mathsf{P}^{tr(J)}(n)$ are distinct subgroups of $\mathsf{GL}(n)$ if $I\neq J$, the result follows. □

We next relate the irreducible submodules of $\mathsf{FL}(n)$ to the direct sum decomposition $\mathsf{FL}(n)=\bigoplus_I\mathsf{FL}_I(n)$ of Theorem 16.5.3.

Proposition 17.6.5 *For $I\subseteq Z[n-1]$,*

(i) *$s_I(\mathsf{FL}(n))$ is the unique irreducible submodule of $\mathsf{FL}_I(n)$,*
(ii) *$\mathsf{FL}_I(n)$ has a unique irreducible quotient, which is isomorphic to $SF_J(n)$ for $J=tr(I)$,*
(iii) *$\mathsf{FL}_I(n)$ is an indecomposable $\mathbb{F}_2\mathsf{GL}(n)$-module.*

Proof By Proposition 16.5.6, $\mathsf{FL}_J(n)$ is the transpose dual of $\mathsf{FL}_I(n)$, so (i) and (ii) are equivalent. Since a module with a unique irreducible submodule is indecomposable, (iii) follows from (i) or (ii).

To prove (i), we use Proposition 17.6.3. An irreducible submodule M of $\mathsf{FL}(n)$ must contain a $\mathsf{B}(n)$-invariant element, which must be of the form $h(\mathsf{B}(n))$ for some nonzero element $h \in \mathsf{H}_0(n)$. The left ideal $h\mathsf{H}_0(n)$ must be 1-dimensional, i.e. $hh_i = h_i$ or 0 for $1 \leq i \leq n-1$, since otherwise some element hh_i generates a proper ideal contained in $h\mathsf{H}_0(n)$ and $hh_i(\mathsf{B}(n))$ is a proper submodule of M. Hence h is in the intersection of the 0- or 1-eigenspaces of h_i for $1 \leq i \leq n-1$. If $I \subseteq Z[n-1]$ and h is in the 1-eigenspace of h_i for $i \in I$ and in the 0-eigenspace of h_j for $j \notin I$, then $h = s_I$ by Theorem 17.6.2. □

Proposition 17.6.6 *Every irreducible* $\mathbb{F}_2\mathsf{GL}(n)$*-module is isomorphic to a quotient module of* $\mathsf{FL}(n)$.

Proof By Proposition 10.1.1, by restriction to $\mathsf{B}(n)$ it follows that an irreducible right $\mathbb{F}_2\mathsf{GL}(n)$-module M is generated by a $\mathsf{B}(n)$-invariant $x \neq 0$. Let $\psi : \mathbb{F}_2\mathsf{GL}(n) \to M$ be the linear map defined by $\psi(A) = x \cdot A$, where $A \in \mathsf{GL}(n)$. Then ψ is a map of $\mathbb{F}_2\mathsf{GL}(n)$-modules, and so M is a quotient of the group algebra $\mathbb{F}_2\mathsf{GL}(n)$. For $B \in \mathsf{B}(n)$, $\psi(BA) = \psi(A)$ since $x \cdot B = x$, so $\psi(A)$ depends only on the coset $\mathsf{B}(n)A$. Hence ψ factors through the quotient map $\mathbb{F}_2\mathsf{GL}(n) \to \mathsf{FL}(n) = \mathbb{F}_2(\mathsf{GL}(n)/\mathsf{B}(n))$, giving a surjection $\mathsf{FL}(n) \to M$ of $\mathbb{F}_2\mathsf{GL}(n)$-modules. □

The preceding results can be summarized as follows.

Theorem 17.6.7 *The modules* $s_I(\mathsf{FL}(n))$ *for* $I \subseteq Z[n-1]$ *are a complete set of inequivalent* $\mathbb{F}_2\mathsf{GL}(n)$*-modules, i.e. every irreducible right* $\mathbb{F}_2\mathsf{GL}(n)$*-module is isomorphic to* $s_I(\mathsf{FL}(n))$ *for one and only one subset* $I \subseteq Z[n-1]$. *Thus there are* 2^{n-1} *isomorphism classes of irreducible* $\mathbb{F}_2\mathsf{GL}(n)$*-modules.*

Proof Let M be an irreducible $\mathbb{F}_2\mathsf{GL}(n)$-module. Then M is isomorphic to a quotient of $\mathsf{FL}(n)$ by Proposition 17.6.6, and so to a submodule of $\mathsf{FL}(n)$ by transpose duality. By Propositions 17.6.5(i) and 17.6.4, $M \cong s_I(\mathsf{FL}(n))$ for a unique $I \subseteq Z[n-1]$. □

The indecomposable summand $\mathsf{FL}_\emptyset(n) = SF_\emptyset(n)$ is the 1-dimensional module generated by the sum of all the flags $s_\emptyset = \sum_{w \in \mathsf{W}(n)} f_w$. The case $n = 3$ of the next result was proved in Proposition 16.6.2.

Theorem 17.6.8 *For* $I = Z[n-1]$, *the direct summand* $\mathsf{FL}_I(n)$ *of* $\mathsf{FL}(n)$ *is an irreducible* $\mathbb{F}_2\mathsf{GL}(n)$*-module of dimension* $2^{n(n-1)/2}$, *the* **Steinberg** *module* $\mathsf{St}(n)$.

Proof For $I = Z[n-1]$, $\mathsf{W}^I(n) = \{I_n\}$ and so $s_I = f_{W_0}$. Thus $s_I(\mathsf{FL}(n))$ is generated by $s_I(\mathsf{B}(n))$, the sum of the flags in the top Schubert cell. On the other hand, $\mathsf{FL}_I(n)$ is generated by $g_I(\mathsf{B}(n))$ where $g_I = e_\emptyset h_I$. Since $e_\emptyset = 1$ is the identity element of $\mathsf{H}_0(n)$ and $h_I = f_{W_0^{-I}} = f_{W_0^\emptyset} = f_{W_0}$, $g_I = s_I$. Hence $\mathsf{FL}_I(n) = s_I(\mathsf{FL}(n))$ is irreducible. By Proposition 16.3.5, $\dim \mathsf{FL}_I(n) = \dim \mathsf{Sch}(W_0) = 2^{\mathrm{len}(W_0)} = 2^{n(n-1)/2}$. □

If $J = \mathrm{tr}(I)$, then it follows from Proposition 17.5.6 that $s_I : \mathsf{FL}(n) \to \mathsf{FL}(n)$ maps the generator $g_J(\mathsf{B}(n))$ of $\mathsf{FL}_J(n)$ to the generator $s_I(\mathsf{B}(n))$ of the socle of $\mathsf{FL}_I(n)$, i.e. its irreducible submodule, and maps $g_K(\mathsf{B}(n))$ to 0 for $K \neq J$. Thus the irreducible quotient, or *head*, of $\mathsf{FL}_J(n)$ is isomorphic to the socle of $\mathsf{FL}_I(n)$, as observed in Proposition 17.6.5.

Proposition 17.5.5 provides an indexing of the irreducible $\mathbb{F}_2\mathsf{GL}(n)$-modules $s_I(\mathsf{FL}(n))$ by subsets $I \subseteq Z[n-1]$. in Section 20.3 we discuss the construction of the irreducible $\mathbb{F}_2\mathsf{GL}(n)$-modules as quotients of Weyl modules $\Delta(\lambda)$, where λ is a column 2-regular partition of length $\leq n-1$.

Definition 17.6.9 Let $I = \{i_1, \ldots, i_r\} \subseteq Z[n-1]$ where $1 \leq i_1 < \cdots < i_r \leq n-1$, and let λ be the column 2-regular partition such that $\lambda^{\mathrm{tr}} = (i_r, \ldots, i_1)$. Then the irreducible quotient of $\mathsf{FL}_I(n)$ is alternatively denoted by $L(\lambda)$.

Each basis element of $\mathsf{H}_0(n)$ given in Theorem 17.4.7 gives an isomorphism between a quotient of $\mathsf{FL}_I(n)$ and a submodule of $\mathsf{FL}_J(n)$ for some I and J. This information can be used to locate certain composition factors in these modules, and for $n \leq 3$ all composition factors are detected in this way. In the case $n = 4$, full composition series can be constructed by using the following additional information, together with the fact that $\mathsf{FL}(4)$ is self-dual with respect to both transpose and contragredient duality (see Proposition 16.5.6): the summand $\mathsf{FL}_{1,3}(4)$ has a quotient of dimension 15 isomorphic to the Weyl module $\Delta(2,1,1)$ (see Proposition 20.4.2), and $\Delta(2,1,1)$ has a 1-dimensional submodule and a 14-dimensional irreducible quotient $L(2,1,1)$). The resulting structure of $\mathsf{FL}(4)$ is shown diagrammatically at the end of this chapter.

17.7 Remarks

Our approach to the representation theory of $\mathbb{F}_2\mathsf{GL}(n)$ follows the methods of [37], which is based on a construction due to Curtis (see [79, Chapter 7]). A detailed account of the Hecke algebra $\mathsf{H}_0(n)$ has been given by Pamela N. Norton [156].

In particular, the results of Section 17.6 provide a description of the irreducible representations of $\mathsf{H}_0(n)$ itself. These are all 1-dimensional, and are given by the characteristic functions of the subsets I of $Z[n-1]$, i.e. the homomorphisms $\chi_I : \mathsf{H}_0(n) \to \mathbb{F}_2$ defined by $\chi_I(h_i) = 1$ if $i \in I$ and $\chi_I(h_i) = 0$ if $i \in -I$. We use the opposite notation to [156] for the subgroups $\mathsf{W}^I(n)$, in that the set I and its complement $-I$ in $Z[n-1]$ are exchanged.

The construction of irreducible modules for $\mathbb{F}_2\mathsf{GL}(n)$ using Weyl modules can be found in [98, Chapter 8]. Since $\mathsf{GL}(n)$ has 2^{n-1} conjugacy classes of odd order elements (see Proposition 18.2.2(ii)), it follows from a standard result in modular representation theory (see e.g. [48, Theorem 83.5]) that the number of inequivalent irreducible $\mathbb{F}_2\mathsf{GL}(n)$-modules is 2^{n-1}.

The Schubert monoid $\mathsf{Sch}(4)$ and the Norton basis for $\mathsf{H}_0(4)$

$I = \text{des}(\rho)$	ρ	reduced words for f_W	$e_{-I}f_W$
$\{1,2,3\}$	$(4,3,2,1)$	$abacba, abcaba, abcbab, acbcab,$ $acbacb, cabcab, cabacb, cbabcb,$ $cbacbc, cbcabc, bcbabc, bcabac,$ $bacbac, bcabca, bacbca, babcba$	$g_{1,2,3}$
$\{1,2\}$	$(4,3,1,2)$	$bacbc, bcabc, babcb, abacb, abcab$	$g_{1,2}cb$
	$(4,2,1,3)$	$abca, abac, babc$	$g_{1,2}c$
	$(3,2,1,4)$	aba, bab	$g_{1,2}$
$\{1,3\}$	$(4,2,3,1)$	$abcba, acbca, cabca, cabac, acbac, cbabc$	$g_{1,3}bac$
	$(4,1,3,2)$	$abcb, acbc, cabc$	$g_{1,3}bc$
	$(3,2,4,1)$	$acba, caba, cbab$	$g_{1,3}ba$
	$(3,1,4,2)$	cab, acb	$g_{1,3}b$
	$(2,1,4,3)$	ac, ca	$g_{1,3}$
$\{2,3\}$	$(3,4,2,1)$	$bacba, bcaba, bcbab, cbcab, cbacb$	$g_{2,3}ab$
	$(2,4,3,1)$	$bcba, cbca, cbac$	$g_{2,3}a$
	$(1,4,3,2)$	bcb, cbc	$g_{2,3}$

$\{1\}$	$(4,1,2,3)$	abc	g_1bc
	$(3,1,2,4)$	ab	g_1b
	$(2,1,3,4)$	a	g_1
$\{2\}$	$(3,4,1,2)$	$bacb, bcab$	g_2acb
	$(2,4,1,3)$	bca, bac	g_2ac
	$(2,3,1,4)$	ba	g_2a
	$(1,4,2,3)$	bc	g_2c
	$(1,3,2,4)$	b	g_2
$\{3\}$	$(2,3,4,1)$	cba	g_3ba
	$(1,3,4,2)$	cb	g_3b
	$(1,2,4,3)$	c	g_3
$\emptyset$	$(1,2,3,4)$	1	$g_\emptyset$

The last column gives the Norton basis element $e_{-I}f_W$, where $W \in \mathsf{W}(4)$ is the permutation matrix corresponding to $\rho \in \Sigma(4)$. The basis elements are written in the form $g_I f_{W'}$, where the elements $g_I = e_{-I} \cdot h_I$ of $\mathsf{H}_0(4)$ such that $g_I(\mathsf{B}(4)) = \mathsf{FL}_I(4)$ are listed below. The table gives all reduced words as products of the generators $a = h_1$, $b = h_2$, $c = h_3$ of $\mathsf{Sch}(4)$.

$$
\begin{aligned}
g_{1,2,3} &= 1 \cdot abacba, & g_1 &= (1+b)(1+c)(1+b)\cdot a,\\
g_{1,2} &= (1+c)\cdot aba, & g_2 &= (1+a)(1+c)\cdot b,\\
g_{1,3} &= (1+b)\cdot ac, & g_3 &= (1+a)(1+b)(1+a)\cdot c,\\
g_{2,3} &= (1+a)\cdot bcb, & g_\emptyset &= (1+a)(1+b)(1+a)(1+c)(1+b)(1+a)\cdot 1.
\end{aligned}
$$

Structure of the flag module FL(4)

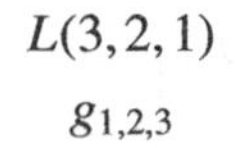

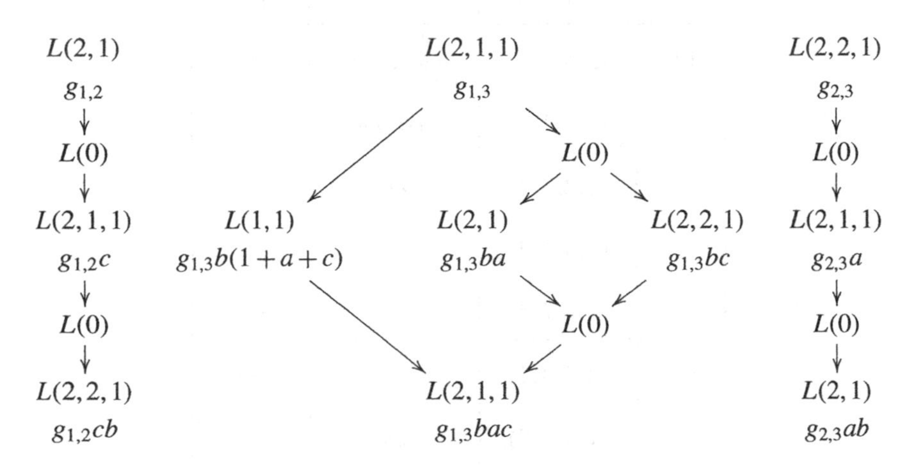

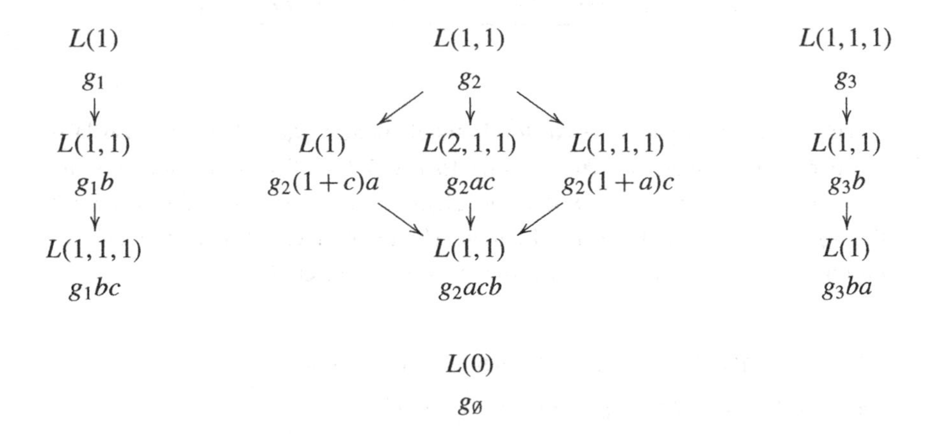

$L(0)$

$g_\emptyset$

18
Idempotents and characters

18.0 Introduction

In Chapter 17 it was shown that there are 2^{n-1} isomorphism classes of irreducible $\mathbb{F}_2\mathsf{GL}(n)$-modules V. In other words, by choosing a $\mathbb{F}_2$-basis for V, $\mathsf{GL}(n)$ has 2^{n-1} equivalence classes of matrix representations over $\mathbb{F}_2$. In order to describe the $\mathbb{F}_2\mathsf{GL}(n)$-module structure of the cohits $\mathsf{Q}^d(n)$ and the Steenrod kernel $\mathsf{K}^d(n)$, we require some of the standard tools of modular representation theory.

Let G be a finite group, K a field of prime characteristic $p > 0$ and KG the group algebra of G over K. Recall that a finite-dimensional right KG-module V is *indecomposable* if it cannot be expressed as the direct sum of two nonzero submodules, and is *irreducible* if it has no submodules other than V itself and 0. Clearly every irreducible module is also indecomposable.

For any field K, every irreducible KG-module V appears both as a submodule and as a quotient module of KG, i.e. of the regular representation of G. For quotients, let $v \in V$ be a generator and observe that $1 \mapsto v$ defines a surjection $KG \to V$ of KG-modules. The result for submodules follows by duality, since KG is isomorphic to its contragredient dual $KG^* = \mathrm{Hom}(KG, K)$, by means of the map $\phi : KG \to KG^*$ defined by $\langle \phi(g), g \rangle = 1$ and $\langle \phi(g), g' \rangle = 0$ if $g' \neq g$, where $g, g' \in G$. In Section 18.1 we discuss decompositions of KG as the direct sum of indecomposable submodules, and relate such decompositions to idempotent elements of KG and to irreducible KG-modules. in order to provide a theoretical framework for the specific results that we use for $G = \mathsf{GL}(n)$, $K = \mathbb{F}_2$.

If $\widetilde{K}$ is an extension field of K, then every KG-module V extends to a $\widetilde{K}G$-module $\widetilde{V} = V \otimes_K \widetilde{K}$. Thus a K-basis of V is a $\widetilde{K}G$-basis of $\widetilde{V}$, and the action of G is given by the same action on the basis elements. We say that $\widetilde{V}$

is obtained from V by *extension of scalars*. The following example shows that irreducibility of V does not imply irreducibility of $\widetilde{V}$.

Example 18.0.1 Let $K = \mathbb{F}_2$, $\widetilde{K} = \mathbb{F}_4$ and let $G \cong C_3$ be the cyclic group of order 3 generated by g. Then $KG = A \oplus B$ where A is the 1-dimensional submodule generated by $1+g+g^2$ and B is the 2-dimensional submodule generated by $1+g$, and A and B are the only non-trivial submodules of KG. Writing ζ for a generator of $\mathbb{F}_4^*$, so that $\zeta^3 = 1$, we find that $\widetilde{K}G$ splits as the direct sum of the 1-dimensional modules generated by $1+g+g^2$, $1+\zeta g+\zeta^2 g^2$ and $1+\zeta^2 g+\zeta g^2$.

An irreducible representation V is *absolutely irreducible* if the extended representation $\widetilde{V}$ is irreducible for all choices of the extension field $\widetilde{K}$, or equivalently if $\widetilde{V}$ is irreducible when $\widetilde{K}$ is algebraically closed. A field K is a *splitting field* for a finite group G if every irreducible KG-module is absolutely irreducible. Example 18.0.1 shows that a splitting field for a group may not be a splitting field for its subgroups, since $\mathsf{GL}(2)$ has a subgroup $G \cong C_3$.

In Section 18.2, we show that $\mathbb{F}_2$ is a splitting field for $\mathsf{GL}(n)$. In effect, this means that every finite-dimensional $\mathbb{F}_2\mathsf{GL}(n)$-module V has the same structure as its scalar extension $\widetilde{V}$. In particular, let $\overline{\mathsf{P}}(n)$ denote the polynomial algebra $\overline{\mathbb{F}}_2[x_1,\dots,x_n]$, graded by degree, with the action of $\mathsf{GL}(n)$ by linear substitutions of the variables. We extend the action of A_2 to $\overline{\mathsf{P}}(n)$ by letting Sq^k act trivially on the coefficients, i.e. $Sq^k(\alpha f) = \alpha Sq^k(f)$ for $\alpha \in \overline{\mathbb{F}}_2$ and $f \in \mathsf{P}(n)$. This action commutes with the right action of $\overline{\mathbb{F}}_2\mathsf{GL}(n)$, but not with linear substitutions such as $x \mapsto \alpha x$ where $\alpha \notin \mathbb{F}_2$.

In Section 18.3 we introduce the Steinberg idempotent $e(n)$ in $\mathbb{F}_2\mathsf{GL}(n)$ and the conjugate idempotent $e'(n)$, and show that they both generate submodules of $\mathbb{F}_2\mathsf{GL}(n)$ isomorphic to the summand $\mathsf{St}(n) = \mathsf{FL}_{Z[n-1]}(n)$ of the flag module $\mathsf{FL}(n)$. The submodule generated by $e(n)$ is invariant under the permutation subgroup $\mathsf{W}(n)$ of $\mathsf{GL}(n)$, while that generated by $e'(n)$ is invariant under the lower triangular subgroup $\mathsf{B}(n)$. We show that $\{e(n)B, B \in \mathsf{B}(n)\}$ is a basis for $e(n)\mathbb{F}_2\mathsf{GL}(n)$ and that $\{e'(n)U, U \in \mathsf{U}(n)\}$ is a basis for $e'(n)\mathbb{F}_2\mathsf{GL}(n)$, where $\mathsf{U}(n)$ is the upper triangular subgroup of $\mathsf{GL}(n)$.

In Section 18.4, we show that the Hecke algebra $\mathsf{H}_0(n)$ can be embedded as a subalgebra in $\mathbb{F}_2\mathsf{GL}(n)$ by mapping the generators $h_1,\dots,h_{n-1}$ in $\mathsf{H}_0(n)$ to the Steinberg idempotents $e_1,\dots,e_{n-1}$ for the 'diagonal' subgroups $G_1,\dots,G_{n-1}$ of $\mathsf{GL}(n)$ which fix all but two of the variables. Under this embedding, the element $f_{W_0} \in \mathsf{H}_0(n)$ given by a reduced word of maximum length in $h_1,\dots,h_{n-1}$ maps to the Steinberg idempotent $e(n)$, and it follows that $e(n)$ is the corresponding product of $e_1,\dots,e_{n-1}$. We conclude Section 18.4 by proving that $\mathsf{St}(n)$ is a projective $\mathbb{F}_2\mathsf{GL}(n)$-module.

By the Jordan–Hölder theorem, the composition factors in every composition series for a finite dimensional $\mathbb{F}_2\mathsf{GL}(n)$-module V are the same multiset of irreducible modules. We conclude this chapter by introducing standard methods relating to composition factors. In Section 18.5 we discuss the Brauer character $\mathrm{br}(V)$, which determines the composition factors of V by lifting the eigenvalues of matrices in an extension field K of $\mathbb{F}_2$ to roots of unity in the complex numbers $\mathbb{C}$. In Section 18.6 we discuss the representation ring $R_2(\mathsf{GL}(n))$ of $\mathsf{GL}(n)$ generated by the irreducible $\mathbb{F}_2\mathsf{GL}(n)$-modules. For a finite-dimensional $\mathbb{F}_2\mathsf{GL}(n)$-module V, the element $[V] \in R_2(\mathsf{GL}(n))$ records the composition factors of V.

18.1 Idempotents and direct sums

An idempotent e in a ring R with identity 1 is an element such that $e^2 = e$. Idempotents are abundant in group algebras: for example, if H is a subgroup of odd order in $\mathsf{GL}(n)$, then $e_H = \sum_{h \in H} h$ is idempotent in $\mathbb{F}_2\mathsf{GL}(n)$. Semigroups such as diagonal matrices over $\mathbb{F}_2$ provide further examples.

Idempotents e_1 and e_2 are *orthogonal* if $e_1e_2 = 0 = e_2e_1$. If e_1 is idempotent, then $e_2 = 1 - e_1$ is also idempotent and is orthogonal to e_1, and R is the direct sum of the right ideals e_1R and e_2R. Conversely, such a decomposition $R = I_1 \oplus I_2$ gives a pair of orthogonal idempotents. To see this, let $1 = e_1 + e_2$ where $e_1 \in I_1$ and $e_2 \in I_2$. Then the equation $e_1 = e_1^2 + e_1e_2$ implies that $e_1 = e_1^2$ and $e_1e_2 = 0$, and similarly the equation $e_2 = e_2e_1 + e_2^2$ implies that $e_2 = e_2^2$ and $e_2e_1 = 0$. Similar results hold for left ideals.

An idempotent $e \neq 0, 1$ is *primitive* if it is not the sum of two nonzero orthogonal idempotents. If $e = e_1 + e_2$ is not primitive, then the preceding argument extends to show that $eR = e_1R \oplus e_2R$ is a direct sum decomposition of the right ideal eR, the summands being nonzero since $e_1 \in e_1R$ and $e_2 \in e_2R$. The converse also extends: if $eR = I_1 \oplus I_2$ where I_1 and I_2 are nonzero right ideals in R, then we can write $e = e_1 + e_2$ where $e_1 \in I_1$ and $e_2 \in I_2$, and verify as above that e_1 and e_2 are nonzero orthogonal idempotents. Thus the right ideal eR generated by an idempotent $e \neq 0, 1$ is *indecomposable*, i.e. not the direct sum of two nonzero right ideals, if and only if e is primitive.

Given a finite group G and a field K, the *(right) regular representation* of G is the group algebra KG itself, regarded as a vector space over K of dimension $|G|$, with the action of G defined by $v \cdot g = vg$ for $v \in KG$ and $g \in G$. In other words, the ring KG acts on itself by right multiplication, and submodules of KG are the same as right ideals. If $KG = U_1 \oplus \cdots \oplus U_m$ is a decomposition of

KG into indecomposable submodules, the summands $U_1, \dots, U_m$ are called the *principal indecomposable modules* of KG.

By the Krull–Schmidt theorem, the modules $U_1, \dots, U_m$ are unique up to isomorphism and the order of the summands. A decomposition $1 = e_1 + \cdots + e_m$ of the identity element of G as a sum of orthogonal primitive idempotents in KG yields a decomposition $KG = e_1 KG \oplus \cdots \oplus e_m KG$ of KG as the direct sum of principal indecomposable modules.

Example 18.1.1 The group $\mathsf{GL}(2)$ is generated by $S = \left(\begin{smallmatrix} 0 & 1 \\ 1 & 0 \end{smallmatrix}\right)$ and $L = \left(\begin{smallmatrix} 1 & 0 \\ 1 & 1 \end{smallmatrix}\right)$, with defining relations $S^2 = I_2$, $L^2 = I_2$, $LSL = SLS$. The elements

$$e_0 = I_2 + SL + LS, \quad e_1 = I_2 + S + L + LS, \quad e_1' = I_2 + S + L + SL$$

are primitive orthogonal idempotents in $\mathbb{F}_2\mathsf{GL}(2)$ with sum I_2, so that $\mathbb{F}_2\mathsf{GL}(2) = U_0 \oplus U_1 \oplus U_1'$ where $U_0 = e_0\mathbb{F}_2\mathsf{GL}(2)$, $U_1 = e_1\mathbb{F}_2\mathsf{GL}(2)$ and $U_1' = e_1'\mathbb{F}_2\mathsf{GL}(2)$ are the corresponding principal indecomposable modules. The module U_0 has dimension 2 with basis $\{e_0, e_0S\}$ and has a 1-dimensional submodule A generated by $e_0(1+S)$, the sum of the elements of $\mathsf{GL}(2)$. The modules A and U_0/A are isomorphic to the trivial module $\mathsf{I}(2)$, while U_1 and U_1' have bases $\{e_1, e_1L\}$ and $\{e_1', e_1'S\}$ respectively, and are isomorphic to the defining module $\mathsf{V}(2)$.

The following result summarizes the relation between decompositions of the identity element 1 and direct sum splittings of KG.

Theorem 18.1.2 (i) *Given a decomposition $1 = e_1 + \cdots + e_r$ of $1 \in KG$ as the sum of primitive orthogonal idempotents, $KG = U_1 \oplus \cdots \oplus U_r$ where $U_i = e_iKG$ is an indecomposable right KG-module for $1 \le i \le r$.*

(ii) *Given a splitting $KG = U_1 \oplus \cdots \oplus U_r$ of KG as the direct sum of indecomposable right KG-modules, the elements $e_i \in U_i$ such that $1 = e_1 + \cdots + e_r$ are primitive orthogonal idempotents for $1 \le i \le r$.*

(iii) *If $1 = e_1 + \cdots + e_r$ and $1 = f_1 + \cdots + f_s$ are two decompositions of 1 as in* (i), *then $r = s$ and with suitable indexing there is an invertible element $u \in KG$ such that $f_i = u^{-1}e_iu$ and $e_iKG = uf_iKG$ for $1 \le i \le r$.*

Proof (i) It follows from the remarks above that a submodule of KG is indecomposable if and only if it is the image $eKG = \{ev, v \in KG\}$ of a primitive idempotent $e \in KG$. Thus a decomposition $1 = e_1 + \cdots + e_r$ yields the stated splitting of KG.

(ii) Let $KG = U_1 \oplus \cdots \oplus U_r$ be a splitting of KG into indecomposable submodules, and let $1 = e_1 + \cdots + e_r$, where $e_i \in U_i$ for $1 \le i \le r$. Then the equations $e_i = \sum_{j=1}^{r} e_ie_j$ imply that $e_1, \dots, e_r$ are orthogonal idempotents. Thus for $x \in U_i$ we have $x = \sum_{j=1}^{r} e_jx$, and again by equating components we obtain

$x = e_i x$. Hence $U_i \subseteq e_i KG$. Since $e_i \in U_i$ and U_i is a right KG-module, we also have $e_i KG \subseteq U_i$, and hence $U_i = e_i KG$. Further, each e_i is primitive, since a decomposition $e_i = e'_i + e''_i$ into orthogonal idempotents gives a decomposition $U_i = e'_i KG \oplus e''_i KG$.

(iii) Let $KG = U_1 \oplus \cdots \oplus U_r$ be the splitting of KG given by the decomposition $1 = e_1 + \cdots + e_r$, so that $U_i = e_i KG$, and let $KG = V_1 \oplus \cdots \oplus V_s$ be another splitting of KG. By the Krull–Schmidt theorem, $r = s$ and we may order the summands so that $U_i \cong V_i$ for $1 \leq i \leq r$. The isomorphism $\bigoplus_{i=1}^r U_i \cong \bigoplus_{i=1}^r V_i$ is then a KG-automorphism of KG itself, regarded as a free left KG-module.

Thus this map must have the form $x \mapsto ux$ for all $x \in KG$, where $u \in KG$ is a unit. Hence $V_i = uU_i$ for $1 \leq i \leq r$. Since V_i is a right KG-module and $e_i \in U_i$, $f_i = ue_i u^{-1} \in V_i$ for $1 \leq i \leq r$. Then $f_1, \dots, f_r$ are primitive orthogonal idempotents, and $1 = f_1 + \cdots + f_r$ is the decomposition of the identity corresponding to the decomposition $KG = V_1 \oplus \cdots \oplus V_r$. It follows that any two decompositions of $1 \in G$ as a sum of primitive orthogonal idempotents have the same number of terms and are conjugate in KG. □

Example 18.1.3 In Example 18.1.1, $e_0 = I_2 + SL + LS$ is in the centre of $\mathbb{F}_2\mathsf{GL}(2)$. Hence all decompositions of 1 as a sum of primitive orthogonal idempotents contain e_0. The elements S,L and SLS are conjugate in $\mathsf{GL}(2)$, and so the idempotents e_1, e'_1 may be replaced by f_1, f'_1 or by g_1, g'_1 where

$$f_1 = I_2 + S + SLS + SL, f'_1 = I_2 + S + SLS + LS,$$
$$g_1 = I_2 + L + SLS + SL, g'_1 = I_2 + L + SLS + LS.$$

There are three decompositions of $\mathbb{F}_2\mathsf{GL}(2) = V_0 \oplus V_1 \oplus V'_1$ into principal indecomposable modules. All contain the same submodule $V_0 = U_0$, but $U_1 \oplus U'_1 \cong \mathsf{V}(2) \oplus \mathsf{V}(2)$ has three submodules isomorphic to $\mathsf{V}(2)$, including the 'diagonal' submodule, and any two of these may be chosen as V_1 and V'_1.

The general relation between the representation theory of a finite group G and primitive idempotents in KG, the group algebra of G over a splitting field K of characteristic $p > 0$, can be summarized as follows. References are given in Section 18.7.

Every principal indecomposable right KG-module U has a unique irreducible quotient module $L = U/\mathrm{rad}(U)$, where the *radical* $\mathrm{rad}(U)$ is the two-sided ideal of U generated by the nilpotent elements. Two principal indecomposable modules are isomorphic if and only if their irreducible quotients are isomorphic, and every irreducible KG-module L is isomorphic to the irreducible quotient of some principal indecomposable module U. Thus there is a 1-1 correspondence between isomorphism classes of irreducible modules and isomorphism classes of principal indecomposable modules.

Given a primitive idempotent $e \in KG$, we may associate to e the principal indecomposable module $U = eKG$ and the irreducible quotient L of U. By Theorem 18.1.2(iii), this defines a 1-1 correspondence between conjugacy classes of primitive idempotents in KG and isomorphism classes of irreducible KG-modules.

The number of conjugacy classes of primitive idempotents, of isomorphism classes of irreducible modules and of principal indecomposable modules is the same as the number of conjugacy classes of elements of G of order prime to p. There are also some important results which count multiplicities in this situation. The multiplicity of U as a direct summand of KG is the dimension of L, and the multiplicity of L as a composition factor of KG is the dimension of U.

We show in Section 18.2 that $\mathbb{F}_2$ is a splitting field for $\mathsf{GL}(n)$. In this case, these results lead to the following statement, where the irreducible modules are indexed as in Definition 17.6.9.

Theorem 18.1.4 *There are 2^{n-1} isomorphism classes of principal indecomposable $\mathbb{F}_2\mathsf{GL}(n)$-modules $U(\lambda)$, indexed by column 2-regular partitions λ of length $\leq n-1$. For each such λ, the number of summands isomorphic to $U(\lambda)$ which occur in any decomposition of $\mathbb{F}_2\mathsf{GL}(n)$ into principal indecomposable modules is the dimension d_λ of the irreducible quotient $L(\lambda)$ of $U(\lambda)$. In any decomposition $1 = e_1 + \cdots + e_r$ of the identity element of $\mathbb{F}_2\mathsf{GL}(n)$ as a sum of primitive orthogonal idempotents, there are d_λ idempotents associated to $L(\lambda)$.* □

18.2 Splitting fields and conjugacy classes

In this section we show that $\mathbb{F}_2$ is a splitting field for $\mathsf{GL}(n)$ and that $\mathsf{GL}(n)$ has 2^{n-1} conjugacy classes of odd order elements, the same as the number of isomorphism classes of $\mathbb{F}_2\mathsf{GL}(n)$-modules.

Proposition 18.2.1 $\mathbb{F}_2$ *is a splitting field for* $\mathsf{GL}(n)$.

Proof We have seen in Theorem 17.6.7 that the submodules $s_I(\mathsf{FL}(n))$ of the flag module $\mathsf{FL}(n)$ are a complete set of 2^{n-1} non-isomorphic $\mathbb{F}_2\mathsf{GL}(n)$-modules. We will show that their scalar extensions $s_I(\mathsf{FL}(n)) \otimes_{\mathbb{F}_2} \overline{\mathbb{F}}_2$ are irreducible $\overline{\mathbb{F}}_2\mathsf{GL}(n)$-modules, where $\overline{\mathbb{F}}_2$ is the algebraic closure of $\mathbb{F}_2$.

We first observe that the counting argument of Proposition 10.1.1 remains valid if $\mathbb{F}_2$ is replaced by an arbitrary finite field $\mathbb{F}_q$ of characteristic 2. Hence if G is a finite 2-group, every nonzero finite-dimensional $\mathbb{F}_qG$-module contains a nonzero G-invariant. If $q = 2^t$, $t \geq 1$, then $\mathbb{F}_q$ has a unique subfield of order $q' = 2^{t'}$ when t' divides t, and we may regard $\overline{\mathbb{F}}_2$ as the union $\bigcup_q \mathbb{F}_q$. Thus if V is a nonzero finite-dimensional $\overline{\mathbb{F}}_2G$-module, V is the scalar extension of a $\mathbb{F}_qG$-module for some q, and so contains a nonzero G-invariant. Arguing as in Proposition 10.1.3, it follows that V is irreducible if the $\mathsf{B}(n)$-invariants in V form a 1-dimensional subspace $\langle v\rangle$ and v generates V.

We can apply this to the scalar extension $\mathsf{FL}(n) \otimes_{\mathbb{F}_2} \overline{\mathbb{F}}_2$. By Proposition 16.3.7, the $\mathsf{B}(n)$-invariants are essentially the same as in $\mathsf{FL}(n)$: they are linear combinations of the sums of all the flags in the same Schubert cell. The argument of Proposition 17.6.4 extends to show that the $\overline{\mathbb{F}}_2\mathsf{GL}(n)$-modules $s_I(\mathsf{FL}(n)) \otimes_{\mathbb{F}_2} \overline{\mathbb{F}}_2$ are irreducible. □

A general result in modular representation theory states that if G is a finite group and K is a splitting field for G of prime characteristic $p > 0$, then the number of isomorphism classes of irreducible KG-modules is equal to the number of conjugacy classes of elements of G whose order is not divisible by p. The next result verifies this in the case $G = \mathsf{GL}(n)$ and $p = 2$ by counting the conjugacy classes of odd order elements.

Proposition 18.2.2 (i) *A matrix $A \in \mathsf{GL}(n)$ is diagonalizable over $\overline{\mathbb{F}}_2$ if and only if A has odd order.*

(ii) *The group $\mathsf{GL}(n)$ has 2^{n-1} conjugacy classes of elements of odd order.*

Proof Since there are 2^{n-1} polynomials of degree n in one variable over $\mathbb{F}_2$, and each is the characteristic polynomial of some matrix $A \in \mathsf{GL}(n)$, there are 2^{n-1} conjugacy classes of diagonalizable elements of $\mathsf{GL}(n)$. Hence (ii) follows from (i).

To prove (i), let $\mathbb{F}_q \subset \overline{\mathbb{F}}_2$ be a finite field over which we can reduce A to its Jordan canonical form. This is a direct sum of elementary Jordan matrices of the form $J = \alpha I_r + E$, where $\alpha \in \mathbb{F}_q^* = \mathbb{F}_q \setminus 0$ is an eigenvalue of A, I_r is the identity matrix and E is the matrix with entries 1 on the superdiagonal and 0 elsewhere. Since $\mathbb{F}_q^*$ has odd order $q - 1$, all eigenvalues α of A have odd order. Hence if A is diagonalizable over $\mathbb{F}_q$, A has odd order.

On the other hand, if A is not diagonalizable over $\mathbb{F}_q$, then some matrix E is nonzero. This matrix E is nilpotent and for $k > 1$ all superdiagonal entries of E^k are 0. By the binomial theorem $J^m = \sum_{k=0}^{m} \binom{m}{k} \alpha^{m-k} E^k$, and so the superdiagonal terms of J^m are given by the term with $k = 1$, which is $m\alpha^{m-1}E$. Thus $J^m = I$ implies that m is even. Hence J has even order, and so does A. □

18.3 The Steinberg idempotents $e(n)$ and $e'(n)$

In Chapter 16, the Steinberg module $\mathsf{St}(n)$ was constructed as a direct summand of the flag module $\mathsf{FL}(n)$. We can identify $\mathsf{St}(n)$ with a submodule of the group algebra $\mathbb{F}_2\mathsf{GL}(n)$ by means of the following observation.

Proposition 18.3.1 *The map* $i : \mathsf{FL}(n) \to \mathbb{F}_2\mathsf{GL}(n)$ *defined by* $i(\mathsf{B}(n)A) = \sum_{B\in\mathsf{B}(n)} BA$, *where* $A \in \mathsf{GL}(n)$, *is an embedding of right* $\mathbb{F}_2\mathsf{GL}(n)$*-modules.* □

Recall from Proposition 17.1.2 that for a permutation matrix $W \in \mathsf{W}(n)$ the basis element f_W of the Hecke algebra $\mathsf{H}_0(n)$ maps $\mathsf{B}(n)$ to the sum of the flags in the Schubert cell $\mathsf{Sch}(W)$, i.e. $f_W(\mathsf{B}(n)) = \sum_A \mathsf{B}(n)A$, where the sum is over all cosets $\mathsf{B}(n)A \subseteq \mathsf{B}(n)W\mathsf{B}(n)$. As the elements of these cosets are the matrices in the Bruhat cell $\mathsf{B}(n)W\mathsf{B}(n)$, the embedding i of Proposition 18.3.1 maps $f_W(\mathsf{B}(n))$ to their sum in $\mathbb{F}_2\mathsf{GL}(n)$. The following notation is convenient for manipulating such sums.

Definition 18.3.2 For a finite group G and a subgroup H of G, $\overline{H} = \sum_{h\in H} h$, the sum of the elements of H in the group algebra $\mathbb{F}_2G$.

Thus the statement above can be written as $i(f_W(\mathsf{B}(n))) = \overline{\mathsf{B}}(n)W\overline{\mathsf{B}}(n)$. Since $\mathsf{B}(n)$ has even order, $\overline{\mathsf{B}}(n)^2 = 0$, and so $i(f_W(\mathsf{B}(n)))^2 = 0$. In the case $W = W_0$, this element is the sum of the matrices in the top Bruhat cell, and we shall show that it generates a submodule of $\mathbb{F}_2\mathsf{GL}(n)$ isomorphic to $\mathsf{St}(n) = \mathsf{FL}_I(n)$, $I = Z[n-1]$.

Definition 18.3.3 The element $e(n) = \overline{\mathsf{B}}(n)\overline{\mathsf{W}}(n) \in \mathbb{F}_2\mathsf{GL}(n)$ is the **Steinberg idempotent** of $\mathsf{GL}(n)$. The **symmetric Steinberg module** is the submodule $e(n)\mathbb{F}_2\mathsf{GL}(n)$ of $\mathbb{F}_2\mathsf{GL}(n)$ generated by $e(n)$. Its **Bruhat generator** is the sum $g(n) = \overline{\mathsf{B}}(n)W_0\overline{\mathsf{B}}(n)$ of the matrices in the top Bruhat cell of $\mathsf{GL}(n)$.

Example 18.3.4 With the matrices S and L as in Examples 18.1.1 and 18.1.3, $e(2) = (I_2 + L)(I_2 + S) = e_1$, and $e(2)\mathbb{F}_2\mathsf{GL}(2)$ has dimension 2 with basis $\{e(2), e(2)L = g_1\}$. The Bruhat generator $g(2) = (I_2 + L)S(I_2 + L) = e(2)(I_2 + L)$.

The next result generalizes this example.

Proposition 18.3.5 (i) $e(n)\overline{\mathsf{B}}(n) = g(n)$, (ii) $g(n)\overline{\mathsf{W}}(n) = e(n)$ *and* (iii) $e(n)$ *is idempotent.*

Proof Since $e(n) = \overline{\mathsf{B}}(n)\overline{\mathsf{W}}(n)$, (iii) follows from (i) and (ii).

(i) follows from the Bruhat decomposition 16.2.7. A matrix $A \in \mathsf{GL}(n)$ can be factored uniquely as $A = BWB'$ where $B \in \mathsf{B}(n)$, $W \in \mathsf{W}(n)$ and $B' \in \mathsf{B}(W)$.

If $W \neq W_0$, then $\mathsf{B}(W)$ is a proper subgroup of $\mathsf{B}(n)$, and an element $A \in \mathsf{B}(n)W\mathsf{B}(n)$ can be written in $[\mathsf{B}(n) : \mathsf{B}(W)]$ ways in the form $A = B_1 W B_2$ for $B_1, B_2 \in \mathsf{B}(n)$. Since $[\mathsf{B}(n) : \mathsf{B}(W)]$ is even, $\overline{\mathsf{B}}(n)W\overline{\mathsf{B}}(n) = 0$ in $\mathbb{F}_2\mathsf{GL}(n)$. Hence $e(n)\overline{\mathsf{B}}(n) = \overline{\mathsf{B}}(n)\overline{\mathsf{W}}(n)\overline{\mathsf{B}}(n) = \overline{\mathsf{B}}(n)W_0\overline{\mathsf{B}}(n) = g(n)$.

(ii) Let $X = \mathsf{B}(n)\overline{\mathsf{W}}(n) = \sum_{W \in \mathsf{W}(n)} \mathsf{B}(n)W \in \mathsf{FL}(n)$ be the sum of the flags represented by permutation matrices, one in each Schubert cell. Thus $i(X) = \overline{\mathsf{B}}(n)\overline{\mathsf{W}}(n) = e(n)$. We claim that X is fixed by the basis element f_W of $\mathsf{H}_0(n)$ for all $W \in \mathsf{W}(n)$, i.e. X is invariant under the action of $\mathsf{Sch}(n)$. In particular $f_{W_0}(X) = X$. Since $i(f_{W_0}(X)) = i(f_{W_0}(\mathsf{B}(n))\overline{\mathsf{W}}(n) = g(n)\overline{\mathsf{W}}(n)$, this implies (ii).

To prove the claim, it is sufficient to show that X is fixed by the generator $h_i = f_{S_i}$ of $\mathsf{Sch}(n)$ for $1 \leq i \leq n-1$, where $S_i \in \mathsf{W}(n)$ is the switch matrix which exchanges x_i and x_{i+1}. Let G_i be the 'diagonal' subgroup of $\mathsf{GL}(n)$ generated by S_i and L_i, where $L_i = T_{i+1,i} \in \mathsf{B}(n)$ is the transvection which adds x_i to x_{i+1} and fixes x_j for $j \neq i$. Then the Schubert cell $\mathsf{Sch}(S_i)$ contains the two flags $\mathsf{B}(n)S_i$ and $\mathsf{B}(n)S_iL_i$, and so $h_i(\mathsf{B}(n)) = \mathsf{B}(n)S_i + \mathsf{B}(n)S_iL_i$ and $h_i(X) = \mathsf{B}(n)S_i\overline{\mathsf{W}}(n) + \mathsf{B}(n)S_iL_i\overline{\mathsf{W}}(n)$.

Since $S_i \in \mathsf{W}(n)$, the first term $\mathsf{B}(n)S_i\overline{\mathsf{W}}(n) = \mathsf{B}(n)\overline{\mathsf{W}}(n) = X$. To show that the second term is 0, we have $\overline{\mathsf{W}}(n) = \overline{\mathsf{W}^+}(n) + S_i\overline{\mathsf{W}^+}(n)$, where $\mathsf{W}^+(n)$ is the (alternating) group of permutation matrices of even length. Since $S_iL_iS_i = L_iS_iL_i$ and $\mathsf{B}(n)L_i = \mathsf{B}(n)$, $\mathsf{B}(n)S_iL_iS_i\overline{\mathsf{W}^+}(n) = \mathsf{B}(n)S_iL_i\overline{\mathsf{W}^+}(n)$, and hence $\mathsf{B}(n)S_iL_i\overline{\mathsf{W}}(n) = 0$. □

Proposition 18.3.6 (i) *The embedding* $i : \mathsf{FL}(n) \to \mathbb{F}_2\mathsf{GL}(n)$ *maps the Steinberg module* $\mathsf{St}(n) = \mathsf{FL}_{Z[n-1]}(n)$ *to the symmetric Steinberg summand* $e(n)\mathbb{F}_2\mathsf{GL}(n)$, *and* (ii) *the* $\mathbb{F}_2\mathsf{GL}(n)$*-modules* $e(n)\mathbb{F}_2\mathsf{GL}(n)$ *and* $\mathsf{St}(n)$ *are isomorphic.*

Proof Clearly (ii) follows from (i). By Theorem 17.6.8, the summand $\mathsf{St}(n) = \mathsf{FL}_{Z[n-1]}(n)$ of $\mathsf{FL}(n)$ is generated by $f_{W_0}(\mathsf{B}(n))$, the sum of the flags in the top Schubert cell $\mathsf{Sch}(W_0)$. These are the flags which are in 'general position' with respect to the reference flag. Hence $i(\mathsf{St}(n))$ is generated by the sum $g(n)$ of the matrices in the top Bruhat cell. By Proposition 18.3.5, $g(n)$ and $e(n)$ generate the same submodule of $\mathbb{F}_2\mathsf{GL}(n)$. □

Proposition 18.3.7 (i) *The* $\mathbb{F}_2\mathsf{GL}(n)$*-module* $e(n)\mathbb{F}_2\mathsf{GL}(n)$ *has* $\mathbb{F}_2$*-basis* $\{e(n)B : B \in \mathsf{B}(n)\}$, *and* (ii) *its restriction to* $\mathsf{B}(n)$ *is the right regular representation of* $\mathsf{B}(n)$.

Proof Since $(e(n)B)(B') = e(n)(BB')$ for all $B, B' \in \mathsf{B}(n)$, (ii) follows from (i). Since $\dim \mathsf{St}(n) = 2^{n(n-1)/2} = |\mathsf{B}(n)|$, it suffices for (i) to show that the elements $e(n)B$, $B \in \mathsf{B}(n)$ are linearly independent. For all $A \in \mathsf{GL}(n)$, $e(n)A =$

$\overline{\mathsf{B}}(n)\overline{\mathsf{W}}(n)A = \sum_{W\in\mathsf{W}(n)} \overline{\mathsf{B}}(n)WA$ is the sum of the elements of $n!$ right cosets of $\mathsf{B}(n)$.

In particular, all elements of the coset $\mathsf{B}(n)W_0A$ appear in the sum $e(n)A$, and for $B \in \mathsf{B}(n)$ the elements of $\mathsf{B}(n)W_0B$ appear in $e(n)B$. They do not appear in $e(n)B' = \overline{\mathsf{B}}(n)\overline{\mathsf{W}}(n)B'$ for $B' \in \mathsf{B}(n)$ and $B' \neq B$, because an equation of the form $W_0B = B''WB'$ implies that $W = W_0$ and $B = B'$, by uniqueness of the Bruhat decomposition for elements of the top Bruhat cell $\mathsf{B}(n)W_0\mathsf{B}(n)$. Hence the elements $e(n)B$ for $B \in \mathsf{B}(n)$ are linearly independent. □

We shall prove in Section 18.4 that the submodule $e(n)\mathbb{F}_2\mathsf{GL}(n)$ is a direct summand of $\mathbb{F}_2\mathsf{GL}(n)$, and hence that it is irreducible and is also a principal indecomposable module. Otherwise stated, $\mathsf{St}(n)$ is a projective module. It follows from the general theory outlined in Section 18.1 that $\mathbb{F}_2\mathsf{GL}(n)$ has a submodule of dimension $2^{n(n-1)}$ which is the direct sum of $2^{n(n-1)/2}$ copies of $\mathsf{St}(n)$, one of which we may take as $e(n)\mathbb{F}_2\mathsf{GL}(n)$. We conclude this section by constructing another of these submodules.

Proposition 18.3.8 *Let G be a finite group and K a field. If $e = \sum_i a_i g_i \in KG$ is idempotent, where $a_i \in K$ and $g_i \in G$, then $e' = \sum_i a_i g_i^{-1}$ is also idempotent.*

Proof Write the equation $e^2 = e$ as $\sum_{i,j} a_i a_j g_i g_j = \sum_k a_k g_k$. By equating coefficients of $g_k \in G$ we obtain $\sum_{i,j} a_i a_j = a_k$, where the sum is over all i and j such that $g_i g_j = g_k$. On the other hand, the equation $(e')^2 = e'$ is equivalent to a similar set of equations $\sum_{j,i} a_j a_i = a_k$, where the sum is over all j and i such that $g_j^{-1} g_i^{-1} = g_k^{-1}$. Since $g_j^{-1} g_i^{-1} = (g_i g_j)^{-1}$ and $a_j a_i = a_i a_j$, this is the same condition on i and j. □

Proposition 18.3.9 *The element $e'(n) = \overline{\mathsf{W}}(n)\overline{\mathsf{B}}(n) \in \mathbb{F}_2\mathsf{GL}(n)$ is idempotent, and $e'(n)$ is conjugate and orthogonal to $e(n)$.*

Proof Since $\mathsf{W}(n)$ and $\mathsf{B}(n)$ are groups and $W^{-1}B^{-1} = (BW)^{-1}$, the element $e'(n) = \sum WB = \sum W^{-1}B^{-1} \in \mathbb{F}_2\mathsf{GL}(n)$ is the sum of the inverses of the matrices in $e(n) = \sum BW$, where the sums are over all $B \in \mathsf{B}(n)$ and $W \in \mathsf{W}(n)$. Hence $e'(n)$ is idempotent by Proposition 18.3.9.

Since $\mathsf{B}(n)$ and $\mathsf{W}(n)$ have even order, $\overline{\mathsf{B}}(n)^2 = 0$ and $\overline{\mathsf{W}}(n)^2 = 0$ in $\mathbb{F}_2\mathsf{GL}(n)$. Let $u = I_n + \overline{\mathsf{B}}(n)$, $v = I_n + \overline{\mathsf{W}}(n)$ and $x = uvu$. Then $u^2 = 1$, $v^2 = 1$ and $x^2 = 1$ in $\mathbb{F}_2\mathsf{GL}(n)$. We claim that $xe(n) = e'(n)x$, so that $e(n)$ and $e'(n)$ are conjugate.

Since $ue(n) = e(n)$, $\overline{\mathsf{B}}(n)e(n) = 0$ and $e(n)$ is idempotent, $xe(n) = uvue(n) = uve(n) = e(n) + \overline{\mathsf{B}}(n)e(n) + \overline{\mathsf{W}}(n)e(n) + e(n)^2 = \overline{\mathsf{W}}(n)e(n)$. Similarly $e'(n)x = e'(n)\overline{\mathsf{W}}(n)$ since $e'(n)u = e'(n)$, $e'(n)\overline{\mathsf{B}}(n) = 0$ and $e'(n)$ is idempotent. Since $\overline{\mathsf{W}}(n)e(n) = e'(n)\overline{\mathsf{W}}(n)$, we obtain $xe(n) = e'(n)x$.

It follows that $e(n)\mathbb{F}_2\mathsf{GL}(n)$ and $e'(n)\mathbb{F}_2\mathsf{GL}(n)$ are isomorphic as right $\mathbb{F}_2\mathsf{GL}(n)$-modules, and so $e'(n)\mathbb{F}_2\mathsf{GL}(n)$ is isomorphic to $\mathsf{St}(n)$. Since $\overline{\mathsf{B}}(n)^2 = 0$ and $\overline{\mathsf{W}}(n)^2 = 0$, $e(n)e'(n) = 0 = e'(n)e(n)$, i.e. $e(n)$ and $e'(n)$ are orthogonal. □

It follows that we have constructed two distinct copies of the Steinberg module $\mathsf{St}(n)$ in $\mathbb{F}_2\mathsf{GL}(n)$, the symmetric summand $e(n)\mathbb{F}_2\mathsf{GL}(n)$ and the $\mathsf{B}(n)$-invariant summand $e'(n)\mathbb{F}_2\mathsf{GL}(n)$.

Proposition 18.3.10 *Let $\mathsf{U}(n) = W_0\mathsf{B}(n)W_0$ be the upper triangular subgroup of $\mathsf{GL}(n)$. Then $\{e'(n)U : U \in \mathsf{U}(n)\}$ is a $\mathbb{F}_2$-basis of $e'(n)\mathbb{F}_2\mathsf{GL}(n)$, and the restriction of $e'(n)\mathbb{F}_2\mathsf{GL}(n)$ to $\mathsf{U}(n)$ is the right regular representation of $\mathsf{U}(n)$.*

Proof The proof is similar to that of Proposition 18.3.7. Since $e'(n)\mathbb{F}_2\mathsf{GL}(n)$ has dimension $2^{n(n-1)/2} = |\mathsf{U}(n)|$, it suffices to prove linear independence of the elements $e'(n)U$ in $e'(n)\mathbb{F}_2\mathsf{GL}(n)$. We write $U = W_0BW_0$ where $B \in \mathsf{B}(n)$. Then $e'(n)U = \overline{\mathsf{W}}(n)\overline{\mathsf{B}}(n)W_0BW_0$ is the sum of the elements of $2^{n(n-1)/2}$ right cosets of $\mathsf{W}(n)$.

In particular, the elements of the coset $\mathsf{W}(n)BW_0$ appear in the sum, and do not appear in $e'(n)U'$ for $U' = W_0B'W_0 \in \mathsf{U}(n)$ and $U' \neq U$, so that $B' \neq B$, because an equation of the form $WB = B''W_0B'$ implies that $W = W_0$ and $B = B'$, by uniqueness of the Bruhat decomposition for elements of $\mathsf{B}(n)W_0\mathsf{B}(n)$. Hence the elements $e(n)U$ for $U \in \mathsf{U}(n)$ are linearly independent. The second statement follows, since $(e(n)U)(U') = e(n)(UU')$ for $U, U' \in \mathsf{U}(n)$. □

18.4 Embedding $\mathsf{H}_0(n)$ in $\mathbb{F}_2\mathsf{GL}(n)$

As in the proof of Proposition 18.3.5, let $G_i \cong \mathsf{GL}(2)$ be the subgroup of $\mathsf{GL}(n)$ generated by the switch matrix S_i and the lower triangular transvection L_i.

Proposition 18.4.1 *There is an algebra map $\iota' : \mathsf{H}_0(n) \to \mathbb{F}_2\mathsf{GL}(n)$ given by $\iota'(h_i) = e_i$, $1 \le i \le n-1$, where $e_i = (I_n + L_i)(I_n + S_i)$ is the Steinberg idempotent for G_i.*

Proof By Example 18.1.1, the elements e_i are idempotents. It suffices to show that e_i and e_j commute if $|i-j| > 1$ and that they satisfy the braid relation for $|i-j| = 1$. The first statement is clear since the subgroups G_i and G_j commute, while the second is effectively the calculation that $e_1e_2e_1 = e(3) = e_2e_1e_2$ in $\mathbb{F}_2\mathsf{GL}(3)$. We omit the details. □

Proposition 18.4.2 *Let $e_{W_0} = \iota'(f_{W_0})$, a word in the elements $e_i \in \mathbb{F}_2\mathsf{GL}(n)$ which corresponds to a word in the switch matrices $S_i \in \mathsf{W}(n)$ for W_0. Then*

(i) *$e_{W_0}\mathbb{F}_2\mathsf{GL}(n)$ is spanned by $\{e_{W_0}B, B \in \mathsf{B}(n)\}$,*
(ii) *$e(n)e_i = e(n) = e_ie(n)$ for $1 \le i \le n-1$,*
(iii) *$e(n)e_{W_0} = e(n) = e_{W_0}e(n)$.*

Proof Since e_{W_0} is a word in the e_i, $1 \le i \le n-1$, (iii) follows from (ii).

(i) Since f_{W_0} is an idempotent in $\mathsf{H}_0(n)$, e_{W_0} is an idempotent in $\mathbb{F}_2\mathsf{GL}(n)$. Using the Bruhat decomposition, we wish to show that for $B_1, B_2 \in \mathsf{B}(n)$ and $W \in \mathsf{W}(n)$, $e_{W_0}B_1WB_2$ is a sum of elements of the form $e_{W_0}B$ where $B \in \mathsf{B}(n)$. Without loss of generality, we may assume that $B_2 = I_n$. By induction on $\mathrm{len}(W)$, it suffices to show that if $e_{W_0}A$ is a sum of the required form, then so is $e_{W_0}AS_i$.

Thus it suffices to consider elements of the form $e_{W_0}BS_i$ where $B \in \mathsf{B}(n)$. Now $\mathsf{B}(n) = \mathsf{B}_i'(n) \cup L_i\mathsf{B}_i'(n)$, where $\mathsf{B}_i'(n)$ is the subgroup of matrices with $(i+1,i)$th entry 0. Since $S_i\mathsf{B}_i'(n) = \mathsf{B}_i'(n)S_i$, $BS_i \in S_i\mathsf{B}_i'(n) \cup L_iS_i\mathsf{B}_i'(n)$. By the case $n=2$ of Proposition 18.3.7, e_iBS_i is in the linear span of e_i and e_iL_i. Hence e_iBS_i is in the linear span of $\{e_iB, B \in \mathsf{B}(n)\}$. Since $e_{W_0} = e_{W_0}e_i$, $e_{W_0}BS_i$ is in the linear span of $\{e_{W_0}B, B \in \mathsf{B}(n)\}$, as required.

(ii) First we prove $e_ie(n) = e(n)$. As in the proof of Proposition 18.3.5, let $\mathsf{W}^+(n)$ be the subgroup of permutation matrices of even length. Then $\overline{\mathsf{B}}(n) = (I_n + L_i)\overline{\mathsf{B}}_i'(n)$, $\overline{\mathsf{W}}(n) = (I_n + S_i)\overline{\mathsf{W}}^+(n)$ and $e_i = (I_n + L_i)(I_n + S_i)$. Since $\overline{\mathsf{B}}_i'(n)$ and S_i commute, we have $e(n) = \overline{\mathsf{B}}(n)\overline{\mathsf{W}}(n) = (I_n + L_i)\overline{\mathsf{B}}_i'(n)(I_n + S_i)\overline{\mathsf{W}}^+(n) = (I_n + L_i)(I_n + S_i)\overline{\mathsf{B}}_i'(n)\overline{\mathsf{W}}^+(n) = e_ix$ where $x = \overline{\mathsf{B}}_i'(n)\overline{\mathsf{W}}^+(n)$. Thus $e_ie(n) = e_ie_ix = e_ix = e(n)$.

It remains to prove that $e(n)e_i = e(n)$. By Proposition 18.3.6, $e(n)$ is the image in $\mathbb{F}_2\mathsf{GL}(n)$ of the symmetrization under $\mathsf{W}(n)$ of the sum $\sum_{B \in \mathsf{B}(n)} \mathsf{B}(n)W_0B$ of the flags in the top Schubert cell $\mathsf{Sch}(W_0)$. This symmetrization is the sum of the flags $\mathsf{B}(n)W$, $W \in \mathsf{W}(n)$. We shall prove that $\sum_{W \in \mathsf{W}(n)} \mathsf{B}(n)We_i = \sum_{W \in \mathsf{W}(n)} \mathsf{B}(n)W$. It follows that $e(n)e_i = e(n)$.

Since $\sum_{W \in \mathsf{W}(n)} \mathsf{B}(n)W = \sum_{W \in \mathsf{W}(n)} \mathsf{B}(n)WS_i$. Hence $\sum_{W \in \mathsf{W}(n)} \mathsf{B}(n)We_i = \sum_{W \in \mathsf{W}(n)} (\mathsf{B}(n)WL_i + \mathsf{B}(n)WL_iS_i)$. We separate the cases (a) $i \notin \mathrm{des}(W^{-1})$ and (b) $i \in \mathrm{des}(W^{-1})$. Since $(WS_i)^{-1} = S_iW^{-1}$, WS_i satisfies (b) if and only if W satisfies (a).

In case (a), $WL_iW^{-1} \in \mathsf{B}(n)$. Hence $\mathsf{B}(n)WL_i = \mathsf{B}(n)W$ and $\mathsf{B}(n)WL_iS_i = \mathsf{B}(n)WS_i$. Since WS_i runs through the matrices of case (b) as W runs through the matrices of case (a), the sum of the flags $\mathsf{B}(n)WL_i$ and $\mathsf{B}(n)WL_iS_i$ over matrices of case (a) is the required sum $\sum_{W \in \mathsf{W}(n)} \mathsf{B}(n)We_i$. In case (b), $WU_iW^{-1} \in \mathsf{B}(n)$ where $U_i = L_iS_iL_i$ is the upper triangular transvection in G_i. Hence $\mathsf{B}(n)WL_i = \mathsf{B}(n)WL_iS_i$ so that these flags cancel in the sum. □

Proposition 18.4.3 $e(n) = e_{W_0}$.

Proof The map $e(n)\mathbb{F}_2\mathsf{GL}(n) \to e_{W_0}\mathbb{F}_2\mathsf{GL}(n)$ given by $x \mapsto e_{W_0}x$ is an inclusion by Proposition 18.4.2(ii), and so by Proposition 18.4.2(i) and Proposition 18.3.7 it is an isomorphism. Hence $e_{W_0} = e(n)y$ for some $y \in \mathbb{F}_2\mathsf{GL}(n)$. Then by Proposition 18.4.2(ii) $e(n) = e(n)e_{W_0} = e(n)(e(n)y) = e(n)y = e_{W_0}$. □

Proposition 18.4.4 *$Me(n) = Me_1 \cap \cdots \cap Me_{n-1}$ for any right $\mathbb{F}_2\mathsf{GL}(n)$-module M.*

Proof Since $e(n) = e(n)e_i$, $Me(n) \subseteq Me_i$. Conversely, if $x \in Me_i$ then $xe_i = x$. Thus $x \in Me_i \cap \cdots \cap Me_{n-1}$ implies that $xe_i = x$ for $1 \le i \le n-1$, and hence $xe_{W_0} = x$, since e_{W_0} is a word in the e_i. Hence $x = xe(n) \in Me(n)$. □

We next show that the map of Proposition 18.4.1 embeds the Hecke algebra $\mathsf{H}_0(n)$ in the group algebra $\mathbb{F}_2\mathsf{GL}(n)$.

Theorem 18.4.5 *The map $\iota' : \mathsf{H}_0(n) \to \mathbb{F}_2\mathsf{GL}(n)$ is injective.*

Proof Assume, for a contradiction, that there is an element $x \neq 0 \in \mathsf{H}_0(n)$ such that $\iota'(x) = 0$. Let $f_{W'}$ be a term of maximal length in the expression for x in the basis $\{f_W : W \in \mathsf{W}(n)\}$ of $\mathsf{H}_0(n)$, and let $W_0 = W'W''$. Then by Proposition 17.2.4, f_{W_0} is a term in the expression for $xf_{W''}$ as a sum of basis elements f_W, and $\iota'(xf_{W''}) = \iota'(x)\iota'(f_{W''}) = 0$.

By induction on $r = \mathrm{len}(W)$, we shall show that the elements $W' \in \mathsf{W}(n)$ which appear as terms in the product $\iota'(f_W)$ are expressible as subwords of a reduced word for W, and so $W' \le W$ in the Bruhat order. Thus all elements of $\mathsf{GL}(n)$ which appear as terms in $\iota'(f_W)$ are in Bruhat cells corresponding to elements $W' < W$. In particular, if $W' \neq W_0$ then no elements of the top Bruhat cell appear. But by Proposition 18.4.3, $\iota'(f_{W_0}) = e(n) = \overline{\mathsf{B}}(n)\overline{\mathsf{W}}(n)$ contains elements BW_0, $B \in \mathsf{B}(n)$, of the top Bruhat cell. Hence $\iota'(xf_{W''}) \neq 0$, contradicting the assumption that $\iota'(x) = 0$.

For the induction, let $W = S_{i_1} \cdots S_{i_r}$ be a reduced word for W. Then $\iota'(f_W) = e_{i_1} \cdots e_{i_r} = (I_n + L_{i_1})(I_n + S_{i_1}) \cdots (I_n + L_{i_r})(I_n + S_{i_r})$. Assume as inductive hypothesis that all terms in the product $e_{i_1} \cdots e_{i_{r-1}}$, when expanded without cancellation, have Bruhat decomposition of the form $B_1W'B_2$ where W' is a subword of $S_{i_1} \cdots S_{i_{r-1}}$. Then all terms A in $\iota'(f_W)$ have the form $B_1W'B_2$, $B_1W'B_2L_{i_r}$, $B_1W'B_2S_{i_r}$ or $B_1W'B_2L_{i_r}S_{i_r}$. If $A = W \in \mathsf{W}(n)$, then $W = W'$ in the first two cases by Theorem 16.2.2. By postmultiplying by S_{i_r}, $WS_{i_r} = W'$ in the remaining two cases, so that $W = W'S_{i_r}$. Since W' is a subword of $S_{i_1} \cdots S_{i_{r-1}}$, W is a subword of $S_{i_1} \cdots S_{i_r}$. This completes the induction. □

There are similar results for the conjugate idempotent $e'(n) = \overline{\mathsf{W}}(n)\overline{\mathsf{B}}(n)$. For $1 \le i \le n-1$, we denote by $e'_i = (1+S_i)(1+L_i)$ the conjugate Steinberg idempotent for the subgroup G_i.

Proposition 18.4.6 *There is an embedding of algebras $\iota'' : \mathsf{H}_0(n) \to \mathbb{F}_2\mathsf{GL}(n)$ defined by $\iota''(h_i) = e'_i$ for $1 \le i \le n-1$. Writing $\iota''(f_{W_0}) = e'_{W_0}$, we have*

(i) *$e'_{W_0}\mathbb{F}_2\mathsf{GL}(n)$ is spanned by $\{e'_{W_0}U,\ U \in \mathsf{U}(n)\}$,*
(ii) *$e'(n)e'_{W_0} = e'(n) = e'_{W_0}e'(n)$,*
(iii) *$e'_{W_0} = e'(n)$ is the longest word in $e'_1,\ldots,e'_{n-1}$,*
(iv) *$Me'(n) = Me'_1 \cap \cdots \cap Me'_{n-1}$ for any right $\mathbb{F}_2\mathsf{GL}(n)$-module M.*

Proof Recall from Proposition 18.3.9 that $e'(n) = \overline{\mathsf{W}}(n)\overline{\mathsf{B}}(n)$ is the sum of the inverses of the matrices in $e(n) = \overline{\mathsf{B}}(n)\overline{\mathsf{W}}(n)$. Thus the identity $e_1e_2e_1 = e(3) = e_2e_1e_2$ gives the corresponding identity $e'_1e'_2e'_1 = e'(3) = e'_2e'_1e'_2$ by taking the inverses of all the matrices involved. All the statements follow by similar arguments from the previous results in this section, together with Proposition 18.3.10 to obtain the basis for the image of $e'(n)$ given in (i). □

Proposition 18.4.7 *The map $\pi : \mathbb{F}_2\mathsf{GL}(n) \to \mathsf{FL}(n)$ defined by $\pi(A) = \mathsf{B}(n)A$ for $A \in \mathsf{GL}(n)$ is a surjection of right $\mathbb{F}_2\mathsf{GL}(n)$-modules which maps the submodule of $\mathbb{F}_2\mathsf{GL}(n)$ generated by $e'(n)$ isomorphically on to the submodule $\mathsf{St}(n)$ of $\mathsf{FL}(n)$.*

Proof Since the action of $\mathsf{GL}(n)$ on $\mathsf{FL}(n)$ is given by $\mathsf{B}(n)A \cdot A' = \mathsf{B}(n)AA'$ for $A, A' \in \mathsf{GL}(n)$, π is a map of right $\mathbb{F}_2\mathsf{GL}(n)$-modules. It is surjective since the flags $\mathsf{B}(n)A$ form a $\mathbb{F}_2$-basis for $\mathsf{FL}(n)$. We shall show that $\pi(e'(n))$ is the standard generator of $\mathsf{St}(n)$, i.e. the sum $f_{W_0}(\mathsf{B}(n)) = \sum_{B \in \mathsf{B}(n)} \mathsf{B}(n)W_0B$ of the flags in the top Schubert cell.

Since $e'_i = (I_n + S_i)(I_n + L_i)$, $\pi(e'_i) = \mathsf{B}(2)(I_n + S_i)(I_n + L_i) = \mathsf{B}(n)S_i + \mathsf{B}(n)S_iL_i = h_i(\mathsf{B}(n)$, the sum of the flags in $\mathsf{Sch}(S_i)$. We claim that $\pi(e'_ie'_j) = h_ih_j(\mathsf{B}(n)$ if $i \neq j$, so that π maps the image of ι'', the subalgebra of $\mathbb{F}_2\mathsf{GL}(n)$ generated by $e'_1,\ldots,e'_{n-1}$, isomorphically on the ring of $\mathsf{B}(n)$-invariants in $\mathsf{FL}(n)$. In particular, it follows from Proposition 18.4.6 that $\pi(e'(n)) = f_{W_0}(\mathsf{B}(n))$.

To prove the claim, we have $\pi(e'_ie'_j) = \mathsf{B}(n)e'_ie'_j = \mathsf{B}(n)(I_n+S_i)(I_n+L_i)(I_n+S_j)(I_n+L_j)$. Since $L_i \in \mathsf{B}(n)$, $\mathsf{B}(n)(I_n+L_i) = 0$, and so this reduces to $\pi(e'_ie'_j) = \mathsf{B}(n)S_i(I_n+L_i)(I_n+S_j)(I_n+L_j)$. On the other hand $h_ih_j(\mathsf{B}(n)) = h_i(\mathsf{B}(n)(S_j + S_jL_j)) = \mathsf{B}(n)(S_i+S_iL_i)(S_j+S_jL_j)$.

To complete the proof, we show that $\mathsf{B}(n)S_i = \mathsf{B}(n)S_iL_j$ and $\mathsf{B}(n)S_iL_iL_j = \mathsf{B}(n)S_iL_i$. These statements follow from Proposition 16.2.5(iii). If $j \neq i$, then $S_iL_jS_i \in \mathsf{B}(n)$ since $L_j \in \mathsf{B}(W_0S_i)$, the subgroup of matrices in $\mathsf{B}(n)$ for which the $(i+1,i)$th entry is 0. Hence $\mathsf{B}(n)S_i = \mathsf{B}(n)S_iL_j$. If $|i-j|>1$, then $L_iL_j = L_jL_i$ and so we also obtain $\mathsf{B}(n)S_iL_iL_j = \mathsf{B}(n)S_iL_jL_i = \mathsf{B}(n)S_iL_i$. The same conclusion holds if $j=i+1$ or $i-1$, as a calculation with 3×3 matrices shows that $L_iL_jL_i \in \mathsf{B}(W_0S_i)$ in these cases also, so that $S_iL_iL_jL_iS_i \in \mathsf{B}(n)$. □

Remark 18.4.8 Note that $\pi(e_i)=0$ for $1\le i\le n-1$ since $e_i = (I_n+L_i)(I_n+S_i)$ and $\mathsf{B}(n)(I_n+L_i)=0$. In particular, the symmetric Steinberg summand $e(n)\mathbb{F}_2\mathsf{GL}(n)$ is in the kernel of π.

It has been shown in the proof of Proposition 18.3.9 that $xe(n)x^{-1} = e'(n)$ where $x = (I_n+\overline{\mathsf{W}}(n))(I_n+\overline{\mathsf{B}}(n))(I_n+\overline{\mathsf{W}}(n))$. Thus $c(e(n)) = e'(n)$ where the map $c:\mathbb{F}_2\mathsf{GL}(n)\to\mathbb{F}_2\mathsf{GL}(n)$ is defined by $c(u)=xux^{-1}$ for $u\in\mathbb{F}_2\mathsf{GL}(n)$.

Proposition 18.4.9 *Let $c:\mathbb{F}_2\mathsf{GL}(n)\to\mathbb{F}_2\mathsf{GL}(n)$ be a conjugation map such that $c(e(n))=e'(n)$. Then*

(i) *the composition* $\mathsf{FL}(n)\xrightarrow{i}\mathbb{F}_2\mathsf{GL}(n)\xrightarrow{c}\mathbb{F}_2\mathsf{GL}(n)\xrightarrow{\pi}\mathsf{FL}(n)$ *restricts to an automorphism of the direct summand* $\mathsf{St}(n)$ *of* $\mathsf{FL}(n)$,

(ii) *the Steinberg module* $\mathsf{St}(n)$ *is a projective* $\mathbb{F}_2\mathsf{GL}(n)$*-module.*

Proof (i) follows from Propositions 18.3.1, 18.3.9 and 18.4.7. Thus $\mathsf{St}(n)$ is a direct summand of the free module $\mathbb{F}_2\mathsf{GL}(n)$, and (ii) follows. □

18.5 Brauer characters

Let G be a finite group and p a prime. Then an element $g\in G$ can be written uniquely as the product of commuting elements g_p and g'_p, where the order of its *p-regular* part g'_p is not divisible by p, and the order of its *p-singular* part g_p is a power of p. We denote by G'_p the set of all p-regular elements of G, the elements whose order is not divisible by p. A conjugacy class C in G is p-regular if $C\subseteq G'_p$. The *p'-exponent* of G is the least common multiple of the orders of elements $g\in G'_p$. By Lagrange's theorem, it divides $|G|$. Thus the 2-regular elements are the odd order elements, and the $2'$-exponent of $\mathsf{GL}(n)$ divides $|\mathsf{GL}(n)|/|\mathsf{B}(n)|$, the dimension of the flag module $\mathsf{FL}(n)$.

Let V be an $\mathbb{F}_pG$-module of finite dimension n, and let $\rho_V: G\to\mathsf{GL}(n,\mathbb{F}_p)$ be the corresponding representation of G by $n\times n$ matrices over $\mathbb{F}_p$, with respect to a given $\mathbb{F}_p$-basis of V. As in Proposition 18.2.2, the matrix $\rho_V(g)$ is diagonalizable over some finite extension field K of $\mathbb{F}_p$ if and only if $g\in G$ is p-regular. The diagonal elements $\alpha_1,\dots,\alpha_n\in K$ are the eigenvalues of $\rho_V(g)$, the roots of the characteristic polynomial $\det(\rho_V(g)-x1_V)\in\mathbb{F}_p[x]$, and their sum in $\mathbb{F}_p$ is the trace of $\rho_V(g)$. This sum, the *ordinary character* of ρ_V, is independent of the choice of basis for V, and is constant on each conjugacy class. However, the ordinary character does not usefully classify modular representations: for example, its value at $g = 1$ is $n \bmod p$. It does not help to include classes which are not p-regular, since the matrices representing an element and its p-regular part have the same eigenvalues.

The Brauer character $\mathrm{br}(V)$ extends the ordinary character theory so as to determine the composition factors of V in the modular case. To construct $\mathrm{br}(V)$, we lift the eigenvalues of $\rho_V(g)$ for p-regular elements to the complex numbers $\mathbb{C}$, as follows. Let m be the p'-exponent of G and let $K = \mathbb{F}_q$, where q is a power of p such that $q = 1 \bmod m$. Then the multiplicative group $K^\times \cong \mathbb{Z}/(q-1)$ has a cyclic subgroup of order m. Given a generator α of $K^\times$, the eigenvalues $\alpha_1, \ldots, \alpha_n$ of $\rho_V(g)$, $g \in G'_p$, may be written as powers of α. Let $\widetilde{\alpha} = e^{2\pi i/m} \in \mathbb{C}$. Then $\widetilde{\alpha} \mapsto \alpha$ defines an isomorphism from the group of mth roots of 1 in $\mathbb{C}$ to the group of mth roots of 1 in K. The *Brauer character* $\mathrm{br}(V)$ is a map $G'_p \to \mathbb{C}$, and we write its value at $g \in G$ as $\mathrm{br}(V, g) = \widetilde{\alpha}_1 + \cdots + \widetilde{\alpha}_n$.

Example 18.5.1 Let $p = 2$, $G = \mathsf{GL}(3)$ and $V = \mathsf{FL}^1(3)$, so that $n = \dim V = 7$. We may take $A = \left(\begin{smallmatrix} 0&1&0 \\ 0&0&1 \\ 1&0&0 \end{smallmatrix}\right)$ to represent the conjugacy class of elements of order 3 in $\mathsf{GL}(3)$. Since A fixes $(1,1,1)$ and permutes the other lines in two 3-cycles, the characteristic polynomial of $\rho_V(A)$ is $(x-1)(x^3-1)^2 \in \mathbb{F}_2[x]$, and $\rho_V(A)$ is diagonalizable over $\mathbb{F}_4 = \{0, 1, \alpha, \alpha^2\}$ with eigenvalues $1, \alpha, \alpha^2$ of multiplicity $3, 2, 2$ respectively. We lift α to $\omega = \exp(2\pi i/3) \in \mathbb{C}$, and so $\mathrm{br}(V, A) = 3 + 2\omega + 2\omega^2 = 1$.

An element of order 7 in $\mathsf{GL}(3)$, such as $B = \left(\begin{smallmatrix} 0&1&0 \\ 0&0&1 \\ 1&0&1 \end{smallmatrix}\right)$, permutes the 7 lines cyclically. The characteristic polynomial of $\rho_V(B)$ is $x^7 - 1$. The two conjugacy classes C_7, C'_7 of elements of order 7 in $\mathsf{GL}(3)$ have eigenvalues ζ, ζ^2, ζ^4 and $\zeta^3, \zeta^5, \zeta^6$ respectively, where ζ generates $\mathbb{F}_8^* \cong \mathbb{Z}/7$. Lifting ζ to $\varepsilon = e^{2\pi i}/7 \in \mathbb{C}$, we have $\gamma = \varepsilon + \varepsilon^2 + \varepsilon^4$, $\overline{\gamma} = \varepsilon^3 + \varepsilon^5 + \varepsilon^6$, and $\mathrm{br}(V, B) = 1 + \gamma + \overline{\gamma} = 0$. Then $\gamma\overline{\gamma} = 2$, so γ and $\overline{\gamma}$ are the roots of $z^2 + z + 2$ in $\mathbb{C}$. Hence $\gamma = (-1 + \sqrt{-7})/2$ and $\overline{\gamma} = (-1 - \sqrt{-7})/2$.

The modules $\mathsf{I}(3)$, $\mathsf{V}(3)$, $\mathsf{V}(3)^*$ and $\mathsf{St}(3)$ are a complete set of irreducible $\mathbb{F}_2\mathsf{GL}(3)$-modules. The Brauer character table is

	C_1	C_3	C_7	C'_7
$\mathsf{I}(3)$	1	1	1	1
$\mathsf{V}(3)$	3	0	γ	$\overline{\gamma}$
$\mathsf{V}(3)^*$	3	0	$\overline{\gamma}$	γ
$\mathsf{St}(3)$	8	-1	1	1

. (18.1)

Thus $\text{br}(\mathsf{FL}^1(3)) = \text{br}(\mathsf{I}(3)) + \text{br}(\mathsf{V}(3)) + \text{br}(\mathsf{V}(3)^*)$, and it follows that the composition factors of $\mathsf{FL}^1(3)$ are $\mathsf{I}(3)$, $\mathsf{V}(3)$ and $\mathsf{V}(3)^*$. The full structure of $\mathsf{FL}^1(3)$ cannot be determined from the Brauer character alone. For example, $\mathsf{FL}^1(3)$ and $\mathsf{FL}^3(3)$ have the same Brauer character, but we have shown in Section 16.6 that they are not isomorphic.

Example 18.5.2 For all $n \geq 1$, the Brauer character of the flag module $\mathsf{FL}(n)$ is $|\mathsf{GL}(n)/\mathsf{B}(n)|$ at the identity I_n and is 0 on all other 2-regular conjugacy classes. To justify this, we consider the corresponding representation Flag(n) over a field F of characteristic 0, i.e. the usual permutation representation of $\mathsf{GL}(n, F)$ on the cosets of the subgroup of lower triangular matrices.

The representing matrices are permutation matrices, with diagonal elements 1 corresponding to fixed flags. But no odd order element $A \in \mathsf{GL}(n)$ other than the identity element I_n can fix a flag, since $\mathsf{B}(n)$ is a 2-group. Hence the character of Flag(n) is 0 on all 2-regular conjugacy classes other than I_n. Under mod 2 reduction, the representative matrices $\tilde{\rho}(A)$ for Flag(n) map to the representative matrices $\rho(A)$ for $\mathsf{FL}(n)$. Hence the characteristic polynomial of $\tilde{\rho}(A)$ in $\mathbb{Z}[t]$ maps to the characteristic polynomial of $\rho(A)$ in $\mathbb{F}_2[t]$, and its eigenvalues in $\mathbb{C}$ are lifts of the eigenvalues of $\rho(A)$ in an extension field of $\mathbb{F}_2$. It follows that the Brauer character of $\mathsf{FL}(n)$ is the restriction of the ordinary character of Flag(n) to 2-regular classes.

The main properties of the Brauer character are as follows.

Theorem 18.5.3 *Let G be a finite group, p a prime and m the p'-exponent of G. Fix an isomorphism between the mth roots of 1 in $\mathbb{C}$ and the mth roots of 1 in a field K of characteristic p. Let V and W be finite-dimensional $\mathbb{F}_pG$-modules and let $br : G'_p \to \mathbb{C}$ be the Brauer character. Then*

(i) *$br(V)$ is a class function, i.e. $br(V, g_1) = br(V, g_2)$ if g_1 and g_2 are conjugate in G;*
(ii) *V and W have the same composition factors, counting multiplicities, if and only if $br(V) = br(W)$;*
(iii) *$br(V) = br(W) + br(V/W)$ if W is a submodule of V;*
(iv) *$br(V \otimes W) = br(V)br(W)$;*
(v) *if K is a splitting field for G, then the Brauer characters of a full set of non-isomorphic irreducible KG-modules form a $\mathbb{C}$-basis for the class functions $G'_p \to \mathbb{C}$.* □

Let $M = \sum_{d \geq 0} M^d$ be a graded KG-module such that M^d is finite dimensional for all $d \geq 0$. We define the Brauer character series of M by

$$\mathrm{br}(M, g, t) = \sum_{d \geq 0} \mathrm{br}(M^d, g) t^d. \tag{18.2}$$

Example 18.5.4 Let $M = \mathsf{P}(3)$ and let A be as in Example 18.5.1. Then A fixes the monomial $x^a y^b z^c$ if $a = b = c$ and permutes a, b, c cyclically otherwise. The fixed monomials contribute 1 to the Brauer character, while the 3-cycles contribute $1 + \omega + \omega^2 = 0$. Hence $\mathrm{br}(\mathsf{P}(3), A, t) = \sum_{r \geq 0} t^{3r} = 1/(1 - t^3)$. This example will be generalized in Theorem 19.1.4.

18.6 The representation ring $R_2(\mathsf{GL}(n))$

For a finite group G, information about the composition factors of $\mathbb{F}_2 G$-modules can be conveniently expressed in terms of the modular representation ring $R_2(G)$. Let p be a prime and K a splitting field for G of characteristic p. The *representation ring* (or *Grothendieck ring*) $R_p(G)$ is a commutative ring with 1 whose elements are formal linear combinations of irreducible KG-modules with integer coefficients. For finite dimensional KG-modules A, B, C, addition in $R_p(G)$ is given by $[A] + [B] = [A \oplus B]$ and $[A][B] = [A \otimes B]$, and an exact sequence $0 \to A \to C \to B \to 0$ gives a relation $[C] = [A] + [B]$ in $R_p(G)$.

By the Jordan–Hölder theorem, the multiset of composition factors appearing in a composition series for a finite dimensional KG-module A is independent of the choice of series. Hence the element $[A] \in R_p(G)$ which records the composition factors of A and their multiplicities is well defined. Thus $R_p(G)$ is the free abelian group with $\mathbb{Z}$-basis given by elements $[L]$ corresponding to isomorphism classes of irreducible KG-modules L, with identity $1 = [\mathsf{I}]$ corresponding to the 1-dimensional trivial representation I.

For a graded KG-module M, we treat the sequence $[M^d]$, $d \geq 0$ of elements of $R_p(G)$ as coefficients in a power series.

Definition 18.6.1 Let $M = \sum_{d \geq 0} M^d$ be a graded KG-module, where M^d is finite dimensional for all d. The **Poincaré series** of M is

$$\Pi(M, t) = \sum_{d \geq 0} [M^d] t^d,$$

a power series in t with coefficients in $R_p(G)$.

It follows from the definitions that if M and N are graded KG-modules then $\Pi(M \oplus N, t) = \Pi(M, t) + \Pi(N, t)$ and $\Pi(M \otimes N, t) = \Pi(M, t)\Pi(N, t)$.

In the case $p = 2$ and $G = \mathsf{GL}(n)$, we may take $K = \mathbb{F}_2$ by Proposition 18.2.1. Then $R_2(\mathsf{GL}(n))$ is the free abelian group of rank 2^{n-1} generated by elements

$[L(\lambda)]$ corresponding to the irreducible $\mathbb{F}_2\mathsf{GL}(n)$-modules $L(\lambda)$, where λ is a column 2-regular partition of length $\leq n-1$. We can express each module $[M^d]$ as a sum $\sum_\lambda m_d(\lambda)[L(\lambda)]$ in $R_2(\mathsf{GL}(n))$, where the coefficients $m_d(\lambda)$ are integers ≥ 0, and write

$$\Pi(M,t) = \sum_\lambda [L(\lambda)]P(M,\lambda,t)$$

where $P(M,\lambda,t) = \sum_{d\geq 0} m_d(\lambda)t^d \in \mathbb{Z}[[t]]$. For all λ, we have $\Pi(M\oplus N,\lambda,t) = \Pi(M,\lambda,t)+\Pi(N,\lambda,t)$ for graded $\mathbb{F}_2\mathsf{GL}(n)$-modules M and N.

It follows from Theorem 18.5.3 that $R_2(\mathsf{GL}(n))$ can be described as the ring of Brauer characters of $\mathsf{GL}(n)$, i.e. the ring of $\mathbb{Z}$-linear combinations of Brauer characters of irreducible $\mathbb{F}_2\mathsf{GL}(n)$-modules, with operations defined pointwise in $\mathbb{C}$. Thus the ring of class functions $G'_p \to \mathbb{C}$ for $G = \mathsf{GL}(n)$ and $p = 2$ can be described as $R_2(\mathsf{GL}(n)) \otimes_{\mathbb{Z}} \mathbb{C}$.

Example 18.6.2 For $G = \mathsf{GL}(3)$, Example 18.5.1 gives the relations $\mathrm{br}(\mathsf{V}(3))^2 = \mathrm{br}(\mathsf{V}(3)) + 2\mathrm{br}(\mathsf{V}(3)^*)$, $\mathrm{br}(\mathsf{V}(3)^*)^2 = \mathrm{br}(\mathsf{V}(3)^*) + 2\mathrm{br}(\mathsf{V}(3))$, $\mathrm{br}(\mathsf{V}(3))\,\mathrm{br}(\mathsf{V}(3)^*) = \mathrm{br}(\mathsf{I}(3)) + \mathrm{br}(\mathsf{St}(3))$. Thus $R_2(\mathsf{GL}(3))$ has an additive basis $\{1,a,b,ab\}$ where $a = [\mathsf{V}(3)]$ and $b = [\mathsf{V}(3)^*]$, with relations $a^2 = a+2b$ and $b^2 = b+2a$. (In the notation of Proposition 18.6.3, $a = [\Lambda^1]$ and $b = [\Lambda^2]$.)

As we do not know the Brauer characters of the irreducible modules of $\mathsf{GL}(n)$ in general, we can determine $R_2(\mathsf{GL}(n))$ in this way only for small values of n. Although the following result describes $R_2(\mathsf{GL}(n))$ for all n, it is not very useful as we cannot locate the classes $[L(\lambda)]$ of the irreducible $\mathbb{F}_2\mathsf{GL}(n)$-modules in terms of the basis $[\Lambda^I]$. In Section 20.1, we shall see what can be done towards dealing with this fundamental difficulty by using the representation theory of the algebraic group $\overline{\mathsf{G}}(n) = \mathsf{GL}(n,\overline{\mathbb{F}}_2)$, where $\overline{\mathbb{F}}_2$ is the algebraic closure of $\mathbb{F}_2$.

For $1 \leq i \leq n-1$, let $\Lambda^i = \Lambda^i(n)$ be the ith exterior power of the defining module $\mathsf{V}(n)$, and let $[\Lambda^i]$ be the corresponding element of $R_2(\mathsf{GL}(n))$. Then $\Lambda^0 = \mathsf{I}(n)$, so that $[\Lambda^0] = 1$. For $I = \{i_1,i_2,\ldots,i_r\} \subseteq Z[n-1]$, let $[\Lambda^I] = [\Lambda^{i_1} \otimes \Lambda^{i_2} \otimes \cdots \otimes \Lambda^{i_r}] = [\Lambda^{i_1}][\Lambda^{i_2}]\cdots[\Lambda^{i_r}]$.

Proposition 18.6.3 *The elements $[\Lambda^I]$, $I \subseteq Z[n-1]$, form a $\mathbb{Z}$-basis of the representation ring $R_2(\mathsf{GL}(n))$. The product is determined by the relations*

$$[\Lambda^i]^2 = [\Lambda^i] + 2\sum_{k=1}^{i}(-1)^{k-1}[\Lambda^{i-k}][\Lambda^{i+k}], \quad 1 \leq i \leq n-1,$$

where $[\Lambda^n] = [\Lambda^0] = 1$, and $[\Lambda^i] = 0$ if $i < 0$ or $i > n$.

When $n=7$, for example, the first three relations are

$$[\Lambda^1]^2 = [\Lambda^1] + 2[\Lambda^2],$$
$$[\Lambda^2]^2 = [\Lambda^2] + 2[\Lambda^1][\Lambda^3] - 2[\Lambda^4],$$
$$[\Lambda^3]^2 = [\Lambda^3] + 2[\Lambda^2][\Lambda^4] - 2[\Lambda^1][\Lambda^5] + 2[\Lambda^6].$$

The remaining three relations are the contragredient duals of these. Since $\Lambda^{n-i} \cong (\Lambda^i)^*$, the ith and $(n-i)$th relations are contragredient duals for all n.

Proof of Proposition 18.6.3 Let $A \in \mathsf{GL}(n)$ be an element of odd order. Let $\{\alpha_1(A),\dots,\alpha_n(A)\}$ be the multiset of eigenvalues of A in a splitting field K for its characteristic polynomial, and let $\{\widetilde{\alpha}_1(A),\dots,\widetilde{\alpha}_n(A)\}$ be the corresponding elements of $\mathbb{C}$ under Brauer lifting. Then the eigenvalues of $\Lambda^i(A)$ are the products of $\widetilde{\alpha}_1(A),\dots,\widetilde{\alpha}_n(A)$, taken i at a time, and the Brauer character $\mathrm{br}(\Lambda^i,A) = e_i(\widetilde{\alpha}_1(A),\dots,\widetilde{\alpha}_n(A))$, where $e_i(u_1,\dots,u_n)$ is the ith elementary symmetric function in the variables $u_1,\dots,u_n$.

Since A is fixed by the Frobenius automorphism F of K and $F(\alpha)=\alpha^2$ for all $\alpha \in K$, the multisets $\{\widetilde{\alpha}_1(A)^2,\dots,\widetilde{\alpha}_n(A)^2\}$ and $\{\widetilde{\alpha}_1(A),\dots,\widetilde{\alpha}_n(A)\}$ are equal. It follows that $e_i(\widetilde{\alpha}_1(A),\dots,\widetilde{\alpha}_n(A)) = e_i(\widetilde{\alpha}_1(A)^2,\dots,\widetilde{\alpha}_n(A)^2)$.

The generating function for the elementary symmetric functions in n variables $u_1,\dots,u_n$ is $\sum_{i\geq 0}(-1)^i e_i(u_1,\dots,u_n)t^i = \prod_{j=1}^n (1-u_jt)$. Substituting u_j^2 for u_j for all j and t^2 for t, this gives $\sum_{i\geq 0}(-1)^i e_i(u_1^2,\dots,u_n^2)t^{2i} = \prod_{j=0}^n(1-u_j^2t^2) = \prod_{j=0}^n(1+u_jt)(1-u_jt) = \prod_{j=0}^n(1+u_jt)\prod_{j=0}^n(1-u_jt) = \sum_{r\geq 0} e_rt^r \sum_{s\geq 0}(-1)^s e_st^s$.

Equating coefficients of t^{2i}, we have $(-1)^i e_i(u_1^2,\dots,u_n^2) = \sum_{r+s=2i}(-1)^s e_re_s$, $1 \leq i \leq n$. We evaluate this in $R_2(G)$ by replacing the variables $u_1,\dots,u_n$ by the eigenvalues $\alpha_1(A),\dots,\alpha_n(A)$. This gives $(-1)^i[\Lambda^i] = \sum_{r+s=2i}(-1)^s[\Lambda^r][\Lambda^s]$, $1 \leq i \leq n$, where $[\Lambda^n] = [\Lambda^0] = 1$ and $[\Lambda^i] = 0$ unless $0 \leq i \leq n$.

It follows from the fundamental theorem of symmetric functions that the elements $[\Lambda^i]$, $i \geq 0$ generate $R_2(\mathsf{GL}(n))$ as a ring. These relations show that the elements $[\Lambda^I]$, $I \subset Z[n-1]$ give an additive spanning set. Since $R_2(\mathsf{GL}(n))$ is additively a free abelian group of rank 2^{n-1}, the products $[\Lambda^I]$ for $I \subseteq Z[n-1]$ satisfy no further relations, and so they give a $\mathbb{Z}$-basis for $R_2(G)$. □

18.7 Remarks

Our general reference for modular representation theory is Curtis and Reiner [48], especially Chapter XII. Here we outline the main results used in the text.

Let G be a finite group, K a field of characteristic $p > 0$ and KG the group algebra of G over K. A finite-dimensional right KG-module V is also called a modular representation of G, and a choice of K-basis for V gives a matrix representation of G, i.e. a homomorphism $\rho_V : G \to \mathsf{GL}(\dim V, K)$. If p does not divide the order of G, then by Maschke's theorem [48, Theorem 10.8] every indecomposable module is irreducible, and so all representations V are direct sums of irreducibles. In particular, this is true for KG if $\mathrm{char}(K) = 0$ or if p does not divide the order of G.

The field K is a splitting field for G if all irreducible KG-modules remain irreducible under field extensions of K. The situation shown in Example 18.0.1 occurs already in characteristic 0, as the map $g \mapsto \left(\begin{smallmatrix} 0 & 1 \\ -1 & -1 \end{smallmatrix}\right)$ defines an irreducible representation of $\mathbb{Z}/3$ over the field of rational numbers $\mathbb{Q}$ (cf. [48, (39.5)]). This splits over $\mathbb{Q}[\omega]$, where $\omega = \exp(2\pi i/3)$. A sufficient condition for K to be a splitting field for G is that K contains a qth root of 1 whenever G has an element of odd order q [48, Section 68].

Proposition 18.2.1 is an important example where this condition is not necessary. When $K = \mathbb{C}$, or any splitting field for G of characteristic 0, there is a one-one correspondence between conjugacy classes of G and isomorphism classes of irreducible KG-modules. The corresponding theorem for characteristic $p > 0$ [95, Theorem 3.9] states that the number of isomorphism classes of irreducible KG-modules is the number of p-regular conjugacy classes of G when K is a splitting field for G. If G is a p-group, this implies that the 1-dimensional trivial representation I is the only irreducible KG-module. When K is finite, this is proved by the argument of Proposition 10.1.1.

Orthogonal sets of idempotents in a ring are discussed in [48, Section 25]. For the Krull–Schmidt theorem, see [48, (14.5)]. For a finite group G and a field K, the relation between primitive idempotents in KG and the representation theory of G (Section 18.1) is developed in [48, Section 54]. By [48, (54.5)]), a submodule of KG is indecomposable if and only if it is the image eKG of a primitive idempotent $e \in KG$. Thus a decomposition $1 = e_1 + \cdots + e_m$ of the identity element of G as a sum of orthogonal primitive idempotents in KG yields a decomposition $KG = e_1KG \oplus \cdots \oplus e_mKG$ of KG as a direct sum of indecomposable modules.

When K is a splitting field for G, [48, (54.16)] provides a means of counting composition factors of KG-modules. Given a right KG-module M and a primitive idempotent $e \in KG$ corresponding to the irreducible module L, the multiplicity of L as a composition factor of M is the dimension of the image Me as a vector space over K. By the proof of [48, (61.13)], the K-vector spaces KGe and eKG are duals, and so have the same dimension. Hence by applying

(54.16) with $M = KG$, it follows that the multiplicity of L as a composition factor of KG is the dimension of the corresponding principal indecomposable module U. It is further proved in [48, (54.16)] that the multiplicity of U as a direct summand of KG is the dimension of the corresponding irreducible module L. These facts are summarized in [48, (83.3)].

Since $\mathsf{B}(n)$ is a Sylow 2-subgroup of $\mathsf{GL}(n)$, a standard result in representation theory (see [79, Chapter 9]) allows us to deduce that the Steinberg module $\mathsf{St}(n)$ is a projective $\mathbb{F}_2\mathsf{GL}(n)$-module directly from Proposition 18.3.7, rather than by the method of Section 18.4. Since $\mathbb{F}_2\mathsf{GL}(n)$ is a Frobenius algebra, it follows from [48, Theorem 62.3] that $\mathsf{St}(n)$ is injective as well as projective. It follows that $\mathsf{St}(n)$ is a direct summand of any $\mathbb{F}_2\mathsf{GL}(n)$-module in which it is a composition factor. In particular, every occurrence of $\mathsf{St}(n)$ in $\mathsf{P}(n)$ is as a direct summand.

Brauer characters and their orthogonality relations are discussed in [48, Sections 82, 84]. For the factorization $g = g_p g'_p$, see e.g. [48, Lemma 40.3]. For the properties of the Brauer character cited in Theorem 18.5.3, see [79, Section 5.5]. Parts (ii), (iii) and (iv) follow from [48, Theorem 30.16], and part (v) then follows from [95, Theorem 1.11(a)]. The observation in Example 18.5.2 that the Brauer character of $\mathsf{FL}(n)$ is the restriction of the ordinary character of Flag(n) to 2-regular classes applies to any permutation representation [48, p.589].

The Steinberg idempotents $e(n)$ and $e'(n)$ were defined in [197] and are treated in [141]. The embedding of the Hecke algebra $\mathsf{H}_0(n)$ in the group algebra $\mathbb{F}_2\mathsf{GL}(n)$ discussed in Section 18.4 is due to N. J. Kuhn [116]. The modular representation ring $R_2(\mathsf{GL}(n))$ is calculated by Carlisle and Kuhn in [28, Section 3].

19

Splitting $\mathsf{P}(n)$ as an A_2-module

19.0 Introduction

In this chapter we study the number of composition factors of the $\mathbb{F}_2\mathsf{GL}(n)$-module $\mathsf{P}^d(n)$ which are in a given isomorphism class. We have seen that there are 2^{n-1} such classes, and that the irreducible $\mathbb{F}_2\mathsf{GL}(n)$-modules $L(\lambda)$ can be indexed by the column 2-regular partitions λ of length $\leq n-1$. Since the Steenrod square $Sq^k : \mathsf{P}^d(n) \to \mathsf{P}^{d+k}(n)$ is a map of $\mathbb{F}_2\mathsf{GL}(n)$-modules, it links composition factors isomorphic to the same $L(\lambda)$ in $\mathsf{P}^d(n)$ and $\mathsf{P}^{d+k}(n)$. Hence $\mathsf{P}(n)$ is the direct sum of graded A_2-modules, each of which is associated to one of the irreducible modules $L(\lambda)$. The number of summands associated to $L(\lambda)$ is $d_\lambda = \dim L(\lambda)$.

For each λ, these d_λ summands are isomorphic as graded A_2-modules, and we denote this isomorphism type by $\mathsf{P}(n,\lambda)$. For all $d \geq 0$, the dimension of $\mathsf{P}^d(n,\lambda)$ as a vector space over $\mathbb{F}_2$ is the number of composition factors of $\mathsf{P}^d(n)$ which are isomorphic to $L(\lambda)$. Finding these numbers is equivalent to expressing the Brauer character of $\mathsf{P}^d(n)$ as the sum of Brauer characters of the irreducible modules $L(\lambda)$. However, the determination of these irreducible Brauer characters for a general value of n remains an open problem.

Every decomposition of the identity element as a sum of primitive orthogonal idempotents in $\mathbb{F}_2\mathsf{GL}(n)$ gives rise to such a splitting of $\mathsf{P}(n)$. In the case $n = 2$, for example, the decomposition $I_2 = e_0 + e_1 + e_1'$ of Example 18.1.1 corresponds to an isomorphism $\mathsf{P}(2) \cong A_0 \oplus A_1 \oplus A_1'$ of A_2-modules, where $A_0 \cong \mathsf{P}(2,(0))$ and $A_1 \cong A_1' \cong \mathsf{P}(2,(1))$. We discuss this splitting in Example 19.1.3.

Although we restrict attention almost entirely to the splitting obtained using $\mathsf{GL}(n)$, a more refined A_2-module decomposition of $\mathsf{P}(n)$ can be obtained using the full matrix semigroup $\mathsf{M}(n)$. In this splitting, each summand $\mathsf{P}(n,\lambda)$ is further split as the direct sum $\mathsf{P}'(n,\lambda) \oplus \mathsf{P}'(n,\lambda')$ of two A_2-modules, and the resulting summands of $\mathsf{P}(n)$ are of 2^n isomorphism types. Here $\mathsf{P}'(n,\lambda)$ and $\mathsf{P}'(n,\lambda')$ correspond to the irreducible $\mathbb{F}_2\mathsf{M}(n)$-modules $L(\lambda)$ and $L(\lambda') = L(\lambda) \otimes \mathsf{det}$, where $\lambda'_i = \lambda_i + 1$ for $1 \le i \le n$. The modules $\mathsf{L}(\lambda)$ and $L(\lambda')$ form a complete set of 2^n isomorphism classes of irreducible $\mathbb{F}_2\mathsf{M}(n)$-modules, indexed by column 2-regular partitions λ of length $< n$ and λ' of length n. We prove in Chapter 21 that the A_2-modules $\mathsf{P}'(n,\lambda)$ and $\mathsf{P}'(n,\lambda')$ are indecomposable, and therefore that the corresponding splitting of $\mathsf{P}(n)$ using representations of $\mathsf{M}(n)$ is maximal.

In Section 19.1 we consider the Poincaré series $P(\mathsf{P}(n,\lambda),t)$, the generating function which gives the number of occurrences of $L(\lambda)$ as a composition factor in $\mathsf{P}^d(n)$ for all $d \ge 0$. We show in Theorem 19.1.4 that $P(\mathsf{P}(n,\lambda),t)$ is a rational function of t with denominator $(1-t)(1-t^3)\cdots(1-t^{2^n-1})$ independent of λ. In Section 19.2 we relate these Poincaré series to the Brauer characters of principal indecomposable $\mathbb{F}_2\mathsf{GL}(n)$-modules.

Section 19.3 introduces the ring of coinvariants $\mathsf{C}(n)$, the quotient of $\mathsf{P}(n)$ by the ideal generated by the Dickson invariants $\mathsf{d}_{n,i}$, $0 \le i \le n-1$. In Theorem 19.3.7, we show that every irreducible $\mathbb{F}_2\mathsf{GL}(n)$-module $L(\lambda)$ appears as a composition factor in $\mathsf{C}(n)$ with the same multiplicity as it has in the regular representation of $\mathsf{GL}(n)$. In particular, every $L(\lambda)$ appears as a composition factor in $\mathsf{C}(n)$, and so also in $\mathsf{P}(n)$.

In Sections 19.4 and 19.5 we show that every $L(\lambda)$ appears as a submodule in $\mathsf{P}(n)$, and that the minimum degree d such that $\mathsf{P}^d(n)$ has such a submodule is $\deg_2 \lambda$. The proofs use a theorem of Huynh Mui on the $\mathsf{B}(n)$-invariants of $\mathsf{P}(n)$, which we discuss in Section 19.4. Results of modular representation theory that we use are stated in Section 19.6.

19.1 The A_2-modules $\mathsf{P}(n,\lambda)$

Our basic splitting theorem follows from the commutativity of Steenrod operations with the action of $\mathsf{GL}(n)$.

Theorem 19.1.1 *There is a direct sum decomposition of graded* A_2*-modules* $\mathsf{P}(n) \cong \bigoplus_\lambda d_\lambda \mathsf{P}(n,\lambda)$, *where the summands* $\mathsf{P}(n,\lambda)$ *are indexed by column 2-regular partitions* λ *of length* $\le n-1$, *and* d_λ *is the dimension of the irreducible* $\mathbb{F}_2\mathsf{GL}(n)$*-module* $L(\lambda)$.

Proof Given a decomposition $1 = e_1 + \cdots + e_r$ of the identity element $1 = \mathsf{I}_n$ of $\mathbb{F}_2\mathsf{GL}(n)$ as a sum of primitive orthogonal idempotents, $\mathsf{P}(n) = \bigoplus_{i=1}^r \mathsf{P}(n)\cdot e_i$ as graded vector spaces over $\mathbb{F}_2$. Since $Sq^k(f\cdot e_i) = Sq^k(f)\cdot e_i$ for $f \in \mathsf{P}^d(n)$, $\mathsf{P}(n)e_i$ is a graded A_2-module for $1 \le i \le r$. By a standard result of modular representation theory (see Section 19.6), the dimension of $P^d(n)\cdot e_i$ is the number of composition factors of $P^d(n)$ which are isomorphic to the irreducible module $L(\lambda)$ associated to the principal indecomposable $\mathbb{F}_2GL(n)$-module $U(\lambda)=e_i\mathbb{F}_2GL(n)$.

If e_i and e_j are associated to the same module $L(\lambda)$, then $P(n)\cdot e_i \cong P(n)\cdot e_j$ as A_2-modules. To see this, by Theorem 18.1.2(iii) we have $e_j = u^{-1}e_iu$ for some invertible element $u \in \mathbb{F}_2GL(n)$. Then for $f \in P(n)$, the map $f\cdot e_i \mapsto f\cdot e_iu = f\cdot ue_j$ is an A_2-module isomorphism from $P(n)\cdot e_i$ to $P(n)\cdot e_j$. Thus we have a decomposition $P(n) = P(n)\cdot e_1 \oplus \cdots \oplus P(n)\cdot e_r$ of $P(n)$ as a direct sum of A_2-modules, with $d_\lambda = \dim L(\lambda)$ summands corresponding to each λ. □

Definition 19.1.2 Given a 2-column regular partition λ of length $\leq n-1$, we denote by $P(n,\lambda)$ the isomorphism class of the A_2-module $P(n)\cdot e$, where e is a primitive idempotent in $\mathbb{F}_2GL(n)$ associated to the irreducible $\mathbb{F}_2GL(n)$-module $L(\lambda)$.

Since the Poincaré series $P(P(n),t) = 1/(1-t)^n$, Theorem 19.1.1 gives a corresponding decomposition

$$\frac{1}{(1-t)^n} = \sum_{\lambda} d_\lambda P(P(n,\lambda),t), \tag{19.1}$$

where $P(P(n,\lambda),t) = P(P(n),\lambda,t) \in \mathbb{Z}[[t]]$ in the notation of Section 18.6.

Example 19.1.3 Let $n=2$ and let e_0, e_1, e_1' be the idempotents in $\mathbb{F}_2GL(2)$ defined in Example 18.1.1. Then $e_0 = I_2+LS+SL$ is the sum of the elements in the cyclic subgroup H_2 of $GL(2)$, and we have shown in Section 15.4 that $P(2)\cdot e_0 = P(2)^{H_2} = D(2)\oplus D(2)g$, where $g = x^3+y^3+x^2y$ and $D(2) = \mathbb{F}_2[d_0,d_1]$ is the Dickson algebra. Thus $P(2,(0)) \cong P(2)\cdot e_0$, the subring of $P(2)$ generated by $d_0 = x^2y+xy^2$, $d_1 = x^2+xy+y^2$ and g, with the defining relation $g^2 = d_0g+d_0^2+d_1^3$.

Since $P(D(2),t) = 1/(1-t^2)(1-t^3)$, $P(P(2,(0)),t) = (1+t^3)/(1-t^2)(1-t^3) = (1-t+t^2)/(1-t)(1-t^3)$. Since $P(P(2),t) = P(P(2,(0)),t) + 2P(P(2,(1)),t)$ by (19.1), it follows that $P(P(2,(1)),t) = t/(1-t)(1-t^3)$. Exceptionally, the summand $P(2,(0))$ is a $\mathbb{F}_2GL(2)$-module, since e_0 is central in $\mathbb{F}_2GL(2)$.

It follows from Theorem 19.1.1 that the hit problem for $P(n)$ splits into a corresponding problem for each summand $P(n,\lambda)$. In the case $n=2$, the diagram following Theorem 1.8.2 shows that a basis for the cohits in $P(2,(1))$ has one element in degrees $2^{t_1}+2^{t_2}-2$ where $t_1 > t_2$, while a basis for the cohits in $P(2,(0))$ has one element in degrees $2^{t_1}+2^{t_2}-2$ where $t_1 = t_2$ or $t_1 > t_2+1$.

Example 19.1.3 shows that the Poincaré series of the summands of $\mathsf{P}(2)$ are rational functions with integer coefficients. We show next that this holds for all n, and that the series are determined by their terms of degree $\leq 2^{n+1}-2n-2$.

Theorem 19.1.4 *Let* $w(n)=\sum_{i=1}^{n}(2^i-1)=2^{n+1}-n-2$. *For all* n *and* λ, $P(\mathsf{P}(n,\lambda),t)=p_n(\lambda,t)/q_n(t)$, *where* $p_n(\lambda,t)\in\mathbb{Z}[t]$ *has degree* $\leq 2w(n-1)=w(n)-n$ *and* $q_n(t)=\prod_{i=1}^{n}(1-t^{2^i-1})$ *has degree* $w(n)$.

Proof Let $A\in\mathsf{GL}(n)$ be an element of odd order. By Theorems 19.1.1 and 18.5.3, the Brauer character $\mathrm{br}(\mathsf{P}^d(n),A)=\sum_\lambda \dim\mathsf{P}^d(n,\lambda)\,\mathrm{br}(L(\lambda),A)$ for $A\in\mathsf{GL}(n)$, since $\dim\mathsf{P}^d(n,\lambda)$ is the number of composition factors of $\mathsf{P}^d(n)$ which are isomorphic to $L(\lambda)$. Hence the Brauer character series (18.2) of $\mathsf{P}(n)$ at A is

$$\mathrm{br}(\mathsf{P}(n),A,t)=\sum_\lambda P(\mathsf{P}(n,\lambda),t)\mathrm{br}(L(\lambda),A). \tag{19.2}$$

The Brauer character $\mathrm{br}(\mathsf{V}(n),A)=\sum_{i=1}^{n}\alpha_i(A)$ of $\mathsf{V}(n)\cong\mathsf{P}^1(n)$ is the sum of the eigenvalues of A in a splitting field K for the characteristic polynomial of A, lifted to $\mathbb{C}$. In general, $\mathrm{br}(\mathsf{P}^d(n),A)=h_d(\alpha_1(A),\ldots,\alpha_n(A))$ for $d\geq 0$, where $h_d(u_1,\ldots,u_n)$ is the dth complete symmetric function in the variables $u_1,\ldots,u_n$. Since $\sum_{d\geq 0}h_d(u_1,\ldots,u_n)t^d=\prod_{i=1}^{n}1/(1-u_it)$, the Brauer character series (18.2) of $\mathsf{P}(n)$ is

$$\mathrm{br}(\mathsf{P}(n),A,t)=\prod_{i=1}^{n}\frac{1}{1-\alpha_i(A)t}. \tag{19.3}$$

Since $A\in\mathsf{GL}(n)$ is fixed by the Frobenius automorphism $F:\alpha\mapsto\alpha^2$ of K, the multiset $\alpha_1(A),\ldots,\alpha_n(A)$ is also fixed by F. Thus we can express the eigenvalues as a union of F-orbits. Let $\{\alpha,\alpha^2,\ldots,\alpha^{2^{\ell-1}}\}$ be such an orbit, where $\alpha^{2^\ell-1}=1$. Since $1-t^{2^\ell-1}=\prod_{j=0}^{2^\ell-2}(1-\alpha^jt)$, the product $\prod_{i=0}^{\ell-1}(1-\alpha^{2^i}t)$ divides $1-t^{2^\ell-1}$ in $K[t]$. If the orbits are all of different lengths ℓ, then $\prod_{i=1}^{n}(1-\alpha_i(g)t)$ divides $\prod_{i=1}^{n}(1-t^{2^i-1})$, since the product of the factors arising from each F-orbit divides a different factor $1-t^{2^i-1}$.

This argument extends to the general case. Let the F-orbit lengths of the eigenvalues of A be $\ell_1,\ldots,\ell_m$ in decreasing order, so that $(\ell_1,\ldots,\ell_m)$ is a partition of n. Since a finite field of order 2^i has a subfield of order 2^j if and only if j divides i, $1-t^{2^j-1}$ divides $1-t^{2^i-1}$ if and only if j divides i. For $1\leq j\leq n$, let j have multiplicity $m(j)$ in the partition $(\ell_1,\ldots,\ell_m)$. Then $m(j)\leq n/j$, and so the factor $1-t^{2^i-1}$ of $\prod_{i=1}^{n}(1-t^{2^i-1})$ is divisible by $1-t^{2^j-1}$ for $i=j,2j,\ldots,m(j)j$.

Thus $\prod_{i=1}^{n}(1-t^{2^i-1})$ is divisible by $(1-t^{2^j-1})^{m(j)}$. Since elements $\alpha_i(A)$ which belong to F-orbits of different lengths j correspond to different factors of $\prod_{i=1}^{n}(1-t^{2^i-1})$, we can conclude that $\prod_{i=1}^{n}(1-t^{2^i-1})$ is divisible by $\prod_{i=1}^{n}(1-$

$\alpha_i(A)t)$, as required. The quotient has degree $\sum_{i=1}^{n}(2^i-1)-n=2^{n+1}-2n-2$. Hence $\prod_{i=1}^{n}(1-\alpha_i(g)t)$ divides $\prod_{i=1}^{n}(1-t^{2^i-1})$ in $K[t]$. □

Example 19.1.5 Theorem 19.1.4 implies that the series $P(\mathsf{P}(2,(0)),t)$ and $P(\mathsf{P}(2,(1)),t)$ are determined by their terms of degree ≤ 2. Since $\mathsf{P}^0(2)\cong L(0)$, $\mathsf{P}^1(2)\cong L(1)$ and $\mathsf{P}^2(2)$ has composition factors $L(1)$ and $L(0)$, $P(\mathsf{P}(2,(0)),t)=1+t^2+$ terms of degree ≥ 3, and $P(\mathsf{P}(2,(1)),t)=t+t^2+$ terms of degree ≥ 3.

Thus $p_2((0),t)=(1-t)(1-t^3)(1+t^2+$ terms of degree $\geq 3)=1-t+t^2$, since $p_2((0),t)\in\mathbb{Z}[t]$ has degree ≤ 2. Similarly $p_2((1),t)=(1-t)(1-t^3)(t+t^2+$ terms of degree $\geq 3)=t$. Hence $P(\mathsf{P}(2,(0)),t)=p_2((0),t)/q_2(t)=(1-t+t^2)/(1-t)(1-t^3)$ and $P(\mathsf{P}(2,(1)),t)=t/(1-t)(1-t^3)$, as shown in Example 19.1.3.

Example 19.1.6 The series in Example 19.1.5 can also be found using Brauer characters. Let C_1 and C_3 denote the conjugacy classes in $\mathsf{GL}(2)$ of the identity I_2 and of an element A of order 3, which has eigenvalues α,α^2 where α generates $\mathbb{F}_4^*$. For the Brauer character, we lift α to $\omega=e^{2\pi i}/3\in\mathbb{C}$. The Brauer character table is

	C_1	C_3
$L(0)\cong\mathsf{I}(2)$	1	1
$L(1)\cong\mathsf{V}(2)$	2	-1

From (19.3) we have $\mathrm{br}(\mathsf{P}(2),I_2)=1/(1-t)^2$, $\mathrm{br}(\mathsf{P}(2),A)=1/(1-\omega t)(1-\omega^2 t)=1/(1+t+t^2)$. Substituting in (19.2), we obtain $1/(1-t)^2=P(\mathsf{P}(2,(0)),t)+2P(\mathsf{P}(2,(1)),t)$ and $1/(1+t+t^2)=P(\mathsf{P}(2,(0)),t)-P(\mathsf{P}(2,(1)),t)$. Using Theorem 19.1.4 we write $P(\mathsf{P}(2,\lambda),t)=p_2(\lambda,t)/(1-t)(1-t^3)$, where $\lambda=(0)$ or (1), and solve for $p_2((0),t)$ and $p_2((1),t)$.

Example 19.1.7 In the case $n=3$, $q_3(t)=(1-t)(1-t^3)(1-t^7)$ and the polynomials $p_3(\lambda,t)$ associated to the irreducible $\mathbb{F}_2\mathsf{GL}(3)$-modules $L(0)\cong\mathsf{I}(3)$, $L(1)\cong\mathsf{V}(3)$, $L(1,1)\cong\mathsf{V}(3)^*$ and $L(2,1)\cong\mathsf{St}(3)$ are $1-t+t^4-t^7+t^8$, $t-t^4+t^5+t^6$, $t^2+t^3-t^4+t^7$ and t^4 respectively. As we show in Example 19.2.5, these can be calculated by the method of Example 19.1.5. The Brauer character table is given in (18.1). From (19.3) we have

$$\mathrm{br}(\mathsf{P}(3),I_3)=1/(1-t)^3,\qquad \mathrm{br}(\mathsf{P}(3),A_3)=1/(1-t^3),$$
$$\mathrm{br}(\mathsf{P}(3),A_7)=1/(1-\gamma t+\overline{\gamma}t^2-t^3),\quad \mathrm{br}(\mathsf{P}(3),A_7')=1/(1-\overline{\gamma}t+\gamma t^2-t^3),$$

where I_3, A_3, A_7, A_7' are elements in the conjugacy classes C_1, C_3, C_7 and C_7' of $\mathsf{GL}(3)$ respectively. By substituting in (19.2) we obtain four equations, which can be solved for the polynomials $p_3(\lambda, t)$.

Example 19.1.8 Given a column 2-regular partition λ of length $\leq n-1$, we have irreducible $\mathbb{F}_2\mathsf{M}(n)$-modules $L(\lambda)$ and $L(\lambda') = L(\lambda) \otimes \det$, where $\lambda_i' = \lambda_i + 1$ for $1 \leq i \leq n$, and these give a complete set of 2^n non-isomorphic $\mathbb{F}_2\mathsf{M}(n)$-modules. The Poincaré series for these can be calculated for the series for $\mathbb{F}_2\mathsf{GL}(n)$-modules.

Let $\varepsilon_n : \mathsf{P}(n) \to \mathsf{P}(n-1)$ be the algebra map defined by $\varepsilon_n(x_i) = x_i$ for $1 \leq i \leq n-1$ and $\varepsilon_n(x_n) = 0$. With λ as above, ε_n maps each $\mathbb{F}_2\mathsf{M}(n)$-composition factor of $\mathsf{P}(n)$ isomorphic to $L(\lambda)$ to a corresponding $\mathbb{F}_2\mathsf{M}(n-1)$-composition factor $L(\lambda)$ of $\mathsf{P}(n-1)$, and maps each $\mathbb{F}_2\mathsf{M}(n)$-composition factor of $\mathsf{P}(n)$ isomorphic to $L(\lambda) \otimes \det$ to 0. This can be seen by considering the action of the idempotent matrix $\left(\begin{smallmatrix} I_{n-1} & \mathbf{0} \\ \mathbf{0} & 0 \end{smallmatrix}\right)$ on the irreducible modules for $\mathsf{M}(n)$. It follows that the series $P(\mathsf{P}'(n,\lambda), t)$ is independent of n for $n \geq \operatorname{len}(\lambda)$.

For example, the four irreducible $\mathbb{F}_2\mathsf{M}(2)$-modules are $L(0) \cong \mathsf{I}(2)$ (where singular matrices also act as the identity), $L(1) \cong \mathsf{V}(2)$, $L(1,1) \cong \det$ and $L(2,1) = L(1) \otimes \det$. The module $L(0)$ occurs only in degree 0. Hence $P(\mathsf{P}'(2,(0)),t) = 1$ and $P(\mathsf{P}'(2,(1)),t) = t/(1-t)$. Using the results of Example 19.1.5, we have $P(\mathsf{P}'(2,(1,1)),t) = P(\mathsf{P}(2,(0)),t) - P(\mathsf{P}'(2,(0)),t) = (1-t+t^2)/(1-t)(1-t^3) - 1 = (t^2+t^3-t^4)/(1-t)(1-t^3)$ and $P(\mathsf{P}'(2,(2,1)),t) = P(\mathsf{P}(2,(1)),t) - P(\mathsf{P}'(2,(1)),t) = t/(1-t)(1-t^3) - t/(1-t) = t^4/(1-t)(1-t^3)$.

Example 19.1.9 The results of Examples 19.1.7 and 19.1.8 can be combined to calculate the Poincaré series for occurrences of the 8 irreducible $\mathbb{F}_2\mathsf{M}(3)$-modules in $\mathsf{P}(3)$, as follows. By Example 19.1.8, the series for $\lambda = (0)$, (1), $(1,1)$ and $(2,1)$ are 1, $t/(1-t)$, $(t^2+t^3-t^4)/(1-t)(1-t^3)$ and $t^4/(1-t)(1-t^3)$ respectively. Using Example 19.1.7, the series for $\lambda = (1,1,1)$, $(2,1,1)$, $(2,2,1)$ and $(3,2,1)$ have denominator $q_3(t) = (1-t)(1-t^3)(1-t^7)$ and numerators $t^3 - t^{10} + t^{11}$, $t^5 + t^6 + t^8 - t^{11}$, $t^7 + t^9 + t^{10} - t^{11}$ and t^{11} respectively. As in Example 19.1.8, these are calculated by using the identity $P(\mathsf{P}'(n,\lambda'),t) = P(\mathsf{P}(n,\lambda),t) - P(\mathsf{P}'(n,\lambda),t)$, which arises from the splitting $\mathsf{P}(n,\lambda) = \mathsf{P}'(n,\lambda) \oplus \mathsf{P}'(n,\lambda')$.

19.2 Poincaré series of $\mathsf{P}(n,\lambda)$

Let G be a finite group, and let K be a splitting field for G of characteristic $p > 0$. Let r be the number of p-regular conjugacy classes in G, let $L_1, \ldots, L_r$ be a full set of irreducible KG-modules and let U_i be the principal indecomposable KG-module corresponding to L_i for $1 \leq i \leq r$. Then the Brauer characters of these modules satisfy the orthogonality relations

$$\frac{1}{|G|}\sum_{g\in G'_p} \mathrm{br}(U_i,g^{-1})\mathrm{br}(L_j,g) = \begin{cases} 1, & \text{if } i=j \\ 0, & \text{if } i\neq j, \end{cases} \tag{19.4}$$

where G'_p is the set of p-regular elements of G.

Given a finite dimensional KG-module M and an irreducible KG-module L, we denote by $[M:L]$ the multiplicity of L as a composition factor in M. Since the Brauer character of M is the sum of the Brauer characters of its composition factors, (19.4) gives

$$[M:L] = \frac{1}{|G|}\sum_{g\in G'_p} \mathrm{br}(U,g^{-1})\,\mathrm{br}(M,g), \tag{19.5}$$

where U is the principal indecomposable module corresponding to L. Let $M = \sum_{d\geq 0} M^d$ be a graded KG-module such that M^d is finite dimensional for all $d \geq 0$, and define the Brauer character series of M by (18.2).

We write the Poincaré series for the multiplicity of L in M, which depends only on the equivalence class $[M]$ of M in $R_p(G)$, as $P([M],L,t) = \sum_{d\geq 0}[M^d : L]t^d$. Then we obtain the following graded form of (19.5).

$$P([M],L,t) = \frac{1}{|G|}\sum_{g\in G'_p} \mathrm{br}(U,g^{-1})\,\mathrm{br}(M,g,t). \tag{19.6}$$

We specialize to the case where $p = 2$, G is a subgroup of $\mathsf{GL}(n)$, K is a splitting field for G of characteristic 2 and $M = \mathsf{P}(n)$. If $A \in G$ has odd order, then as $M^1 \cong \mathsf{V}(n)$, $\mathrm{br}(M^1,A) = \sum_{i=1}^n \alpha_i(A)$, where $\alpha_1(A),\cdots,\alpha_n(A)$ are the eigenvalues of A lifted to $\mathbb{C}$. Then $\mathrm{br}(\mathsf{P}(n),A,t) = \prod_{i=1}^n 1/(1-\alpha_i(A)t)$ by (19.3), and so (19.6) gives the following result.

Proposition 19.2.1 *Let G be a subgroup of $\mathsf{GL}(n)$, let K be a splitting field for G of characteristic 2, and let L be an irreducible KG-module. Then*

$$P([\mathsf{P}(n)],L,t) = \frac{1}{|G|}\sum_{A\in G'_2} \frac{br(U,A^{-1})}{\prod_{i=1}^n (1-\alpha_i(A)t)},$$

where U is the principal indecomposable module corresponding to L. □

Given a representation of a finite group G over a field K of characteristic 0, Molien's theorem gives the dimension of the K-space of invariant polynomials in degree d as the coefficient of t^d in the series $(1/|G|)\sum_{g\in G} 1/\det(1-gt)$. In a sense, Proposition 19.2.1 is a modular analogue of Molien's formula, which counts all occurrences of L as a composition factor in $\mathsf{P}^d(n)$, and not only submodules.

Example 19.2.2 In the case $n=2$, $U(0)$ has two composition factors $\cong L(0)$ and $U(1)\cong L(1)$. Using the Brauer character table (see Example 19.1.6), Proposition 19.2.1 gives

$$P(\mathsf{P}(2,(0)),t)=\frac{1}{6}\left(\frac{2}{(1-t)^2}+\frac{2\cdot 2}{(1-\omega t)(1-\omega^2 t)}\right)=\frac{1-t+t^2}{(1-t)(1-t^3)},$$
$$P(\mathsf{P}(2,(1)),t)=\frac{1}{6}\left(\frac{2}{(1-t)^2}+\frac{2\cdot -1}{(1-\omega t)(1-\omega^2 t)}\right)=\frac{t}{(1-t)(1-t^3)},$$

where $\omega=(-1+\sqrt{-3})/2$.

Example 19.2.3 In the case $n=3$, $[U(0)]=2[L(0)]+[L(1)]+[L(1,1)]$, $[U(1)]=[L(0)]+3[L(1)]+2[L(1,1)]$, $[U(1,1)]=[L(0)]+2[L(1)]+3[L(1,1)]$ and $[U(2,1)]=[L(2,1)]$. The congruence classes C_1,C_3,C_7,C_7' have $1,56,24,24$ elements respectively. Using the Brauer character table (18.1), Proposition 19.2.1 gives the formula

$$\frac{1}{168}\left(\frac{8}{(1-t)^3}+\frac{56\cdot 2}{1-t^3}+\frac{24\cdot 1}{1-\gamma t+\overline{\gamma}t^2-t^3}+\frac{24\cdot 1}{1-\overline{\gamma}t+\gamma t^2-t^3}\right)$$

for $P(\mathsf{P}(3),L(0),t)$, where $\gamma=(-1+\sqrt{-7})/2$. Using the relations $\gamma+\overline{\gamma}=-1$ and $\gamma\overline{\gamma}=2$, this reduces to $(1-t+t^4-t^7+t^8)/(1-t)(1-t^3)(1-t^7)$. Similar calculations give the Poincaré series for $L(1)$, $L(1,1)$ and $L(2,1)$ as in Example 19.1.7.

We next relate the Poincaré series of the summands $\mathsf{P}(n,\lambda)$ and $\mathsf{P}(n,\lambda^*)$ of $\mathsf{P}(n)$, where $L(\lambda^*)$ is the contragredient dual of $L(\lambda)$. By Proposition 17.6.5, $\lambda_i^*=\lambda_1-\lambda_{n+1-i}$ for $1\le i\le n$. Using Proposition 19.2.1 we have

$$P(\mathsf{P}(n,\lambda),t)=\frac{1}{|\mathsf{GL}(n)|}\sum_A\frac{\mathrm{br}(U(\lambda),A^{-1})}{\prod_{i=1}^n(1-\alpha_i(A)t)}$$

where $U(\lambda)$ is the principal indecomposable module corresponding to $L(\lambda)$, and the sum is over all elements A of odd order in $\mathsf{GL}(n)$.

As in Theorem 19.1.4, we write $w(n)=\deg q_n(t)=2^{n+1}-n-2$, where $q_n(t)=\prod_{i=1}^n(1-t^{2^i-1})$, and we observe that $t^{w(n)}q_n(t^{-1})=(-1)^n q_n(t)$.

Proposition 19.2.4 *Let λ be a column 2-regular partition of length $<n$. Then the numerators of the Poincaré series of* $\mathsf{P}(n,\lambda)$ *and* $\mathsf{P}(n,\lambda^*)$ *are related by the equation $p_n(\lambda^*,t)=t^m p_n(\lambda,t^{-1})$, where $m=2w(n-1)=2^{n+1}-2n-2$.*

Proof With sums over elements A of odd order in $\mathsf{GL}(n)$,

$$P(\mathsf{P}(n,\lambda^*),t) = \frac{1}{|\mathsf{GL}(n)|}\sum_A \frac{\mathrm{br}(U(\lambda^*),A)}{\prod_{i=1}^n(1-\alpha_i(A^{-1})t)}$$

since A^{-1} and A have the same order. Since $A \mapsto (A^{\mathrm{tr}})^{-1}$ is an automorphism of $\mathsf{GL}(n)$, $\mathrm{br}(U(\lambda^*),A) = \mathrm{br}(U(\lambda),A^{-1})$. As

$$P(\mathsf{P}(n,\lambda),t^{-1}) = \frac{1}{|\mathsf{GL}(n)|}\sum_A \frac{\mathrm{br}(U(\lambda),A^{-1})}{\prod_{i=1}^n(1-\alpha_i(A)t^{-1})}$$

and $\prod_{i=1}^n(1-\alpha_i(A^{-1})t) = (-t)^n\prod_{i=1}^n(1-\alpha_i(A)t^{-1})$, it follows that $p(\lambda,t^{-1}) = (-t)^n p(\lambda^*,t)$. The result follows, since $p_n(\lambda^*,t) = q_n(t)p(\lambda^*,t)$ and $p_n(\lambda,t^{-1}) = q_n(t^{-1})p(\lambda,t^{-1})$, using the fact that $t^{w(n)}q_n(t^{-1}) = (-1)^n q_n(t)$. □

We can use Proposition 19.2.4 to simplify the calculation of the polynomials $p_n(\lambda,t)$. It is sufficient to determine the coefficients of the Poincaré series in degrees $\leq w(n-1)$, and so we need only determine the number of composition factors of each isomorphism type in $\mathsf{P}^d(n)$ for $d \leq w(n-1) = 2^n - n - 1$.

Example 19.2.5 In the case $n = 3$, $L(1) \cong \mathsf{V}(3)$ and $L(1,1) \cong \mathsf{V}(3)^*$ are contragredient duals, and $p_3((1),t) = t - t^4 + t^5 + t^6$, $p_3((1,1),t) = t^2 + t^3 - t^4 + t^7 = t^8(t^{-1} - t^{-4} + t^{-5} + t^{-6})$, while $L(0) \cong \mathsf{I}(3)$ and $L(2,1) \cong \mathsf{St}(3)$ are self-dual and $p_3(\lambda,t)$ is 'palindromic' for $\lambda = (0)$ and $\lambda = (2,1)$, i.e. $p_3(\lambda,t) = t^8 p_3(\lambda,t^{-1})$.

The number of composition factors of $\mathsf{P}^d(3)$ for $d \leq 4$ of each isomorphism class $L(\lambda)$ is given by

$\lambda \setminus d$	0	1	2	3	4
(0)	1	0	0	1	1
(1)	0	1	1	1	1
(1,1)	0	0	1	2	1
(2,1)	0	0	0	0	1

This table can be calculated using the lifted eigenvalues for the conjugacy classes C_1, C_3, C_7, C_7' of $\mathsf{GL}(3)$, which are $(1,1,1)$, $(1,\omega,\omega^2)$, $(\varepsilon,\varepsilon^2,\varepsilon^4)$ and $(\varepsilon^3,\varepsilon^5,\varepsilon^6)$ respectively, where $\omega = \exp(2\pi i/3)$ and $\varepsilon = \exp(2\pi i/7)$. We evaluate the complete symmetric function h_d for these eigenvalues and use the Brauer character table (18.1) to find the composition factors of $\mathsf{P}^d(3)$. The columns of this table give the coefficients a_i for $0 \leq i \leq 4$ in the series $P(\mathsf{P}(n,\lambda),t) = \sum_{i\geq 0} a_i t^i$. Since $P(\mathsf{P}(n,\lambda),t) = p_3(\lambda,t)/q_3(t)$ where $q_3(t) =$

$(1-t)(1-t^3)(1-t^7)$, $p_3(\lambda,t)$ has the form $p_3(\lambda,t) = a_0 + (a_1 - a_0)t + (a_2 - a_1)t^2 + (a_3 - a_2 + a_0)t^3 + (a_4 - a_3 - a_1 + a_0)t^4 +$ higher terms. Thus $p_3((0),t) = 1 - t + t^4 + \cdots$, $p_3((1),t) = t - t^4 + \cdots$, $p_3((1,1),t) = t^2 + t^3 - t^4 + \cdots$, $p_3((2,1),t) = t^4 + \cdots$, and the calculation is completed using Proposition 19.2.4.

Example 19.2.6 We show that e_{H_3} is a primitive idempotent in $\mathbb{F}_2\mathsf{GL}(3)$ associated to the irreducible $\mathsf{GL}(3)$-module $L(0) = \mathsf{I}(3)$, where H_3 is the subgroup of order 21 in $\mathsf{GL}(3)$ defined in Section 15.5. If $e_{H_3} = e + f$ where e and f are orthogonal idempotents in $\mathbb{F}_2\mathsf{GL}(3)$, then $\mathsf{P}(3)e_{H_3} = \mathsf{P}(3)e \oplus \mathsf{P}(3)f$, and so the Poincaré series of $\mathsf{P}(3)e_{H_3}$ is the sum of the Poincaré series for $\mathsf{P}(3)e$ and $\mathsf{P}(3)f$. By Examples 19.1.7 and 19.2.5, if e is such an idempotent then the Poincaré series of $\mathsf{P}(3)e$ is $(1 - t + t^4 - t^7 + t^8)/(1-t)(1-t^3)(1-t^7)$. By Proposition 15.5.1, the Poincaré series of $\mathsf{P}(3)^{H_3}$ is $(1 + t^3 + t^5 + t^6 + t^8 + t^9 + t^{11} + t^{14})/(1-t^4)(1-t^6)(1-t^7)$. Since these series are equal, $\mathsf{P}(3)e_{H_3} = \mathsf{P}(3)e$. It follows that e_{H_3} is primitive. Hence $\mathbb{F}_2$-basis elements of $\mathsf{P}(3)^{H_3}$ correspond to occurrences of $\mathsf{I}(3)$ as composition factors in $\mathsf{P}(3)$.

19.3 The coinvariant algebra $\mathsf{C}(n)$

Recall from Chapter 15 that the Dickson algebra $\mathsf{D}(n)$ is the polynomial algebra over $\mathbb{F}_2$ generated by the Dickson invariants $\mathsf{d}_{n,i}$, $0 \le i \le n-1$, where $\mathsf{d}_{n,i}$ is a homogeneous polynomial of degree $2^n - 2^i$.

Definition 19.3.1 Let $\mathsf{D}(n)^+$ be the set of all $\mathsf{GL}(n)$-invariant polynomials with zero constant term, i.e. the maximal ideal in $\mathsf{D}(n)$ generated by $\mathsf{d}_{n,i}$, $0 \le i \le n-1$. The quotient $\mathsf{C}(n) = \mathsf{P}(n)/\mathsf{D}(n)^+\mathsf{P}(n)$ is the algebra of **coinvariants** $\mathsf{C}(n)$.

We begin by showing that $\mathsf{C}(n)$ has dimension $|\mathsf{GL}(n)| = \prod_{i=0}^{n-1}(2^n - 2^i)$ as a vector space over $\mathbb{F}_2$, and giving a basis for it.

Proposition 19.3.2 *A vector space basis for the coinvariant algebra* $\mathsf{C}(n)$ *is given by the monomials* $x_1^{a_1}\cdots x_n^{a_n} \in \mathsf{P}(n)$ *such that* $0 \le a_{i+1} < 2^n - 2^i$ *for* $0 \le i \le n-1$.

Proof Let $f \in \mathsf{P}(n)$. By Proposition 15.1.5, f can be written uniquely in the form $f = \sum f_{a_1,\ldots,a_n} x_1^{a_1}\cdots x_n^{a_n}$, where the sum is over all sequences of exponents $(a_1,\ldots,a_n)$ such that $0 \le a_{i+1} < 2^n - 2^i$ for $0 \le i \le n-1$, and the coefficients are in $\mathsf{D}(n)$. Hence the equivalence class of f in $\mathsf{P}(n)/\mathsf{D}(n)^+\mathsf{P}(n)$ has a unique representative as a sum of the monomials $x_1^{a_1}\cdots x_n^{a_n}$. □

Proposition 19.3.3 *The coinvariant algebra* $\mathsf{C}(n)$ *has dimension* $|\mathsf{GL}(n)| = \prod_{i=0}^{n-1}(2^n - 2^i)$ *as a vector space over* $\mathbb{F}_2$. *As a graded algebra the Poincaré polynomial of* $\mathsf{C}(n)$ *is* $P(\mathsf{C}(n),t) = P(\mathsf{P}(n),t)/P(\mathsf{D}(n),t) = (1-t)^{-n}\prod_{i=0}^{n-1}(1-t^{2^n-2^i})$. □

The ideal $\mathsf{D}(n)^+$ is a $\mathbb{F}_2\mathsf{GL}(n)$-submodule of $\mathsf{P}(n)$, since if $f = \mathsf{d}_{n,i}g$ is a polynomial divisible by a Dickson invariant, then for all $A \in \mathsf{GL}(n)$, $f \cdot A = (\mathsf{d}_{n,i} \cdot A)(g \cdot A) = \mathsf{d}_{n,i}(g \cdot A)$ is divisible by the same Dickson invariant. Since $Sq^r(\mathsf{d}_{n,i}) \in \mathsf{D}(n)$, $Sq^k(\mathsf{d}_{n,i}g) = \sum_{r+s=k} Sq^r(\mathsf{d}_{n,i})Sq^s(g)$ is again in $\mathsf{D}(n)^+$, and so $\mathsf{D}(n)^+$ is also an A_2-submodule of $\mathsf{P}(n)$. Hence the quotient algebra $\mathsf{C}(n)$ has commuting actions of $\mathsf{GL}(n)$ and A_2, inherited from the corresponding actions on $\mathsf{P}(n)$. The next result shows that the multiplicity of each irreducible module $L(\lambda)$ as a composition factor in each degree for $\mathsf{P}(n)$ is determined by the corresponding information for $\mathsf{C}(n)$.

Recall from Definition 18.6.1 that the Poincaré series of a graded $\mathbb{F}_2\mathsf{GL}(n)$-module $M = \sum_{d\geq 0} M^d$ is $\Pi(M,t) = \sum_{d\geq 0}[M^d]t^d$, where $[M^d]$ is the equivalence class of M^d in the representation ring $R_2(\mathsf{GL}(n))$.

Proposition 19.3.4 *The graded* $\mathbb{F}_2\mathsf{GL}(n)$*-modules* $\mathsf{P}(n)$ *and* $\mathsf{D}(n) \otimes \mathsf{C}(n)$ *have the same composition factors in each degree, i.e.*

$$\Pi(\mathsf{P}(n),t) = \Pi(\mathsf{D}(n),t)\ \Pi(\mathsf{C}(n),t).$$

Proof Let f_i, $1 \leq i \leq |\mathsf{GL}(n)|$ be homogeneous polynomials in $\mathsf{P}(n)$ which are coset representatives for a $\mathbb{F}_2$-basis of $\mathsf{C}(n)$, and for each exponent sequence $I = (i_0, \ldots, i_{n-1})$ of integers ≥ 0 let $\mathsf{d}_n^I = \mathsf{d}_{n,0}^{i_0} \cdots \mathsf{d}_{n,n-1}^{i_{n-1}}$ be the Dickson monomial indexed by I. Then the products $f_i\mathsf{d}_n^I$ form a vector space basis for $\mathsf{P}(n)$, and so $\mathsf{P}^d(n)$ has a basis given by the union over $j \geq 0$ of the sets of products $f_i\mathsf{d}_n^I$, where $\deg f_i = j$ and $\deg \mathsf{d}_n^I = d - j$. The products with $j = 0$ span a submodule which is a direct sum of copies of the trivial module $\mathsf{I}(n)$. More generally, for $k \geq 0$ the products $f_i\mathsf{d}_n^I$ such that $\deg f_i = j \leq k$ and $\deg \mathsf{d}_n^I = d - j \geq d - k$ span a submodule S_j of $\mathsf{P}^d(n)$. This gives a filtration of $\mathsf{P}^d(n)$ with quotients S_j/S_{j-1} isomorphic to a direct sum of copies of $\mathsf{C}^j(n)$ indexed by Dickson monomials of degree $d - j$. □

The Poincaré series of $\mathsf{D}(n)$, $\Pi(\mathsf{D}(n),t) = \prod_{i=0}^{n-1} 1/(1 - t^{2^n-2^i}) = P(\mathsf{D}(n),t)$ since $[\mathsf{I}(n)]$ is the identity element of $R_2(\mathsf{GL}(n))$. Hence we can use Proposition 19.3.4 to determine either $\Pi(\mathsf{P}(n),t)$ or $\Pi(\mathsf{C}(n),t)$ when the other is known. In the notation of Section 19.1,

$$\Pi(\mathsf{P}(n),t) = \sum_{\lambda}[L(\lambda)]P(\mathsf{P}(n,\lambda),t) = \sum_{\lambda}[L(\lambda)]p_n(\lambda,t)/q_n(t)$$

where the sum is over all column 2-regular partitions λ of length $\leq n-1$. Now setting $T = t^{2^i-1}$, $T^{2^k} = t^{2^{k+i}-2^k}$, and so $(1-T^{2^r})/(1-T) = \sum_{j=0}^{2^r-1} T^j = \prod_{k=0}^{r-1}(1+T^{2^k}) = \prod_{k=0}^{r-1}(1+t^{2^{k+i}-2^k})$. Hence

$$\frac{1}{q_n(t)P(\mathsf{D}(n),t)} = \prod_{i=1}^{n} \frac{1-t^{2^n-2^{n-i}}}{1-t^{2^i-1}} = \prod_{i=1}^{n} \prod_{k=0}^{n-i-1} (1+t^{2^{k+i}-2^k}) = \prod_a (1+t^a),$$

where the product is over all 2-atomic numbers $a < 2^{n-1}$ (Definition 12.5.1). By (12.5) this is the Poincaré polynomial of the subalgebra $\mathsf{A}_2(n-2)$ of A_2. Hence

$$\Pi(\mathsf{C}(n),t) = P(\mathsf{A}_2(n-2),t)\sum_\lambda [L(\lambda)]p_n(\lambda,t). \tag{19.7}$$

Example 19.3.5 By Example 19.1.5, $p_2((0),t) = 1-t+t^2$ and $p_2((1),t) = t$, and $P(\mathsf{A}_2(0),t) = 1+t$. Hence $\Pi(\mathsf{C}(2),t) = (1+t)((1-t+t^2)[L(0)] + tL[(1)]) = (1+t^3)[L(0)] + (t+t^2)[L(1)]$. In other words, $\mathsf{C}(2)$ has a copy of the trivial module $L(0)$ in degrees 0 and 3, and a copy of the defining module $L(1)$ in degrees 1 and 2. This can also be shown directly by counting composition factors of $\mathsf{P}^d(2)$ for $d \leq 3$. Then we can use Proposition 19.3.4 to determine the composition factors of $\mathsf{P}^d(2)$ for all $d \geq 0$. This gives a method for determining these composition factors without using Brauer characters.

Since $P(\mathsf{C}(n),t) = \prod_{i=1}^{n}(1-t^{2^n-2^i})/(1-t)$ has degree $(n-1)(2^n-1)$, the argument of Example 19.3.5 shows that the composition factors of $\mathsf{P}^d(n)$ for all $d \geq 0$ are determined by those for $d \leq (n-1)(2^n-1)$.

Example 19.3.6 For $n=3$, we can use the results of Example 19.1.7 to calculate $\Pi(\mathsf{C}(3),t)$ by the same method. For $\lambda = (0)$, (1), $(1,1)$ and $(2,1)$ we multiply the polynomials $p_3(\lambda,t)$ of Example 19.1.7 by $P(\mathsf{A}_2(1),t) = (1+t)(1+t^2)(1+t^3)$ to obtain $\Pi(\mathsf{C}(3),t)$. The table below gives the multiplicity of each $L(\lambda)$ in $\mathsf{C}^d(3)$ for all d. The last column gives the total number of occurrences of $L(\lambda)$ in $\mathsf{C}(3)$. By Theorem 19.3.7, this is also the multiplicity of $L(\lambda)$ in the regular representation of $\mathsf{GL}(3)$.

$\lambda \setminus d$	0	1	2	3	4	5	6	7	8	9	10	11	12	13	14	
$L(0)$	1	0	0	1	0	1	1	0	1	1	0	1	0	0	1	8
$L(1)$	0	1	1	1	1	1	2	1	2	2	1	2	1	0	0	16
$L(1,1)$	0	0	1	2	1	2	2	1	2	1	1	1	1	1	0	16
$L(2,1)$	0	0	0	0	1	1	1	2	1	1	1	0	0	0	0	8

The following result implies that every irreducible $\mathbb{F}_2\mathsf{GL}(n)$-module $L(\lambda)$ appears as a composition factor in $\mathsf{P}^d(n)$ for infinitely many degrees d.

Theorem 19.3.7 *In the representation ring $R_2(\mathsf{GL}(n))$, $[\mathsf{C}(n)] = [\mathbb{F}_2\mathsf{GL}(n)]$, the equivalence class of the regular representation.*

Proof We use Brauer characters. By a standard result of modular representation theory (see Section 19.6), the multiplicity of $L(\lambda)$ in $\mathbb{F}_2\mathsf{GL}(n)$ is the dimension of the projective indecomposable module $U(\lambda)$ associated to $L(\lambda)$. Thus we have to show that the same is true for $\mathsf{C}(n)$. By (19.3) the Brauer character of $\mathsf{P}(n)$ is given by

$$\mathrm{br}(\mathsf{P}(n),A) = \prod_{i=1}^{n} \frac{1}{1-\alpha_i(A)t},$$

where A is an element of odd order in $\mathsf{GL}(n)$ and $\alpha_i(A)$, $1 \le i \le n$ are the eigenvalues of A, lifted to $\mathbb{C}$. Applying Proposition 19.2.1 with $L = L(\lambda)$, we obtain

$$P(\mathsf{P}(n,\lambda),t) = \frac{1}{|\mathsf{GL}(n)|}\sum_{A} \frac{\mathrm{br}(U(\lambda),A^{-1})}{\prod_{i=1}^{n}(1-\alpha_i(A)t)},$$

where the sum is over all odd order elements $A \in \mathsf{GL}(n)$. By Proposition 19.3.4, $P([\mathsf{C}(n)],L(\lambda),t) = P(\mathsf{P}(n,\lambda),t)\cdot\prod_{i=1}^{n}(1-t^{2^n-2^i})$, and so

$$P([\mathsf{C}(n)],L(\lambda),t) = \frac{1}{|\mathsf{GL}(n)|}\sum_{A} \mathrm{br}(U(\lambda),A^{-1})\cdot\prod_{i=1}^{n}\frac{1-t^{2^n-2^i}}{1-\alpha_i(A)t}.$$

Since $\mathsf{C}(n)$ is finite dimensional and $P([\mathsf{C}(n)],L(\lambda),t) = \sum_{d\ge 0} m_d t^d$, where m_d is the multiplicity of $L(\lambda)$ in $\mathsf{C}^d(n)$, the multiplicity of $L(\lambda)$ in $\mathsf{C}(n)$ is given by evaluating this series at $t = 1$. If $A \ne I_n$ then A is diagonalizable over some finite extension of $\mathbb{F}_2$ since it has odd order, and so has at least one eigenvalue $\alpha_i(A) \ne 1$. Hence $\prod_{i=1}^{n}(1-t^{2^n-2^i})/(1-\alpha_i(A)t) = 0$ when $t = 1$. Hence the multiplicity of $L(\lambda)$ in $\mathsf{C}(n)$ is $(1/|\mathsf{GL}(n)|)\mathrm{br}(U(\lambda),1)\prod_{i=0}^{n-1}(2^n-2^i) = \mathrm{br}(U(\lambda),1) = \dim U(\lambda)$ as required, since $|\mathsf{GL}(n)| = \prod_{i=0}^{n-1}(2^n-2^i)$. □

Note that Theorem 19.3.7 gives no information about the grading of the composition factors of $\mathsf{C}(n)$ which are isomorphic to a given $L(\lambda)$.

19.4 B(*n*)-invariants and irreducible modules

In Chapter 17 a complete set of irreducible $\mathbb{F}_2\mathsf{GL}(n)$-modules $L(\lambda)$ was constructed, both as quotient modules and as submodules of the flag module

$\mathsf{FL}(n)$. In this section we give a construction of the modules $L(\lambda)$ as submodules of $\mathsf{P}(n)$. We begin by finding the invariants of the lower triangular subgroup $\mathsf{B}(n)$ in $\mathsf{P}(n)$. By Proposition 15.1.3, the Dickson invariants are the nonzero coefficients of the polynomial $f_n(t) = \prod_{x \in \mathsf{P}^1(n)}(x+t) = \sum_{i=0}^{n} \mathsf{d}_{n,i} t^{2^i}$, where $\mathsf{d}_{n,n} = 1$.

Definition 19.4.1 For $1 \le k \le n$, the kth **Mui invariant** V_k is the product $f_{k-1}(x_k) = \prod(x + x_k) = \sum_{i=0}^{k} \mathsf{d}_{k-1,i} x_k^{2^i}$, where the product is over all $x \in \mathsf{P}^1(k-1)$.

Thus V_k is a homogeneous polynomial of degree 2^{k-1} in $x_1,\dots,x_k$. When $n = 3$, for example, the Mui invariants in $\mathbb{F}_2[x,y,z]$ are $V_1 = x$, $V_2 = y(x+y)$ and $V_3 = z(x+z)(y+z)(x+y+z) = z^4 + \mathsf{d}_{2,1}z^2 + \mathsf{d}_{2,0}z$. The top Dickson invariant in $\mathsf{P}(n)$ is the product $\mathsf{d}_{n,0} = V_1 V_2 \cdots V_n$ of the Mui invariants.

Theorem 19.4.2 *The Mui invariants* $V_1,\dots,V_n$ *are algebraically independent in* $\mathsf{P}(n)$, *and the ring of invariants of* $\mathsf{B}(n)$ *in* $\mathsf{P}(n)$ *is the* **Mui algebra**, *the polynomial algebra* $V(n) = \mathbb{F}_2[V_1,\dots,V_n]$.

Proof Since an element of $\mathsf{B}(n)$ permutes the factors $a_1x_1 + \ldots + a_{k-1}x_{k-1} + x_k$ of V_k, V_k is $\mathsf{B}(n)$-invariant. For the same reason, if $f \in \mathsf{P}(n)$ is $\mathsf{B}(n)$-invariant and x_k divides f, then all the factors of V_k must divide f, and so V_k divides f.

We show by induction on n that the Mui invariants $V_1,\dots,V_n$ generate the ring of $\mathsf{B}(n)$-invariants in $\mathsf{P}(n)$. The case $n = 1$ is trivial. Thus we assume that the result holds for $\mathsf{P}(n-1)$ and for all $\mathsf{B}(n)$-invariant polynomials in $\mathsf{P}(n)$ of degree $< d$. We may write a $\mathsf{B}(n)$-invariant polynomial $f \in \mathsf{P}^d(n)$ uniquely as $f = g + x_n h$, where $g \in \mathsf{P}(n-1)$. Then if $A \in \mathsf{B}(n-1) \subset \mathsf{B}(n)$, so that $x_n \cdot A = x_n$, we have $f = f \cdot A = g \cdot A + x_n(h \cdot A)$. Then $g \cdot A = g$ by uniqueness of the decomposition of f, and so g is invariant under $\mathsf{B}(n-1)$.

Since g does not involve x_n, g is invariant under $\mathsf{B}(n)$, and hence $x_n h$ is also $\mathsf{B}(n)$-invariant. As observed above, it follows that V_k divides $x_n h$, and so we can write $x_n h = V_k f'$, where $f' \in \mathsf{P}(n)$ is $\mathsf{B}(n)$-invariant and has degree $< d$. By the induction hypothesis, g can be written as a polynomial in $V_1,\dots,V_{n-1}$ and f' as a polynomial in $V_1,\dots,V_n$. This completes the inductive step.

To prove that $V_1,\dots,V_n$ are algebraically independent, we show that the Jacobian determinant $\partial(V_1,\dots,V_n)/\partial(x_1,\dots,x_n) \neq 0$. Since V_k is a polynomial in $x_1,\dots,x_k$, the determinant is lower triangular. Since $V_k = \sum_{i=0}^{k-1} \mathsf{d}_{k-1,i} x_k^{2^i}$, $\partial V_k/\partial x_k = \mathsf{d}_{k-1,0}$, and so $\partial(V_1,\dots,V_n)/\partial(x_1,\dots,x_n) = \prod_{k=1}^{n} \mathsf{d}_{k-1,0} \neq 0$. □

Given a column 2-regular partition λ, let I be the set of parts of the conjugate partition λ^{tr}. Then by Definition 17.6.9 $L(\lambda)$ is the head, or irreducible quotient, of the indecomposable $\mathbb{F}_2\mathsf{GL}(n)$-module $\mathsf{FL}_I(n)$. Let $I = \{i_1,\dots,i_r\}$ where $1 \le i_1 \le \cdots \le i_r \le n-1$, and let $i_0 = 0$, $i_{r+1} = n$. We define $\Delta_I = \Delta_{i_1} \cdots \Delta_{i_r}$, where $\Delta_k = \mathrm{d}_{k,0} = \det(x_j^{2^{i-1}})$, $1 \le i,j \le k$ is the top Dickson invariant for $\mathsf{P}(k)$. Then $\Delta_I \in \mathsf{P}^d(n)$, where $d = \deg_2 \lambda$. We regard Δ_k as a polynomial in all the variables $x_1,\dots,x_n$, although only $x_1,\dots,x_k$ are present. In particular, Δ_k is invariant under all lower triangular transvections for $1 \le i \le n$, and so it is invariant under $\mathsf{B}(n)$. Thus Δ_I is invariant under $\mathsf{B}(n)$. Since Δ_k is the product of all nonzero elements of $\mathsf{P}^1(k)$, $\Delta_k = V_1 \cdots V_k$, the product of the first k Mui invariants. Hence $\Delta_I = V_1^{\lambda_1} \cdots V_{n-1}^{\lambda_{n-1}}$ in $\mathsf{P}^d(n)$, where $d = \deg_2 \lambda$.

The polynomial Δ_I is an example of a *bideterminant*, i.e. a product of determinants of various sizes in $\mathsf{P}(n)$. We shall prove that Δ_I is the unique $\mathsf{B}(n)$-invariant element of the $\mathbb{F}_2\mathsf{GL}(n)$-submodule $\langle \Delta_I \rangle$ of $\mathsf{P}^d(n)$ that it generates. To do this, we show that Δ_I is the unique $\mathsf{B}(n)$-invariant element in the submodule of $\mathsf{P}^d(n)$ spanned by a family of similar bideterminants obtained by independent substitutions of variables in the factors of Δ_I.

Definition 19.4.3 Given $I = \{i_1,\dots,i_r\} \subseteq Z[n-1]$ and λ as above with $d = \deg_2 \lambda$, the **dual Weyl module** $\nabla(\lambda,n)$ is the $\mathbb{F}_2\mathsf{GL}(n)$-submodule of $\mathsf{P}^d(n)$ spanned as a vector space over $\mathbb{F}_2$ by the bideterminants $\Delta_0(J_1)\cdots\Delta_0(J_r)$ in $\mathsf{P}^d(n)$, where $J_1,\dots,J_r$ are subsets of $Z[n-1]$ such that $|J_1| = i_1,\dots,|J_r| = i_r$, where $\Delta_0(I)$ is the Vandermonde determinant

$$\Delta_0(I) = \begin{vmatrix} x_{i_1} & x_{i_2} & \cdots & x_{i_r} \\ x_{i_1}^2 & x_{i_2}^2 & \cdots & x_{i_r}^2 \\ \vdots & \vdots & \ddots & \vdots \\ x_{i_1}^{2^{r-1}} & x_{i_2}^{2^{r-1}} & \cdots & x_{i_r}^{2^{r-1}} \end{vmatrix}. \tag{19.8}$$

Example 19.4.4 Let $n = 4$ and $\lambda = (2,1,1)$, so that $I = \{1,3\}$. Then $\nabla(\lambda,n)$ is the subspace of $\mathsf{P}^8(4)$ spanned by the 16 bideterminants

$$\Delta_0(J_1)\Delta_0(J_2) = x_i \begin{vmatrix} x_j & x_k & x_\ell \\ x_j^2 & x_k^2 & x_\ell^2 \\ x_j^4 & x_k^4 & x_\ell^4 \end{vmatrix}$$

where $J_1 = \{i\}, J_2 = \{j,k,\ell\} \subseteq Z[4]$. It is the cyclic $\mathbb{F}_2\mathsf{GL}(4)$-module generated by any of the 4 bideterminants which involve all four variables. As the sum of these 4 elements is 0, $\dim \nabla(\lambda,n) = 15$. The submodule generated by the bideterminant $\Delta_I = x_1\mathrm{d}_{3,0}$ has dimension 14 and is isomorphic to $L(\lambda)$.

Proposition 19.4.5 *Δ_I is the unique nonzero $\mathsf{B}(n)$-invariant in $\nabla(\lambda,n)$.*

Proof Since the maximum degree of any variable x_j in $\Delta_0(J)$ is $2^{|J|-1}$, the degree of x_j in a polynomial $f \in \nabla(\lambda,n)$ is $\le 2^{i_1-1} + \cdots + 2^{i_r-1} < 2^{i_r}$.

If f is $\mathsf{B}(n)$-invariant, it follows that $f \in \mathbb{F}_2[V_1,\dots,V_{i_r}]$, since x_j has 2^{j-1} in V_j. Hence f is a polynomial in $x_1,\dots,x_{i_r}$ only.

Let π be the projection of $\nabla(\lambda,n)$ which maps x_j to 0 for all $j > i_r$. By the preceding remarks, $\pi(f) = f$. Since π maps all bideterminants involving variables x_j with $j > i_r$ to 0, f can be written as a sum of bideterminants involving only $x_1,\dots,x_{i_r}$. Since $|J_r| = i_r$, $J_r = \{1,\dots,i_r\}$, and so all these bideterminants have a factor Δ_{i_r}. Hence $f = \Delta_{i_r} g$, where g is a $\mathsf{B}(n)$-invariant, and is a sum of bideterminants corresponding to $I' = \{i_1,\dots,i_{r-1}\} \subset I$. It follows by induction on r that $g = \Delta_{I'}$, and so $f = \Delta_I$. □

Proposition 19.4.6 *The cyclic module* $\langle \Delta_I \rangle$ *is an irreducible* $\mathbb{F}_2\mathsf{GL}(n)$*-module.*

Proof This follows from Proposition 19.4.5 using the criterion of Proposition 10.1.3. □

Proposition 19.4.7 *The modules* $\langle \Delta_I \rangle$, $I \subseteq Z[n-1]$, *are a complete set of* 2^{n-1} *non-equivalent irreducible* $\mathbb{F}_2\mathsf{GL}(n)$*-modules.*

Proof Suppose that there exists an isomorphism $\iota : \langle \Delta_I \rangle \to \langle \Delta_{I'} \rangle$, where $I, I' \subseteq Z[n-1]$. Then $\iota(\Delta_I) = \Delta_{I'}$, since these are the unique $\mathsf{B}(n)$-invariant elements in the two modules. We consider the effect on Δ_I of the switch S_k of adjacent variables x_k and x_{k+1} for $1 \le k \le n-1$. If $k \in I$, then there is a factor of Δ_I which involves x_k but not x_{k+1}, and so $S_k(\Delta_I) \neq \Delta_I$. On the other hand, if $k \notin I$, then every factor of Δ_I involves both x_k and x_{k+1} or neither, and so $S_k(\Delta_I) = \Delta_I$. Since ι commutes with S_k for all k, the equation $\iota(\Delta_I) = \Delta_{I'}$ implies that $I = I'$. □

We shall show that $\langle \Delta_I \rangle \cong L(\lambda)$, thus relating the above construction of the irreducible $\mathbb{F}_2\mathsf{GL}(n)$ modules to the construction in Section 17.6.

Definition 19.4.8 Let $\theta_\lambda : \mathsf{FL}(n) \to \mathsf{P}^d(n)$ be the $\mathbb{F}_2\mathsf{GL}(n)$-module map defined by $\theta_\lambda(\mathsf{B}(n)A) = \Delta_I(u_1,\dots,u_n)$, where $u_1,\dots,u_n$ are the rows of $A \in \mathsf{GL}(n)$.

Thus $\theta_\lambda(\mathsf{B}(n)A)$ is the bideterminant obtained by substituting u_i for x_i in Δ_I for $1 \le i \le n$. The map θ_λ is well defined, because $\Delta_I(u_1,\dots,u_n)$ is invariant under the action of the lower triangular subgroup $\mathsf{B}(n)$ on the representative matrix A for the flag. Clearly θ_λ is map of $\mathbb{F}_2\mathsf{GL}(n)$ modules, and the image of θ_λ is contained in $\langle \Delta_I \rangle$. Since $\theta_\lambda \neq 0$, it follows from Proposition 19.4.6 that the image of θ_λ is $\langle \Delta_I \rangle$.

Proposition 19.4.9 θ_λ *defines a surjection from the indecomposable summand* $\mathsf{FL}_I(n)$ *of* $\mathsf{FL}(n)$ *on to* $\langle \Delta_I \rangle$.

Proof We show that θ_λ is nonzero on the partial flag module $\mathsf{FL}^I(n) \subseteq \mathsf{FL}(n)$. Let R^I be the reference partial flag of type I, i.e. the flag consisting of the subspaces $\mathsf{V}(i)$ for $i \in I$. Then R^I generates the cyclic module $\mathsf{FL}^I(n)$, and the image of R^I under the standard embedding of $\mathsf{FL}^I(n)$ in $\mathsf{FL}(n)$ is the sum of the complete flags X extending R^I. The number of such flags X is odd, since it is the index of $\mathsf{B}(n)$ in the parabolic subgroup $\mathsf{P}^I(n)$. We claim that $\theta_\lambda(X) = \Delta_I$ for all X. This follows from the fact that each factor Δ_i, $i \in I$ is invariant under $\mathsf{GL}(i) \subset \mathsf{GL}(n)$, and so depends only on the subspace of dimension i in the flag X. Since this subspace is $\mathsf{V}(i)$, $\theta_\lambda(X)$ takes the same value for all X extending R^I, and we may choose X as the reference flag R, for which it is clear that $\theta_\lambda(X) = \Delta_I$. □

Proposition 19.4.10 *Let λ be a column 2-regular partition of length $\leq n-1$, and let I be the set of parts of the conjugate partition λ^{tr}. Then $\langle \Delta_I \rangle \cong L(\lambda)$ as $\mathbb{F}_2\mathsf{GL}(n)$-modules.*

Proof It follows from Proposition 19.4.9 that $\langle \Delta_I \rangle$ is isomorphic to a quotient module of $\mathsf{FL}^I(n)$. Now the possible irreducible quotients of $\mathsf{FL}^I(n)$ are the heads of its indecomposable summands $\mathsf{FL}_J(n)$ where $J \subseteq I$, and these are all distinct. Since the irreducible modules $\langle \Delta_I \rangle$ are also all distinct, we can match them by induction on $|I|$ to see that $\langle \Delta_I \rangle$ is isomorphic to the head $L(\lambda)$ of $\mathsf{FL}_I(n)$. □

19.5 Irreducible submodules of P(*n*)

In this section we consider the irreducible $\mathbb{F}_2\mathsf{GL}(n)$-submodules of $\mathsf{P}(n)$. We recall from Definition 1.2.6 that $S_{i,j}$ denotes the matrix which switches the variables x_i and x_j, and $T_{i,j}$ denotes the transvection which adds x_j to x_i.

Proposition 19.5.1 *Let f be a homogeneous polynomial in the Mui algebra $V(n) = \mathbb{F}_2[V_1, \dots, V_n]$. If $f \cdot T_{i,j} = f + f \cdot S_{i,j}$ for some i and j such that $1 \leq i < j \leq n$, then V_i divides f.*

Proof Let $g(u,v) = f(x_1, \dots, x_{i-1}, u, x_{i+1}, \dots, x_{j-1}, v, x_{j+1}, \dots, x_n)$. Since $f \cdot T_{i,j} = f + f \cdot S_{i,j}$, we have $g(u+v, v) = g(u,v) + g(v,u)$, which we regard as a polynomial identity in variables u, v with coefficients involving x_k, $k \neq i,j$. Setting $u = 0$ gives $g(v,v) = g(0,v) + g(v,0)$. Next, since $f \in V(n)$ and $i < j$, f is invariant under $T_{j,i}$, we have $g(u, u+v) = g(u,v)$ for all u, v, and setting $u = v$ gives $g(v,0) = g(v,v)$. Hence $g(0,v) = 0$, i.e. $f(x_1, \dots, x_{i-1}, 0, x_{i+1}, \dots, x_n) = 0$. Hence x_i divides f. Since f is $\mathsf{B}(n)$-invariant, it follows that V_i divides f. □

Example 19.5.2 Let $n=4$ and $\lambda=(2,2,1)$, so that $I=\{2,3\}$. Then

$$\Delta_I = \Delta_2\Delta_3 = \begin{vmatrix} x_1 & x_2 \\ x_1^2 & x_2^2 \end{vmatrix} \begin{vmatrix} x_1 & x_2 & x_3 \\ x_1^2 & x_2^2 & x_3^2 \\ x_1^4 & x_2^4 & x_3^4 \end{vmatrix},$$

which we regard as a function of x_1,x_2,x_3,x_4. Applying the transvections $T_{1,3},T_{2,3}$ and $T_{3,4}$, we obtain

$$\Delta_I(x_1+x_3,x_2,x_3,x_4) = \Delta_I(x_1,x_2,x_3,x_4)+\Delta_I(x_3,x_2,x_1,x_4),$$
$$\Delta_I(x_1,x_2+x_3,x_3,x_4) = \Delta_I(x_1,x_2,x_3,x_4)+\Delta_I(x_1,x_3,x_2,x_4),$$
$$\Delta_I(x_1,x_2,x_3+x_4,x_4) = \Delta_I(x_1,x_2,x_3,x_4)+\Delta_I(x_1,x_2,x_4,x_3).$$

If $f \in \mathsf{P}^d(4)$ a homogeneous polynomial such that $f \mapsto \Delta_I$ by an isomorphism of $\mathbb{F}_2\mathsf{GL}(4)$-modules, then f also satisfies the above three identities. By Proposition 19.5.1, it follows that f is divisible by V_1, V_2 and V_3, and hence by $V_1V_2V_3 = \Delta_3$.

Proposition 19.5.3 *Let* $\lambda = (\lambda_1,\ldots,\lambda_{n-1})$ *be a column 2-regular partition, and let* $\lambda_n = 0$*. Let* $M \subseteq \mathsf{P}^d(n)$ *be a* $\mathbb{F}_2\mathsf{GL}(n)$*-submodule isomorphic to* $L(\lambda)$*, and let* f *be the unique nonzero* $\mathsf{B}(n)$*-invariant element of* M*. Then* f *is divisible by* Δ_I*, where* I *is the set of parts of* λ^{tr}*.*

Proof We argue by induction on $r=|I|=\lambda_1$. As above, $\Delta_I = \Delta_{i_1}\cdots\Delta_{i_r}$, where $I=\{i_1,\ldots,i_k\}$ and $1 \le i_1 < \cdots < i_r \le n-1$. We first prove that Δ_{i_r} divides f, using the method illustrated by Example 19.5.2.

Let $i_0 = 0$, $i_{r+1} = n$, and for $1 \le i \le i_r$, choose $j > i$ such that $i_s < j \le i_{s+1}$ if $i_{s-1} < i \le i_s$. Then $\Delta_{i_s}(x_1,\ldots,x_i,\ldots,x_n)\cdot T_{i,j} = \Delta_{i_s}(x_1,\ldots,x_i + x_j,\ldots,x_n) = \Delta_{i_s}(x_1,\ldots,x_i,\ldots,x_n) + \Delta_{i_s}(x_1,\ldots,x_j,\ldots,x_n)$ and $\Delta_{i_k}\cdot T_{i,j} = \Delta_{i_k}$ for $k \neq s$. Hence $\Delta_I(x_1,\ldots,x_i,\ldots,x_n)\cdot T_{i,j} = \Delta_I(x_1,\ldots,x_i+x_j,\ldots,x_n) = \Delta_I(x_1,\ldots,x_i,\ldots,x_j,\ldots,x_n) + \Delta_I(x_1,\ldots,x_j,\ldots,x_i,\ldots,x_n)$. Since f maps to Δ_I by an isomorphism of $\mathbb{F}_2\mathsf{GL}(n)$-modules, this identity holds for f in place of Δ_I. By Proposition 19.5.1, it follows that V_i divides f, and since this has been proved for all i such that $1 \le i \le i_r$, it follows that $V_1\cdots V_{i_r} = \Delta_{i_r}$ divides f.

Thus we can write $f = \Delta_{i_r}g$, where g is invariant under $\mathsf{B}(n)$. On restriction to $\mathbb{F}_2\mathsf{GL}(i_r)$-modules, f and g generate isomorphic submodules, since Δ_{i_r} generates the trivial module $\mathsf{I}(i_r)$, and these submodules are both isomorphic to the irreducible $\mathbb{F}_2\mathsf{GL}(i_r)$-module $\langle\Delta_{I'}\rangle$, where $I' = \{i_1,\ldots,i_{r-1}\}$. By the induction hypothesis on r, $\Delta_{I'}$ divides g. Hence Δ_I divides f. □

Since Δ_I has degree $\deg_2 \lambda$, Proposition 19.5.3 determines the first occurrence of each irreducible $\mathbb{F}_2\mathsf{GL}(n)$-module $L(\lambda)$ as a submodule in $\mathsf{P}(n)$, as follows.

Theorem 19.5.4 *Let λ be a column 2-regular partition of length $\leq n-1$. Then $\mathsf{P}^d(n)$ has no $\mathbb{F}_2\mathsf{GL}(n)$-submodule isomorphic to $L(\lambda)$ if $d < \deg_2 \lambda$, and if $d = \deg_2 \lambda$, $\mathsf{P}^d(n)$ has a unique such submodule generated by Δ_I.* □

This result determines the Poincaré series for the Steinberg summand of $\mathsf{P}(n)$.

Proposition 19.5.5 *Let $\lambda = (n-1, n-2, \ldots, 1)$, so that $L(\lambda) = \mathsf{St}(n)$ is the Steinberg module for $\mathbb{F}_2\mathsf{GL}(n)$. Then*

$$P(\mathsf{P}(n,\lambda),t) = \frac{t^{w(n-1)}}{\prod_{i=1}^{n}(1-t^{2^i-1})},$$

where $w(n-1) = \sum_{i=1}^{n-1}(2^i-1) = 2^n - n - 1$.

Proof By Theorem 19.1.4, $P(\mathsf{P}(n,\lambda),t) = p_n(\lambda,t)/\prod_{i=1}^{n}(1-t^{2^i-1})$, where $p_n(\lambda,t)$ is a polynomial of degree $\leq 2w(n-1) = 2^{n+1} - 2n - 2$. By Theorem 19.5.4 the first occurrence of $\mathsf{St}(n)$ as a submodule in $\mathsf{P}(n)$ is in degree $w(n-1) = \deg_2 \lambda^{\mathrm{tr}}$ where $\lambda^{\mathrm{tr}} = \lambda$. By Proposition 18.4.9(ii), the module $\mathsf{St}(n)$ is projective, and hence it is a submodule of every $\mathbb{F}_2\mathsf{GL}(n)$-module in which it is a composition factor. Hence the first occurrence of $\mathsf{St}(n)$ as a composition factor in $\mathsf{P}(n)$ is in degree $w(n-1)$, and so $p_n(\lambda,t)$ has no terms of degree $< w(n-1)$. As $\mathsf{St}(n)$ occurs only once as a submodule of $\mathsf{P}^{w(n-1)}(n)$, the coefficient of $t^{w(n-1)}$ in $p_n(\lambda,t)$ is 1. Finally, since $\mathsf{St}(n)$ is its own contragredient dual, it follows from Proposition 19.2.4 that $p_n(\lambda,t)$ has no terms of degree $> w(n-1)$. Hence $p_n(\lambda,t) = t^{w(n-1)}$. □

19.6 Remarks

For the results of modular representation theory used to prove Theorems 19.1.1 and 19.3.7, see [48, Theorem (83.3)]. Given a finite group G, a splitting field K for G, and a decomposition $1 = e_1 + \cdots + e_m$ of the identity element of KG as the sum of primitive orthogonal idempotents, the number of composition factors of a KG-module M which are isomorphic to a given irreducible KG-module F is the dimension of Me_i as a vector space over K, where $U = e_iKG$ is the principal indecomposable KG-module associated to F. The multiplicity of F in KG is the dimension of U as a vector space over K. In our application, $G = \mathsf{GL}(n)$, $K = \mathbb{F}_2$, $M = \mathsf{P}^d(n)$, $F = L(\lambda)$ and $U = U(\lambda)$. For the orthogonality relations (19.4) for Brauer characters, see ([48, (84.11)]). Our proof of Theorem 19.1.4 follows [26, 139]. Proposition 19.2.1 was proved by S. A. Mitchell [139, (1.5)]. For Molien's theorem in invariant theory, see [190].

The splitting 19.1.1 of $\mathsf{P}(n)$ as an A_2-module using idempotents in $\mathbb{F}_2\mathsf{GL}(n)$ can be realized topologically as a stable splitting of the product $\mathbb{R}P^\infty \times \cdots \times \mathbb{R}P^\infty$ of n copies of infinite real projective space, in which there are $\dim L(\lambda)$ copies of a space $\mathbb{R}P(\lambda)$ such that $H^*(\mathbb{R}P(\lambda);\mathbb{F}_2) = \mathsf{P}(n,\lambda)$. If $\mathbb{F}_2\mathsf{M}(n)$ is used in place of $\mathbb{F}_2\mathsf{GL}(n)$, then each space $\mathbb{R}P(\lambda)$ splits further into two spaces corresponding to the irreducible $\mathbb{F}_2\mathsf{M}(n)$-modules $L(\lambda)$ and $L(\lambda') = L(\lambda)\otimes \det$.

For every irreducible $\mathbb{F}_2\mathsf{M}(n)$-module $L(\lambda)$, where λ is column 2-regular of length $\leq n$, Theorem 19.3.7 implies that $L(\lambda)$ occurs as a composition factor in $\mathsf{P}(n)$, and with the exception of $L(0)$, which occurs only in degree 0, $L(\lambda)$ occurs in $\mathsf{P}^d(n)$ for infinitely many degrees d. It is proved in [28] that $L(\lambda)$ first occurs in degree $d(\lambda) = \deg_2 \lambda^{\mathrm{tr}}$, so that in the notation of Section 19.1, the Poincaré series $P(\mathsf{P}'(n,\lambda),t) = p_n'(\lambda,t)/q_n(t)$ where $d(\lambda)$ is the minimum exponent of the polynomial $p_n(\lambda',t)$. Since a polynomial $v(\lambda)$ which generates such a minimal composition factor cannot be hit, $L(\lambda)$ also occurs as a composition factor in the cohit module $\mathsf{Q}^{d(\lambda)}(n)$. Such a polynomial $v(\lambda)$ is given in [218], together with an operation $\theta_\lambda \in \mathsf{A}_2$ such that $\theta_\lambda(v(\lambda)) = \Delta_I$, the generator of the first submodule occurrence of $L(\lambda)$.

These results were first established in the case of the Steinberg module $\mathsf{St}(n)$ by S. A. Mitchell and S. B. Priddy [141], who used symmetric products to construct the stable summand $\mathbb{R}P(\lambda)$, where $L(\lambda) = \mathsf{St}(n)$. As $\mathbb{R}P^\infty \times \cdots \times \mathbb{R}P^\infty$ is the classifying space of the additive group $(\mathbb{Z}/2)^n = \mathbb{Z}/2 \times \cdots \times \mathbb{Z}/2$ of $\mathsf{V}(n)$, this is an important example in the topological problem of stable splitting of classifying spaces of finite groups. The multiplicity of $\mathsf{St}(n)$ as a composition factor in $\mathsf{P}^d(n)$ was also determined in [141]. We return to this topic in Chapter 22.

A deeper connection between the $\mathsf{M}(n)$-splitting of $\mathsf{P}(n)$ and the structure of A_2 was found by Carlisle and Kuhn [28]: for every column 2-regular partition λ of length $\leq n$, $\mathsf{P}'(n,\lambda)$ is free over the subalgebra $\mathsf{A}_2(k)$ of A_2 if and only if $\lambda_1 > k$.

Proposition 19.2.4 is proved in [30], where some further results on the Poincaré series for occurrences in $\mathsf{P}(n)$ of irreducible modules for $\mathbb{F}_2\mathsf{GL}(n)$ and $\mathbb{F}_2\mathsf{M}(n)$ can also be found. The ring of coinvariants $\mathsf{C}(n)$ is discussed in [190], mainly in the non-modular case, and by Mitchell [139]. Steinberg's basis for $\mathsf{C}(n)$ (Proposition 19.3.2) appears in [199]. Theorem 19.5.4 is proved in [177, 214, 137]. Our proof uses Ton That Tri's construction of a complete set of irreducible $\mathbb{F}_2\mathsf{GL}(n)$-modules as submodules of $\mathsf{P}(n)$ [213]. Proposition 19.5.1 is also due to Tri [214]. Theorem 19.4.2 is due to Huynh Mui [152].

20

The algebraic group $\overline{\mathsf{G}}(n)$

20.0 Introduction

In this chapter we apply the representation theory of the algebraic group $\overline{\mathsf{G}}(n) = \mathsf{GL}(n,\overline{\mathbb{F}}_2)$ to the hit problem for $\mathsf{P}(n)$, where $\overline{\mathbb{F}}_2$ is the algebraic closure of $\mathbb{F}_2$. These applications are given in Sections 20.5 and 20.6, while the rest of the chapter gives an outline of the results on algebraic groups that we use to relate $\overline{\mathsf{G}}(n)$-modules to $\mathbb{F}_2\mathsf{GL}(n)$-modules.

In Section 20.1 we define polynomial representations of $\overline{\mathsf{G}}(n)$ and their formal characters. We use Steinberg's tensor product theorem for algebraic groups to obtain results for $\mathsf{GL}(n)$ by restriction of $\overline{\mathsf{G}}(n)$-modules.

In Section 20.2 we discuss the structure of the polynomial algebra $\overline{\mathsf{P}}(n) = \overline{\mathbb{F}}_2[x_1,\ldots,x_n]$ as a graded representation of $\overline{\mathsf{G}}(n)$. When the action is restricted to $\mathsf{GL}(n) \subseteq \overline{\mathsf{G}}(n)$, the composition factors of $\overline{\mathsf{P}}^d(n)$ correspond to the filtration of $\mathsf{P}^d(n)$ by ω-sequences. We show that the filtration quotients $\mathsf{P}^\omega(n)$ can also be obtained by restriction of tensor products of exterior powers of the defining $\overline{\mathsf{G}}(n)$-module $\overline{\mathsf{V}}(n) \cong \overline{\mathsf{P}}^1(n)$. These modules belong to the important family of 'tilting' modules for $\overline{\mathsf{G}}(n)$.

In Section 20.3 we introduce two standard families of $\overline{\mathsf{G}}(n)$-modules, the Weyl modules $\overline{\Delta}(\lambda,n)$ and their transpose duals $\overline{\nabla}(\lambda,n)$, and consider the 'restricted' modules $\Delta(\lambda,n)$ and $\nabla(\lambda,n)$ obtained by regarding them as $\mathbb{F}_2\mathsf{GL}(n)$-modules. In Section 20.4, we show that when the partition λ is column 2-regular, the restricted Weyl module $\Delta(\lambda,n)$ is indecomposable, with a unique irreducible quotient isomorphic to $L(\lambda)$.

In Sections 20.5 and 20.6 we apply these results to the hit problem. In Section 20.5 we show that when μ is strictly decreasing and $\lambda = \mu^{\mathrm{tr}}$, the cohit module $\mathsf{Q}^\mu(n)$ is isomorphic to the restricted dual Weyl module $\nabla(\lambda,n)$, and the Steenrod kernel $\mathsf{K}^\mu(n)$ to $\Delta(\lambda,n)$. In Section 20.6 we show that the first occurrence of $L(\lambda)$ as a composition factor in $\mathsf{P}(n)$ is in degree $\deg_2\mu$.

This complements Theorem 19.5.4, which is concerned with submodules $L(\lambda)$ rather than composition factors.

20.1 Polynomial representations of $\overline{\mathsf{G}}(n)$

In this section we show that the filtration of $\mathsf{P}^d(n)$ by n-bounded sequences ω such that $\deg_2 \omega = d$ has a natural interpretation in terms of its module structure. To do this, we consider the corresponding situation over $\overline{\mathbb{F}}_2$, the algebraic closure of $\mathbb{F}_2$. If $q = 2^t$ then the finite field $\mathbb{F}_q$ of order q has a unique subfield of order 2^s when s divides t, and we may regard the infinite field $\overline{\mathbb{F}}_2$ as the union (or direct limit) $\bigcup_q \mathbb{F}_q$. We denote by $\overline{\mathsf{G}}(n)$ the group $\mathsf{GL}(n,\overline{\mathbb{F}}_2)$ of invertible $n \times n$ matrices over $\overline{\mathbb{F}}_2$. We show that the corresponding filtration quotients $\overline{\mathsf{P}}^{\omega}(n)$ of the polynomial algebra $\overline{\mathsf{P}}(n) = \overline{\mathbb{F}}_2[x_1,\ldots,x_n]$ are the irreducible composition factors of the module $\overline{\mathsf{P}}^d(n)$ given by the right action of $\overline{\mathsf{G}}(n)$ on homogeneous polynomials of degree d over $\overline{\mathbb{F}}_2$.

The group $\overline{\mathsf{G}}(n)$ is an *algebraic group*. This means that (i) it is an algebraic variety, and (ii) the group operations are polynomial maps. For (i), we introduce a new variable Δ and observe that $\overline{\mathsf{G}}(n)$ can be defined by using the polynomial equation $\det(A) \cdot \Delta = 1$ of degree $n+1$ in n^2+1 variables over $\overline{\mathbb{F}}_2$ in place of the inequality $\det(A) \neq 0$ in the n^2 entries of $A \in \overline{\mathsf{G}}(n)$. For (ii), we observe that the entries of AB are polynomials in the entries of A and B, and that the entries of A^{-1} are polynomials in the entries of A and Δ. For algebraic groups over infinite fields of prime characteristic, there is a theory of root systems, weights and characters analogous to that for Lie groups, in which $\overline{\mathsf{G}}(n)$ is of type A_{n-1}.

In particular, a *polynomial representation* of $\overline{\mathsf{G}}(n)$ of dimension $m \geq 1$ is a homomorphism $\rho : \overline{\mathsf{G}}(n) \to \overline{\mathsf{G}}(m)$ such that the entries of $\rho(A)$ are polynomials in the entries of A. If V is a finite dimensional vector space over $\overline{\mathbb{F}}_2$ with a right action of $\overline{\mathsf{G}}(n)$, then V is a *polynomial right* $\overline{\mathsf{G}}(n)$*-module* if the map ρ obtained by choosing a basis for V is a polynomial representation of $\overline{\mathsf{G}}(n)$.

More generally, ρ is a *rational representation* of $\overline{\mathsf{G}}(n)$ when the entries of $\rho(A)$ are rational functions in the entries of A. The defining representation is the polynomial representation given by the identity map of $\overline{\mathsf{G}}(n)$. The corresponding module $\overline{\mathsf{V}}(n)$ is defined by extending Definition 1.2.1 to coefficients in $\overline{\mathbb{F}}_2$. The trivial 1-dimensional representation and the determinant representation $A \mapsto \det(A)$ are polynomial representations, while the representation $A \mapsto \det^{-1}(A)$ is rational but not polynomial. Every polynomial representation ρ of $\overline{\mathsf{G}}(n)$ gives rise to an infinite sequence of non-isomorphic polynomial representations by multiplying $\rho(A)$ by $\det^k(A)$ for $k \geq 0$.

Let $\overline{\mathsf{P}}(n) = \sum_{d\geq 0} \overline{\mathsf{P}}^d(n)$ denote the polynomial algebra $\overline{\mathbb{F}}_2[x_1,\ldots,x_n]$, graded by degree, with the action of $\overline{\mathsf{G}}(n)$ by linear substitutions of the variables. Then $\overline{\mathsf{P}}^d(n)$ is a polynomial right $\overline{\mathsf{G}}(n)$-module. In particular, $\overline{\mathsf{P}}^1(n)$ gives the defining representation $\overline{\mathsf{V}}(n)$. We can obtain further examples of polynomial representations by forming direct sums, tensor products, submodules and quotients. We denote the kth exterior power of $\overline{\mathsf{V}}(n)$ by $\overline{\Lambda}^k(n)$, and note that $\overline{\Lambda}^n(n) \cong \det$, the determinant representation. Every finite dimensional rational representation is isomorphic to the product of a polynomial representation and a power of $\det^{-1}$, i.e. the entries of $\rho(A)$ are polynomials in the entries of A and the variable $\Delta = \det^{-1}(A)$.

The irreducible polynomial representations $\overline{L}(\lambda)$ of $\overline{\mathsf{G}}(n)$ are in one-one correspondence with partitions λ of length $\leq n$. This correspondence is given by the *highest weight*. To explain this, let $\overline{\mathsf{D}}(n)$ be the algebraic subgroup of diagonal matrices in $\overline{\mathsf{G}}(n)$. The 1-dimensional rational representations of $\overline{\mathsf{D}}(n)$ are given by $A \mapsto a_{1,1}^{\lambda_1}\cdots a_{n,n}^{\lambda_n}$, where $a_{1,1},\ldots,a_{n,n}$ are the diagonal entries of $A \in \overline{\mathsf{D}}(n)$, and $\lambda_1,\ldots,\lambda_n$ are integers.

Given a finite dimensional rational $\overline{\mathsf{G}}(n)$-module V, by restricting the action to $\overline{\mathsf{D}}(n)$ we can write V as the direct sum of its λ-*weight spaces* V_λ, where V_λ is the set of elements $v \in V$ such that $v \cdot A = a_{1,1}^{\lambda_1}\cdots a_{n,n}^{\lambda_n} v$ for all $A \in \overline{\mathsf{D}}(n)$. The sequence λ is a *weight* of V if $V_\lambda \neq 0$. If V is a polynomial $\overline{\mathsf{G}}(n)$-module, then $\lambda_i \geq 0$ for $1 \leq i \leq n$, and we associate to V its *formal character* $\mathrm{ch}(V) = \sum_\lambda (\dim V_\lambda) t_1^{\lambda_1}\cdots t_n^{\lambda_n}$, a polynomial in $t_1,\ldots,t_n$.

For polynomial $\overline{\mathsf{G}}(n)$-modules V and W, $\mathrm{ch}(V) = \mathrm{ch}(W)$ if and only if V and W have the same irreducible composition factors, counting multiplicities, i.e. $[V] = [W]$ in the representation ring $R_2(\overline{\mathsf{G}}(n))$. The formal character also respects sums and products, i.e. $\mathrm{ch}(V \oplus W) = \mathrm{ch}(V) + \mathrm{ch}(W)$ and $\mathrm{ch}(V \otimes W) = \mathrm{ch}(V)\mathrm{ch}(W)$. In particular $\mathrm{ch}(\overline{\Lambda}^k(n)) = e_k$, the kth elementary symmetric function in $t_1,\ldots,t_n$, and $\mathrm{ch}(\overline{\mathsf{P}}^d(n)) = h_d$, the sum of all monomials of degree d in $t_1,\ldots,t_n$. Thus $\mathrm{ch}(V)$ is determined by the dimensions of the weight spaces V_λ when λ is a decreasing sequence, i.e. a partition.

Weights for $\overline{\mathsf{G}}(n)$ can be identified with exponent sequences of monomials in $t_1,\ldots,t_n$, and we order them by dominance. The diagram below shows the dominance order on the weights of the defining module $\overline{L}(1) = \overline{\mathsf{V}}(3)$ for $\overline{\mathsf{G}}(3)$, $\overline{L}(1,1) = \overline{\Lambda}^2(3)$ and the Steinberg module $\overline{L}(2,1)$. These modules restrict to the corresponding irreducible modules for $\mathsf{GL}(3)$.

$$
\begin{array}{ccccccc}
 & & & & & (2,1,0) & \\
 & & & & \swarrow & & \searrow \\
(1,0,0) & & (1,1,0) & & (2,0,1) & & (1,2,0) \\
\downarrow & & \downarrow & & & \searrow \quad \swarrow & \\
(0,1,0)\ , & & (1,0,1)\ , & & & (1,1,1) & \\
\downarrow & & \downarrow & & \swarrow & & \searrow \\
(0,0,1) & & (0,1,1) & & (1,0,2) & & (0,2,1) \\
 & & & & \searrow & & \swarrow \\
 & & & & & (0,1,2) &
\end{array}
$$

Here the $(1,1,1)$ weight space of $\overline{L}(2,1)$ has dimension 2 and the others have dimension 1. Thus the corresponding formal characters are $e_1 = t_1 + t_2 + t_3$, $e_2 = t_1t_2 + t_1t_3 + t_2t_3$ and $t_1^2t_2 + t_1t_2^2 + t_1^2t_3 + t_1t_3^2 + t_2^2t_3 + t_2t_3^2 + 2t_1t_2t_3$.

A polynomial module for $\overline{\mathsf{G}}(n)$ has a *highest weight* if the partially ordered set of its weights has a unique maximal element. If the corresponding weight space V_λ is 1-dimensional, then a nonzero vector in V_λ is called a *highest weight vector*. A module generated by a highest weight vector is called a *highest weight module*, as in the examples above.

We next describe the set of irreducible polynomial $\overline{\mathsf{G}}(n)$-modules. For each partition λ of length $\leq n$, there is a corresponding irreducible polynomial $\overline{\mathsf{G}}(n)$-module $\overline{L}(\lambda)$, which is a highest weight module with highest weight λ. In particular, the kth exterior power $\overline{\Lambda}^k(n)$ is irreducible with highest weight $(1,\ldots,1)$ of length k. All weights ν of $\overline{L}(\lambda)$ satisfy $|\nu| = |\lambda|$, $\mathrm{len}(\nu) \leq n$ and $\nu \preceq \lambda$. Hence the formal character of $\overline{L}(\lambda)$ is of the form

$$\mathrm{ch}(\overline{L}(\lambda)) = m(\lambda) + \sum_{\nu \prec \lambda} a_{\lambda,\nu} m(\nu), \tag{20.1}$$

where the sum is over partitions ν of length $\leq n$ which are dominated by λ, $m(\nu)$ is the monomial symmetric function containing $t_1^{\nu_1} \cdots t_n^{\nu_n}$ and $a_{\lambda,\nu}$ is an integer ≥ 0. The determination of the coefficients $a_{\lambda,\nu}$ for general n, λ and ν is an important unsolved problem.

Recall from Proposition 18.2.1 that $\mathbb{F}_2$ is a splitting field for $\mathsf{GL}(n)$. Given a $\mathbb{F}_2\mathsf{GL}(n)$-module M, we can construct a corresponding $\overline{\mathbb{F}}_2\mathsf{GL}(n)$-module M' by extension of scalars, i.e. $M' \cong M \otimes_{\mathbb{F}_2} \overline{\mathbb{F}}_2$, and every $\overline{\mathbb{F}}_2\mathsf{GL}(n)$-module is obtained in this way. In particular, we lose no information by studying the action of $\mathsf{GL}(n)$ on $\overline{\mathsf{P}}(n)$ rather than on $\mathsf{P}(n)$. Recall that $\mathbb{F}_2\mathsf{GL}(n)$ has 2^{n-1} inequivalent irreducible modules $L(\lambda)$ indexed by column 2-regular partitions λ with $\mathrm{len}(\lambda) \leq n-1$. Thus $\overline{\mathbb{F}}_2\mathsf{GL}(n)$ has 2^{n-1} inequivalent irreducible modules $L(\lambda) \otimes_{\mathbb{F}_2} \overline{\mathbb{F}}_2$. To simplify notation, we shall denote this module also by $L(\lambda)$.

If the partition λ is column 2-regular, then the $\overline{\mathsf{G}}(n)$-module $\overline{L}(\lambda)$ restricts to the corresponding irreducible $\overline{\mathbb{F}}_2\mathsf{GL}(n)$-module $L(\lambda)$. These particular modules $\overline{L}(\lambda)$ determine all other irreducible polynomial $\overline{\mathsf{G}}(n)$-modules via the tensor product theorem 20.1.1. We do not prove this result, but only explain the notation required to state it. The Frobenius map F of $\overline{\mathbb{F}}_2$ is the automorphism $a \mapsto a^2$. Any $\overline{\mathsf{G}}(n)$-module V gives rise to a 'twisted' $\overline{\mathsf{G}}(n)$-module $V^{(1)}$ by means of the composition $\overline{\mathsf{G}}(n) \xrightarrow{F} \overline{\mathsf{G}}(n) \to \mathsf{GL}(V)$. By iterating this construction, we obtain a module $V^{(k)}$ for all $k \geq 1$. In particular, if V is irreducible then all the modules $V^{(k)}$ are irreducible.

Given a partition λ and a positive integer m, let $m\lambda$ denote the partition obtained by multiplying each part of λ by m. In particular, the weights of the kth Frobenius twist $V^{(k)}$ of V are obtained by multiplying the weights of V by $m=2^k$, and so its formal character $\mathrm{ch}(V^{(k)})$ is obtained by substituting $t_i^{2^k}$ for t_i in $\mathrm{ch}(V)$ for $1 \leq i \leq n$. By combining the binary expansions of $\lambda_i - \lambda_{i+1}$ for all i, we see that a partition λ has a unique binary expansion as the sum $\lambda = \lambda(1) + 2\lambda(2) + \cdots + 2^r\lambda(r)$, where the partitions $\lambda(i)$, $1 \leq i \leq r$, are column 2-regular. For example, if $\lambda = (13,7,3)$, the difference sequence is $(6,4,3) = (0,0,1) + 2(1,0,1) + 4(1,1,0)$, and so $(13,7,3) = (1,1,1) + 2(2,1,1) + 4(2,1,0)$.

The following result is a special case of Steinberg's tensor product theorem, a fundamental result in the representation theory of algebraic groups.

Theorem 20.1.1 (R. Steinberg) *Let λ be a partition of length $\leq n$ with binary expansion as above. Then* $\overline{L}(\lambda) \cong \overline{L}(\lambda(1)) \otimes \overline{L}(\lambda(2))^{(1)} \otimes \cdots \otimes \overline{L}(\lambda(r))^{(r)}$.

The finite group $\mathsf{GL}(n)$ is the fixed point set of $\overline{\mathsf{G}}(n)$ under the Frobenius map F. Hence the Frobenius twists disappear on restriction of $\overline{\mathsf{G}}(n)$-modules to $\mathbb{F}_2\mathsf{GL}(n)$-modules, and so $\overline{L}(\lambda)$ restricts to $L(\lambda(1)) \otimes L(\lambda(2)) \otimes \cdots \otimes L(\lambda(r))$.

We note that Theorem 20.1.1 reduces the problem of calculating the formal characters of the representations $\overline{L}(\lambda)$ to the case where λ is column 2-regular. As $\overline{L}(\lambda_1, \ldots, \lambda_{n-1}, 1) = \overline{L}(\lambda_1 - 1, \ldots, \lambda_{n-1} - 1) \otimes \det$, we can further reduce to the case $\mathrm{len}(\lambda) \leq n-1$. These 2^{n-1} modules $\overline{L}(\lambda)$ restrict to the full set of irreducible $\mathbb{F}_2\mathsf{GL}(n)$-modules $L(\lambda)$. In this sense, the problem of determining the irreducible modular representations of $\mathsf{GL}(n)$ and the problem of determining the irreducible polynomial representations of $\overline{\mathsf{G}}(n)$ are equivalent.

20.2 The $\overline{\mathsf{G}}(n)$-module $\overline{\mathsf{P}}(n)$

In this section we consider the structure of $\overline{\mathsf{P}}(n)$ as a $\overline{\mathsf{G}}(n)$-module. As in Section 6.2, we denote by $\overline{\mathsf{P}}^{\omega}(n)$ the $\overline{\mathbb{F}}_2$-subspace of $\overline{\mathsf{P}}^{d}(n)$ spanned by monomials f such that $\omega(f) = \omega$, where $\deg_2 \omega = d$, and define $\overline{\mathsf{P}}^{<_l\omega}(n)$ and $\overline{\mathsf{P}}^{\leq_l\omega}(n)$ similarly. The group $\overline{\mathsf{G}}(n)$ is generated by permutation matrices and the transvections $U_\alpha, \alpha \in \overline{\mathbb{F}}_2$, defined by $U_\alpha(x_1) = x_1 + \alpha x_2$ and $U_\alpha(x_i) = x_i$ for $i > 1$.

As in Sections 6.1 and 6.2, $\omega(f \cdot A) \leq \omega(f)$ for $A \in \overline{\mathsf{G}}(n)$ in any ordering on ω-sequences which extends 2-dominance. Hence $\overline{\mathsf{P}}^{<_l\omega}(n)$ and $\overline{\mathsf{P}}^{\leq_l\omega}(n)$ are $\overline{\mathsf{G}}(n)$-submodules of $\overline{\mathsf{P}}^{d}(n)$, and $\overline{\mathsf{P}}^{\omega}(n) \cong \overline{\mathsf{P}}^{\leq_l\omega}(n)/\overline{\mathsf{P}}^{<_l\omega}(n)$. Thus the modules $\overline{\mathsf{P}}^{\omega}(n)$ are the quotients of the filtration on $\overline{\mathsf{P}}^{d}(n)$ defined using the left order, and so $[\overline{\mathsf{P}}^{d}(n)] = \sum_\omega [\overline{\mathsf{P}}^{\omega}(n)]$ in $R_2(\overline{\mathsf{G}}(n))$. Each monomial in $\overline{\mathsf{P}}^{\omega}(n)$ spans a 1-dimensional weight space, and so the formal character $\mathrm{ch}(\overline{\mathsf{P}}^{\omega}(n))$ is the sum of all monomials f in $t_1, \ldots, t_n$ with $\omega(f) = \omega$.

We associate to ω the exponent sequence λ of the $\leq_l$-maximal monomial f in $\overline{\mathsf{P}}^{\omega}(n)$. Thus $\lambda = (\lambda_1, \ldots, \lambda_n)$ is a partition of $d = \deg_2 \omega$, and $f = x_1^{\lambda_1} \cdots x_n^{\lambda_n} = \prod_{i=1}^{r} (x_1 \cdots x_{\omega_i})^{2^{i-1}}$, where $\omega = (\omega_1, \ldots, \omega_r)$. Thus $\lambda_j = \sum_{\omega_i \geq j} 2^{i-1}$ for $1 \leq j \leq n$. For example, if $\omega = (1,3)$ then $\deg_2 \omega = 7$ and $\lambda = (3,2,2)$. The monomial f is a highest weight vector in $\overline{\mathsf{P}}^{\omega}(n)$, with weight $t^\lambda = t_1^{\lambda_1} \cdots t_n^{\lambda_n}$.

Example 20.2.1 The block $B = \begin{smallmatrix} 1 & 0 & 1 & 1 \\ 1 & 0 & 0 & 1 \\ 1 & & & \end{smallmatrix}$ represents the $\leq_l$-maximal monomial $x_1^{13} x_2^{9} x_3$ in $\overline{\mathsf{P}}^{23}(3)$ with ω-sequence $\omega = (3,0,1,2)$. The binary expansion $\lambda = (13,9,1) = (1,1,1) + 4(1,0,0) + 8(1,1,0)$ can be read off from the columns of B.

It is clear from this example that the partitions λ which correspond to the highest weights of the modules $\overline{\mathsf{P}}^{\omega}(n)$ are of a special type, as their binary expansions involve only the partitions (1^k) for $0 \leq k \leq n$. The blocks representing the corresponding highest weight monomials are characterized by the property that no entry 1 lies below an entry 0 in the same column. For example, the partition $(13,7,3)$ corresponds to the block $\begin{smallmatrix} 1 & 0 & 1 & 1 \\ 1 & 1 & 1 & \\ 1 & 1 & & \end{smallmatrix}$, and so it is not of this form.

The next result describes the composition factors of the $\overline{\mathsf{G}}(n)$-modules $\overline{\mathsf{P}}^{d}(n)$. They are multiplicity-free, i.e. no two of the modules $\overline{\mathsf{P}}^{\omega}(n)$ are isomorphic. Not all the irreducible modules $\overline{L}(\lambda)$ appear in $\mathsf{P}(n)$, and those which appear do so just once, in degree $d = |\lambda|$. The restriction of $\overline{\mathsf{P}}^{\omega}(n)$ to $\mathbb{F}_2\mathsf{GL}(n)$ is $\mathsf{P}^{\omega}(n) \cong \Lambda^{\omega_1}(n) \otimes \Lambda^{\omega_2}(n) \otimes \cdots \otimes \Lambda^{\omega_r}(n)$.

Proposition 20.2.2 *For all $\omega \in \mathsf{Seq}_d(n)$, $\overline{\mathsf{P}}^{\omega}(n)$ is an irreducible $\overline{\mathsf{G}}(n)$-module $\overline{L}(\lambda)$ of highest weight λ. More precisely, if $\omega = (\omega_1, \ldots, \omega_r)$ then $\overline{\mathsf{P}}^{\omega}(n)$ is isomorphic to the tensor product $\overline{\Lambda}^{\omega_1}(n) \otimes (\overline{\Lambda}^{\omega_2}(n))^{(1)} \otimes \cdots \otimes (\overline{\Lambda}^{\omega_r}(n))^{(r-1)}$ of Frobenius twisted exterior powers of the defining $\overline{\mathsf{G}}(n)$-module $\overline{\mathsf{V}}(n)$.*

Example 20.2.3 The dominance order on ω-sequences of 2-degree 7 is shown below, with the corresponding highest weight monomials in $\overline{\mathsf{P}}^7(n)$ and irreducible $\overline{\mathsf{G}}(n)$-modules $\overline{L}(\lambda)$. Here $\overline{L}(\lambda) = 0$ if $\mathrm{len}(\lambda) > n$. The formal character $\mathrm{ch}(\overline{L}(\lambda))$ is the sum of all monomials in $t_1, \ldots, t_n$ with ω-sequence ω.

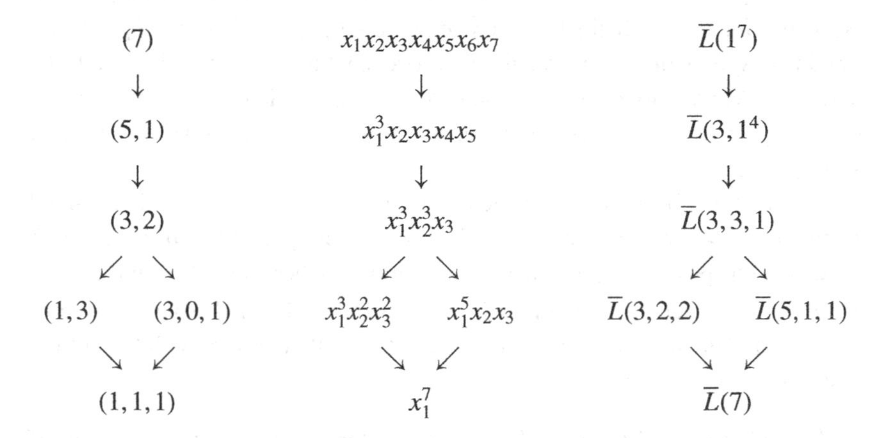

Every $\overline{\mathsf{G}}(n)$-submodule of $\overline{\mathsf{P}}^d(n)$ corresponds to a downward-closed subset S of ω-sequences in the dominance order, and is spanned by monomials f with $\omega(f) \in S$. Thus the lattice of ω-sequences describes the join-irreducible submodules. For example, for $n \geq 7$ there are 8 submodules of $\mathsf{P}^7(n)$, the two which are not visible in the ω-lattice being the zero submodule and the submodule spanned by monomials f with $\omega(f) = (1,3)$, $(3,0,1)$ or $(1,1,1)$.

Proof of Proposition 20.2.2 In the case $\omega = (k)$, it is easily seen that the map $\phi : \overline{\mathsf{P}}^{\omega}(n) \to \overline{\Lambda}^k(n)$ defined by $x_{i_1} \cdots x_{i_k} \mapsto v_{i_1} \wedge \cdots \wedge v_{i_k}$ is an isomorphism of $\overline{\mathsf{G}}(n)$-modules. More generally, for $\omega = (0, \ldots, 0, k)$, with k in the rth place, the map $\phi^{(r-1)} : \overline{\mathsf{P}}^{\omega}(n) \to (\overline{\Lambda}^k(n))^{r-1}$ defined by $(x_{i_1} \cdots x_{i_k})^{2^{r-1}} \mapsto (v_{i_1} \wedge \cdots \wedge v_{i_k})^{(r-1)}$ is an isomorphism of $\overline{\mathsf{G}}(n)$-modules.

For the general case, let $\omega = (\omega_1, \ldots, \omega_r)$ and write a monomial $f \in \overline{\mathsf{P}}^{\omega}(n)$ in the form $f = f_1 f_2^2 \cdots f_r^{2^{r-1}}$, where each f_j is a product of distinct variables x_i. Then the map $\overline{\mathsf{P}}^{\omega}(n) \to \overline{\Lambda}^{\omega_1}(n) \otimes (\overline{\Lambda}^{\omega_2}(n))^{(1)} \otimes \cdots \otimes (\overline{\Lambda}^{\omega_r}(n))^{(r-1)}$ given by $f \mapsto \phi(f_1) \otimes \phi^{(1)}(f_2) \otimes \cdots \otimes \phi^{(r-1)}(f_r)$ gives the required isomorphism. It follows from Theorem 20.1.1 that $\overline{\mathsf{P}}^{\omega}(n)$ is an irreducible $\overline{\mathsf{G}}(n)$-module.

The irreducibility of $\overline{\mathsf{P}}^{\omega}(n)$ can also be proved directly as follows. Since the weight vectors in $\overline{\mathsf{P}}^{\omega}(n)$ are the monomials f such that $\omega(f) = \omega$, it follows by restriction to the diagonal subgroup $\overline{\mathsf{D}}(n)$ that every $\overline{\mathsf{G}}(n)$-submodule M of $\overline{\mathsf{P}}^{\omega}(n)$ contains a monomial $f = x_1^{b_1} \cdots x_n^{b_n}$. We shall show that M contains all monomials in $\overline{\mathsf{P}}^{\omega}(n)$.

For $1 \leq i,j \leq n$, $i \neq j$, and a nonzero element $\alpha \in \overline{\mathbb{F}}_2$, let $T_{i,j}(\alpha) \in \overline{\mathsf{G}}(n)$ be the transvection which maps x_i to $x_i + \alpha x_j$ and fixes the other variables. Then $(T_{i,j}(\alpha))(f) = x_1^{b_1} \cdots (x_i + \alpha x_j)^{b_i} \cdots x_n^{b_n} = \sum_{s=0}^{b_i} \alpha^s f(i,j,s)$ where $f(i,j,s) = \binom{b_i}{s} x_1^{b_1} \cdots x_i^{b_i - s} \cdots x_j^{b_j + s} \cdots x_n^{b_n}$, a linear combination of the $b_i + 1$ terms $f(i,j,s)$. By choosing $b_i + 1$ distinct nonzero elements $\alpha_r \in \overline{\mathbb{F}}_2$, $1 \leq r \leq b_i + 1$, we obtain a system of linear equations whose coefficient matrix is invertible, since the Vandermonde determinant $\det(\alpha_r^s) \neq 0$. Hence we can express each $f(i,j,s)$ as a linear combination of the elements $(T_{i,j}(\alpha))(f) \in M$, and so $f(i,j,s) \in M$.

Let $g \in \overline{\mathsf{P}}^{\omega}(n)$ be a monomial such that the block representing g is obtained from the block representing f by exchanging an entry 1 in position (i,t) with an entry 0 in position (j,t). Then $\binom{b_i}{s} = 1 \bmod 2$ where $s = 2^{t-1}$, and so $g = f(i,j,s) \in M$. Since all monomials in $\overline{\mathsf{P}}^{\omega}(n)$ can be obtained from any one of them by a sequence of such moves, M contains all such monomials, and the result follows. □

We study the $\mathbb{F}_2\mathsf{GL}(n)$-module structure of the filtration quotients $\mathsf{P}^{\omega}(n)$ of $\mathsf{P}^d(n)$ by restriction of the action of $\overline{\mathsf{G}}(n)$ on $\overline{\mathsf{P}}(n)$ to $\mathsf{GL}(n)$. Since $\mathbb{F}_2$ is a splitting field for $\mathsf{GL}(n)$, results for $\overline{\mathbb{F}}_2\mathsf{GL}(n)$-modules are equivalent to results for $\mathbb{F}_2\mathsf{GL}(n)$modules. We shall therefore abuse notation by writing $\mathsf{P}(n)$ for the $\overline{\mathbb{F}}_2\mathsf{GL}(n)$-module $\mathsf{P}(n) \otimes_{\mathbb{F}_2} \overline{\mathbb{F}}_2$, and similarly for $\mathsf{P}^{\omega}(n) \otimes_{\mathbb{F}_2} \overline{\mathbb{F}}_2$, $\Lambda^k(n) \otimes_{\mathbb{F}_2} \overline{\mathbb{F}}_2$, etc.

Proposition 20.2.4 *For an n-bounded sequence* $\omega = (\omega_1, \ldots, \omega_r)$,

$$\mathsf{P}^{\omega}(n) \cong \Lambda^{\omega_1}(n) \otimes \cdots \otimes \Lambda^{\omega_r}(n).$$

Proof This follows from Proposition 20.2.2 by restriction to $\mathsf{GL}(n)$, since all iterated Frobenius twists of $\overline{\Lambda}^k(n)$ restrict to $\Lambda^k(n)$. Let the monomial $f \in \mathsf{P}^{\omega}(n)$ be written as a product $f = f_1 f_2^2 \cdots f_r^{2^{r-1}}$, each f_j being a product of distinct variables x_i. Then the required isomorphism $\mathsf{P}^{\omega}(n) \to \Lambda^{\omega_1}(n) \otimes \cdots \otimes \Lambda^{\omega_r}(n)$ is given by $f \mapsto \phi(f_1) \otimes \phi(f_2) \otimes \cdots \otimes \phi(f_r)$, where $\phi(x_{i_1} \cdots x_{i_k}) = v_{i_1} \wedge \cdots \wedge v_{i_k}$. □

Since all iterated Frobenius twists of $\overline{\Lambda}^k(n)$ restrict to $\Lambda^k(n)$, the $\mathbb{F}_2\mathsf{GL}(n)$-module $\mathsf{P}^{\omega}(n)$ can be regarded as the restriction of many different $\overline{\mathsf{G}}(n)$-modules.

In particular, $\mathsf{P}^{\omega}(n)$ is the restriction of $\overline{\Lambda}^{\omega}(n) = \overline{\Lambda}^{\omega_1}(n) \otimes \overline{\Lambda}^{\omega_2}(n) \otimes \cdots \otimes \overline{\Lambda}^{\omega_r}(n)$, where no Frobenius twists appear.

Example 20.2.5 Let $n \geq 7$. Then the left hand diagram below shows the modules $\overline{\Lambda}^{\omega} = \overline{\Lambda}^{\omega}(n)$ for $\omega \in \mathsf{Seq}_2(7)$ and the right hand diagram shows the irreducible composition factors $\overline{\mathsf{P}}^{\omega}(n)$ of $\overline{\mathsf{P}}^{7}(n)$. Note that the first diagram does *not* represent the $\overline{\mathsf{G}}(n)$-module $\overline{\mathsf{P}}^{7}(n)$. On restriction to $\mathsf{GL}(n)$, both diagrams give the same 'approximation' to the submodule lattice for $\mathsf{P}^{7}(n)$.

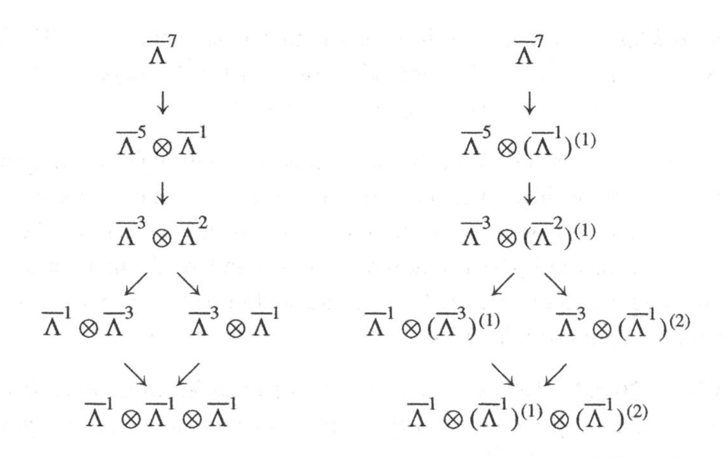

The modules $\overline{\Lambda}^{\omega}(n)$ in the left hand diagram are examples of a special type of polynomial $\overline{\mathsf{G}}(n)$-module called a 'tilting' module (see Section 20.7). The next result gives some information about these modules.

Proposition 20.2.6 *Let $\mu = (\mu_1, \ldots, \mu_r)$ be a partition with $\mu_1 \leq n$, and let $\lambda = \mu^{tr}$. Then $\overline{\Lambda}^{\mu}(n) = \overline{\Lambda}^{\mu_1}(n) \otimes \cdots \otimes \overline{\Lambda}^{\mu_r}(n)$ has one composition factor isomorphic to the irreducible $\overline{\mathsf{G}}(n)$-module $\overline{L}(\lambda)$, all others being isomorphic to $\overline{L}(\nu)$ for some partition ν such that $\nu \prec \lambda$.*

Proof Let $\lambda = (\lambda_1, \ldots, \lambda_n)$ where $\lambda_i = 0$ if $i > \mu_1$, and let $m(\lambda)$ be the monomial symmetric function in $t_1, \ldots, t_n$ containing $t_1^{\lambda_1} \cdots t_n^{\lambda_n}$. The formal character $\mathrm{ch}(\overline{L}(\lambda))$ is given by (20.1). Since ch is multiplicative and $\mathrm{ch}(\overline{\Lambda}^k(n)) = e_k$, the kth elementary symmetric function in $t_1, \ldots, t_n$, $\mathrm{ch}(\overline{\Lambda}^{\mu}(n)) = e_\mu = e_{\mu_1} \cdots e_{\mu_r}$. By a standard result for symmetric functions, e_μ can be written in the same form as $\mathrm{ch}(\overline{L}(\lambda))$, i.e. $e_\mu = m(\lambda) + \sum_{\nu \prec \lambda} b_{\lambda,\nu} m(\nu)$, with coefficients $b_{\lambda,\nu} \geq 0$. The result follows by induction on the dominance order. □

20.3 Weyl modules

Since $v_1 \wedge \cdots \wedge v_k$ is a highest weight vector for $\overline{\Lambda}^k(n)$, a highest weight vector for $\overline{\Lambda}^\mu(n)$ is given by $(v_1 \wedge \cdots \wedge v_{\mu_1}) \otimes \cdots \otimes (v_1 \wedge \cdots \wedge v_{\mu_r})$. This element corresponds to the spike monomial $s^\lambda = x_1^{2^{\lambda_1}-1} \cdots x_n^{2^{\lambda_n}-1} \in \overline{\mathsf{P}}^\mu(n)$ under the isomorphism ϕ of Proposition 20.2.2. It follows from Proposition 20.2.6 that if M is a submodule of $\overline{\mathsf{P}}^\mu(n)$ which contains s^λ, then M has a composition factor $\overline{L}(\lambda)$, and all other composition factors of M are of the form $\overline{L}(\nu)$ for some partition ν such that $\nu \prec \lambda$. We consider the smallest such submodule M.

Definition 20.3.1 Let λ be a partition of length $\leq n$, and let $\mu = \lambda^{\mathrm{tr}}$. The **Weyl module** $\overline{\Delta}(\lambda, n)$ for $\overline{\mathsf{G}}(n)$ is the submodule of $\overline{\Lambda}^\mu(n) = \overline{\Lambda}^{\mu_1}(n) \otimes \cdots \otimes \overline{\Lambda}^{\mu_r}(n)$ generated by $(v_1 \wedge \cdots \wedge v_{\mu_1}) \otimes \cdots \otimes (v_1 \wedge \cdots \wedge v_{\mu_r})$.

We recall from Section 7.4 that a Young tableau is a filling of the diagram of a partition λ with positive integers, and that a Young tableau is semi-standard if its entries increase strictly on columns and weakly on rows. The Weyl module $\overline{\Delta}(\lambda, n)$ has an $\overline{\mathbb{F}}_2$-basis indexed by semi-standard Young tableaux with diagram λ and distinct entries $1, 2, \ldots, n$, and so $\dim \overline{\Delta}(\lambda, n) = \tau(\lambda, n)$ is given by the hook-length formula (7.1).

Proposition 20.3.2 *The Weyl module $\overline{\Delta}(\lambda, n)$ is indecomposable. It has a unique irreducible quotient isomorphic to $\overline{L}(\lambda)$ and all other composition factors are isomorphic to $\overline{L}(\nu)$ for some $\nu \prec \lambda$.*

Proof Proposition 20.2.6 implies that the composition factors of $\overline{\Delta}(\lambda, n)$ are as stated. Since the spike monomial s^λ is a highest weight vector and generates $\overline{\Delta}(\lambda, n)$, $\overline{\Delta}(\lambda, n)$ has a quotient isomorphic to $\overline{L}(\lambda)$. If $\overline{\Delta}(\lambda, n)$ has a quotient isomorphic to $\overline{L}(\nu)$ with $\nu \neq \lambda$, then by considering weight vectors in a composition series for $\overline{\Delta}(\lambda, n)$ with highest quotient $\overline{L}(\nu)$ it follows that s^λ is an element of a proper submodule of $\overline{\Delta}(\lambda, n)$, a contradiction. In particular, $\overline{\Delta}(\lambda, n)$ is indecomposable. □

Dually, we consider the action of $\mathsf{GL}(n)$ on the divided polynomial algebra $\overline{\mathsf{DP}}(n)$ in n variables $v_1, \ldots, v_n$ over $\overline{\mathbb{F}}_2$, in place of its action on $\overline{\mathsf{P}}(n)$. As for the $\mathbb{F}_2\mathsf{GL}(n)$-modules $\mathsf{P}(n)$ and $\mathsf{DP}(n)$, the $\overline{\mathsf{G}}(n)$-module $\overline{\mathsf{DP}}^d(n)$ is the transpose dual of $\overline{\mathsf{P}}^d(n)$. Thus the polynomial $\overline{\mathsf{G}}(n)$-module $\overline{\mathsf{DP}}^d(n)$ has the same irreducible composition factors as $\overline{\mathsf{P}}(n)$, but composition series are reversed. For ω-sequences, this corresponds to reversal of the left or right order.

In particular, for each n-bounded sequence ω with $\deg_2 \omega = d$ we have a corresponding filtration quotient $\overline{\mathsf{DP}}^\omega(n) = \overline{\mathsf{DP}}^{\geq \omega}(n)/\overline{\mathsf{DP}}^{> \omega}(n)$ which is the transpose dual of $\overline{\mathsf{P}}^\omega(n)$. Since $\overline{\mathsf{P}}^\omega(n)$ is irreducible by Proposition 20.2.2,

$\overline{\mathsf{DP}}^{\omega}(n) \cong \overline{\mathsf{P}}^{\omega}(n)$ as a $\overline{\mathsf{G}}(n)$-module, and so by restriction of the action $\mathsf{DP}^{\omega}(n) \cong \mathsf{P}^{\omega}(n)$ as $\mathbb{F}_2\mathsf{GL}(n)$-modules. By the results of Section 9.4, these modules are also transpose dual, and hence $\mathsf{P}^{\omega}(n)$ is isomorphic to its transpose dual.

Since transpose duality commutes with tensor products, $\overline{\Lambda}^{\omega}(n) = \overline{\Lambda}^{\omega_1}(n) \otimes \cdots \otimes \overline{\Lambda}^{\omega_r}(n)$ is also isomorphic to its transpose dual, and restriction to $\mathsf{GL}(n)$ gives another proof that $\mathsf{DP}^{\omega}(n)$ and $\mathsf{P}^{\omega}(n)$ are isomorphic $\mathbb{F}_2\mathsf{GL}(n)$-modules. Since a canonical isomorphism between $\overline{\mathsf{DP}}^{(k)}(n)$ and $\overline{\Lambda}^{k}(n)$ is given by associating the d-monomial $v_{i_1} \cdots v_{i_k}$ to $v_{i_1} \wedge \cdots \wedge v_{i_k}$, the map which associates the monomial in $\overline{\mathsf{P}}^{\omega}(n)$ and the d-monomial in $\overline{\mathsf{DP}}^{\omega}(n)$ which are represented by the same block is a $\mathbb{F}_2\mathsf{GL}(n)$-isomorphism.

Hence all weight vectors in $\overline{\mathsf{P}}^{\omega}(n)$ and $\overline{\mathsf{DP}}^{\omega}(n)$ are represented by the same blocks. In particular, for a partition μ with $\mu_1 \leq n$, $\overline{\mathsf{DP}}^{\mu}(n)$ has highest weight $\lambda = \mu^{\mathrm{tr}}$, and the d-spike monomial $s_\lambda = v_1^{(2^{\lambda_1}-1)} \cdots v_n^{(2^{\lambda_n}-1)}$ is a highest weight vector. It follows that the submodule of $\overline{\mathsf{DP}}^{\mu}(n)$ generated by s_λ is isomorphic to $\overline{\Delta}(\lambda, n)$.

Definition 20.3.3 Let λ be a partition of length $\leq n$. The **dual Weyl module** $\overline{\nabla}(\lambda, n)$ for $\overline{\mathsf{G}}(n)$ with highest weight λ is the transpose dual $\overline{\Delta}(\lambda, n)^{\mathrm{tr}}$.

Since transpose duality fixes irreducible modules and reverses composition series, it follows from Proposition 20.3.2 that $\overline{\nabla}(\lambda, n)$ is an indecomposable $\overline{\mathsf{G}}(n)$-module with socle isomorphic to $\overline{L}(\lambda)$, all other composition factors being isomorphic to $\overline{L}(\nu)$ where $\nu \prec \lambda$. Since $\overline{\mathsf{P}}^{\mu}(n)$ and $\overline{\mathsf{DP}}^{\mu}(n)$ are transpose duals, the smallest quotient module of $\overline{\mathsf{DP}}^{\mu}(n)$ containing s_λ is isomorphic to $\overline{\nabla}(\lambda, n)$. Since there is an isomorphism from $\overline{\mathsf{P}}^{\mu}(n)$ and $\overline{\mathsf{DP}}^{\mu}(n)$ which maps s^λ to s_λ, the smallest quotient module of $\overline{\mathsf{P}}^{\mu}(n)$ containing s^λ is isomorphic to $\overline{\nabla}(\lambda, n)$, and the submodule of $\overline{\mathsf{DP}}^{\mu}(n)$ generated by s_λ is isomorphic to $\overline{\Delta}(\lambda, n)$.

We next define Weyl modules and dual Weyl modules for the finite group $\mathsf{GL}(n)$.

Definition 20.3.4 Let λ be a partition of length $\leq n$. The $\mathbb{F}_2\mathsf{GL}(n)$-modules $\Delta(\lambda, n)$ and $\nabla(\lambda, n)$ given by restriction of $\overline{\Delta}(\lambda, n)$ and $\overline{\nabla}(\lambda, n)$ to $\mathsf{GL}(n)$, or the corresponding $\mathbb{F}_2\mathsf{GL}(n)$-modules, are the **(restricted) Weyl module** $\Delta(\lambda, n)$ and the **(restricted) dual Weyl module** $\nabla(\lambda, n)$ corresponding to λ.

If λ is column 2-regular, then the irreducible $\overline{\mathsf{G}}(n)$-module $\overline{L}(\lambda)$ restricts to the irreducible $\mathbb{F}_2\mathsf{GL}(n)$-module $L(\lambda)$, so $\Delta(\lambda, n)$ has a quotient isomorphic to $L(\lambda)$, and dually $\nabla(\lambda, n)$ has a submodule isomorphic to $L(\lambda)$.

Example 20.3.5 Let $n = 4$ and $\mu = (3,1)$, so that $\lambda = (2,1,1)$. The $\overline{\mathsf{G}}(4)$-module $\overline{\Lambda}^3 \otimes \overline{\Lambda}^1$ has composition series

$$\begin{array}{cc} \overline{L}(1^4) & (v_1 \wedge v_2 \wedge v_3) \otimes v_4 \\ \downarrow & \\ \overline{L}(2,1,1) & (v_1 \wedge v_2 \wedge v_3) \otimes v_1 \\ \downarrow & \\ \overline{L}(1^4) & \sum_4 (v_1 \wedge v_2 \wedge v_3) \otimes v_4 \end{array}$$

with generators (chosen as highest weight vectors) as shown for the composition factors, where the notation $\sum_4$ means the 4-term symmetrization. The Weyl module $\overline{\Delta}(\lambda,4)$ has dimension 15 and is given by the lower two factors, and its dual $\overline{\nabla}(\lambda,4)$ by the upper two factors. The irreducible module $\overline{L}(2,1,1)$ has dimension 14, while $\overline{L}(1^4) \cong \det$ has dimension 1. The weights of $\overline{\Delta}(\lambda,4)$ and $\overline{\nabla}(\lambda,4)$ are the 12 permutations of $(2,1,1,0)$ (weight spaces of dimension 1) and $(1^4) = (1,1,1,1)$ (weight space of dimension 3).

On restriction to $\mathsf{GL}(4)$, the irreducible modules $\overline{L}(2,1,1)$ and $\overline{L}(1^4)$ restrict to the irreducible $\overline{\mathbb{F}}_2\mathsf{GL}(4)$-modules $L(2,1,1)$ and $L(0) \cong \mathsf{I}(4)$ respectively. Hence the diagram also shows the structure of $\mathsf{P}^\mu(4)$ or of $\mathsf{DP}^\mu(4)$. The spike monomial $s^\lambda = x_1^3x_2x_3$ generates the submodule $\Delta(\lambda,4)$ of $\mathsf{P}^\mu(4)$, and the d-spike monomial $s_\lambda = v_1^{(3)}v_2v_3$ generates the submodule $\Delta(\lambda,4)$ of $\mathsf{DP}^\mu(4)$. A basis for $\Delta(\lambda,4) \subset \mathsf{P}^\mu(4)$ is given by the 12 spikes obtained from s_λ by permuting the variables, together with the polynomials $(x_1+x_i)x_1x_2x_3x_4$, $i = 2,3,4$. The corresponding d-polynomials form a basis for $\Delta(\lambda,4) \subset \mathsf{DP}^\mu(4)$.

The submodule $L(0)$ of $\mathsf{P}^\mu(4)$ is generated by the hit polynomial $Sq^1(\mathsf{c}(4)) = \sum_4 x_1^2x_2x_3x_4$, and so $\mathsf{Q}^\mu(4)$ is the quotient module $\nabla(\lambda,4)$. The quotient $\nabla(\lambda,4)$ has a monomial basis given by omitting any one of the monomials with α-sequence $(1,1,1,1)$. If we omit $x_1^2x_2x_3x_4$, then the basis corresponds to the 15 semi-standard Young tableaux for λ via the bijective correspondence of Section 7.4. The submodule $L(0)$ of $\mathsf{DP}^\mu(4)$ is generated by $\sum_4 v_1^{(2)}v_2v_3v_4$, and $\mathsf{J}^\mu(4) = \mathsf{K}^\mu(4)$ is the submodule $\Delta(\lambda,4) \subset \mathsf{DP}^\mu(4)$. A correspondence between semi-standard Young tableaux and basis elements for $\Delta(\lambda,4) \subset \mathsf{DP}^\mu(4)$ is given by row-symmetrizing the tableau and taking the sum of the corresponding monomials.

20.4 Weyl modules and flag modules

The following result will be useful in generalizing Example 20.3.5 to all strictly decreasing ω-sequences.

Proposition 20.4.1 *Let λ be a column 2-regular partition. Then the restricted Weyl module $\Delta(\lambda,n)$ has a unique irreducible quotient $L(\lambda)$. In particular, $\Delta(\lambda,n)$ is an indecomposable $\overline{\mathbb{F}}_2\mathsf{GL}(n)$-module.*

Since $\nabla(\lambda,n)$ is the transpose dual of $\Delta(\lambda,n)$, we have the equivalent statement that when λ is column 2-regular $\nabla(\lambda,n)$ is an indecomposable $\overline{\mathbb{F}}_2\mathsf{GL}(n)$-module with unique irreducible submodule $L(\lambda)$.

Since $\mathsf{FL}_I(n)$ has a unique irreducible quotient isomorphic to $L(\lambda)$ by Proposition 17.6.5, the following result implies Proposition 20.4.1.

Proposition 20.4.2 *Let λ be a column 2-regular partition, and $I \subseteq Z[n-1]$ the set of parts of $\mu = \lambda^{tr}$. Then the restricted Weyl module $\Delta(\lambda,n)$ is isomorphic to a quotient of the indecomposable summand $\mathsf{FL}_I(n)$ of the flag module $\mathsf{FL}(n)$.*

We begin by defining a map from the partial flag module $\mathsf{FL}^I(n)$ to $\Delta(\lambda,n)$. Let $I = \{i_1, i_2, \ldots, i_r\}$, so that a partial flag of type I is a sequence of subspaces $X : 0 \in X_{i_1} \subset X_{i_2} \subset \cdots \subset X_{i_r} \subset \mathsf{V}(n)$, where $\mathsf{V}(n)$ is the defining $\mathbb{F}_2\mathsf{GL}(n)$-module. The partial flags X form a $\mathbb{F}_2$-basis of $\mathsf{FL}^I(n)$. Let $\Lambda^I = \Lambda^I(n)$ denote the tensor product $\Lambda^{i_r} \otimes \Lambda^{i_{r-1}} \otimes \cdots \otimes \Lambda^{i_1}$, where $\Lambda^i = \Lambda^i(n)$ is the ith exterior power of $\mathsf{V}(n)$, and let $v_1, \ldots, v_n$ be the standard basis of $\mathsf{V}(n)$. The Weyl module $\Delta(\lambda,n)$ is the cyclic submodule of Λ^I generated by the element $v_\lambda = (v_1 \wedge \cdots \wedge v_{i_r}) \otimes (v_1 \wedge \cdots \wedge v_{i_{r-1}}) \otimes \cdots \otimes (v_1 \wedge \cdots \wedge v_{i_1})$.

Definition 20.4.3 The map p_λ is defined as follows. Let $u_1, \ldots, u_n$ be a basis for $\mathsf{V}(n)$ adapted to the flag X, so that $u_1, \ldots, u_{i_j}$ is a basis for X_{i_j} for $1 \leq j \leq r$. Then $p_\lambda(X) = (u_1 \wedge \cdots \wedge u_{i_r}) \otimes (u_1 \wedge \cdots \wedge u_{i_{r-1}}) \otimes \cdots \otimes (u_1 \wedge \cdots \wedge u_{i_1})$.

Since the product $u_1 \wedge \cdots \wedge u_i \in \Lambda^i(n)$ is independent of the basis for X_i, the map p_λ is well defined.

Proposition 20.4.4 *The map $p_\lambda : \mathsf{FL}^I(n) \to \Delta(\lambda,n)$ is a surjection of $\mathbb{F}_2\mathsf{GL}(n)$-modules.*

Proof It is clear that p_λ is a map of $\mathbb{F}_2\mathsf{GL}(n)$-modules, since if $A \in \mathsf{GL}(n)$ maps the subspace X of $\mathsf{V}(n)$ to the subspace Y, then it maps a basis for X to a basis for Y. Since p_λ maps the reference flag R^I, which corresponds to the parabolic subgroup $\mathsf{P}^I(n)$ of $\mathsf{GL}(n)$, to the generator v_λ, p_λ is a surjection. □

The partial flag module $\mathsf{FL}^I(n) \cong \oplus_{J \subseteq I} \mathsf{FL}_J(n)$ is decomposable. We recall from Proposition 17.4.8 that the indecomposable summand $\mathsf{FL}_I(n)$ is the cyclic module generated by $g_I(\mathsf{B}(n))$, where $\mathsf{B}(n)$ corresponds to the reference flag R and $g_I = e_{-I} h_I$ is the element of the 0-Hecke algebra $\mathsf{H}_0(n)$ defined in

Definition 17.4.1. Then the following result implies that the restriction of p_λ to $\mathsf{FL}_I(n)$ is a surjection, and completes the proof of Proposition 20.4.2.

Proposition 20.4.5 $p_\lambda(g_I(\mathsf{B}(n))) = v_\lambda$.

Proof By Proposition 17.4.6, $g_I = h_I +$ elements $f_{W'}$ such that $\mathrm{len}(W') > \mathrm{len}(h_I)$. We shall prove (i) $p_\lambda(h_I(\mathsf{B}(n))) = v_\lambda$, and (ii) $p_\lambda(f_{W'}(\mathsf{B}(n))) = 0$ for all the other terms $f_{W'}$ in g_I.

We first consider $p_\lambda(h_i(\mathsf{B}(n)))$ where $h_i = f_{S_i}$ is the generator of $\mathsf{H}_0(n)$ which maps the reference flag $\mathsf{B}(n)$ to the sum $\mathsf{B}(n)S_i + \mathsf{B}(n)S_iL_i$ of the flags in $\mathsf{Sch}(S_i)$, where L_i is the lower triangular transvection obtained by adding row i of the identity matrix I_n to row $i+1$, and $p_\lambda(h_i(\mathsf{B}(n))) = p_\lambda(\mathsf{B}(n)S_i) + p_\lambda(\mathsf{B}(n)S_iL_i)$.

By definition, the jth factor of $p_\lambda(\mathsf{B}(n))$ is $v_1 \wedge \cdots v_{i_j}$ for $1 \le j \le r$. The corresponding factor of $p_\lambda(\mathsf{B}(n)S_i)$ is obtained by exchanging v_i and v_{i+1}, and the corresponding factor of $p_\lambda(\mathsf{B}(n)S_iL_i)$ is obtained by replacing v_i by $v_1 + v_{i+1}$ and v_{i+1} by v_i. We separate the cases (1) $i_j < i$, (2) $i_j = i$ and (3) $i_j > i$.

In case (1), neither v_i nor v_{i+1} appears in $v_1 \wedge \cdots v_{i_j}$, and so this factor is the same in $p_\lambda(\mathsf{B}(n))$, $p_\lambda(\mathsf{B}(n)S_i)$ and $p_\lambda(\mathsf{B}(n)S_iL_i)$.

In case (2), v_i appears in $v_1 \wedge \cdots v_{i_j}$, but v_{i+1} does not. Hence the corresponding factor of $p_\lambda(\mathsf{B}(n)S_i)$ is $v_1 \wedge \cdots v_{i+1}$, the corresponding factor of $p_\lambda(\mathsf{B}(n)S_iL_i)$ is $v_1 \wedge \cdots v_{i-1} \wedge (v_i + v_{i+1})$, and their sum is $v_1 \wedge \cdots v_i$.

In case (3), both v_i and v_{i+1} appear in $v_1 \wedge \cdots v_{i_j}$. Hence the corresponding factor of $p_\lambda(\mathsf{B}(n)S_i)$ is $v_1 \wedge \cdots v_{i+1} \wedge v_i \wedge \cdots \wedge v_{i_j}$ and the corresponding factor of $p_\lambda(\mathsf{B}(n)S_iL_i)$ is $v_1 \wedge \cdots v_{i-1} \wedge (v_i + v_{i+1}) \wedge v_i \wedge \cdots \wedge v_{i_j}$. Since $(v_i + v_{i+1}) \wedge v_i = v_i \wedge v_{i+1}$, these are equal.

Applying these observations to all the factors of v_λ, it follows that

$$p_\lambda(h_i(\mathsf{B}(n))) = p_\lambda(\mathsf{B}(n)S_i) + p_\lambda(\mathsf{B}(n)S_iL_i) = \begin{cases} p_\lambda(\mathsf{B}(n)), & \text{if } i \in I, \\ 0, & \text{if } i \notin I. \end{cases}$$

More generally we can replace the basis vectors $v_1, \ldots, v_n$ by arbitrary elements $u_1, \ldots, u_n$ of $\mathsf{V}(n)$ in the three cases above. It follows that $p_\lambda(h_i(\mathsf{B}(n)A)) = p_\lambda(\mathsf{B}(n)A)$ for all $A \in \mathsf{GL}(n)$ if $i \in I$ and $p_\lambda(h_i(\mathsf{B}(n)A)) = 0$ if $i \notin I$. Summing over representative matrices A for the cosets in the Schubert cell $\mathsf{Sch}(W)$, we obtain $p_\lambda(h_if_W(\mathsf{B}(n))) = p_\lambda(f_W(\mathsf{B}(n)))$ if $i \in I$ and $p_\lambda(h_if_W(\mathsf{B}(n))) = 0$ if $i \notin I$.

Now $h_I = f_{W_0^{-I}}$ is a product of the generators h_i of $\mathsf{H}_0(n)$ such that $i \in I$, and all other terms $f_{W'}$ of g_I are products involving generators h_i with $i \notin I$. Thus $p_\lambda(g_I(\mathsf{B}(n))) = p_\lambda(h_I(\mathsf{B}(n))) = p_\lambda(\mathsf{B}(n)) = v_\lambda$. □

20.5 Weyl modules and the hit problem

In this section, we return to the hit problem for the action of the Steenrod algebra A_2 on $\mathsf{P}(n)$, and show that the restricted dual Weyl modules whose socles are a complete set of irreducible $\mathbb{F}_2\mathsf{GL}(n)$-modules $L(\lambda)$ are quotients of the corresponding cohit modules $\mathsf{Q}^\mu(n)$, where $\mu = \lambda^{\text{tr}}$. In Section 30.2 we interpret the module $\mathsf{J}^\mu(n)^{\text{tr}}$ as the quotient of $\mathsf{Q}^d(n)$ by the submodule $\mathsf{SF}^d(n)$ of 'strongly spike-free' polynomials.

Theorem 20.5.1 *Let $\mu = \lambda^{tr}$ be a strictly decreasing sequence with $\mu_1 \le n$, and let $\mathsf{P}^\mu(n)$ and $\mathsf{DP}^\mu(n)$ be defined using an ordering which refines 2-dominance, such as $\le_l$ or $\le_r$. Then the submodule $\mathsf{J}^\mu(n) \subseteq \mathsf{K}^\mu(n)$ generated by the d-spike $s_\lambda = v_1^{(2^{\lambda_1}-1)} \cdots v_n^{(2^{\lambda_n}-1)}$ is isomorphic to the restricted Weyl module $\Delta(\lambda,n)$ as a $\mathbb{F}_2\mathsf{GL}(n)$-module. The quotient module $\mathsf{J}^\mu(n)^{tr}$ of the cohit module $\mathsf{Q}^\mu(n)$ is isomorphic to the restricted dual Weyl module $\nabla(\lambda,n)$; its socle is generated by the spike $s^\lambda = x_1^{2^{\lambda_1}-1} \cdots x_n^{2^{\lambda_n}-1}$ and is isomorphic to $L(\lambda)$. In particular,* $\dim \mathsf{Q}^\mu(n) = \dim \mathsf{K}^\mu(n) \ge \tau(\lambda,n)$, *the hook number.*

Proof Let $\rho = (2^{\lambda_1}-1, \cdots, 2^{\lambda_n}-1)$, so that $\rho^{\text{tr}} = (1^{\mu_1}) + 2(1^{\mu_2}) + \cdots + 2^{r-1}(1^{\mu_r})$. By Proposition 20.2.2 we have isomorphisms of irreducible $\overline{\mathsf{G}}(n)$-modules $\overline{L}(\rho)$

$$\overline{\Lambda}^{\mu_1}(n) \otimes (\overline{\Lambda}^{\mu_2}(n))^{(1)} \otimes \cdots \otimes (\overline{\Lambda}^{\mu_r}(n))^{(r-1)} \cong \overline{\mathsf{P}}^\mu(n) \cong \overline{\mathsf{DP}}^\mu(n)$$

such that the highest weight vector $v = (v_1 \wedge \cdots \wedge v_{\mu_1}) \otimes \cdots \otimes (v_1 \wedge \cdots \wedge v_{\mu_r})$ corresponds to the highest weight vectors s^λ in $\overline{\mathsf{P}}^\mu(n)$ and s_λ in $\overline{\mathsf{DP}}^\mu(n)$.

If we change the action of $\overline{\mathsf{G}}(n)$ on the tensor product of exterior powers by removing the Frobenius twists, then we obtain isomorphisms of $\overline{\mathbb{F}}_2$-spaces

$$\overline{\Lambda}^{\mu_1}(n) \otimes \overline{\Lambda}^{\mu_2}(n) \otimes \cdots \otimes \overline{\Lambda}^{\mu_r}(n) \cong \overline{\mathsf{P}}^\mu(n) \cong \overline{\mathsf{DP}}^\mu(n)$$

such that v still corresponds to s^λ and to s_λ, but the map is no longer an isomorphism of $\overline{\mathsf{G}}(n)$-modules. On restriction to $\mathsf{GL}(n)$, however, we obtain isomorphisms of $\mathbb{F}_2\mathsf{GL}(n)$-modules

$$\Lambda^{\mu_1}(n) \otimes \Lambda^{\mu_2}(n) \otimes \cdots \otimes \Lambda^{\mu_r}(n) \cong \mathsf{P}^\mu(n) \cong \mathsf{DP}^\mu(n).$$

As an element of $\overline{\Lambda}^{\mu_1}(n) \otimes \overline{\Lambda}^{\mu_2}(n) \otimes \cdots \otimes \overline{\Lambda}^{\mu_r}(n)$, v is still a highest weight vector, but the highest weight is now λ and not ρ, and the submodule generated by v is the Weyl module $\overline{\Delta}(\lambda,n)$. Since λ is column 2-regular, the irreducible head $\overline{L}(\lambda)$ of $\overline{\Delta}(\lambda,n)$ gives $L(\lambda)$ on restriction to $\mathsf{GL}(n)$, and so the submodule of $\Lambda^{\mu_1}(n) \otimes \Lambda^{\mu_2}(n) \otimes \cdots \otimes \Lambda^{\mu_r}(n)$ generated by v is a submodule M of $\Delta(\lambda,n)$. By Proposition 20.4.1, $\Delta(\lambda,n)$ has irreducible head $L(\lambda)$, and hence $M = \Delta(\lambda,n)$.

It follows that the submodules of $\mathsf{P}^\mu(n)$ and $\mathsf{DP}^\mu(n)$ generated by s^λ and s_λ respectively are also isomorphic to $\Delta(\lambda,n)$. Since transpose duality between $\mathsf{P}^\mu(n)$ and $\mathsf{DP}^\mu(n)$ maps s^λ to s_λ, it follows that the smallest quotient modules of $\mathsf{P}^\mu(n)$ and $\mathsf{DP}^\mu(n)$ containing s^λ and s_λ respectively are isomorphic to $\nabla(\lambda,n)$. Since s^λ is not hit and $s_\lambda \in \mathsf{K}^\mu(n)$, it follows that the quotient module $\nabla(\lambda,n)$ containing s^λ is a quotient of $\mathsf{Q}^\mu(n)$, and the submodule $\Delta(\lambda,n)$ generated by s_λ is a submodule of $\mathsf{K}^\mu(n)$. □

From the discussion in Section 20.3, it follows that the $\overline{\mathsf{G}}(n)$-modules $\overline{\mathsf{P}}^\omega(n)$ and $\overline{\mathsf{DP}}^\omega(n)$ are canonically isomorphic via the map which sends each monomial to the d-monomial represented by the same block. Hence we may combine the lower bound on $\mathsf{Q}^\omega(n)$ given by Theorem 20.5.1 with the upper bound given by semi-standard monomials (Proposition 7.4.5). Note that this result requires the use of an extra condition on the ordering of ω-sequences of 2-degree $\deg_2 \mu$, which may not be satisfied by the standard left and right orderings.

Theorem 20.5.2 *Let μ be strictly decreasing, with $\lambda = \mu^{tr}$, and let $\leq$ be an ordering giving preference to μ as in Proposition 7.4.5. Then $\mathsf{Q}^\mu(n) \cong \nabla(\lambda,n)$ as a $\overline{\mathbb{F}}_2\mathsf{GL}(n)$-module, with a $\overline{\mathbb{F}}_2$-basis given by semi-standard monomials. Dually, $\mathsf{K}^\mu(n) \cong \Delta(\lambda,n)$ as a $\overline{\mathbb{F}}_2\mathsf{GL}(n)$-module, and $\mathsf{K}^\mu(n)$ can be identified with the cyclic submodule of $\mathsf{DP}^\mu(n)$ generated by $s^\lambda = v_1^{(2^{\lambda_1}-1)} \cdots v_n^{(2^{\lambda_n}-1)}$. In particular, $\dim \mathsf{Q}^\mu(n) = \dim \mathsf{K}^\mu(n) = \tau(\lambda,n)$.* □

The following example extends Example 7.4.7 to a case where $\omega = \mu$ is strictly decreasing.

Example 20.5.3 Let $n = 6$ and $\mu = (4,3)$. We have $\omega <_l (6,0,1)$ but $\omega >_r (6,0,1)$, so Proposition 7.4.5 applies only to the right order, and gives $\dim \mathsf{Q}^\omega(6) \leq 210$. As $Sq^2(x_1^3x_2x_3x_4x_5x_6) = x_1^5x_2x_3x_4x_5x_6 \bmod \mathsf{P}^\omega(5)$, a monomial in $\mathsf{P}^{(6,0,1)}(6)$ is left reducible, but it is not right reducible since it does not appear in any other hit equation. Hence $\dim \mathsf{Q}^{(6,0,1)}(6) = 0$ for the left order and $\dim \mathsf{Q}^{(6,0,1)}(6) = 6$ for the right order.

The only other hit equations involving $\mathsf{P}^\omega(6)$ are 60 independent equations in 5 variables obtained from $Sq^1(x_1^3x_2^3x_3x_4x_5)$ and 30 equations in 6 variables obtained from $Sq^1(x_1^3x_2^2x_3x_4x_5x_6)$ by permuting the variables. Of these 30, only 24 are linearly independent. Since $\dim \mathsf{P}^\omega(6) = 300$, we obtain $\dim \mathsf{Q}^\omega(6) = 216$ for the left order. For the right order, using the 6 hit equations involving Sq^2 to reduce elements of $\mathsf{P}^\omega(6)$ to $\mathsf{P}^{(6,0,1)}(6)$ we obtain $\dim \mathsf{Q}^\omega(6) = 210$.

20.6 First occurrences of irreducibles in $\mathsf{P}(n)$

Since $Sq^k : \mathsf{P}^d(n) \to \mathsf{P}^{d+k}(n)$ links occurrences of the same irreducible $\mathbb{F}_2\mathsf{GL}(n)$-module $L(\lambda)$ as a composition factor of $\mathsf{P}(n)$ in different degrees, a polynomial which generates the lowest degree occurrence of $L(\lambda)$ as a composition factor in $\mathsf{P}(n)$ is not hit. Thus information about this minimal degree $d(\lambda)$ is relevant to the hit problem. In Theorem 20.6.2 we show that $d(\lambda) = \deg_2 \mu$ where $\mu = \lambda^{\text{tr}}$ for all column 2-regular partitions λ of length $\leq n-1$, and that $\mathsf{P}^d(n)$ has just one composition factor isomorphic to $L(\lambda)$ when $d = d(\lambda)$.

Given an n-bounded sequence $\omega = (\omega_1, \dots, \omega_r)$, an isomorphism $\phi : \mathsf{P}^\omega(n) \cong \Lambda^{\omega_1}(n) \otimes \cdots \otimes \Lambda^{\omega_r}(n)$ of $\mathbb{F}_2\mathsf{GL}(n)$-modules is defined in Proposition 20.2.4. Since we require only the lowest degree occurrences of irreducible composition factors, we may permute factors so as to minimize $\deg_2 \omega$, and so we take the terms of ω in decreasing order. Then $\omega = \mu$ is a partition, and it suffices to study the filtration quotients

$$\phi : \mathsf{P}^\mu(n) \cong \Lambda^\mu(n) = \Lambda^{\mu_1}(n) \otimes \cdots \otimes \Lambda^{\mu_r}(n), \tag{20.2}$$

where $d = \deg_2 \mu$. Since $\Lambda^n(n) = \mathsf{I}(n)$, we may assume that $\mu_1 \leq n-1$.

Proposition 20.6.1 *Let μ be a partition with len$(\mu) \leq n$ and $\deg_2 \mu = d$, and let $\lambda = \mu^{tr}$. If μ is strictly decreasing, then one composition factor of $\mathsf{P}^\mu(n)$ is isomorphic to $L(\lambda)$, and all other composition factors are isomorphic to $L(\sigma)$ for some column 2-regular partition σ such that either*

(i) *$|\sigma| = |\lambda|$ and $\sigma \prec \lambda$, or*
(ii) *$|\sigma| < |\lambda|$ and $L(\sigma)$ occurs as a composition factor of $\mathsf{P}^\rho(n)$ for some partition ρ such that $d' = \deg_2 \rho < d$, so that $\mathsf{P}^{d'}(n)$ has a composition factor isomorphic to $L(\sigma)$.*

If μ is not strictly decreasing, then all composition factors of $\mathsf{P}^\mu(n)$ are isomorphic to $L(\sigma)$ where σ satisfies (i) *or* (ii).

Proof First we assume that μ is strictly decreasing, so that λ is column 2-regular. By Proposition 20.2.6, $\overline{L}(\lambda)$ is a composition factor in $\overline{\Lambda}^\mu(n)$. Hence on restriction to $\mathsf{GL}(n)$, $L(\lambda)$ is a composition factor in $\Lambda^\mu(n)$. By (20.2), $\Lambda^\mu(n) \cong \mathsf{P}^\mu(n)$.

Next we consider the other composition factors of $\mathsf{P}^\mu(n)$. By Proposition 20.2.6, the other composition factors of $\overline{\Lambda}^\mu(n)$ have the form $\overline{L}(\nu)$, where $|\nu| = |\lambda|$ and $\nu \prec \lambda$. If ν is also column 2-regular, then $\overline{L}(\nu)$ restricts to the irreducible $\mathbb{F}_2\mathsf{GL}(n)$-module $L(\nu)$, and so (i) holds, with $\sigma = \nu$.

However, ν may be column 2-singular. In this case, we can use Theorem 20.1.1 to write $\overline{L}(\nu)$ as a tensor product $\overline{L}(\nu(1)) \otimes \overline{L}(\nu(2))^{(1)} \otimes \cdots \otimes \overline{L}(\nu(r))^{(r)}$, where $\nu(i)$ is column 2-regular for $1 \leq i \leq r$. On restriction to $\mathsf{GL}(n)$, this gives $L(\nu(1)) \otimes L(\nu(2)) \otimes \cdots \otimes L(\nu(r))$. Thus we have to consider the composition factors $L(\sigma)$ of this tensor product of irreducible $\mathbb{F}_2\mathsf{GL}(n)$-modules.

We have already seen that if λ is column 2-regular then $L(\lambda)$ is a composition factor in $\Lambda^{\mu}(n)$, where $\mu = \lambda^{\mathrm{tr}}$. Hence every composition factor of $L(\nu(1)) \otimes \cdots \otimes L(\nu(r))$ is a composition factor of $\Lambda^{\nu(1)^{\mathrm{tr}}}(n) \otimes \cdots \otimes \Lambda^{\nu(r)^{\mathrm{tr}}}(n)$. But this module is isomorphic to $\mathsf{P}^{\rho}(n)$, where ρ is the partition obtained by arranging all the parts of the partitions $\nu(i)^{\mathrm{tr}}$, $1 \leq i \leq r$, in decreasing order.

In terms of Ferrers diagrams of partitions, the passage from ν to ρ^{tr} is carried out by removal of certain duplicated columns. For example, if $\nu = (8,6)$,

$$\nu = \begin{array}{l} 1\,1\,1\,1\,1\,1\,1\,1 \\ 1\,1\,1\,1\,1\,1 \end{array}, \quad \rho^{\mathrm{tr}} = \begin{array}{l} 1\,1\,1 \\ 1\,1 \end{array},$$

since Theorem 20.1.1 gives $\overline{L}(\nu) \cong \overline{L}(2,1)^{(1)} \otimes \overline{L}(1,1)^{(2)}$, and so $\nu(1) = (0), \nu(2) = (2,1)$ and $\nu(3) = (1,1)$. Thus the nonzero parts of the transposed partitions are $2,1,2$, giving $\rho = (2,2,1)$. As this example shows, ρ^{tr} may again be column 2-singular.

Let $d' = \deg_2 \rho$. Then $d' < \deg_2 \nu^{\mathrm{tr}}$, since removal of any term of a sequence lowers its 2-degree. By Propositions 5.2.4 and 5.2.5, $\deg_2 \nu^{\mathrm{tr}} < \deg_2 \mu = d$ since $\nu \prec \lambda$, and so $d' < d$. Thus every composition factor $L(\sigma)$ of $L(\nu(1)) \otimes L(\nu(2)) \otimes \cdots \otimes L(\nu(r))$ occurs as a composition factor in $\mathsf{P}^{d'}(n)$ for some $d' < d$, so that (ii) holds.

Finally we consider the case where μ is not strictly decreasing, so that $\lambda = \mu^{\mathrm{tr}}$ is column 2-singular. Then the composition factor $L(\lambda)$ of $\overline{\Lambda}^{\mu}(n)$ is treated by applying the preceding argument with $\nu = \lambda$, to show that restriction to $\mathsf{GL}(n)$ gives only composition factors which satisfy (ii). The other composition factors of the $\overline{\mathsf{G}}(n)$-module $\overline{\Lambda}^{\mu}(n)$ are treated as before, leading to the conclusion that all composition factors of their restriction to $\mathsf{GL}(n)$ satisfy either (i) or (ii). □

Theorem 20.6.2 *Let λ be a column 2-regular partition of length $\leq n-1$, and let $\mu = \lambda^{tr}$. Then $\mathsf{P}^d(n)$ has no $\mathbb{F}_2\mathsf{GL}(n)$-composition factors isomorphic to $L(\lambda)$ if $d < \deg_2 \mu$, and has exactly one such factor if $d = \deg_2 \mu$.*

Proof Proposition 20.6.1 reduces the problem to the local problem for the composition factors of $\mathsf{P}^{\mu}(n)$ for all partitions μ with $\mathrm{len}(\mu) \leq n-1$ and $\deg_2 \mu = d$. Combining cases (i) and (ii) of Proposition 20.6.1, it follows that a composition factor in $\mathsf{P}^d(n)$ is either the unique composition factor $L(\lambda)$

corresponding to a strictly decreasing ω-sequence $\mu = \lambda^{\text{tr}}$ of 2-degree d, or else it is isomorphic to a composition factor of $\mathsf{P}^{d'}(n)$ for some $d' < d$, and the result follows. □

In particular, it follows from Theorem 20.6.2 that the degree of first occurrence of the Steinberg module $\mathsf{St}(n)$ in $\mathsf{P}(n)$ is $w(n-1) = 2^n - 1 - n$, and that for all other irreducible $\mathbb{F}_2\mathsf{GL}(n)$-modules $L(\lambda)$ there is some $d < w(n-1)$ such that $L(\lambda)$ appears in $\mathsf{P}^d(n)$.

20.7 Remarks

The $\overline{\mathsf{G}}(n)$-modules $\overline{\Lambda}^{\omega}(n)$ are examples of 'tilting' modules. We include here some explanation of this term, although the results are not needed in our arguments. For further information on tilting modules we refer to [53].

For each partition λ of length $\leq n$, there is a corresponding $\overline{\mathsf{G}}(n)$-module $\overline{T}(\lambda,n)$, called the (partial) tilting module of highest weight λ, with the following (defining) properties: (i) $\overline{T}(\lambda,n)$ is indecomposable; (ii) $\overline{T}(\lambda,n)$ has a λ-weight space of dimension 1, and all other weights $\leq_l \lambda$; (iii) $\overline{T}(\lambda,n)$ has an increasing filtration with quotients isomorphic to $\overline{\Delta}(\nu,n)$, where all $\nu \preceq \lambda$; (iv) $\overline{T}(\lambda,n)$ is self-dual under transpose duality, so that it also has an increasing filtration with quotients isomorphic to $\overline{\nabla}(\mu,n)$, where all $\mu \preceq \lambda$.

In general, a tilting module is a direct sum of partial tilting modules. Since each exterior power $\overline{\Lambda}^k(n)$ is a tilting module, and the tensor product of tilting modules is a tilting module, the $\overline{\mathsf{G}}(n)$-module $\overline{\Lambda}^{\omega}(n) \cong \overline{\Lambda}^{\omega_1}(n) \otimes \ldots \otimes \overline{\Lambda}^{\omega_r}(n)$ is a tilting module, where $\omega = (\omega_1,\ldots,\omega_r)$. In addition, $\overline{\Lambda}^{\omega}(n)$ is a highest weight module. If $\omega = \mu$ is weakly decreasing, the highest weight λ of $\overline{\Lambda}^{\mu}(n)$ is $\lambda = \mu^{\text{tr}}$, and in general λ is the conjugate of the partition μ obtained by sorting the components of ω in decreasing order. It follows that $\overline{\Lambda}^{\omega}(n)$ contains $\overline{T}(\lambda,n)$ as a direct summand, and all other indecomposable summands of $\overline{\Lambda}^{\omega}(n)$ are isomorphic to $\overline{T}(\nu,n)$ for some $\nu \prec \lambda$.

The degree of first occurrence of $L(\lambda)$ in $\mathsf{P}(n)$ was determined by Lionel Schwartz in the 1980s by studying categories of A_2-modules [59, Section 7], [177, Theorem 4.2.4]. A more algebraic proof was given by David Carlisle and Nicholas J. Kuhn [28] using results of [98] concerning a Lie algebra called the 'hyperalgebra'. Our proof in Section 20.6 avoids Lie algebras at the cost of introducing algebraic group machinery.

Further material on formal characters and Steinberg's tensor product theorem 20.1.1 can be found in [79, Section 2.7]. Theorem 20.2.2 was proved by S. R. Doty [54]. Theorem 20.4.1 is due to W. J. Wong (see [226, (2D)] or [79, Section 5.9]).

21

Endomorphisms of $\mathsf{P}(n)$ over A_2

21.0 Introduction

Given graded left A_2-modules M_1 and M_2, we denote by $\mathrm{Hom}_{\mathsf{A}_2}(M_1,M_2)$ the $\mathbb{F}_2$-vector space of A_2-module maps $L : M_1 \to M_2$ which preserve the grading. When $M_1 = M_2 = M$, $\mathrm{Hom}_{\mathsf{A}_2}(M_1,M_2) = \mathrm{End}_{\mathsf{A}_2}(M)$, the algebra of A_2-endomorphisms of M with multiplication defined by composition. We are concerned with the case $M_1 = \mathsf{P}(m)$, $M_2 = \mathsf{P}(n)$. For $f \in \mathsf{P}(m)$, the right action of $L \in \mathrm{Hom}_{\mathsf{A}_2}(\mathsf{P}(m),\mathsf{P}(n))$ is denoted by $f \mapsto f * L$.

We denote the set of rectangular (m,n)-matrices over $\mathbb{F}_2$ by $\mathsf{M}(m,n)$, and write $\mathbb{F}_2\mathsf{M}(m,n)$ for the $\mathbb{F}_2$-vector space spanned by $\mathsf{M}(m,n)$. A matrix $A \in \mathsf{M}(m,n)$ acts on the right of $f \in \mathsf{P}^d(m)$ by linear substitution $f \cdot A$, giving a linear map $\mathsf{P}^d(m) \to \mathsf{P}^d(n)$ which commutes with the action of A_2, and this is extended by linearity to $\mathbb{F}_2\mathsf{M}(m,n)$.

Definition 21.0.1 Let $\psi : \mathbb{F}_2\mathsf{M}(m,n) \to \mathrm{Hom}_{\mathsf{A}_2}(\mathsf{P}(m),\mathsf{P}(n))$ be the linear map induced by matrix substitution, i.e. $\psi(A)(f) = f \cdot A$.

The object of this chapter is to prove that ψ is an isomorphism of vector spaces. In particular, the algebra $\mathrm{End}_{\mathsf{A}_2}(\mathsf{P}(n))$ is isomorphic to $\mathbb{F}_2\mathsf{M}(n)$. Injectivity of ψ is proved in Section 21.1, and surjectivity in Section 21.6.

Example 21.0.2 We show that $\psi : \mathbb{F}_2\mathsf{M}(1) \to \mathrm{End}_{\mathsf{A}_2}(\mathsf{P}(1))$ is an isomorphism. The elements of $\mathbb{F}_2\mathsf{M}(1)$ can be written as $a_0A_0 + a_1A_1$, where $A_0 = (0)$ and $A_1 = (1)$, and $a_0, a_1 \in \mathbb{F}_2$. The identity matrix A_1 acts as the identity on $\mathsf{P}(1)$, while the zero matrix A_0 acts by $x^k \cdot A_0 = 0$ if $k \geq 1$ and $1 \cdot A_0 = 1$. It follows that $\psi(a_0A_0 + a_1A_1) = 0$ if and only if $a_0 = a_1 = 0$, i.e. ψ is injective.

To prove that ψ is surjective, let $L \in \mathrm{End}_{\mathsf{A}_2}(\mathsf{P}(1))$ be given by $x^k * L = b_k x^k$ where $b_k \in \mathbb{F}_2$, $k \geq 0$. We show that all the coefficients b_k are equal for $k > 0$. From the solution of the hit problem for $\mathsf{P}(1)$, we know that the powers x^{2^r-1} for $r \geq 0$ form a minimal generating set of $\mathsf{P}(1)$ as a module over A_2. Hence for $k > 0$, $x^k = \theta(x^{2^r-1})$ for some $\theta \in \mathsf{A}_2$ and $r > 0$. Then $x^k * L = \theta(x^{2^r-1} * L)$, and hence $b_k = b_{2^r-1}$. Now consider the formula $x^{2^r-1} * L = b_{2^r-1}x^{2^r-1}$, and apply Sq^1 to obtain $x^{2^r} * L = b_{2^r-1}x^{2^r}$, which shows that $b_{2^r-1} = b_{2^r}$. Finally, $x^{2^r} = Sq^{2^{r-1}} \cdots Sq^2Sq^1(x)$. Hence $x^{2^r} * L = b_1x^{2^r}$, and $b_{2^r} = b_1$ for all $r \geq 0$.

We conclude that $x^k * L = b_1x^k$ for all $k > 0$, where $b_1 = 0$ or 1 and $1 * L = 0$ or 1. Hence L is given by the action of an element $a_0A_0 + a_1A_1 \in \mathbb{F}_2\mathsf{M}(1)$ on $\mathsf{P}(1)$, where $a_1 = b_1$.

The guiding principle of the proof that ψ is surjective is to characterize properties of the action of a matrix A on $\mathsf{P}(m)$ in terms of Steenrod operations, thus showing that these properties also belong to $L \in \mathrm{Hom}_{\mathsf{A}_2}(\mathsf{P}(m), \mathsf{P}(n))$. For example, if $f \in \mathsf{P}^d(m)$ then $f^2 \cdot A = (f \cdot A)^2$ by definition of the action of A, while $f^2 * L = Sq^d(f) * L = Sq^d(f * L) = (f * L)^2$ because L commutes with the action of A_2. Hence A and L act in the same way on squares of polynomials. Although $\mathbb{F}_2\mathsf{M}(m,n)$ is finite dimensional, it is not clear *a priori* that this is true for $\mathrm{Hom}_{\mathsf{A}_2}(\mathsf{P}(m), \mathsf{P}(n))$. This is proved in Section 21.4.

We recall from Section 1.4 the direct sum decomposition $\mathsf{P}(m) = \bigoplus_Y \mathsf{P}(Y)$ of A_2-modules, where $Y \subseteq Z[m]$ and $\mathsf{P}(Y)$ is spanned by monomials which are divisible by exactly the variables x_i for $i \in Y$. In particular, $\mathsf{P}(Z[m])$ is the A_2-submodule of $\mathsf{P}(m)$ consisting of polynomials divisible by $x_1 \cdots x_m$. We note that $\mathsf{P}(Y) \cong \mathsf{P}(Z[r])$ as A_2-modules where $r = |Y|$. We then have the direct sum decomposition $\mathrm{Hom}_{\mathsf{A}_2}(\mathsf{P}(m), \mathsf{P}(n)) = \bigoplus_{Y \subseteq Z[m]} \mathrm{Hom}_{\mathsf{A}_2}(\mathsf{P}(Y), \mathsf{P}(n))$, where $\mathsf{P}(\emptyset) = \mathsf{P}^0(n) = \mathbb{F}_2$, and $L \in \mathrm{Hom}_{\mathsf{A}_2}(\mathsf{P}(m), \mathsf{P}(n))$ is determined by its restriction to $\mathsf{P}(Y)$ for each Y.

In Section 21.5, we show that an element of $\mathrm{Hom}_{\mathsf{A}_2}(\mathsf{P}(Y), \mathsf{P}(n))$ is determined by its value on one special monomial in $\mathsf{P}(Y)$. There are many choices for such a special monomial, but we single out one, which we call the 'key' monomial. In the case $Y = Z[m]$, the key monomial is $x_1^{d_1} \cdots x_m^{d_m}$, where $d_j = (2^n - 1)2^{(n+1)(j-1)}$ for $1 \leq j \leq m$. For a general Y the exponents are the same and the variables are x_i for $i \in Y$. For example, when $m = 3$ and $n = 2$ the block corresponding to the key monomial in $\mathsf{P}(Z[3])$ is

$$\begin{matrix} 1\,1 \\ 0\,0\,0\,1\,1 \\ 0\,0\,0\,0\,0\,0\,1\,1 \end{matrix} .$$

The elements of $\mathrm{Hom}_{\mathsf{A}_2}(\mathsf{P}(m), \mathsf{P}(n))$ are determined by their values on the 2^m key monomials, one for each subset Y of $Z[m]$.

21.1 Embedding $\mathbb{F}_2\mathsf{M}(m,n)$ in $\mathrm{Hom}_{\mathsf{A}_2}(\mathsf{P}(m),\mathsf{P}(n))$

Proposition 21.1.1 *The map* $\psi : \mathbb{F}_2\mathsf{M}(m,n) \to Hom_{\mathsf{A}_2}(\mathsf{P}(m),\mathsf{P}(n))$ *is injective.*

Proof We show by induction on m that a nonzero element of $\mathbb{F}_2\mathsf{M}(m,n)$ cannot annihilate $\mathsf{P}(m)$. Let $m = 1$ and let c denote an irredundant formal sum of $(1,n)$-matrices. Choose one matrix A in c for which the number t of entries 1 in A is maximal. If $A = 0$, then $1 \cdot A \neq 0$. Otherwise let $Y = \{i_1,\ldots,i_t\}$ be the set of $j \in Z[n]$ such that $A_{1,i_j} = 1$. Then the irredundant expansion of $x^{2^t-1} \cdot A$ contains the term $\prod_{k=1}^{t} x_{i_k}^{2^{t-k}}$, which cannot appear in the expansion of $x^{2^t-1} \cdot A'$ for any A' in c with $A' \neq A$.

Assume that the result is true for numbers less than $m > 1$, and consider an irredundant formal sum of matrices $c \in \mathbb{F}_2\mathsf{M}(m,n)$. Let A be a matrix in c whose first row has the maximal number $t > 0$ of entries 1, and let $Y = \{i_1,\ldots,i_t\}$ be the set of $j \in Z[n]$ such that $A_{1,i_j} = 1$. Let b denote the formal sum of all matrices in c with the same first row as A, and let $a \in \mathbb{F}_2(m-1,n)$ be obtained by deleting the first row of each matrix in b. Then a is an irredundant formal sum of $(m-1) \times n$ matrices. By the inductive hypothesis, there exists a monomial g in $x_2,\ldots,x_m$ such that $g \cdot a \neq 0$. Let $f = x_1^{2^s(2^t-1)} g$, where s is chosen so that $2^s > \deg g$. Then $f \cdot c \neq 0$, because the expression $(\prod_{k=1}^{t} x_{i_k}^{2^{t-k}})^{2^s}(g \cdot a)$ for $i_k \in Y$ appears once in $f \cdot b = (x_{i_1} + \cdots + x_{i_t})^{2^s(2^t-1)}(g \cdot a)$, but nowhere else in $f \cdot c$. This completes the inductive step. □

21.2 Detecting squares and Dickson invariants

We begin by finding Steenrod algebra criteria for detecting squares of polynomials and certain Dickson invariants. It is clear from the results of Section 1.3 that any element in the two-sided ideal in A_2 generated by Sq^1 annihilates the square of a polynomial. In particular, this is true for the primitive Milnor basis elements $Q_r = Sq(0,\ldots,0,1)$, $r \geq 1$, as shown in Proposition 3.5.10(ii). We consider some converse statements.

Proposition 21.2.1 *If* $f \in \mathsf{P}^d(n)$ *and* $Sq^k(f) = 0$ *for all odd k, then* f *is a square.*

Proof The result is true if d is odd, since $f^2 = Sq^d(f) = 0$ implies that $f = 0$. Let d be even, and let $Sq^k(f) = 0$ for all odd k. The result is true for $n = 1$, since $f = (x^{d/2})^2$, and so we argue by induction on n. Let $n > 1$ and suppose the result is true for $\mathsf{P}(n-1)$. Let $x = x_n$, and write $f = \sum_{i=s}^{d} x^i g_i$, for some s such that $0 \leq s \leq d$, where $g_i \in \mathsf{P}(n-1)$. If $g_s \neq 0$, then the term containing the least power of x in the expansion of $Sq^k(f)$ by the Cartan formula is $x^s Sq^k(g_s)$. Hence $Sq^k(g_s) = 0$ if k is odd. By the induction hypothesis, g_s is a square. Hence $\deg g_s$ is even, and since d is even, s is even and $x^s g_s$ is a square, which may be subtracted from f. Upward induction on s, which is bounded by d, then establishes the inductive step for n. □

Proposition 21.2.2 *If $f \in \mathsf{P}^d(n)$ and $Q_r(f) = 0$ for $1 \le r \le n$, then f is a square.*

Proof Since $Q_r(f) = 0$ for $1 \le r \le n$, by Proposition 15.3.4 $Q_r(f) = 0$ for all $r \ge 1$. We argue by contradiction. If f is not a square, then there is a variable $x = x_i$ and an odd integer s such that $f = x^s g + h$, where $g \ne 0$ is independent of x and all odd exponents of x in h are $< s$. Applying Q_r and using Proposition 3.5.10, we have $0 = Q_r(f) = x^{s-1+2^r} g + h'$, where $h' = Q_r(h) + x^s Q_r(g)$. By the choice of s, the odd exponents of x in h' are $< s - 1 + 2^r$. If r is chosen sufficiently large, then $s - 1 + 2^r$ is greater than any even exponent of x in h', because $Q_r(g)$ does not involve x and the application of Q_r does not change even exponents. Then $g = 0$, which is a contradiction. Hence f has no variable with odd exponent, and so it is a square. □

Proposition 21.2.3 *If $f \in \mathsf{P}^d(n)$ for $1 \le d < 2^k - 1$ and $Q_r(f) = 0$ for $1 \le r < k$, then f is a square.*

Proof We argue by induction on n and s, where s is the highest even exponent of $x = x_n$ in f. For $\mathsf{P}(1) = \mathbb{F}_2[x]$ the result is true for all d, since even powers of x are detected by the vanishing of Sq^1. We assume the result is true for $\mathsf{P}(n-1)$, where $n > 1$, and prove it for $\mathsf{P}(n)$ by induction on s.

Let $f \in \mathsf{P}^d(n)$, where $1 \le d < 2^k - 1$ and $Q_r(f) = 0$ for $1 \le r < k$. If all exponents of x in f are even, then $f = \sum_{i \le d/2} x^{2i} g_i$, where $g_i \in \mathsf{P}(n-1)$ for all i. Hence $Q_r(f) = \sum_{i \le d/2} x^{2i} Q_r(g_i)$ by Proposition 3.5.10(i), and so $Q_r(g_i) = 0$ for $1 \le r < k$. Since $\deg g_i \le d < 2^k - 1$, it follows by the induction hypothesis on n that g_i is a square for all i, and so f is a square.

Hence we may assume that $x = x_n$ has an odd exponent in f. Let $t \ge 1$ be the maximum odd exponent of x in f, and let $s \ge 0$ be the maximum even exponent of x in f. Then we can write $f = x^t g_1 + x^s g_2 + h$, where $g_1, g_2 \in \mathsf{P}(n-1)$, $g_1 \ne 0$ and the exponents of x in h are $< t$ if they are odd and are $< s$ if they are even. By Proposition 3.5.10(i), $Q_r(f) = x^{t+2^r-1} g_1 + x^t Q_r(g_1) + x^s Q_r(g_2) + Q_r(h)$ for $r \ge 1$. If $s < t + 2^k - 1$, then $Q_r(f) = x^{t+2^r-1} g_1 +$ terms of lower even exponent in x, and so $Q_r(f) \ne 0$. But $Q_r(f) = 0$ for $1 \le r < k$. Hence $s \ge t + 2^{k-1} - 1$. In particular, $s > 0$ and $g_2 \ne 0$.

We assume as induction hypothesis that the result is true for all $f \in \mathsf{P}(n)$ such that the maximum even exponent of x in f is $< s$. By the argument above, we may assume that $s \geq t + 2^{k-1} - 1$ and that $g_2 \neq 0$. Since $f \in \mathsf{P}^d(n)$ and $d < 2^k - 1$, $\deg g_2 = d - s < (2^k - 1) - (t + 2^{k-1} - 1) = 2^{k-1} - t$. Since $t \geq 1$, $\deg g_2 < 2^{k-1} - 1$. For $1 \leq r < k - 1$, $s > t + 2^r - 1$, and so $Q_r(f) = x^s Q_r(g_2)+$ terms of lower even exponent in x. Since $Q_r(f) = 0$, this implies that $Q_r(g_2) = 0$.

Thus the induction hypothesis on n, with k replaced by $k - 1$, applies to g_2. Hence g_2 is a square, and by subtracting $x^s g_2$ from f it suffices to prove that $f_1 = x^t g_1 + h$ is a square. Since $\deg f_1 = d$, $Q_r(f_1) = Q_r(f) = 0$ for $1 \leq r < k$, and all exponents of x_n in f_1 are $< s$, the induction hypothesis on s applies to f_1. This completes the induction on s, and thus the inductive step on n. □

Proposition 21.2.3 explains the nature of a polynomial of degree $d < 2^k - 1$ which is annihilated by Q_r for $1 \leq r < k$. We next investigate the situation where $d = 2^k - 1$.

The Dickson invariant $\mathrm{d}_{n,0} = \Delta_n$ (15.1.1) is the Vandermonde determinant in the variables $x_1, \ldots, x_n$ with exponents 2^j, $0 \leq j \leq n - 1$. Given $Y \subseteq Z[n]$ with $|Y| = r$, we denote by $\Delta(Y)$ the Dickson invariant of degree $2^r - 1$ in the variables x_i for $i \in Y$, i.e. if $Y = \{i_1, \ldots, i_r\}$ then

$$\Delta(Y) = \begin{vmatrix} x_{i_1} & x_{i_2} & \cdots & x_{i_r} \\ x_{i_1}^2 & x_{i_2}^2 & \cdots & x_{i_r}^2 \\ \vdots & \vdots & \ddots & \vdots \\ x_{i_1}^{2^{r-1}} & x_{i_2}^{2^{r-1}} & \cdots & x_{i_r}^{2^{r-1}} \end{vmatrix}. \tag{21.1}$$

In particular, $\Delta_n = \Delta(Z[n])$ for $n \geq 1$. It is clear from Proposition 3.5.10(i) that Δ_k is annihilated by Q_r for $1 \leq r < k$, because this operation effectively equates two rows of the determinant. In the 3-variable case, for example,

$$\Delta_3 = \begin{vmatrix} x_1 & x_2 & x_3 \\ x_1^2 & x_2^2 & x_3^2 \\ x_1^4 & x_2^4 & x_3^4 \end{vmatrix}, \quad Q_2(\Delta_3) = \begin{vmatrix} x_1^4 & x_2^4 & x_3^4 \\ x_1^2 & x_2^2 & x_3^2 \\ x_1^4 & x_2^4 & x_3^4 \end{vmatrix} = 0.$$

In general Δ_n is the product of all the linear forms in the variables. For example, $\Delta_3 = x_1 x_2 x_3 (x_1 + x_2)(x_1 + x_3)(x_2 + x_3)(x_1 + x_2 + x_3)$.

Proposition 21.2.4 *For $n > 0$ and $d = 2^k - 1$, let $f \in \mathsf{P}^d(n)$ be a polynomial annihilated by Q_r for $1 \leq r < k$. Then $f = \sum_Y a_Y \Delta(Y)$, where $a_Y \in \mathbb{F}_2$ and Y ranges over subsets of $Z[n]$ of cardinality k.*

Proof Using the decomposition $\mathsf{P}(n) = \bigoplus_{Y \subseteq Z[n]} \mathsf{P}(Y)$ of A_2-modules, the problem reduces to the case $n = k$ and showing that if $f \neq 0 \in \mathsf{P}^d(Z[n])$ satisfies the hypotheses of the theorem then $f = \Delta(Z[n])$. If $k > n$ then f is a square by Proposition 21.2.3, but since the degree of f is odd, $f = 0$. Suppose $k \leq n$ and that f is annihilated by Q_r for $1 \leq r < k$. We show that f is divisible by every nonzero linear form in $x_1,\ldots,x_n$, so that $k = n$ and $f = \Delta_n$.

The case $n = 1$ is straightforward. Taking $n > 1$ we have $f = x_1 \cdots x_n g$ for some $g \in \mathsf{P}^{d-n}(n)$ by hypothesis. We argue by induction on the number of variables in a linear form. Assume that f is divisible by all linear forms in $< k$ of the variables. This is true for $k = 2$. Let x be a linear form in k variables. Without loss of generality we may assume, after possibly permuting the variables, that $x = x_1 + \cdots + x_k$. Let $y = x - x_1$ and substitute $x_1 = y$ in f.

By the inductive hypothesis, the resulting expression g already contains the factor y, and it now contains another y in place of x_1. Hence we can write $g = y^2 h$ for some polynomial h. Since substitution commutes with the action of A_2, we see by Proposition 3.5.10(i) that $y^2 Q_j(h) = Q_j(y^2 h) = Q_j(f) = 0$ for $1 \leq j < k$. Hence $Q_j(h) = 0$. Furthermore $\deg h < 2^k - 1$. Hence h is a square by Proposition 21.2.3. However h has odd degree and is therefore 0. We conclude that f vanishes on substitution of $x_1 = y$, and so it is divisible by x. This establishes the inductive step, and shows that f is divisible by all nonzero linear forms in $x_1,\ldots,x_n$. Hence $f = \Delta_n$. □

21.3 Kernels of squaring operations

The vector space of squares of polynomials in $\mathsf{P}^d(m)$ is a $\mathbb{F}_2\mathsf{M}(m)$-submodule of $\mathsf{P}^{2d}(m)$ isomorphic to $\mathsf{P}^d(m)$. It is spanned by blocks with zero first column, and Proposition 21.2.1 identifies it as the intersection of the kernels of all Sq^t with t odd. We wish to investigate the subspace of $\mathsf{P}^d(m)$ spanned by blocks of the form $F|G$ where F is a (m,t)-block of degree $2^{t-1} - 1$. In particular, column t of F is zero.

We begin by considering a more general situation. Recall that $\mathsf{P}(m) \otimes \mathsf{P}(m)$ is a right $\mathbb{F}_2\mathsf{M}(m)$-module with the diagonal action $(f \otimes g) \cdot A = (f \cdot A) \otimes (g \cdot A)$, where $f, g \in \mathsf{P}(m)$ and $A \in \mathsf{M}(m)$.

Proposition 21.3.1 *Let $\phi_t^m : \mathsf{P}^{d_1}(m) \otimes \mathsf{P}^{d_2}(m) \to \mathsf{P}^{d_1 + 2^t d_2}(m)$ be the linear map defined by $\phi_t^m(g \otimes h) = gh^{2^t}$, where $g \in \mathsf{P}^{d_1}(m)$ and $h \in \mathsf{P}^{d_2}(m)$. Then ϕ_t^m is a $\mathbb{F}_2\mathsf{M}(m)$-module map, and is injective if $d_1 < 2^t$.*

Proof Since $A \in \mathsf{M}(m)$ acts on $\mathsf{P}(m)$ by $(gh^{2^t}) \cdot A = (g \cdot A)(h \cdot A)^{2^t}$, ϕ_t^m is a map of $\mathbb{F}_2\mathsf{M}(m)$-modules. A nonzero element of $\mathsf{P}^{d_1}(m) \otimes \mathsf{P}^{d_2}(m)$ has the form $f = \sum_{j=1}^r g_j \otimes h_j$ for some $r > 0$, where $g_1,\ldots,g_r$ are distinct monomials in $\mathsf{P}^{d_1}(m)$ and $h_1,\ldots,h_r$ are nonzero polynomials in $\mathsf{P}^{d_2}(m)$. If $d_1 < 2^t$, then $\phi_t^m(f) = \sum_{j=1}^r g_j h_j^{2^t}$ is represented by a sum of concatenated blocks $G_j|H_j$, where G_j is the (m,t)-block corresponding to g_j, and H_j is a formal sum of blocks corresponding to h_j. Since $G_1,\ldots,G_r$ are distinct, there is no cancellation. Hence $\phi_t^m(f) \neq 0$. □

In the 1-variable case, $\phi_t^1(x^{d_1} \otimes x^{d_2}) = x^d$, where $d = d_1 + 2^t d_2$. Hence ϕ_t^1 is an isomorphism for all d_1, d_2 and t. It is clear for dimensional reasons that ϕ_t^m is not injective in general when $m > 1$. We next characterize the image of ϕ_t^m in terms of the Steenrod algebra when $d_1 < 2^{t-1}$.

Proposition 21.3.2 *If* $d_1 < 2^{t-1}$, *then* ϕ_t^m *is an isomorphism from* $\mathsf{P}^{d_1}(m) \otimes \mathsf{P}^{d_2}(m)$ *to the intersection of the kernels of all* Sq^s *with* $2^{t-1} \in \mathrm{bin}(s)$, *acting on* $\mathsf{P}^{d_1+2^t d_2}(m)$.

Proof Let $f = gh^{2^t}$ where $g \in \mathsf{P}^{d_1}(m)$ and $h \in \mathsf{P}^{d_2}(m)$. By the Cartan formula and Proposition 1.3.3,

$$Sq^s(f) = \sum_{i+2^t j=s} Sq^i(g)(Sq^j(h))^{2^t}.$$

If $2^{t-1} \in \mathrm{bin}(s)$ and $i + 2^t j = s$, then $2^{t-1} \in \mathrm{bin}(i)$. Thus if $d_1 < 2^{t-1}$, then $i > d_1 = \deg g$, and so $Sq^i(g) = 0$. It follows that $Sq^s(f) = 0$. Hence the image of ϕ_t^m lies in the intersection of the kernels of the squares Sq^s, where $2^{t-1} \in \mathrm{bin}(s)$.

Conversely, let $f \in \mathsf{P}^d(m)$, where $d = d_1 + 2^t d_2$ with $d_1 < 2^t$ and assume that $Sq^s(f) = 0$ whenever $2^{t-1} \in \mathrm{bin}(s)$. If $2^{t-1} \in \mathrm{bin}(d)$, then by choosing $s = d$ we obtain $f^2 = Sq^d(f) = 0$ and $f = 0$. Hence we may assume that $2^{t-1} \notin \mathrm{bin}(d)$, and write $d = d_1 + 2^t d_2$ where $d_1 < 2^{t-1}$. We shall show by induction on m that $f \in \mathsf{P}^d(m)$ is of the form $f = \sum_i g_i h_i^{2^t}$, where $g_i \in \mathsf{P}^{d_1}(m)$ and $h_i \in \mathsf{P}^{d_2}(m)$.

This is clear when $m = 1$, since $x^d = x^{d_1}(x^{d_2})^{2^t}$. For $m > 1$, we assume the result for $\mathsf{P}(m-1)$. Given $f \in \mathsf{P}(m)$, let $f = f_0 x^d + f_1 x^{d-1} + \cdots + f_{d-k} x^k$ for some $k \leq d$, where $x = x_m$, $f_i \in \mathsf{P}^i(m-1)$ for $i \leq d-k$ and $f_{d-k} \neq 0$. Expanding $Sq^s(f)$ by the Cartan formula, the term of lowest degree in x is $x^k Sq^s(f_{d-k})$. Hence the given condition $Sq^s(f) = 0$ implies that $Sq^s(f_{d-k}) = 0$. Since $Sq^s(f) = 0$ for all s such that $2^{t-1} \in \mathrm{bin}(s)$, the same is true for f_{d-k}. Hence $2^{t-1} \notin \mathrm{bin}(d-k)$, and so $f_{d-k} = \sum_i g_i'(h_i')^{2^t}$ by the inductive hypothesis, where $g_i' \in \mathsf{P}^{d_1}(m)$, $h_i' \in \mathsf{P}^{d_2}(m)$, and $d - k = d_1' + 2^t d_2'$ where $d_1' < 2^{t-1}$.

Let $s = d - k + 2^{t-1}$. Since $2^{t-1} \notin \mathrm{bin}(d-k)$, $2^{t-1} \in \mathrm{bin}(s)$, and so $Sq^s(f) = 0$. Since $Sq^i(f_r) = 0$ if $i > r$, the term of lowest degree in x when f is written as above and $Sq^s(f)$ is expanded by the Cartan formula is $Sq^{d-k}(f_{d-k})Sq^{2^{t-1}}(x^k) = f_{d-k}^2 Sq^{2^{t-1}}(x^k)$. Since $Sq^s(f) = 0$ and $f_{d-k} \neq 0$, it follows that $Sq^{2^{t-1}}(x^k) = 0$, and so $2^{t-1} \notin \mathrm{bin}(k)$.

Thus let $k = d_1'' + 2^t d_2''$, where $d_1'' < 2^{t-1}$. Then $d = (d-k)+k = (d_1' + d_1'') + 2^t(d_2' + d_2'')$. Since $d_1' + d_1'' < 2^t$ and $2^{t-1} \notin \mathrm{bin}(d)$, $d_1' + d_1'' < 2^{t-1}$. Hence $d_1 = d_1' + d_1''$ and $d_2 = d_2' + d_2''$, and

$$x^k f_{d-k} = \sum_i (x^{d_1'} g_i')(x^{d_2'} h_i')^{2^t},$$

where $\deg(x^{d_1'} g_i') = d_1$ and $\deg(x^{d_2'} h_i') = d_2$. Hence we may cancel this term from the expansion of f, to obtain a polynomial divisible by x^{k+1}. The inductive step for m is completed by upward induction on k, which is bounded above by d. □

21.4 Root generation

The action of the operations $Q_r = Sq(0,\ldots,0,1)$ on monomials can be described in terms of blocks, generalizing the description for $Sq^1 = Q_1$ given in Section 6.1. In the case $n = 1$, $Q_r(x^d) = x^{2^r+d-1}$ when d is odd by Proposition 3.5.10(iii). Thus $f = x^d$ is represented by a 1-block F with $F_{(1,1)} = 1$, and $g = Q_r(f)$ by a block G formed from F by putting the first entry equal to 0 and adding 1 arithmetically at position $r+1$ in F. If $F_{1,r+1} = 0$, then this amounts to moving the initial 1 of F into position $r+1$.

Example 21.4.1 The equations $Q_2(x^{13}) = x^{16}$ and $Q_3(x^{21}) = x^{28}$ are written in block notation as $Q_2(1\ 0\ 1\ 1) = 0\ 0\ 0\ 0\ 1$ and $Q_3(1\ 0\ 1\ 0\ 1) = 0\ 0\ 1\ 1\ 1$.

More generally, the effect of Q_r on a block F with a single entry 1 in its first column, in row i say, is to form a new block G from F by putting $G_{i,1} = 0$ and adding 1 arithmetically to row i at position $(i, r+1)$. Then G represents a squared monomial $g = h^2$. We consider processes which involve compositions of the operations Q_r and the square root operation $h^2 \mapsto h$.

Definition 21.4.2 For $f, g \in \mathrm{P}(n)$, f is **root-generated** by g if $f = \Phi(g)$, where Φ is a composition of operations Q_r and taking square roots.

For example, x^5 is root-generated by x^3, because $Q_3(x^3) = x^{10} = (x^5)^2$. The set of variables which divide a monomial is not changed by Q_r or by the square root operation, so $\mathrm{P}(Y) \subset \mathrm{P}(n)$ is preserved by root generation for all $Y \subseteq Z[n]$.

Proposition 21.4.3 *If $f \in \mathrm{P}^d(m)$ is root-generated by g and $g * L = 0$ for $L \in Hom_{\mathrm{A}_2}(\mathrm{P}(m), \mathrm{P}(n))$, then $f * L = 0$.*

Proof It suffices to prove that if $g * L = 0$, then $f * L = 0$ if $f = Q_r(g)$ or if $f^2 = g$. In the first case, $f * L = Q_r(g) * L = Q_r(g * L) = Q_r(0) = 0$. In the second case, $f*L \in \mathsf{P}^d(n)$ since L preserves grading. Hence $(f*L)^2 = Sq^d(f*L) = Sq^d(f)*L = f^2*L = g*L = 0$. It follows that $f*L = 0$. □

The main result of this section is as follows.

Theorem 21.4.4 *If* $L \in Hom_{\mathsf{A}_2}(\mathsf{P}(\mathsf{Z}[m]), \mathsf{P}(n))$ *and* $f*L = 0$ *for all monomials* $f = x_1^{d_1}\cdots x_m^{d_m}$ *with* $1 \le d_i < 2^n$ *for* $1 \le i \le m$*, then* $L = 0$.

Proof Given a monomial $f = x_1^{d_1}\cdots x_m^{d_m}$ with $d_i \ge 1$ for $1 \le i \le m$, let $G(f)$ denote the set of monomials $g = x_1^{a_1}\cdots x_m^{a_m}$ such that

1. if d_i is even, then $a_i < d_i$;
2. if d_i is odd and $d_i > 2^n - 1$, then $a_i < d_i$;
3. if d_i is odd and $d_i \le 2^n - 1$, then $a_i \le 2^n - 1$.

The proof is divided into three steps. In Steps 1 and 2, we show that $f*L = 0$ if $g*L = 0$ for all $g \in G(f)$. This statement is true if there are no odd exponents d_i, because the square root g of f then belongs to $G(f)$ and $g*L = 0$ implies $f*L = 0$.

The exponent of x_i in a monomial in $G(f)$ is $< d_i$ if $d_i > 2^n - 1$, and is $\le 2^n - 1$ otherwise. Thus Step 3 consists of iterating Steps 1 and 2 on $G(f)$ until all exponents are $\le 2^n - 1$.

The following example illustrates the procedure.

Example 21.4.5 Let $n = m = 2$, let $f = x^5y$ and let $L \in \mathrm{End}_{\mathsf{A}_2}(\mathsf{P}(2))$. In Step 1, starting with the variable x, we have

$$f^2 = x^{10}y^2 = Q_1(x^9y^2), \quad Q_2(x^9y^2) = x^{12}y^2, \quad Q_3(x^9y^2) = x^{16}y^2.$$

Let $S = \{x^6y, x^8y\}$. From Proposition 15.3.4 we have the relation $Q_3 = \mathsf{d}_0^2Q_1 + \mathsf{d}_1^2Q_2$, where d_0, d_1 are the Dickson invariants in x and y. It follows that if $Q_2(x^9y^2*L) = 0$ and $Q_3(x^9y^2*L) = 0$ then $Q_1(x^9y^2*L) = 0$, i.e. $f^2*L = 0$ and so $f*L = 0$. Thus if L vanishes on S then L vanishes on f. In Step 2 we carry out Step 1 on each element of S, concentrating on the variable y. Thus

$$(x^6y)^2 = x^{12}y^2 = Q_1(x^{12}y), \quad Q_2(x^{12}y) = x^{12}y^4, \quad Q_3(x^{12}y) = x^{12}y^8,$$
$$(x^8y)^2 = x^{16}y^2 = Q_1(x^{16}y), \quad Q_2(x^{16}y) = x^{16}y^4, \quad Q_3(x^{16}y) = x^{16}y^8.$$

Taking square roots twice, we obtain the set $T = \{x^3y, x^3y^2, x^4y, x^4y^2\}$ such that if L vanishes on T then L vanishes on S. Thus $T \subset G(f)$. Replacing x^4y^2 by x^2y we have the set $T' = \{x^3y, x^3y^2, x^4y, x^2y\}$, where all exponents are < 4 except for x^4y, to which we apply Steps 1 and 2 once again, to obtain the set $\{xy, xy^2\}$. The final set $\{x^3y, x^3y^2, x^2y, xy^2, xy\}$ has no exponent > 3, and $f*L = 0$ if L is 0 on this set.

To complete the proof of Theorem 21.4.4, we continue by describing Steps 1 and 2 in the general case.

Step 1. Suppose that one of the variables $x = x_i$ has odd exponent d, and let $f = x^d h$. Then $f^2 = x^{2d}h^2 = Q_1(x^{2d-1}h^2)$. Taking $k = n+1$ in Proposition 15.3.4, $Q_{n+1} = \mathrm{d}_0^2 Q_1 + \mathrm{d}_1^2 Q_2 + \cdots + \mathrm{d}_{n-1}^2 Q_n$. Applying this operator to the polynomial $x^{2d-1}h^2 * L$, we obtain $\mathrm{d}_0^2 Q_1(x^{2d-1}h^2 * L) = \mathrm{d}_1^2 Q_2(x^{2d-1}h^2 * L) + \cdots + \mathrm{d}_{n-1}^2 Q_n(x^{2d-1}h^2 * L) + Q_{n+1}(x^{2d-1}h^2 * L)$, and so $\mathrm{d}_0^2(f^2 * L) = \mathrm{d}_1^2(Q_2(x^{2d-1}h^2) * L) + \cdots + \mathrm{d}_{n-1}^2(Q_n(x^{2d-1}h^2) * L) + Q_{n+1}(x^{2d-1}h^2) * L$. Hence $f^2 * L = 0$ if $Q_{r+1}(x^{2d-1}h^2) * L = 0$ for $1 \leq r \leq n$. By Proposition 3.5.10, $Q_{r+1}(x^{2d-1}h^2) = x^{2^{r+1}+2d-2}h^2$. Taking square roots, $f * L = 0$ if $(x^{2^r+d-1}h) * L = 0$ for $1 \leq r \leq n$. The above procedure replaces $f = x^d h$ by a set S of monomials $x^{2^r+d-1}h$, in each of which the original odd exponent d of x is replaced by an even exponent $2^r + d - 1$ for $1 \leq r \leq n$, while the other exponents are unaltered. The set S has the property that $f * L = 0$ if $g * L = 0$ for all $g \in S$.

Step 2. We iterate the procedure of Step 1 on the monomials in S, one variable at a time, until all exponents are even. At each stage, the original even exponents of f remain unaltered, and no new exponent of x_i exceeds $d_i + 2^n - 1$. Following this, we take square roots of the monomials to obtain a set T such that $f * L = 0$ if $g * L = 0$ for all $g \in T$. We shall show that $T \subseteq G(f)$. If d_i is even, then the new exponent is $d_i/2 < d_i$. If d_i is odd and $d_i > 2^n - 1$, then the new exponent is $\leq (d_i + 2^n - 1)/2 < d_i$. If d_i is odd and $d_i \leq 2^n - 1$, then the new exponent is $\leq (d_i + 2^n - 1)/2 \leq 2^n - 1$. Hence $T \subseteq G(f)$. □

Proposition 21.4.6 *The $\mathbb{F}_2$-vector space $Hom_{\mathsf{A}_2}(\mathsf{P}(m), \mathsf{P}(n))$ is finite dimensional.*

Proof By the decomposition $\mathrm{Hom}_{\mathsf{A}_2}(\mathsf{P}(m), \mathsf{P}(n)) = \bigoplus_{Y \subseteq Z[m]} \mathrm{Hom}_{\mathsf{A}_2}(\mathsf{P}(Y), \mathsf{P}(n))$ of Section 21.0, it suffices to prove that $\mathrm{Hom}_{\mathsf{A}_2}(\mathsf{P}(Z[m]), \mathsf{P}(n))$ is finite dimensional. By Theorem 21.4.4, an element $L \in \mathrm{Hom}_{\mathsf{A}_2}(\mathsf{P}(Z[m]), \mathsf{P}(n))$ is determined by its values on a finite set of monomials $g \in \mathsf{P}(Z[m])$. Since each $g * L$ is a polynomial in $\mathsf{P}(n)$ of the same degree as g, it is determined by a finite set of coefficients in $\mathbb{F}_2$. □

The preceding argument does not yield a satisfactory estimate for the dimension of $\mathrm{Hom}_{\mathsf{A}_2}(\mathsf{P}(m), \mathsf{P}(n))$. We need to take into account further constraints imposed by A_2 on the values of $L \in \mathrm{Hom}_{\mathsf{A}_2}(\mathsf{P}(m), \mathsf{P}(n))$ on certain 'key' monomials, which we introduce in Section 21.5.

Example 21.4.7 By Theorem 21.4.4, an A_2-module map $L : \mathsf{P}(m) \to \mathsf{P}(1)$ is determined by its values on the square free monomials in x_i, $1 \leq i \leq m$. Thus, for each $Y \subseteq Z[m]$, L is determined by its value on just one monomial, the product of the variables x_i for $i \in Y$. In particular, there are four possible elements $L \in \mathrm{End}_{\mathsf{A}_2}(\mathsf{P}(1))$, which map $(1,x)$ to $(0,0)$, $(1,x)$, $(1,0)$ and $(0,x)$ respectively. These are realized by the four elements $0, (1), (0), (0)+(1)$ of $\mathbb{F}_2\mathsf{M}(1)$ as shown in Example 21.0.2.

21.5 Key monomials

In this section we show that $L \in \mathrm{Hom}_{\mathsf{A}_2}(\mathsf{P}(\mathsf{Z}[m]), \mathsf{P}(n))$ is determined by its value on a single suitably chosen monomial. We recall from Definition 3.3.7 that the row sum sequence of the block associated to a monomial $f = x_1^{d_1} \cdots x_m^{d_m}$ is $\alpha(f) = (\alpha(d_1), \ldots, \alpha(d_m))$. We associate to each sequence $\rho \in \mathsf{Seq}$ two monomials with α-sequence ρ, the 'key' monomial $\mathsf{k}(\rho)$ and the 'subkey' monomial $\mathsf{k}'(\rho)$. Although $\mathsf{k}'(\rho)$ appears to be the more natural choice, we prefer to work with $\mathsf{k}(\rho)$, because $\mathsf{k}(\rho) * L$ lies in a subspace of $\mathsf{P}^d(n)$ which can be identified as the intersection of the kernels of Sq^s for certain values of s which are determined by ρ. This will be proved in Section 21.6.

Definition 21.5.1 Given a sequence $\rho = (r_1, \ldots, r_m)$ of integers $r_i > 0$, let $s_1 = 0$ and $s_i = i - 1 + r_1 + \cdots + r_{i-1}$ for $2 \le i \le m$. Define $d_i = 2^{r_i} - 1$. Then $\mathsf{k}(\rho) = x_1^{2^{s_1} d_1} \cdots x_m^{2^{s_m} d_m}$ is the **key monomial** associated with ρ. The **subkey monomial** $\mathsf{k}'(\rho)$ is defined similarly, with s_i replaced by $s_i' = r_1 + \cdots + r_{i-1}$.

The key monomial $\mathsf{k}(\rho)$ lies in $\mathsf{P}^d(m)$, where $d = \sum_{i=1}^m 2^{s_i}(2^{r_i} - 1)$. The blocks corresponding to $\mathsf{k}(\rho)$ and $\mathsf{k}'(\rho)$ are denoted by $\mathsf{K}(\rho)$ and $\mathsf{K}'(\rho)$. A subkey block is a concatenation $F_1|F_2 \cdots |F_m$, where all rows of F_i are zero except the ith row, which has length r_i and all entries equal to 1. A key block differs from a subkey block in having a column of 0s separating the subblocks F_i.

Example 21.5.2

$$\mathsf{K}'(2,3,1) = \begin{matrix} 1\,1 \\ 0\,0\,1\,1\,1 \\ 0\,0\,0\,0\,0\,1 \end{matrix}\;, \quad \mathsf{K}(2,3,1) = \begin{matrix} 1\,1\,0 \\ 0\,0\,0\,1\,1\,1\,0 \\ 0\,0\,0\,0\,0\,0\,0\,1 \end{matrix}\;.$$

Proposition 21.5.3 *Let $f \in \mathsf{P}(\mathsf{Z}[m])$ be a monomial with $\alpha(f) = \rho$. Then f is root-generated by the subkey monomial $\mathsf{k}'(\rho)$.*

Proof Let F be the block representing $f \in \mathsf{P}(\mathsf{Z}[m])$ and let s be the length of row m of F, which is the last row of F and is nonzero by assumption. If $s > 1$ then we form the block G from F by moving all columns forward one place, and then moving the last entry $F_{m,s} = 1$ into position $(m,1)$ of G. For example

$$F = \begin{matrix} 101 \\ 1101 \\ 011 \end{matrix}, \quad G = \begin{matrix} 0101 \\ 01101 \\ 101 \end{matrix},$$

where $s = 3$. In general $f^2 = Q_s(g)$ and $\alpha(f) = \alpha(g)$. The procedure is iterated on the last row of G until all entries 1 have been moved to the left and are then contiguous. The whole process is repeated on each row from bottom to top of the block. The final block for the above example is

$$H = \begin{matrix} 11 \\ 00111 \\ 0000011 \end{matrix}.$$

In general, the final outcome H is a subkey block. Since at each stage of the process the square of the original block can be recovered by applying Q_r for some r, f is root-generated by h. □

In Definition 21.5.1 we have chosen to define a subkey or key block with respect to a particular order of the F_i in the concatenation $\mathsf{K}'(\rho) = F_1|F_2\cdots|F_m$. We see that the block F formed by a permutation of the blocks F_i can be produced by carrying out the procedure in the above proof on the rows of $\mathsf{K}'(\rho)$ in a suitable order. Hence F also serves as a root-generator of all monomials with α-sequence ρ. In fact, more is true.

Proposition 21.5.4 *Let $F = F_1|F_2\cdots|F_m$ be a concatenated block such that the only nonzero row of F_i is row i. Then the associated monomial f of F is a root-generator of all monomials in $\mathsf{P}(\mathsf{Z}[m])$ with α-sequence $\alpha(f)$.*

Proof For $\rho = (2,3,2)$, an example of such a block is

$$F = \begin{array}{c|c|c} 101 & 0000 & 0000 \\ 000 & 1101 & 0000 \\ 000 & 0000 & 0101 \end{array}.$$

It is enough to prove that F is a root-generator of the subkey block $\mathsf{K}'(\rho)$, where $\rho = \alpha(F)$. To do this we start with the top row of F, take iterated square roots of f, if necessary, to bring an entry 1 of this row into the first column and then apply Q_r, for suitable r, to move that entry into the next after last column of the block. For the above example F, we obtain

$$\begin{array}{cc|cccc|cccc|c} 0&1&0&0&0&0&0&0&0&0&1\\ 0&0&1&1&0&1&0&0&0&0&0\\ 0&0&0&0&0&0&0&1&0&1&0 \end{array}.$$

The process is iterated until all the entries 1 in the first row of F are moved into contiguous places on the right, and is then repeated on the rows of the new blocks in turn. Taking further square roots if necessary, we finally arrive at the subkey block $\mathsf{K}'(\rho)$. The result is

$$\begin{array}{cc|ccc|cc} 1&1&0&0&0&0&0\\ 0&0&1&1&1&0&0\\ 0&0&0&0&0&1&1 \end{array}$$

for the example F above. □

From Proposition 21.5.4 we deduce in particular that the key monomial $\mathsf{k}(\rho)$, where ρ is a positive sequence of length m, root-generates all monomials $f \in \mathsf{P}(\mathsf{Z}[m])$ with $\alpha(f) = \rho$. Using the following partial order on positive sequences, we widen the scope of Proposition 21.5.3.

Definition 21.5.5 Let $\rho = (r_1, \ldots, r_m)$ and $\sigma = (s_1, \ldots, s_m)$ be positive sequences of length m. Then $\sigma \leq_p \rho$ if $s_i \leq r_i$ for all $1 \leq i \leq m$.

If $\sigma <_p \rho$, then σ can be obtained from ρ through a chain of inequalities $\sigma' <_p \rho'$, where $s'_i = r'_i$ for $1 \leq i \leq n$, $i \neq j$ for some j where $s'_j = r'_j - 1$.

Proposition 21.5.6 *Let ρ be a positive sequence of length m. The key monomial $\mathsf{k}(\rho)$ root-generates every monomial $f \in \mathsf{P}(\mathsf{Z}[m])$ with $\alpha(f) \leq_p \rho$.*

Proof It is enough to show that $\mathsf{k}(\rho)$ root-generates $k(\sigma)$ in the case where ρ and σ differ only in one position j, where $s_j = r_j - 1 > 0$. We deal with the case $j = 1$, as the other cases follow the same pattern. Let $\rho = (r_1, \ldots, r_m)$ where $r_1 > 1$. We note that $F = Q_{r_1 - 1}(\mathsf{K}(\rho))$ has α-sequence $(r_1 - 1, r_2, \ldots, r_m)$. Further, F is a block to which Proposition 21.5.4 applies, as illustrated below for $\rho = (3, 3, 2)$.

$$\mathsf{K}(\rho) = \begin{array}{c} 1110000000\\ 0000111000\\ 0000000011 \end{array}, \quad F = Q_2(\mathsf{K}(\rho)) = \begin{array}{c} 0101000000\\ 0000111000\\ 0000000011 \end{array}.$$

In particular, F root-generates all blocks with α-sequence σ. A similar argument applies to $j > 1$ after switching F_1 and F_j in the concatenation $\mathsf{K}(\rho) = F_1|F_2\cdots|F_m$. □

By Theorem 21.4.4, $L \in \mathrm{Hom}_{\mathsf{A}_2}(\mathsf{P}(Z[m]), \mathsf{P}(n))$ is determined by its values on monomials $f = x_1^{d_1}\cdots x_m^{d_m}$, where $0 < d_i \leq 2^n - 1$ for $1 \leq i \leq m$. Since $\alpha(f) \leq_p (n,\ldots,n)$, Definition 21.5.1 and Proposition 21.5.6 lead to the following conclusion.

Theorem 21.5.7 *An element $L \in \mathrm{Hom}_{\mathsf{A}_2}(\mathsf{P}(Z[m]), \mathsf{P}(n))$ is determined by its value on the key monomial $\mathsf{k}(\rho)$, where $\rho = (n,\ldots,n)$ is a constant sequence of length m.* □

Explicitly, $\mathsf{k}(\rho) = x_1^{d_1}\cdots x_m^{d_m}$ where $d_{i+1} = (2^n - 1)2^{(n+1)i}$ for $0 \leq i < m$. We note that L is also determined by its value on the subkey monomial $\mathsf{k}'(\rho) = x_1^{d'_1}\cdots x_m^{d'_m}$ where $d'_{i+1} = (2^n - 1)2^{ni}$ for $0 \leq i < m$. Since $\mathsf{P}(Y) \cong \mathsf{P}(Z[m])$ where $|Y| = m$, we can extend Definition 21.5.1 so as to define key monomials in $\mathsf{P}(Y)$.

Example 21.5.8 By Theorem 21.4.4, an element $L \in \mathrm{Hom}_{\mathsf{A}_2}(\mathsf{P}(Z[m]), \mathsf{P}(1))$ is determined by its value on $x_1\cdots x_m$. Theorem 21.5.7 states that L is determined by its value on the key monomial $x_1x_2^2\cdots x_m^{2^{m-1}}$. The subkey monomial is not always of least possible degree for the determination of L. On the other hand $L \in \mathrm{Hom}_{\mathsf{A}_2}(\mathsf{P}(Z[1]), \mathsf{P}(n))$ is determined by its value on the subkey monomial x^{2^n-1}. Thus an A_2-map $L : \mathsf{P}(1) \to \mathsf{P}(n)$ is determined by its values on 1 and x^{2^n-1}, and there is no lower power of x with this property.

Example 21.5.9 Let $L \in \mathrm{Hom}_{\mathsf{A}_2}(\mathsf{P}(Z[1]), \mathsf{P}(2))$ be the sum of the linear substitutions $x \mapsto x$, $x \mapsto y$ and $x \mapsto x + y$. Then $x * L = 0$, $x^2 * L = 0$ but $x^3 * L = x^2y + xy^2$. Hence L is not determined by its values on x and x^2 alone.

Example 21.5.10 An element of $\mathrm{End}_{\mathsf{A}_2}(\mathsf{P}(2))$ is determined by its values on the four subkey monomials $1, x^3, y^3, x^3y^{12}$. For $\mathrm{Hom}_{\mathsf{A}_2}(\mathsf{P}(1), \mathsf{P}(2))$, the subkey monomials are $1, x^3$, and for $\mathrm{Hom}_{\mathsf{A}_2}(\mathsf{P}(2), \mathsf{P}(1))$, they are $1, x, y, xy^2$.

21.6 Values on key monomials

In this section we find constraints imposed by A_2 on the values of elements in $\mathrm{Hom}_{\mathsf{A}_2}(\mathsf{P}(m), \mathsf{P}(n))$ on key monomials in $\mathsf{P}(m)$. We first generalize the map ϕ_t^m defined in Proposition 21.3.1.

Definition 21.6.1 Given $\sigma = (s_1, \ldots, s_k) \in \mathsf{Seq}$, the map

$$\phi_\sigma^m : \mathsf{P}^{d_1}(m) \otimes \cdots \otimes \mathsf{P}^{d_k}(m) \to \mathsf{P}^d(m)$$

is defined by $\phi_\sigma^n(f_1 \otimes \cdots \otimes f_k) = f_1^{2^{s_1}} \cdots f_k^{2^{s_k}}$, where $f_i \in \mathsf{P}^{d_i}(m)$ and $d = \sum_{i=1}^k 2^{s_i} d_i$.

Since matrices in $\mathsf{M}(m)$ act multiplicatively on polynomials, ϕ_σ^n is a map of $\mathbb{F}_2\mathsf{M}(m)$-modules, where $\mathbb{F}_2\mathsf{M}(m)$ acts diagonally on the tensor product. For the same reason, $(f_1^{2^{s_1}} \cdots f_k^{2^{s_k}}) \cdot A = (f_1 \cdot A)^{2^{s_1}} \cdots (f_k \cdot A)^{2^{s_k}}$ for $A \in \mathsf{M}(m,n)$. Consequently A maps the image of ϕ_σ^m into the image of ϕ_σ^n. We wish to prove the corresponding statement for $L \in \mathrm{Hom}_{\mathsf{A}_2}(\mathsf{P}(m), \mathsf{P}(n))$, under suitable conditions on σ. We begin by generalizing Proposition 21.3.1.

Proposition 21.6.2 *If $k = 1$ or if $k \geq 2$ and $2^{s_i} > \sum_{j=1}^{i-1} 2^{s_j} d_j$ for $2 \leq i \leq k$, then ϕ_σ^m is injective, where $\sigma = (s_1, \ldots, s_k)$.*

Proof The case $k = 1$ is true because the squaring map is an isomorphism on to its image, and for the same reason the general case follows from the case $s_1 = 0$. The case $k = 2$ is true by Proposition 21.3.1. We argue by induction on k. Let $k > 2$ and assume the proposition is true for numbers less than k. Let $M = \mathsf{P}^{d_1}(m) \otimes \cdots \otimes \mathsf{P}^{d_{k-1}}(m)$, and let $d' = \sum_{i=1}^{k-1} 2^{s_i} d_i$. Then $\sigma' = (s_1, \ldots, s_{k-1})$ satisfies $2^{s_i} > \sum_{j=1}^{i-1} 2^{s_j} d_j$ for $2 \leq i \leq k-1$, and the inductive hypothesis implies that $\phi_{\sigma'}^m : M \to \mathsf{P}^{d'}(m)$ is injective. Also $2^{s_k} > d'$ by the given conditions. By Proposition 21.3.1, $\phi_{s_k}^m : \mathsf{P}^{d'}(m) \otimes \mathsf{P}^{d_k}(m) \to \mathsf{P}^d(m)$ is injective. Since $\phi_\sigma^m = \phi_{s_k}^m(\phi_{\sigma'}^m \otimes 1)$, we see that ϕ_σ^m is injective. This completes the inductive step. □

We can use the same method to generalize Proposition 21.3.2.

Proposition 21.6.3 *Let $\sigma = (s_1, \ldots, s_k)$. If $2^{s_i - 1} > \sum_{j=1}^{i-1} 2^{s_j} d_j <$ for $2 \leq i \leq k$, then ϕ_σ^m is injective and its image is the intersection of the kernels of Sq^s for all s such that $2^{s_i - 1} \in bin(s)$ for some i in the range $2 \leq i \leq k$. In particular, this holds if σ is strictly increasing and $d_i = 2^{s_{i+1} - s_i - 1} - 1$ for $1 \leq i \leq k$.*

Proof We argue by induction on k for $k \geq 2$. Since $d_1 < 2^{s_2 - 1}$, the case $k = 2$ is given by setting $t = s_2$ in Proposition 21.3.2. Let $k > 2$ and assume that the proposition is true for numbers less than k. Let $M = \mathsf{P}^{d_1}(m) \otimes \cdots \otimes \mathsf{P}^{d_{k-1}}(m)$, $d' = \sum_{i=1}^{k-1} 2^{s_i} d_i$ and $\sigma' = (s_1, \ldots, s_{k-1})$. By the inductive hypothesis, the image of $\phi_{\sigma'}^m : M \to \mathsf{P}^{d'}(m)$ is the subspace of $\mathsf{P}^{d'}(m)$ characterized as the intersection of the kernels of Sq^s where $2^{s_i - 1} \in \mathrm{bin}(s)$ for at least one i in the range $2 \leq i \leq k-1$. Since $2^{s_k - 1} > d'$, Proposition 21.3.2 identifies the image of $\phi_t^m : \mathsf{P}^{d'}(m) \otimes \mathsf{P}^{d_k}(m) \to \mathsf{P}^d(m)$ as the intersection of the kernels of those Sq^s such that $2^{s_k - 1} \in \mathrm{bin}(s)$. Since $\phi_\sigma^m = \phi_{s_k}^m(\phi_{\sigma'}^m \otimes 1)$, we see that ϕ_σ^m is injective and its image is identified as the subspace of $\mathsf{P}^d(m)$ annihilated by all Sq^s such that $2^{s_i-1} \in \text{bin}(s)$ for at least one i in the range $2 \leq i \leq k$.

For the last statement, we show that $\sum_{j=1}^{i-1} 2^{s_j} d_j < 2^{s_i-1}$ for $2 \leq i \leq k$. Again we use induction on i. For $i = 2$ the statement is $2^{s_1}(2^{s_2-s_1-1} - 1) = 2^{s_2-1} - 2^{s_1} < 2^{s_2-1}$. Assume that the statement is true for some $i < k$. Then $\sum_{j=1}^{i} 2^{s_j} d_j < 2^{s_i-1} + 2^{s_i} d_i = 2^{s_i-1} + (2^{s_{i+1}-1} - 2^{s_i}) = 2^{s_{i+1}-1} - 2^{s_i-1} < 2^{s_{i+1}-1}$, so the statement is true for $i+1$, completing the induction. □

An element $L \in \mathrm{Hom}_{\mathsf{A}_2}(\mathsf{P}(m), \mathsf{P}(n))$ preserves kernels of Steenrod squares. Hence the next result follows immediately from Proposition 21.6.3.

Proposition 21.6.4 *Let $\sigma = (s_1, \ldots, s_k) \in \mathsf{Seq}$, where $2^{s_i-1} > \sum_{j=1}^{i-1} 2^{s_j} d_j$ for $2 \leq i \leq k$ and $d = \sum_{i=1}^{k} 2^{s_i} d_i$. Then $L \in Hom_{\mathsf{A}_2}(\mathsf{P}(m), \mathsf{P}(n))$ maps the image of ϕ_σ^m into the image of ϕ_σ^n.* □

The key monomial $\mathsf{k}(\rho)$ is preferred over the subkey monomial $\mathsf{k}'(\rho)$ because $\mathsf{k}(\rho)$ is located in the image of ϕ_σ^m, a space characterized by Steenrod squares according to Proposition 21.6.3. We recall the Dickson invariant $\Delta(Y)$ (21.1) and the partial order $<_p$ of Definition 21.5.5.

Proposition 21.6.5 *Let $\rho = (r_1, \ldots, r_m)$ be a positive sequence. Suppose that $L \in Hom_{\mathsf{A}_2}(\mathsf{P}(m), \mathsf{P}(n))$ annihilates all key monomials $\mathsf{k}(\rho')$ for $\rho' <_p \rho$. Then $\mathsf{k}(\rho) * L$ is a sum of terms of the form $\Delta(Y_1)^{2^{s_1}} \cdots \Delta(Y_m)^{2^{s_m}}$, where $Y_i \subseteq Z[n]$, with $|Y_i| = r_i$ and $s_i = i - 1 + r_1 + \cdots + r_{i-1}$ for $1 \leq i \leq m$.*

Proof When $m = 1$, $\rho = (r_1)$ and $\mathsf{k}(\rho) = x^{2^{r_1}-1}$. Writing $\rho' = \alpha(Q_t(\mathsf{k}(\rho)))$, we note that $\rho' <_p \rho$ for $1 \leq t < r_1$, and so $\mathsf{k}(\rho')$ root-generates $Q_t(\mathsf{k}(\rho))$ by Proposition 21.5.6. Hence $Q_t(\mathsf{k}(\rho)) * L = 0$ if L annihilates all key monomials $\mathsf{k}(\rho')$ for $\rho' <_p \rho$. Since L commutes with Q_t, $\mathsf{k}(\rho) * L$ is in the kernel of Q_t for $1 \leq t < r_1$. Since $\deg \mathsf{k}(\rho) = 2^{r_1} - 1$, it follows from Proposition 21.2.4 that $\mathsf{k}(\rho) * L$ is a linear combination of Dickson invariants $\Delta(Y)$, where $Y \subseteq Z[n]$ and $|Y| = r_1$.

For $m > 1$ we use the operations $Q_t^k = Sq_t^{2^k} = Sq(0, \ldots, 0, 2^k)$. In particular, $Q_t^0 = Q_t$. The key monomial $\mathsf{k}(\rho) = x_1^{2^{s_1} d_1} \cdots x_m^{2^{s_m} d_m}$ is in the image of ϕ_σ^m : $\mathsf{P}^{d_1}(m) \otimes \cdots \otimes \mathsf{P}^{d_m}(m) \to \mathsf{P}^d(m)$, where $\sigma = (s_1, \ldots, s_m)$, $d_i = 2^{r_i} - 1$ and $d = \sum_{i=1}^{m} 2^{s_i} d_i$. By Proposition 12.2.11 $Q_t^{s_i}(\mathsf{k}(\rho)) = x_1^{2^{s_1} d_1} \cdots (Q_t(x^{d_i}))^{2^{s_i}} \cdots x_m^{2^{s_m} d_m}$, where only the ith factor of $\mathsf{k}(\rho)$ is affected. As in the case $m = 1$, $\alpha(Q_t^{s_i}(\mathsf{k}(\rho))) <_p \rho$ for $1 \leq t < r_i$. Hence $\mathsf{k}(\rho) * L$ is in the kernel of $Q_t^{s_i}$ for $1 \leq t < r_i$.

By Proposition 21.6.4, $\mathsf{k}(\rho) * L$ lies in the image of ϕ_σ^n. It can therefore be written as a linear combination $\sum_\epsilon a_\epsilon f_{e_1}^{2^{s_1}} \cdots f_{e_i}^{2^{s_i}} \cdots f_{e_m}^{2^{s_m}}$, where $a_\epsilon \in \mathbb{F}_2$, $f_{e_i} \in \mathsf{P}^{d_i}(n)$ and the summation runs over sequences $\epsilon = (e_1, \ldots, e_m) \in \mathsf{Seq}$. We can organize the summation in such a way that for a given i the polynomials f_{e_j} are linearly independent in $\mathsf{P}^{d_j}(n)$ for each $j \neq i$, and the sequence $(e_1,\ldots,e_{i-1},e_{i+1},\ldots,e_m)$, where e_i is omitted, appears at most once.

By Proposition 12.2.11, $Q_t^{s_i}\mathsf{k}(\rho)*L = \sum_\epsilon a_\epsilon f_{e_1}^{2^{s_1}}\cdots(Q_t(f_{e_i}))^{2^{s_i}}\cdots f_{e_m}^{2^{s_m}}$. Since ϕ_σ^n is an injection of a tensor product, we see from the way we have organized this summation that if $Q_t^{s_i}(\mathsf{k}(\rho))*L = 0$ then $Q_t(f_{e_i}) = 0$ for all e_i and $1 \le t < r_i$.

As in the case $m = 1$, it follows that each f_{e_i} is a linear combination of Dickson invariants $\Delta(Y)$, where $Y \subseteq Z[n]$ has cardinality r_i. As Y ranges over distinct subsets of $Z[n]$, the polynomials $\Delta(Y)$ are linearly independent in $\mathsf{P}^{2^{r_i}-1}(n)$, and may be taken as the polynomials f_{e_i} in the original summation. The result follows by applying the above process iteratively to each i in turn for $1 \le i \le m$. □

Proposition 21.6.6 *Given a positive sequence $\rho = (r_1,\ldots,r_m)$, let $s_1 = 0$ and $s_i = i-1+r_1+\cdots+r_{i-1}$ for $2 \le i \le m$. Let $Y_i \subseteq Z[n]$ with $|Y_i| = r_i$ for $1 \le i \le m$. Let $c \in \mathbb{F}_2\mathsf{M}(m,n)$ be the sum of all $A = (a_{i,j})$ such that $a_{i,j} = 0$ if $j \notin Y_i$ for $1 \le i \le m$, $1 \le j \le n$. Then $\mathsf{k}(\rho)\cdot c = \prod_{i=1}^m \Delta(Y_i)^{2^{s_i}}$, and $\mathsf{k}(\rho')\cdot c = 0$ if $\rho' = (r_1',\ldots,r_m')$ is a positive sequence such that $r_i' < r_i$ for some i.*

Proof The key monomial $\mathsf{k}(\rho)$ is the product of factors $x_i^{d_i 2^{s_i}}$, where $d_i = 2^{r_i}-1$. The element c substitutes for x_i the linear forms x in the variables x_j for $j \in Y_i$, and gives the sum $\sum_x x^{d_i 2^{s_i}}$. By Proposition 15.1.8 the result is $\Delta(Y_i)^{2^{s_i}}$, because $|Y_i| = r_i$. Applying this to each variable x_i produces the required expression $\prod_{i=1}^m \Delta(Y_i)^{2^{s_i}}$. For the last statement, $\mathsf{k}(\rho')$ has a factor $x_i^{d_i' 2^{s_i}}$, where $d_i' < 2^{r_i}-1$, and so by Proposition 15.1.8 the corresponding expression $\sum_x x^{d_i 2^{s_i}}$ is 0. Hence $k(\rho')\cdot c = 0$. □

The element c in Proposition 21.6.6 depends on the choice of $Y_1,\ldots,Y_m$. By adding them for suitable choices of these sets and using Proposition 21.6.5, we obtain the following result.

Proposition 21.6.7 *For a positive sequence ρ, let $L \in Hom_{\mathsf{A}_2}(\mathsf{P}(m),\mathsf{P}(n))$ annihilate $\mathsf{k}(\rho')$ for all positive sequences ρ' such that $\rho' <_p \rho$. Then $\mathsf{k}(\rho)*L = \mathsf{k}(\rho)\cdot c$ for some $c \in \mathbb{F}_2\mathsf{M}(m,n)$.* □

Finally we remove the constraint that L annihilates $\mathsf{k}(\rho')$ for all positive sequences ρ' such that $\rho' <_p \rho$.

Proposition 21.6.8 $\psi : \mathbb{F}_2\mathsf{M}(m,n) \to Hom_{\mathsf{A}_2}(\mathsf{P}(m),\mathsf{P}(n))$ *is surjective.*

Proof It is enough to show that, for $L \in \mathrm{Hom}_{\mathsf{A}_2}(\mathsf{P}(Z[m]),\mathsf{P}(n))$, there is an element $c \in \mathbb{F}_2\mathsf{M}(m,n)$ such that $f\cdot c = f*L$ for all $f \in \mathsf{P}(Z[m])$. By Theorem 21.5.7, L is determined by its value on one particular key monomial.

The argument is by induction on $|\rho|$, where $\rho = (r_1, \ldots, r_m)$. The $<_p$-minimal positive sequence ρ of length m is $\rho = (1, \ldots, 1)$, for which $|\rho| = m$. By Proposition 21.6.7, we can find $c \in \mathbb{F}_2\mathsf{M}(m,n)$ such that $\mathsf{k}(\rho) \cdot c = \mathsf{k}(\rho) * L$ in this case.

We assume as inductive hypothesis that, for some given $q > m$, there is an element $c \in \mathbb{F}_2\mathsf{M}(m,n)$ such that $\mathsf{k}(\rho') \cdot c = \mathsf{k}(\rho') * L$ for all positive sequences ρ' of length m with $|\rho'| < q$. Let ρ be a sequence of length m with modulus $|\rho| = q$, and let $L' = L - \psi(c)$. If $\rho' <_p \rho$ then $|\rho'| < q$, and so by the inductive hypothesis $\mathsf{k}(\rho') * L = 0$ for all $\rho' <_p \rho$. By Propositions 21.6.6 and 21.6.5, we can find $c' \in \mathbb{F}_2\mathsf{M}(m,n)$ such that $\mathsf{k}(\rho) \cdot c' = \mathsf{k}(\rho) * L'$. Further, if $\rho' <_p \rho$ or $|\rho'| = q$ and $\rho' \neq \rho$, then $r_i < r_i'$ for at least one i. By Proposition 21.6.6 it follows that $\mathsf{k}(\rho') \cdot c' = 0$. Hence by taking the sum $\bar{c}$ of the elements $c' \in \mathbb{F}_2\mathsf{M}(m,n)$, one for each ρ such that $|\rho| = q$, we obtain $\mathsf{k}(\rho) \cdot \bar{c} = \mathsf{k}(\rho) * L'$ for every ρ with $|\rho| < q+1$. Therefore $L = L' + \psi(c)$ agrees with the element $\bar{c} + c \in \mathbb{F}_2\mathsf{M}(m,n)$ on key monomials $\mathsf{k}(\rho)$ with $|\rho| < q+1$. This completes the inductive step. □

Combining Propositions 21.1.1 and 21.6.8, we obtain our main result.

Theorem 21.6.9 $\psi : \mathbb{F}_2\mathsf{M}(m,n) \to Hom_{\mathsf{A}_2}(\mathsf{P}(m), \mathsf{P}(n))$ *is an isomorphism of vector spaces and* $\psi : \mathbb{F}_2\mathsf{M}(n) \to End_{\mathsf{A}_2}(\mathsf{P}(n))$ *is an isomorphism of algebras.* □

The significance of this result is that any decomposition of $\mathsf{P}(n)$ into a direct sum of A_2-submodules $M_1 \oplus \cdots \oplus M_r$, given by projection operators $e_i : \mathsf{P}(n) \to M_i$ for $1 \leq i \leq r$ in $\mathrm{End}_{\mathsf{A}_2}(\mathsf{P}(n))$, is realized by a decomposition of the identity matrix I_n into a sum of orthogonal idempotents in $\mathbb{F}_2\mathsf{M}(n)$. Since a decomposition of I_n into primitive idempotents corresponds to a decomposition of the module into indecomposable summands, we obtain the following result.

Theorem 21.6.10 *A splitting of* $\mathsf{P}(n)$ *into indecomposable* A_2*-submodules is given by a splitting of the identity element of* $\mathbb{F}_2\mathsf{M}(n)$ *into primitive orthogonal idempotents.*

Example 21.6.11 Let $E_1 = \left(\begin{smallmatrix} 1 & 0 \\ 0 & 0 \end{smallmatrix}\right), E_2 = \left(\begin{smallmatrix} 0 & 0 \\ 0 & 1 \end{smallmatrix}\right) \in \mathsf{M}(2)$, and let I and O be the zero and identity matrices in $\mathsf{M}(2)$. Then $e_\emptyset = O$, $e_{\{1\}} = E_1 + O$, $e_{\{2\}} = E_2 + O$ and $e_{\{1,2\}} = I + E_1 + E_2 + O$ form a set of orthogonal idempotents giving a decomposition of the identity $I \in \mathsf{M}(2)$.

More generally, for $Y \subseteq Z[n]$ there is a corresponding idempotent $e_Y \in \mathbb{F}_2\mathsf{M}(n)$ given by the sum of all diagonal matrices in $\mathsf{M}(n)$ with diagonal entries 1 in rows in Y and 0 in other rows, and the set of idempotents e_Y provides a primitive orthogonal decomposition of the identity I_n with respect to the diagonal monoid of $\mathsf{M}(n)$. The corresponding A_2-module decomposition of $\mathsf{P}(n)$ is the decomposition $\mathsf{P}(n) = \bigoplus_{Y \subseteq Z[n]} \mathsf{P}(Y)$ of Section 1.4.

21.7 Examples

Theorem 19.1.1 describes the splitting of $\mathsf{P}(n)$ as a direct sum of A_2-modules obtained using the 2^{n-1} irreducible $\mathbb{F}_2\mathsf{GL}(n)$-modules $L(\lambda)$, where λ is a column 2-regular partition of length $\leq n-1$. As explained in Section 19.0, this decomposition can be refined: we can split each summand $\mathsf{P}(n,\lambda)$ into two summands $\mathsf{P}'(n,\lambda)$ and $\mathsf{P}'(n,\lambda')$ by using the irreducible $\mathbb{F}_2\mathsf{M}(n)$-modules $\mathsf{L}(\lambda)$ and $L(\lambda')$, where $\lambda'_i = \lambda_i + 1$ for $1 \leq i \leq n$ and $L(\lambda') = L(\lambda) \otimes \det$. We shall denote a complete set of 2^n irreducible representations of $\mathsf{M}(n)$ by $\mathsf{L}(\lambda)$, where λ is column 2-regular of length $\leq n$. By adapting the proof to use a maximal set of orthogonal primitive idempotents in $\mathbb{F}_2\mathsf{M}(n)$, we can refine Theorem 19.1.1 as follows.

Theorem 21.7.1 *There is a direct sum decomposition of graded* A_2*-modules* $\mathsf{P}(n) \cong \bigoplus_\lambda d_\lambda \mathsf{P}(n,\lambda)$*, where the summands* $\mathsf{P}(n,\lambda)$ *are indexed by column 2-regular partitions* λ *of length* $\leq n$*, and* d_λ *is the dimension of the irreducible* $\mathbb{F}_2\mathsf{M}(n)$*-module* $L(\lambda)$*.*

It follows from Theorem 21.6.10 that this is a maximal direct sum splitting of $\mathsf{P}(n)$, so that the summands $\mathsf{P}(n,\lambda)$ are indecomposable A_2-modules. In this section we describe A_2-maps $L : \mathsf{P}(m) \to \mathsf{P}(n)$ in terms of the direct sum decomposition of Theorem 21.7.1 in the special cases where $m,n = 1$ or 2. We write $\mathsf{P}(1) = \mathbb{F}_2[x]$ and $\mathsf{P}[2] = \mathbb{F}_2[x,y]$.

For $n = 1$, $\lambda = (0)$ or (1) and $d_\lambda = 1$ in both cases. Thus $\mathsf{P}(1) = \mathsf{P}(1,(0)) \oplus \mathsf{P}(1,(1))$ where $\mathsf{P}(1,(0)) = \mathbb{F}_2$ and $\mathsf{P}(1,(1))$ is spanned by $\{x^k, k \geq 1\}$. By Example 21.6.11, the splitting $\mathsf{P}(1) = \mathsf{P}(1,(0)) \oplus \mathsf{P}(1,(1))$ is effected by the idempotents $e_\emptyset = (0)$ and $e_{\{1\}} = (1) + (0)$ in $\mathbb{F}_2\mathsf{M}(1)$, which are central since this ring is commutative. By Example 21.0.2, $\mathrm{End}_{\mathsf{A}_2}(\mathsf{P}(1))$ has dimension 2, with basis given by the projections on the summands $\mathsf{P}(1,(0))$ and $\mathsf{P}(1,(1))$, given by the action of $e_\emptyset$ and $e_{\{1\}}$.

For $n = 2$, $d_\lambda = 1$ for $\lambda = (0)$ and $(1,1)$ and $d_\lambda = 2$ for $\lambda = (1)$ or $(2,1)$. Hence $\mathsf{P}(2) \cong \mathsf{P}(2,(0)) \oplus 2\mathsf{P}(2,(1)) \oplus \mathsf{P}(2,(1,1)) \oplus 2\mathsf{P}(2,(2,1))$. In terms of the decomposition $\mathsf{P}(n) = \bigoplus_{Y \subseteq Z[n]} \mathsf{P}(Y)$, the two copies of $\mathsf{P}(2,(1))$ correspond to $\mathsf{P}(Y)$ for $Y = \{1\}$ and $Y = \{2\}$, with $\mathbb{F}_2$-bases $\{x^k, k \geq 1\}$ and $\{y^k, k \geq 1\}$, and $\mathsf{P}(\mathsf{Z}[2]) = \mathsf{P}(2,(1,1)) \oplus 2\mathsf{P}(2,(2,1))$. There is also a 'diagonal' copy of $\mathsf{P}(2,(1))$ with $\mathbb{F}_2$-basis $\{x^k + y^k, k \geq 1\}$.

The Poincaré series $P(\mathsf{P}(2,(1,1)),t) = (1+t^3)/(1-t^2)(1-t^3) - 1 = (t^2 + t^3 - t^4)/(1-t)(1-t^3)$, by Example 19.1.8. We can identify $\mathsf{P}(2,(1,1))$ with the positive degree part of the A_2-module of cyclic invariants $\mathsf{P}(2)^{H_2}$ of Section 15.4. Thus $\mathsf{P}(2,(1,1))$ has an $\mathbb{F}_2$-basis given by monomials of positive degree in $\mathsf{d}_1 = x^2+xy+y^2$, $\mathsf{d}_0 = x^2y+xy^2$ and $g = x^3+y^3+x^2y$, subject to the relation $g^2 = g\mathsf{d}_0 + \mathsf{d}_0^2 + \mathsf{d}_1^3$. Projection on to $\mathsf{P}(Z[2])$ gives an isomorphic copy of $\mathsf{P}(2,(1,1))$, replacing d_1 by xy and g by x^2y.

The (Steinberg) summands $\mathsf{P}(2,(2,1))$ are discussed in Chapter 22. The Poincaré series $P(\mathsf{P}(2,(2,1)),t) = t^4/(1-t)(1-t^3)$, by Example 19.1.8. We may take the polynomials $x\mathsf{d}_0 = x^3y + x^2y^2$ and $y\mathsf{d}_0 = xy^3 + x^2y^2$ as their lowest degree elements. The symmetric Steinberg summand is the 'diagonal' summand with lowest degree element $(x+y)\mathsf{d}_0 = x^3y + xy^3$.

There is a unique splitting of the identity element $1 = z_0 + z_1 + z_2$ as the sum of three primitive orthogonal central idempotents $z_0 = A_0$, $z_1 = A_0 + A_1 + A_3 + A_5 + A_7 + A_8 + A_9 + A_{10} + A_{12} + A_{14}$, $z_2 = A_1 + A_3 + A_5 + A_7 + A_8 + A_{10} + A_{12} + A_{14}$. These give the projections of $\mathsf{P}(2)$ on to the summands $\mathsf{P}(2,(0))$, $2\mathsf{P}(2,(1)) \oplus \mathsf{P}(2,(1,1))$ and $2\mathsf{P}(2,(2,1))$ respectively. The central idempotents z_1 and z_2 can be decomposed (not uniquely) as the sums of non-central primitive orthogonal idempotents in $\mathbb{F}_2\mathsf{M}(2)$ which give the projections of $\mathsf{P}(2)$ on to its indecomposable A_2-summands. We give details in Example 21.7.4.

Example 21.7.2 The $\mathbb{F}_2$-space $\mathrm{Hom}_{\mathsf{A}_2}(\mathsf{P}(1),\mathsf{P}(2))$ has dimension 4 and is discussed in Example 21.5.9. We may project $\mathsf{P}(1)$ on the summand $\mathsf{P}(2,(0))$ of $\mathsf{P}(2)$ or on the summand $\mathsf{P}(2,(1))$ generated by x, y or $x+y$. Writing $A_0 = (0\ 0)$, $A_1 = (1\ 0)$, $\mathsf{A}_2 = (0\ 1)$, $A_3 = (1\ 1)$ for the elements of $\mathsf{M}(1,2)$, these four maps $\mathsf{P}(1) \to \mathsf{P}(2)$ are given by the action of A_0, $A_1 + A_0$, $A_2 + A_0$ and $A_3 + A_0$ respectively. As these elements form a basis of $\mathbb{F}_2\mathsf{M}(1,2)$, the four maps span $\mathrm{Hom}_{\mathsf{A}_2}(\mathsf{P}(1),\mathsf{P}(2))$. We note that if $f * L = f \cdot (A_1 + A_2 + A_3)$ then $x * L = 0$, $x^2 * L = 0$ but $x^3 * L = \mathsf{d}_0$, as in Example 21.5.9. The kernel of L is the A_2-submodule of $\mathsf{P}(1)$ spanned by $\{x^{2^j}, j \geq 0\}$ and the image of L in degrees $d > 0$ with $\alpha(d) > 1$ is spanned by $x^k + y^k + (x+y)^k$. As this polynomial is invariant under permutations of x, y and $x+y$ it lies in the Dickson algebra $\mathsf{D}(2)$. In particular, if $\alpha(d) = 2$ and $d = 2^a + 2^b$ then $x^d * L = x^{2^a}y^{2^b} + x^{2^b}y^{2^a}$.

More generally, $\mathrm{Hom}_{\mathsf{A}_2}(\mathsf{P}(1),\mathsf{P}(n))$ has dimension 2^n, with basis given by the projection on $\mathsf{P}(n,(0))$ and the projection on $\mathsf{P}(n,(1))$ followed by the one of the $2^n - 1$ maps $x^d \mapsto u^d$ for $d \geq 1$, where $u \neq 0 \in \mathsf{P}^1(n)$. The sum L of these $2^n - 1$ projections has kernel spanned by $\{x^a, \alpha(a) < n\}$ and image in the Dickson algebra $\mathsf{D}(n)$. In particular, $x^{2^n-1} * L = \Delta_n = \mathsf{d}_{n,0}$ is the sum of the (2^n-1)th powers of the elements $u \in \mathsf{P}^1(n)$.

Example 21.7.3 By Example 21.5.10, $\dim \mathrm{Hom}_{\mathsf{A}_2}(\mathsf{P}(2),\mathsf{P}(1)) = 4$, and its elements are determined by their values on the subkey monomials 1, x, y,

xy^2. Writing $A_0 = \binom{0}{0}$, $A_1 = \binom{1}{0}$, $A_2 = \binom{0}{1}$, $A_3 = \binom{1}{1}$ for the elements of $\mathsf{M}(2,1)$, the corresponding maps $\mathsf{P}(2) \to \mathsf{P}(1)$ are the projections of $\mathsf{P}(2)$ on $\mathsf{P}(2,(0))$, $\mathsf{P}(Y)$ for $Y = \{1\}$ and $Y = \{2\}$, and the 'diagonal' summand of $\mathsf{P}(2)$ with basis $\{(x+y)^k, k \geq 0\}$, followed by identification of the image with $\mathsf{P}(1,(0))$ in the case of A_0 and with $\mathsf{P}(1)$ in the other three cases. We consider the map $L : \mathsf{P}(2) \to \mathsf{P}(1)$ given by $f * L = f \cdot (A_0 + A_1 + A_2 + A_3)$. Evaluating L on the subkey monomials, $1 * L = 0$, $x * L = 0$, $y * L = 0$ and $xy^2 * L = x^3$. The kernel of L is given by $\mathsf{P}(Y)$ for $Y = \emptyset$, $\{1\}$ and $\{2\}$, together with the $\mathbb{F}_2$-subspace of $\mathsf{P}(Z[2])$ spanned by polynomials with an even number of terms. The image of L is spanned by x^d for $d \geq 2$. The Steinberg summands $\mathsf{P}(2,(2,1))$ are mapped to 0 by L, as is the ideal $\langle \mathsf{d}_0 \rangle \subset \mathsf{P}(2,(1,1))$. The quotient of the $L(1,1) = \det$ summand of $\mathsf{P}(2)$ by $\langle \mathsf{d}_0 \rangle$ has basis $\{\mathsf{d}_1, g, \mathsf{d}_1^2, \mathsf{d}_1 g, \ldots\}$ and is mapped isomorphically to the A_2-submodule $\mathrm{Im}(L) = \langle x^2, x^3, \ldots \rangle \subset \mathsf{P}(1)$.

More generally, $\dim \mathrm{Hom}_{\mathsf{A}_2}(\mathsf{P}(n), \mathsf{P}(1)) = 2^n$. There is a basis given by following the projections $\mathsf{P}(n) \to \mathsf{P}(Y)$ for $Y \subseteq Z[n]$ with the map $\mathsf{P}(Y) \to \mathsf{P}(1)$ which maps each variable x_i, $i \in Y$, to the generator x of $\mathsf{P}(1)$. If $|Y| = k$, then the image of the corresponding A_2-map $L(Y) : \mathsf{P}(n) \to \mathsf{P}(1)$ is spanned by x^d for $d \geq k$. In particular, $x_1 \cdots x_n * L = x^n$, where $L = L(Z[n])$.

In the case $n = 3$, by comparing Poincaré series as in Example 19.2.6 we can identify the summand of $\mathsf{P}(3)$ corresponding to the det representation of $\mathsf{M}(3)$ with the elements of positive degree in the invariant subalgebra $\mathsf{P}(3)^{H_3}$ of Section 15.5. The action of the map $L = L(Z[3])$ on $\mathsf{P}(3)^{H_3}$ can be described as follows. The kernel of L is given by $\mathsf{P}(Y)$ for $Y \neq Z[3]$, together with the $\mathbb{F}_2$-subspace of $\mathsf{P}(Z[3])$ spanned by polynomials with an even number of terms, while the image of L is spanned by x^d for $d \geq 3$. Thus the Dickson invariants d_0 and d_1 are in the kernel of L, but $\mathsf{d}_2 * L = x^4$ and the generators h and k map to x^3 and x^5. We also have $hk * L = x^8 = \mathsf{d}_2^2 * L$ and $h^3 * L = x^9 = \mathsf{d}_2 k * L$. The quotient of the $L(1,1,1) = \det$ summand of $\mathsf{P}(3)$ by the ideal generated by d_0, d_1, $hk + \mathsf{d}_2^2$ and $h^3 + \mathsf{d}_2 k$ is mapped isomorphically to $\mathrm{Im}(L) = \langle x^3, x^4, \ldots \rangle \subset \mathsf{P}(1)$.

Example 21.7.4 The $\mathbb{F}_2$-algebra $\mathrm{End}_{\mathsf{A}_2}(\mathsf{P}(2))$ has dimension 16 and is discussed in Example 21.5.10, where it is shown that its elements are determined by their values on the four subkey monomials 1, x^3, y^3, x^3y^{12}. We encode the elements of $\mathsf{M}(2)$ as $A_j = \begin{pmatrix} a_0 & a_1 \\ a_2 & a_3 \end{pmatrix}$, where $j = \sum_{\{i : a_i = 1\}} 2^i$, $0 \leq j \leq 15$, i.e. $A_0 = \begin{pmatrix} 0 & 0 \\ 0 & 0 \end{pmatrix}$, $A_1 = \begin{pmatrix} 1 & 0 \\ 0 & 0 \end{pmatrix}$, $\ldots$, $A_{15} = \begin{pmatrix} 1 & 1 \\ 1 & 1 \end{pmatrix}$. Thus by Example 21.6.11 the projections of $\mathsf{P}(2)$ on the summands $\mathsf{P}(Y)$ are given by the action of the idempotents $e_\emptyset = A_0$, $e_{\{1\}} = A_0 + A_1$, $e_{\{2\}} = A_0 + A_8$ and $e_{\{1,2\}} = A_0 + A_1 + A_8 + A_9$.

The central projection on the summands $2\mathsf{P}(2,(1)) \oplus \mathsf{P}(2,(1,1))$ is given by $z_1 = A_0 + A_1 + A_3 + A_5 + A_7 + A_8 + A_9 + A_{10} + A_{12} + A_{14}$, so the projection $e_{1,1}$

of $\mathsf{P}(2)$ on the summand $\mathsf{P}(2,(1,1))$ is given by $e_{1,1} = A_0 + A_3 + A_5 + A_7 + A_9 + A_{10} + A_{12} + A_{14}$.

The central projection on the Steinberg summands $2\mathsf{P}(2,(2,1))$ is given by $z_2 = A_1 + A_3 + A_5 + A_7 + A_8 + A_{10} + A_{12} + A_{14}$. We can split z_2 as the sum of the two primitive orthogonal idempotents $e_{2,1} = A_8 + A_9 + A_{10} + A_{11} + A_{12} + A_{13} + A_{14} + A_{15}$ and $e'_{2,1} = A_1 + A_3 + A_5 + A_7 + A_9 + A_{11} + A_{13} + A_{15}$. Then $x\mathrm{d}_0 \cdot e_{2,1} = x\mathrm{d}_0$, $y\mathrm{d}_0 \cdot e_{2,1} = 0$ while $x\mathrm{d}_0 \cdot e'_{2,1} = 0$, $y\mathrm{d}_0 \cdot e'_{2,1} = y\mathrm{d}_0$, so $e_{2,1}$ and $e'_{2,1}$ give projections on the Steinberg summands which are invariant under the lower and upper triangular subgroups respectively.

Thus a maximal set of primitive orthogonal idempotents $e \in \mathbb{F}_2\mathsf{M}(2)$ is given by e_0, e_1, e'_1, $e_{1,1}$, $e_{2,1}$ and $e'_{2,1}$. We expand this list to a basis of $\mathbb{F}_2\mathsf{M}(2)$ by taking the union of bases for the ideals $e\mathbb{F}_2\mathsf{M}(2)$. The table below gives this basis $\{c_i, 0 \leq i \leq 15\}$ and a brief description of the corresponding maps $\mathsf{P}(2) \to \mathsf{P}(2)$ as the projection on one of the summands, followed by the further map indicated. In the table, $\mathsf{P}(x)$ is the summand $\mathsf{P}(\{1\})$, $\mathsf{P}(y)$ is $\mathsf{P}(\{2\})$ and $\mathsf{P}(x+y)$ is the 'diagonal' copy of $\mathsf{P}(2,(1))$, while $\mathsf{P}(x\mathrm{d}_0)$, $\mathsf{P}(y\mathrm{d}_0)$ denote the summands $\mathsf{P}(2,(2,1))$ with lowest degree element $x\mathrm{d}_0$, $y\mathrm{d}_0$. Further notation for elements follows the table.

i	c_i	$f \to f \cdot c_i$
0	$e_\emptyset = A_0$	projection on $\mathsf{P}(2,(0))$
1	$e_{\{1\}} = A_0 + A_1$	projection on $\mathsf{P}(x)$
2	$A_0 + A_2$	$\mathsf{P}(x) \to \mathsf{P}(y)$
3	$A_0 + A_3$	$\mathsf{P}(x) \to \mathsf{P}(x+y)$
4	$e_{\{2\}} = A_0 + A_8$	projection on $\mathsf{P}(y)$
5	$A_0 + A_4$	$\mathsf{P}(y) \to \mathsf{P}(x)$
6	$A_0 + A_{12}$	$\mathsf{P}(y) \to \mathsf{P}(x+y)$
7	$e_{1,1}$	projection on $\mathsf{P}(2,(1,1))$
8	$\tilde{e}_{1,1}$	automorphism $x \leftrightarrow y$ of $\mathsf{P}(2,(1,1))$
9	$A_0 + A_1 + A_4 + A_5$	$\mathsf{P}(2,(1,1))/\langle \mathrm{d}_0 \rangle \to \mathsf{P}(x)$
10	$A_0 + A_2 + A_8 + A_{10}$	$\mathsf{P}(2,(1,1))/\langle \mathrm{d}_0 \rangle \to \mathsf{P}(y)$
11	$A_0 + A_3 + A_{12} + A_{15}$	$\mathsf{P}(2,(1,1))/\langle \mathrm{d}_0 \rangle \to \mathsf{P}(x+y)$
12	$e_{2,1}$	projection on $\mathsf{P}(x\mathrm{d}_0)$
13	$f_{2,1}$	$\mathsf{P}(x\mathrm{d}_0) \to \mathsf{P}(y\mathrm{d}_0)$
14	$e'_{2,1}$	projection on $\mathsf{P}(y\mathrm{d}_0)$
15	$f'_{2,1}$	$\mathsf{P}(y\mathrm{d}_0) \to \mathsf{P}(x\mathrm{d}_0)$

$$e_{1,1} = A_0 + A_3 + A_5 + A_7 + A_9 + A_{10} + A_{12} + A_{14},$$
$$\tilde{e}_{1,1} = A_0 + A_3 + A_5 + A_6 + A_{10} + A_{11} + A_{12} + A_{13},$$
$$e_{2,1} = A_8 + A_9 + A_{10} + A_{11} + A_{12} + A_{13} + A_{14} + A_{15},$$
$$e'_{2,1} = A_1 + A_3 + A_5 + A_7 + A_9 + A_{11} + A_{13} + A_{15},$$
$$f_{2,1} = A_4 + A_5 + A_6 + A_7 + A_{12} + A_{13} + A_{14} + A_{15},$$
$$f'_{2,1} = A_2 + A_3 + A_6 + A_7 + A_{10} + A_{11} + A_{14} + A_{15}.$$

We next consider the effect of replacing the choice of the summand $\mathsf{P}(2,(1,1))$ by $\mathsf{P}(2)^{H_2}$, with lowest degree class $\mathsf{d}_1 = x^2 + xy + y^2$, rather than the submodule of $\mathsf{P}(Z[2])$ with lowest degree class xy. This changes the idempotents giving the projections on to $\mathsf{P}(\{1\})$ and $\mathsf{P}(\{2\})$ as well as the projection on to $\mathsf{P}(2,(1,1))$. The projection of $\mathsf{P}(2)$ on to $\mathsf{P}(x)$ with respect to this maximal splitting is given by the action of $e'_1 = A_4 + A_5$, and not by the action of $e_{\{1\}} = A_0 + A_1$, which maps d_1 to x^2 and g to x^3, and so gives a surjection $\mathsf{P}(2,(1,1)) \to P(x)$. Similarly, the projection of $\mathsf{P}(2)$ on to $\mathsf{P}(y)$ is given by the action of $e'_2 = A_2 + A_{10}$. As the central projection on the Steinberg block is given by z_2 for all choices of maximal splitting, the projection $e'_{1,1}$ of $\mathsf{P}(2)$ on the summand $\mathsf{P}(2)^{H_2} \cong \mathsf{P}(2,(1,1))$ is given by the sum of the identity A_9, the projections $e_0 = A_0$, e'_1, e'_2 and z_2, i.e. $e'_{1,1} = A_0 + A_1 + A_2 + A_3 + A_4 + A_7 + A_8 + A_9 + A_{12} + A_{14}$.

Remark 21.7.5 By Theorem 21.6.9, the table above also describes the decomposition of the regular representation of $\mathbb{F}_2\mathsf{M}(2)$ as the sum of principal indecomposable modules $e\mathbb{F}_2\mathsf{M}(2)$, where $e = e_\emptyset, e_{\{1\}}, e_{\{2\}}, e_{1,1}, e_{2,1}, e'_{2,1}$ with bases given by the corresponding elements c_i of dimensions 1, 3, 3, 5, 2 and 2 and unique irreducible quotients $L(0), L(1), L(1), L(1,1), L(2,1)$ and $L(2,1)$ respectively. The modules $e_{\{1\}}\mathbb{F}_2\mathsf{M}(2)$ and $e_{\{2\}}\mathbb{F}_2\mathsf{M}(2)$ have submodules $L(1,1)$ generated by $A_0 + A_1 + A_2 + A_3$ and $A_0 + A_2 + A_8 + A_{10}$ respectively, while $e_{1,1}\mathbb{F}_2\mathsf{M}(2)$ has two submodules $L(1,1)$ generated by the sums of the singular and non-singular elements of $\mathsf{M}(2)$, and a composition factor $L(1)$ generated by $A_0 + A_1 + A_4 + A_5$. We note that, as for group algebras, the multiplicity of each irreducible module as a composition factor in the regular representation is the dimension of the corresponding principal indecomposable module.

Remark 21.7.6 The Steinberg idempotents $e = \overline{\mathsf{B}}(2)\overline{\mathsf{W}}(2) = A_9 + A_6 + A_{13} + A_{14}$ and $e' = \overline{\mathsf{W}}(2)\overline{\mathsf{B}}(2) = A_9 + A_6 + A_{13} + A_7$ in $\mathbb{F}_2\mathsf{GL}(2)$ are not primitive in $\mathbb{F}_2\mathsf{M}(2)$, but if we replace the subgroup $\overline{\mathsf{W}}(2)$ by the 'rook monoid' (matrices with at most one entry 1 in each row or column) and $\overline{\mathsf{B}}(2)$ by the monoid of all lower triangular matrices, then the corresponding products $f = e + A_1 + A_2 + A_5 + A_{10}$ and $f' = e' + A_2 + A_3 + A_8 + A_{12}$ are primitive orthogonal idempotents giving projections of $\mathsf{P}(2)$ on the symmetric Steinberg summand and the lower triangular invariant Steinberg summand respectively, and $f+f'=z_2$. It would be interesting to know if the corresponding result holds for $n>2$ variables.

21.8 Remarks

Theorem 21.6.9 is a special case of the Adams–Gunawardena–Miller theorem [2]. In [2] the extension groups $\mathrm{Ext}^{p,q}_{\mathsf{A}_2}(\mathsf{P}(m),\mathsf{P}(n))$ are evaluated using sophisticated homological algebra, whereas in this chapter we give an elementary and constructive proof based on [228] of the case $\mathrm{Ext}^{0,0}_{\mathsf{A}_2}(\mathsf{P}(m),\mathsf{P}(n)) = \mathrm{Hom}_{\mathsf{A}_2}(\mathsf{P}(m),\mathsf{P}(n))$.

It was proved in [74] that a maximal decomposition of the A_2-module $\mathsf{P}(n)$ as a direct sum of indecomposable A_2-modules is obtained from a maximal set of orthogonal idempotents in $\mathbb{F}_2\mathsf{M}(n)$. The idempotents given in Example 21.7.4 appear in [234, Example 3.2], while [68] contains much information about the structure of $\mathsf{P}(2)$ as a module over $\mathbb{F}_2\mathsf{M}(2)$ as well as over $\mathbb{F}_2\mathsf{GL}(2)$.

In principle, Theorem 21.6.10 reduces the problem of splitting $\mathsf{P}(n)$ into a direct sum of indecomposable A_2-submodules to matrix algebra. Hence the hit problem could be treated for each indecomposable summand separately. In practice it is difficult to gain a sufficiently good grip on the idempotents to make this approach workable except in the Steinberg case (Chapter 22). Less refined, but more useful, decompositions of $\mathsf{P}(n)$ can be obtained from the group algebras of suitable subgroups of $\mathsf{GL}(n)$. We consider the case of cyclic groups of odd order in Chapters 27 and 28.

22
The Steinberg summands of $\mathsf{P}(n)$

22.0 Introduction

The Steinberg module $\mathsf{St}(n)$ is the irreducible $\mathbb{F}_2\mathsf{GL}(n)$-module $L(\lambda)$ corresponding to the partition $\lambda = (n-1, n-2, \dots, 1)$. In Chapter 17, $\mathsf{St}(n)$ has appeared as an indecomposable summand of the flag module $\mathsf{FL}(n)$. By Theorem 16.5.3, $\dim \mathsf{St}(n) = 2^{n(n-1)/2}$, the number of flags in the top Schubert cell $\mathrm{Sch}(W_0)$. By Proposition 16.5.6, $\mathsf{St}(n)$ is self-dual with respect to contragredient duality.

We have also constructed $\mathsf{St}(n)$ in Chapter 19 as the submodule of $\mathsf{P}(n)$ generated by Δ_I where $I = Z[n-1]$, a polynomial of degree $\deg_2 \lambda = w(n-1) = 2^n - 1 - n$ formed by the product of the top Dickson invariants $\Delta_i \in \mathsf{D}(i)$, $1 \leq i \leq n-1$. By Proposition 19.5.5 this is both the first occurrence of $\mathsf{St}(n)$ as a composition factor in $\mathsf{P}(n)$ and its first occurrence as a submodule. As observed in Section 18.7, the $\mathbb{F}_2\mathsf{GL}(n)$-module $\mathsf{St}(n)$ is both projective and injective, and hence every occurrence of $\mathsf{St}(n)$ in $\mathsf{P}(n)$ is as a direct summand.

By Theorem 19.1.1 $\mathsf{P}(n)$ is a direct sum of A_2-modules, with $d_\lambda = \dim L(\lambda)$ isomorphic summands $\mathsf{P}(n,\lambda)$ corresponding to each irreducible $\mathbb{F}_2\mathsf{GL}(n)$-module $L(\lambda)$. Thus $\mathsf{P}(n)$ has $2^{n(n-1)/2}$ summands isomorphic to $\mathsf{P}(\mathsf{St}(n))$, where we write $\mathsf{P}(\mathsf{St}(n))$ for $\mathsf{P}(n,\lambda)$ in the case of the Steinberg module. By Proposition 19.5.5, the multiplicity of $\mathsf{St}(n)$ as a direct summand of $\mathsf{P}^d(n)$ is the coefficient of t^d in $P(\mathsf{P}(\mathsf{St}(n)),t) = t^{w(n-1)} / \prod_{i=1}^{n}(1-t^{2^i-1})$.

As discussed in Section 18.1, a primitive idempotent $e \in \mathbb{F}_2\mathsf{GL}(n)$ is associated to an irreducible $\mathbb{F}_2\mathsf{GL}(n)$-module $L(\lambda)$, which is isomorphic both to the head and to the socle of the principal indecomposable $\mathbb{F}_2\mathsf{GL}(n)$-module $U(\lambda) = e\mathbb{F}_2\mathsf{GL}(n)$. The idempotent $e(n)$ of Section 18.3 was used by Steinberg to construct $\mathsf{St}(n)$. In the case $\lambda = (n-1, n-2, \dots, 1)$, the summand $U(\lambda) = L(\lambda) = \mathsf{St}(n)$ appears in $\mathbb{F}_2\mathsf{GL}(n)$ with multiplicity $\dim \mathsf{St}(n) = 2^{n(n-1)/2}$, and $e(n)\mathbb{F}_2\mathsf{GL}(n)$ is an example of such a summand. Thus the A_2-module $\mathsf{P}(n)e(n)$

has isomorphism type $\mathsf{P}(\mathsf{St}(n))$, and $\mathsf{P}^d(n)e(n)$ is a vector space over $\mathbb{F}_2$ of dimension a_d for $d \geq 0$.

The Steinberg idempotent $e(n) \in \mathbb{F}_2\mathsf{GL}(n)$ is the sum of the products BW where $B \in \mathsf{B}(n)$ is lower triangular and $W \in \mathsf{W}(n)$ is a permutation matrix. Thus $\mathsf{P}(n)e(n)$ is contained in the subalgebra of symmetric polynomials in $\mathsf{P}(n)$, and so we call $\mathsf{P}(n)e(n)$ the **symmetric Steinberg summand**. The idempotent $e'(n)$ of Section 18.3 is the sum of the corresponding products WB. Since $e(n)$ and $e'(n)$ are orthogonal and conjugate in $\mathbb{F}_2\mathsf{GL}(n)$, $\mathsf{P}(n)e'(n)$ is also an A_2-module isomorphic to $\mathsf{P}(\mathsf{St}(n))$, and $\mathsf{P}(n)e'(n)$ is contained in the Mui algebra $V(n)$ of $\mathsf{B}(n)$-invariant polynomials (Section 19.4).

As we explain in Section 22.7, this chapter is centred on some fundamental results of Stephen A. Mitchell and Stewart B. Priddy. A remarkable property of the A_2-module $\mathsf{P}(\mathsf{St}(n))$ is that it can be constructed as a subquotient module of A_2 itself. To describe this construction, let $\mathsf{Ad}_{\geq n}$ denote the vector subspace of A_2 spanned by the admissible monomials Sq^A of length $\geq n$. It follows from the Adem relations that $\mathsf{Ad}_{\geq n}$ is preserved by the left action of A_2, and that the quotient module $\mathsf{Ad}_{\geq n}/\mathsf{Ad}_{\geq n+1}$ has a $\mathbb{F}_2$-basis given by the admissible monomials Sq^A of length n. With the shifted grading defined by giving Sq^A grading $|A|-n$, we call the module $\mathsf{Ad}_{\geq n}/\mathsf{Ad}_{\geq n+1}$ the **Mitchell–Priddy** module $\mathsf{MP}(n)$.

The module $\mathsf{MP}(n)$ is isomorphic to $\mathsf{P}(\mathsf{St}(n))$ as a graded A_2-module. In particular, $Sq^{2^{n-1}}Sq^{2^{n-2}}\cdots Sq^2Sq^1$ is the admissible monomial of minimum degree of length n. It has degree 2^n-1 as an element of A_2, but its shifted grading in $\mathsf{MP}(n)$ is $w(n-1) = 2^n-1-n$, in agreement with the first occurrence degree of $\mathsf{St}(n)$. In the case $n=1$, $\mathsf{P}(\mathsf{St}(1)) = \mathsf{P}(1)$ and $\mathsf{MP}(1)$ has basis $\{Sq^j : j \geq 1\}$. The correspondence $Sq^j \leftrightarrow x^{j-1}$ is a map of A_2-modules, because the Adem relations give $Sq^iSq^j = \binom{j-1}{i}Sq^{i+j}$ modulo admissibles of length 2, matching the formula $Sq^i(x^{j-1}) = \binom{j-1}{i}x^{i+j-1}$ in $\mathsf{P}(1)$.

In Section 22.1, we extend the action of A_2 on polynomials to an action on rational functions over $\mathbb{F}_2$. This leads to the construction of the important A_2-module $\mathsf{T}(n)$ in Section 22.2 and to that of $\mathsf{MP}(n)$ in Section 22.3.

In Section 22.4, we show that the admissible monomials $Sq^{2^{r_1}}\cdots Sq^{2^{r_n}}$, where $r_1 \geq \cdots \geq r_n \geq 0$, form a minimal generating set for $\mathsf{MP}(n)$, and it follows that $\mathsf{P}(\mathsf{St}(n))$ has a minimal generating set with one element in each degree d such that $\alpha(d+n) = n$.

In Sections 22.5 and 22.6, we apply this result to obtain minimal generating sets for the symmetric Steinberg summand $\mathsf{P}(n)e(n)$ and for the $\mathsf{B}(n)$-invariant Steinberg summand $\mathsf{P}(n)e'(n)$. In terms of the splitting of Theorem 19.1.1, the module $L(\lambda) = \mathsf{St}(n)$ is the only irreducible $\mathbb{F}_2\mathsf{GL}(n)$-module for which the hit problem for $\mathsf{P}(n,\lambda)$ has been solved for all n.

22.1 Steenrod operations on rational functions

In this section we extend the action of A_2 on $P(n)$ to an action on the algebra of rational functions $R(n) = \mathbb{F}_2(x_1,\dots,x_n)$, whose elements are quotients of polynomials of the form f/g, where $f,g \in P(n)$ and $g \neq 0$. For applications to the Steinberg summand $P(St(n))$ of $P(n)$, we shall need only the case where g is a product of linear combinations of $x_1,\dots,x_n$. We begin with the case $n=1$, where the only nonzero linear form is $x_1 = x$.

Example 22.1.1 The ring of Laurent polynomials $\mathbb{F}_2[x,x^{-1}]$ is the vector space over $\mathbb{F}_2$ with basis $\{x^d\}$ where $d \in \mathbb{Z}$ is an integer. We define squaring operations on $\mathbb{F}_2[x,x^{-1}]$ by extending the formula of Proposition 1.1.8

$$Sq^k(x^d) = \binom{d}{k} x^{d+k} \tag{22.1}$$

to all integers d, where by (2.2) $\binom{-a}{b} = (-1)^b\binom{a+b-1}{b}$ for $a > 0$ and $b \geq 0$. We shall see below that the Adem relations hold in this wider context: for example, $Sq^3(x^{-5}) = \binom{7}{3}x^{-2} = x^{-2}$ and $Sq^1Sq^2(x^{-5}) = \binom{6}{2}Sq^1(x^{-3}) = x^{-2}$. Thus $\mathbb{F}_2[x,x^{-1}]$ is a module over A_2.

The element x^{-1} occupies a special place in the A_2-module $\mathbb{F}_2[x,x^{-1}]$. Since $\binom{-1}{k} = (-1)^k$, $Sq^k(x^{-1}) = x^{k-1}$ for all $k \geq 0$, so the identity element 1 and all positive powers of x, including spikes, are hit in $\mathbb{F}_2[x,x^{-1}]$ by squaring operations on x^{-1}, but as $Sq^k(x^{-(k+1)}) = \binom{2k}{k}x^{-1} = 0$ for all $k > 0$, x^{-1} is not hit. Further, x^{-1} is the only cohit Laurent polynomial, in the usual sense that the hit elements form an A_2-submodule $H(1)$ of $\mathbb{F}_2[x,x^{-1}]$, with quotient $Q(1) = \mathbb{F}_2[x,x^{-1}]/H(1) \cong \mathbb{F}_2$ generated by the equivalence class of x^{-1}. This can be seen as follows. If $k > 1$, then $k = 2^r + 1 \bmod 2^r$ for some $r \geq 0$. Then $Sq^{2^r}(x^{-k-2^r}) = \binom{-k-2^r}{2^r}x^{-k} = \binom{k+2^{r+1}-1}{2^r}x^{-k} = x^{-k}$, so x^{-k} is hit.

The A_2-submodule $T(1)$ of $\mathbb{F}_2[x,x^{-1}]$ generated by x^{-1} is of particular importance. It has $\mathbb{F}_2$-basis $\{x^{-1},1,x,x^2,x^3,\dots\}$. More generally, we shall study the A_2-submodule $T(n)$ of the Laurent polynomial ring over $\mathbb{F}_2$ in $x_1,\dots,x_n$ generated by $c(n)^{-1}$ where $c(n) = x_1\cdots x_n$. The Laurent polynomial ring has as $\mathbb{F}_2$-basis the set of all monomials $x_1^{d_1}\cdots x_n^{d_n}$, where the exponents $d_1,\dots,d_n$ are integers which may be positive, negative or 0. The action of A_2 is defined on the generators $x_1,\dots,x_n$ by (22.1), and is extended to monomials by the Cartan formula and to all Laurent polynomials by linearity.

We observe that the action of $GL(n)$ on $P(n)$ does not extend to an action on Laurent polynomials if $n > 1$, since the substitution $x \mapsto x+y$ maps x^{-1} to $(x+y)^{-1}$, and this is not a Laurent polynomial in x and y since $(x+1)^{-1}$ is not a Laurent polynomial in x. However, the group of permutation matrices

$\mathsf{W}(n)$ does act on the Laurent polynomial ring, and the action commutes with that of A_2.

Since the generator $\mathsf{c}(n)^{-1}$ of $\mathsf{T}(n)$ is fixed by $\mathsf{W}(n)$, $\mathsf{T}(n)$ consists of symmetric Laurent polynomials, and so $\mathsf{T}(n) \cap \mathsf{P}(n)$ is an A_2-submodule of the algebra of symmetric polynomials $\mathsf{S}(n)$. In Proposition 22.2.7 we show that $\mathsf{T}(n) \cap \mathsf{P}(n)$ can be identified with the symmetric Steinberg summand $\mathsf{P}(n)e(n)$ of $\mathsf{P}(n)$.

Example 22.1.2 In the case $n = 2$, we have $Sq^2Sq^1(x^{-1}y^{-1}) = Sq^2(x^{-1} + y^{-1}) = x+y$, $Sq^4Sq^2(x^{-1}y^{-1}) = Sq^4(xy^{-1} + 1 + x^{-1}y) = (x^3y + x^2y^2) + (xy^3 + x^2y^2) = x^3y + xy^3$. Hence $x+y$ and x^3y+xy^3 are elements of $\mathsf{T}(2) \cap \mathsf{P}(2)$. Since

$$e(2) = \overline{\mathsf{B}}(2)\overline{\mathsf{W}}(2) = \begin{pmatrix} 1 & 0 \\ 0 & 1 \end{pmatrix} + \begin{pmatrix} 1 & 0 \\ 1 & 1 \end{pmatrix} + \begin{pmatrix} 0 & 1 \\ 1 & 0 \end{pmatrix} + \begin{pmatrix} 0 & 1 \\ 1 & 1 \end{pmatrix},$$

$(x+y)e(2) = (x+y) + (x+x+y) + (y+x) + (y+x+y) = x+y$ and $(x^3y + xy^3)e(2) = (x^3y + xy^3) + (x^3(x+y) + x(x+y)^3) + (y^3x + x^3y) + (y^3(x+y) + y(x+y)^3) = x^3y + xy^3$. Thus $x+y$ and x^3y+xy^3 are fixed by $e(2)$, and so they lie in the direct summand $\mathsf{P}(2)e(2)$ of the A_2-module $\mathsf{P}(2)$.

Note that Sq^2Sq^1 and Sq^4Sq^2 are admissible monomials of length 2. In fact, we shall show that the elements $Sq^A(\mathsf{c}(n)^{-1})$, where Sq^A is an admissible monomial of length n in A_2, form an $\mathbb{F}_2$-basis for $\mathsf{T}(n) \cap \mathsf{P}(n)$.

We begin by extending the A_2 action on polynomials to rational functions. Note that if $g \in \mathsf{P}(n)$ and $g \neq 0$, then the total Steenrod square Sq is nonzero on g since $Sq^0(g) = g$.

Definition 22.1.3 Given $f, g \in \mathsf{P}(n)$ with $g \neq 0$, let $\mathsf{Sq}(f/g) = \mathsf{Sq}(f)/\mathsf{Sq}(g)$.

To see that Sq is well defined on $\mathsf{R}(n)$, note that if $f_1/g_1 = f_2/g_2$ then $f_1g_2 = f_2g_1$ in $\mathsf{P}(n)$, and so $\mathsf{Sq}(f_1)\mathsf{Sq}(g_2) = \mathsf{Sq}(f_2)\mathsf{Sq}(g_1)$ and $\mathsf{Sq}(f_1)/\mathsf{Sq}(g_1) = \mathsf{Sq}(f_2)/\mathsf{Sq}(g_2)$. Hence $\mathsf{Sq}(f_1/g_1) = \mathsf{Sq}(f_2/g_2)$. Similarly it is straightforward to verify that for $f_1/g_1, f_2/g_2 \in \mathsf{R}(n)$ we have $\mathsf{Sq}(f_1/g_1 \cdot f_2/g_2) = \mathsf{Sq}(f_1/g_1)\mathsf{Sq}(f_2/g_2)$. Thus Sq is an injective ring homomorphism $\mathsf{Sq} : \mathsf{R}(n) \to \mathsf{R}(n)$. It is not surjective, since the composition inverse of $\mathsf{Sq} : \mathsf{P}(n) \to \mathsf{P}(n)$ is the conjugate total squaring operation Xq, and $\mathsf{Xq}(x) = x + x^2 + x^4 + x^8 + \dots$ is not a rational function of x.

As well as the total squaring operation Sq, we can define operations Sq^k on rational functions, as in the special case of Example 22.1.1. In general we can use the Cartan formula

$$Sq^k(f) = \sum_{i+j=k} Sq^i(f/g)Sq^j(g)$$

to define $Sq^k(f/g)$ by induction on k, starting as usual with $Sq^0 = 1$. For example $Sq^1(f/g) = (Sq^1(f)g + fSq^1(g))/g^2$, which corresponds to the 'quotient rule' of differential calculus.

Proposition 22.1.4 *With the operations $Sq^k : \mathsf{R}(n) \to \mathsf{R}(n)$ defined as above, the field of rational functions $\mathsf{R}(n) = \mathbb{F}_2(x_1,\ldots,x_n)$ is an A_2-module.*

Proof We must show that the operations Sq^k on $\mathsf{R}(n)$ satisfy the Adem relations. For this, we use the Bullett–Macdonald identity, Proposition 4.2.3. We extend the generalized total squaring operations $\mathsf{Sq}[u]$ for $u \in \mathbb{F}_2[[t]]$ from $\mathsf{P}(n)$ to $\mathsf{R}(n)$ by $\mathsf{Sq}[u](f/g) = \mathsf{Sq}[u](f)/\mathsf{Sq}[u](g)$, so that $\mathsf{Sq}[t + t^2]\mathsf{Sq}[1](f/g) = \mathsf{Sq}[t+t^2]\mathsf{Sq}[1](f)/\mathsf{Sq}[t+t^2]\mathsf{Sq}[1](g) = \mathsf{Sq}[1+t]\mathsf{Sq}[t^2](f)/\mathsf{Sq}[1+t]\mathsf{Sq}[t^2](g) = \mathsf{Sq}[1+t]\mathsf{Sq}[t^2](f/g)$. The result follows from Proposition 4.2.4. □

Proposition 22.1.5 *The right action of $\mathbb{F}_2\mathsf{M}(n)$ on $\mathsf{R}(n)$ commutes with the left action of A_2, i.e. for $\theta \in \mathsf{A}_2$, $f \in \mathsf{P}(n)$, $g \neq 0 \in \mathsf{P}(n)$ and $A \in \mathsf{M}(n)$,*

$$\theta(fg^{-1}) \cdot A = \theta(fg^{-1} \cdot A).$$

Proof It suffices to prove the result for $\theta = Sq^k$, and hence for the total square Sq. First let $f = 1$. Then since $\mathsf{Sq} : \mathsf{R}(n) \to \mathsf{R}(n)$ is multiplicative, $\mathsf{Sq}(g^{-1} \cdot A)\mathsf{Sq}(g \cdot A) = \mathsf{Sq}(g^{-1}g \cdot A) = \mathsf{Sq}(1 \cdot A) = \mathsf{Sq}(1) = 1$, and $(\mathsf{Sq}(g^{-1}) \cdot A)(\mathsf{Sq}(g) \cdot A) = \mathsf{Sq}(g^{-1})\mathsf{Sq}(g) \cdot A = \mathsf{Sq}(g^{-1}g) \cdot A = \mathsf{Sq}(1) \cdot A = 1 \cdot A = 1$. Since $\mathsf{Sq}(g \cdot A) = \mathsf{Sq}(g) \cdot A$, it follows that $\mathsf{Sq}(g^{-1} \cdot A) = \mathsf{Sq}(g^{-1}) \cdot A$. The general case follows since $\mathsf{Sq}(fg^{-1} \cdot A) = \mathsf{Sq}((f \cdot A)(g^{-1} \cdot A)) = \mathsf{Sq}(f \cdot A)\mathsf{Sq}(g^{-1} \cdot A) = (\mathsf{Sq}(f) \cdot A)(\mathsf{Sq}(g^{-1}) \cdot A) = \mathsf{Sq}(f)\mathsf{Sq}(g^{-1}) \cdot A = \mathsf{Sq}(fg^{-1}) \cdot A$. □

We note that Proposition 2.5.1 depends only on the Cartan formula, and so the χ-trick is valid in $\mathsf{R}(n)$. We make use of this in the proof of Proposition 22.2.6.

The action of A_2 on the Laurent polynomial ring $\mathsf{P}(n)[x_1^{-1},\ldots,x_n^{-1}]$ is determined by (22.1) and the Cartan formula, and every Laurent polynomial can be written in the form $f\mathsf{c}(n)^{-m}$ where $f \in \mathsf{P}(n)$ and $m \geq 0$. Hence we have the following result.

Proposition 22.1.6 *The Laurent polynomial ring $\mathsf{P}(n)[x_1^{-1},\ldots,x_n^{-1}]$ is an A_2-submodule of the field of rational functions $\mathsf{R}(n)$. For $m \geq 0$, the set of Laurent polynomials of the form $f\mathsf{c}(n)^{-m}$, where $f \in \mathsf{P}(n)$, is an A_2-submodule $\mathsf{P}(n)\mathsf{c}(n)^{-m}$ of $\mathsf{P}(n)[x_1^{-1},\ldots,x_n^{-1}]$.* □

22.2 The cyclic A_2-module $\mathsf{T}(n)$

Definition 22.2.1 $\mathsf{T}(n)$ is the cyclic A_2-submodule of the Laurent polynomial ring $\mathsf{P}(n)[x_1^{-1},\dots,x_n^{-1}]$ generated by $\mathsf{c}(n)^{-1} = x_1^{-1}\cdots x_n^{-1}$.

The main result of this section, Proposition 22.2.7, is that the polynomials in $\mathsf{T}(n)$ are precisely those which are in $\mathsf{P}(n)\cdot e(n)$, the symmetric Steinberg summand of $\mathsf{P}(n)$. It was shown in Section 22.1 that all Laurent polynomials $a_{-1}x^{-1} + a_0 + a_1x + \cdots + a_mx^m$ are in $\mathsf{T}(1)$, and so $\mathsf{P}(1) \subset \mathsf{T}(1)$. Thus the results in this section are easily proved for $n = 1$. We begin by showing that the generator $\mathsf{c}(n)^{-1}$ of $\mathsf{T}(n)$ is fixed by the Steinberg idempotent $e(n)$.

Proposition 22.2.2 $\mathsf{c}(n)^{-1}\cdot e(n) = \mathsf{c}(n)^{-1}$.

Proof In the case $n = 2$, $e(2)$ is given by Example 22.1.2, and so, writing $x_1 = x$ and $x_2 = y$, $(x^{-1}y^{-1})\cdot e(2) = x^{-1}y^{-1} + y^{-1}x^{-1} + x^{-1}(x+y)^{-1} + y^{-1}(x+y)^{-1} = (x^{-1}+y^{-1})(x+y)^{-1} = x^{-1}y^{-1}$. Let the idempotents $e_1,\dots,e_{n-1}$ be as in Section 18.4, so that $e(n)$ is the longest word in $e_1,\dots,e_{n-1}$. Then all the matrices which appear in e_i fix x_j for $j \neq i, i+1$, and so, by the above calculation with $x = x_i$, $y = x_{i+1}$ and e_i replacing $e(2)$, we have $\mathsf{c}(n)^{-1}\cdot e_i = \mathsf{c}(n)^{-1}$ for $i \le i \le n-1$. □

Proposition 22.2.3 $\mathsf{T}(n) \subseteq \mathsf{P}(n)\mathsf{c}(n)^{-1}\cdot e(n)$.

Proof As $\mathsf{Sq}(\mathsf{c}(n)^{-1}) = \mathsf{Sq}(\mathsf{c}(n))^{-1} = (x_1+x_1^2)^{-1}\cdots(x_n+x_n^2)^{-1} = \mathsf{c}(n)^{-1}(1+x_1)^{-1}\cdots(1+x_n)^{-1}$, $Sq^k(\mathsf{c}(n)^{-1}) \in \mathsf{P}(n)\mathsf{c}(n)^{-1}$ for $k \ge 0$. By Propositions 22.1.5 and 22.2.2, it follows that $\theta(\mathsf{c}(n)^{-1}) = \theta(\mathsf{c}(n)^{-1}\cdot e(n)) = \theta(\mathsf{c}(n)^{-1})\cdot e(n) \in \mathsf{P}(n)\mathsf{c}(n)^{-1}\cdot e(n)$ for any $\theta \in \mathsf{A}_2$. Hence $\mathsf{T}(n) \subseteq \mathsf{P}(n)\mathsf{c}(n)^{-1}\cdot e(n)$. □

To prove the reverse inclusion, we need some preliminary results.

Proposition 22.2.4 *For any admissible monomial* $Sq^A \in \mathsf{A}_2$*,*

$$(x_n^{-1}Sq^A(\mathsf{c}(n-1)^{-1}))\cdot e_{n-1} = Sq^A(\mathsf{c}(n)^{-1}).$$

Proof Using the Cartan formula, $x_n^{-1}Sq^A(\mathsf{c}(n-1)^{-1}) = x_n^{-1}Sq^A(\mathsf{c}(n)^{-1}x_n) = x_n^{-1}(Sq^A(\mathsf{c}(n)^{-1})x_n + Sq^B(\mathsf{c}(n)^{-1})x_n^2) = Sq^A(\mathsf{c}(n)^{-1}) + Sq^B(\mathsf{c}(n)^{-1})x_n$, where B is the admissible sequence obtained by reducing the last term of A by 1. Since $\mathsf{c}(n)^{-1}$ is fixed by e_{n-1}, $Sq^A(\mathsf{c}(n)^{-1})\cdot e_{n-1} = Sq^A(\mathsf{c}(n)^{-1})$. We shall show that $f\cdot e_{n-1} = 0$, where $f = Sq^B(\mathsf{c}(n)^{-1})x_n$.

Since $e_{n-1} = (I_n + L_{n-1})(I_n + S_{n-1})$, $f\cdot e_{n-1} = f\cdot I_n + f\cdot S_{n-1} + f\cdot L_{n-1} + f\cdot L_{n-1}S_{n-1}$, and if the matrix M is one of the four terms of e_{n-1} then $f\cdot M = (Sq^B(\mathsf{c}(n)^{-1})\cdot M)(x_n\cdot M) = Sq^B(\mathsf{c}(n)^{-1}\cdot M)(x_n\cdot M)$ by Proposition 22.1.5. Hence $f\cdot I_n + f\cdot S_{n-1} = Sq^B(\mathsf{c}(n)^{-1})(x_n + x_{n-1})$, since S_{n-1} fixes $\mathsf{c}(n)$.

Since $x_{n-1}^{-1}(x_{n-1}+x_n)^{-1} + x_n^{-1}(x_n+x_{n-1})^{-1} = x_{n-1}^{-1}x_n^{-1}$ and $\mathsf{c}(n-2)^{-1}$ $x_{n-1}^{-1}x_n^{-1} = \mathsf{c}(n)^{-1}$, the two remaining terms

$$f \cdot L_{n-1} = Sq^B\left(\mathsf{c}(n-2)^{-1}x_{n-1}^{-1}(x_{n-1}+x_n)^{-1}\right)(x_{n-1}+x_n)$$
$$f \cdot L_{n-1}S_{n-1} = Sq^B\left(\mathsf{c}(n-2)^{-1}x_n^{-1}(x_n+x_{n-1})^{-1}\right)(x_n+x_{n-1})$$

also have sum $Sq^B(\mathsf{c}(n)^{-1})(x_n+x_{n-1})$. It follows that $f \cdot e_{n-1} = 0$. □

The next result is similar, but the idempotent e_{n-1} is replaced by the Steinberg idempotent $e(n)$.

Proposition 22.2.5 *For any admissible monomial* $Sq^A \in \mathsf{A}_2$,

$$(x_n^{-1}Sq^A(\mathsf{c}(n-1)^{-1})) \cdot e(n) = Sq^A(\mathsf{c}(n)^{-1}).$$

Proof Since $c(n-1)^{-1}$ is fixed by $e(n-1)$ by (22.2.2) and since x_n is also fixed by $e(n-1)$, $x_n^{-1}Sq^A(\mathsf{c}(n-1)^{-1})$ is fixed by $e(n-1)$. By Proposition 18.4.1, the idempotent $e(n)$ factors as $e(n) = e(n-1)e_{n-1}\cdots e_1$. Hence

$$\begin{aligned}(x_n^{-1}Sq^A(\mathsf{c}(n-1)^{-1}))\cdot e(n) &= (x_n^{-1}Sq^A(\mathsf{c}(n-1)^{-1}))\cdot e(n-1)e_{n-1}\cdots e_1,\\ &= (x_n^{-1}Sq^A(\mathsf{c}(n-1)^{-1})\cdot e(n-1))\cdot e_{n-1}\cdots e_1,\\ &= (x_n^{-1}Sq^A(\mathsf{c}(n-1)^{-1}))\cdot e_{n-1}\cdots e_1,\\ &= ((x_n^{-1}Sq^A(\mathsf{c}(n-1)^{-1}))\cdot e_{n-1})e_{n-2}\cdots e_1,\\ &= Sq^A(\mathsf{c}(n)^{-1})\cdot e_{n-2}\cdots e_1\end{aligned}$$

by Proposition 22.2.4. Using Proposition 22.1.5, this reduces to $Sq^A(\mathsf{c}(n)^{-1})$ since $\mathsf{c}(n)^{-1}$ is fixed by e_i for $1 \le i \le n-2$. □

The next result is the key step in the proof of Proposition 22.2.7.

Proposition 22.2.6 $\mathsf{T}(n) = \mathsf{P}(n)\mathsf{c}(n)^{-1} \cdot e(n)$.

Proof By Proposition 22.2.3, it remains to show that

$$\mathsf{P}(n)\mathsf{c}(n)^{-1} \cdot e(n) \subseteq \mathsf{T}(n). \tag{22.2}$$

For this we use induction on n. In the case $n = 1$, we have $e(1) = 1$ and $\mathsf{P}(1)c(1)^{-1} = \mathsf{P}(1)x^{-1} = \mathsf{T}(1)$. We may write a monomial $f \in \mathsf{P}(n)\mathsf{c}(n)^{-1}$ as $f = x_n^{k-1}g$, where $k \ge 0$ and $g \in \mathsf{P}(n-1)\mathsf{c}(n-1)^{-1}$. Since $x_n^{k-1} = Sq^k(x_n^{-1})$, using Proposition 2.5.1 (which is valid, as noted above, for the action of A_2 on rational functions), we have

$$f = Sq^k(x_n^{-1})g = x_n^{-1}Xq^k(g) + \sum_{i=1}^{k} Sq^i(x_n^{-1}Xq^{k-i}(g)). \tag{22.3}$$

Thus to prove (22.2) it suffices to prove that

(i) $x_n^{-1}Xq^k(g)\cdot e(n) \in \mathsf{T}(n)$ for $k \geq 0$,
(ii) $Sq^i(x_n^{-1}Xq^{k-i}(g))\cdot e(n) \in \mathsf{T}(n)$ for $1 \leq i \leq k$.

By Proposition 22.1.5, $Sq^i(x_n^{-1}Xq^{k-i}(g))\cdot e(n) = Sq^i(x_n^{-1}Xq^{k-i}(g)\cdot e(n))$. Since $\mathsf{T}(n)$ is closed under the action of A_2, (ii) follows from (i).

For (i), since $e(n) = e(n-1)e(n)$, $(x_n^{-1}Xq^k(g))\cdot e(n) = (x_n^{-1}Xq^k(g)\cdot e(n-1))\cdot e(n) = (x_n^{-1}\cdot e(n-1))(Xq^k(g)\cdot e(n-1))\cdot e(n)$. Since $g \in \mathsf{P}(n-1)\mathsf{c}(n-1)^{-1}$, $Xq^k(g)\cdot e(n-1) \in \mathsf{P}(n-1)\mathsf{c}(n-1)^{-1}\cdot e(n-1)$. Hence $Xq^k(g)\cdot e(n-1) \in \mathsf{T}(n-1)$ by the induction hypothesis, and so by definition of $\mathsf{T}(n-1)$ it is a sum of terms of the form $Sq^A(c(n-1)^{-1})$ where A is an admissible sequence. Since also x_n^{-1} is fixed by $e(n-1)$, (i) follows from (22.2.5). □

Proposition 22.2.7 $\mathsf{T}(n) \cap \mathsf{P}(n) = \mathsf{P}(n)\cdot e(n)$.

Proof Since $\mathsf{P}(n)\cdot e(n) \subseteq \mathsf{P}(n)(\mathsf{c}(n)^{-1})\cdot e(n) = \mathsf{T}(n)$ by Proposition 22.2.6, $\mathsf{P}(n)\cdot e(n) \subseteq \mathsf{P}(n)\cap\mathsf{T}(n)$. Conversely, if $f \in \mathsf{P}(n)\cap\mathsf{T}(n)$ then again by Proposition 22.2.6 we can write $f = g\cdot e(n)$, where $g \in \mathsf{P}(n)(\mathsf{c}(n)^{-1})$. Then $f\cdot e(n) = g\cdot e(n)^2 = g\cdot e(n) = f$, since $e(n)$ is idempotent. Hence f is fixed by $e(n)$, and so $f \in \mathsf{P}(n)\cdot e(n)$. □

We next use the left lexicographic order on monomials in $\mathsf{P}(n)(\mathsf{c}(n)^{-1})$ to study the A_2-module $\mathsf{T}(n)$ in more detail.

Proposition 22.2.8 *Let $A = (a_1,\ldots,a_r)$ be an admissible sequence of length $r > 0$, and let $a_i = 0$ for $i > r$. Then*

(i) *if $r > n$ then $Sq^A(\mathsf{c}(n)^{-1}) = 0$,*
(ii) *if $r \leq n$ then $Sq^A(\mathsf{c}(n)^{-1}) = x_1^{a_1-1}\cdots x_n^{a_n-1} +$ lower terms,*
(iii) *the elements $Sq^A(\mathsf{c}(n)^{-1})$, where $r \leq n$, form a basis for $\mathsf{T}(n)$,*
(iv) *the elements $Sq^A(\mathsf{c}(n)^{-1})$, where $r = n$, form a basis for $\mathsf{T}(n)\cap\mathsf{P}(n)$.*

Proof (i) Let $A_s = (a_s, a_{s+1},\ldots,a_r)$ where $1 < s \leq r$. Then $Sq^{A_s}(\mathsf{c}(n)^{-1})$ has degree $a_s + a_{s+1} + \cdots + a_r - n < a_{s-1}$, since A is admissible. Let f be a monomial which appears in $Sq^{A_s}(\mathsf{c}(n)^{-1})$, and let g be a monomial which appears in $Sq^{A_{s-1}}(f)$. Since $Sq^{a_{s-1}}$ annihilates all polynomials in $\mathsf{P}(n)$ of degree $< a_{s-1}$, f must have more exponents -1 than g has, and (i) follows.

(ii) We use induction on n. By the Cartan formula

$$Sq^A(\mathsf{c}(n)^{-1}) = \sum_{A=B+C} Sq^B(x_1^{-1})Sq^C(x_2^{-1}\cdots x_n^{-1}) \tag{22.4}$$

where the sum is over all sequences B and C with sum $A = (a_1,\ldots,a_n)$, admissible or not. By choosing $B = (a_1,0,\ldots,0)$ and $C = (0,a_2,\ldots,a_n)$

and using the induction hypothesis, it is clear that $x_1^{a_1-1}\cdots x_n^{a_n-1}$ is the highest monomial in the expansion of the term $Sq^B(x_1^{-1})Sq^C(x_2^{-1}\cdots x_n^{-1})$ in the summation 22.4.

Clearly the term with $B=0$, $C=A$ in the summation gives only monomials where the exponent of x_1 is -1. Otherwise $B=(b_1,\dots,b_s)$ where $b_s>0$ and $s\le r$. If $Sq^B(x_1^{-1})\ne 0$, then $b_{s-1}\le b_s-1$, $b_{s-2}\le b_{s-1}+b_s-1,\dots,b_1\le b_2+\cdots+b_s-1$, and if $Sq^B(x_1^{-1})=x_1^a$ then $a\le b_1+b_2+\cdots+b_s-1$. Then $a\le 2(b_2+\cdots+b_s-1)\le 4(b_3+\cdots+b_s-1)\le\cdots\le 2^{s-1}(b_s-1)$. Now $b_s\le a_s$ since $A=B+C$, and $a_1\ge 2a_2\ge\cdots\ge 2^{s-1}a_s$ since A is admissible. Hence $a\le a_1-2^{s-1}<a_1-1$ if $s>1$. In the case $s=1$, $a=b_1-1$ and so $a=a_1-1$ only in the case $B=(a_1,0,\dots,0)$ considered above. This completes the induction.

(iii) follows at once from (i) and (ii), since these elements span $\mathsf{T}(n)$ by (i), and are linearly independent by (ii).

(iv) The proof of (i) shows that $Sq^A(\mathsf{c}(n)^{-1})$ is in $\mathsf{P}(n)$ if the admissible sequence A has length n. If A has length $r<n$, then $a_n=0$ and $Sq^A(\mathsf{c}(n)^{-1})$ is not in $\mathsf{P}(n)$ by (ii). □

Proposition 22.2.8 gives an alternative proof of Proposition 19.5.5, by calculating the number of admissible monomials Sq^A of given length n and degree d. The vector subspace $\mathsf{Ad}_{\le n}$ of A_2 spanned by admissible monomials of length $\le n$ is also spanned by Milnor basis elements $Sq(R)$ of length $\le n$, and is the subcoalgebra of A_2 dual to the subalgebra $\mathsf{Ad}^*_{\le n}$ of A_2^* generated by ξ_i, $1\le i\le n$. Since $\mathsf{Ad}^*_{\le n}$ is a polynomial algebra on generators of degree 2^i-1, $1\le i\le n$, its Poincaré series $P(\mathsf{Ad}^*_{\le n},t)=\prod_{i=1}^n(1-t^{2^i-1})^{-1}$. Hence

$$P(\mathsf{Ad}_{\le n}/\mathsf{Ad}_{\le n-1},t)=\prod_{i=1}^{n}(1-t^{2^i-1})^{-1}-\prod_{i=1}^{n-1}(1-t^{2^i-1})^{-1}=t^{2^n-1}\prod_{i=1}^{n}(1-t^{2^i-1})^{-1}.$$

By Proposition 22.2.8(iv), the Poincaré series of $\mathsf{T}(n)\cap\mathsf{P}(n)=\mathsf{P}(n)\cdot e(n)$ is obtained from that of $\mathsf{Ad}_{\le n}/\mathsf{Ad}_{\le n-1}$ by shifting the grading by $\deg(\mathsf{c}(n)^{-1})=-n$. Hence $P(\mathsf{P}(\mathsf{St}(n)),t)=t^{w(n-1)}\prod_{i=1}^n(1-t^{2^i-1})^{-1}$, where $w(n-1)=2^n-1-n$.

22.3 The Mitchell–Priddy module $\mathsf{MP}(n)$

The admissible basis gives a decreasing filtration

$$\mathsf{A}_2=\mathsf{Ad}_{\ge 0}\supset\mathsf{Ad}_{\ge 1}\supset\mathsf{Ad}_{\ge 2}\supset\cdots\supset\mathsf{Ad}_{\ge n}\supset\cdots$$

of A_2, where $\mathsf{Ad}_{\ge n}$ is the $\mathbb{F}_2$-subspace of A_2 spanned by admissible monomials Sq^A of length $\mathrm{len}(A)\ge n$. We shall show that $\mathsf{Ad}_{\ge n}$ is preserved by the left action of A_2, so that it is an A_2-module. The equivalence classes of admissible monomials of length n give a $\mathbb{F}_2$-basis for the quotient module $\mathsf{Ad}_{\geq n}/\mathsf{Ad}_{\geq n+1}$, with the grading $|A|$ inherited from A_2. We write the equivalence class of Sq^A as $[Sq^A]$ and define the **Mitchell–Priddy module** $\mathsf{MP}(n)$ by regrading $\mathsf{Ad}_{\geq n}/\mathsf{Ad}_{\geq n+1}$ so that $[Sq^A]$ has grading $|A|-n$. The aim of this section is to show that $\mathsf{MP}(n)$ is isomorphic as an A_2-module to the Steinberg summand $\mathsf{P}(\mathsf{St}(n))$ of $\mathsf{P}(n)$.

We start with some observations about the Adem relations. It is convenient to say that the term $Sq^{a+b-j}Sq^j$ 'appears' in the Adem relation (3.1) if the mod 2 binomial coefficient $\binom{b-j-1}{a-2j}=1$.

Proposition 22.3.1 (i) *When the Adem relation (3.1) is used to replace the left hand side Sq^aSq^b by the right hand side, and the resulting expression is written as a sum of monomials in A_2, the length of index sequences is never increased.*

(ii) *All terms on the right of the Adem relation (3.1) have length 2 except possibly the first term $j=0$, which can be Sq^{a+b}. If this term appears, we say that a* **contraction** *occurs.*

(iii) *If the term $Sq^{a+b-j}Sq^j$ appears, then $j>a-b$. In particular, a contraction does not occur when $a\geq b$.*

(iv) *When a is even, the last term $Sq^{b+a/2}Sq^{a/2}$ appears.*

(v) *If a is odd and b is even, then no term with $a+b-j$ even and j odd appears.*

Proof All these statements follow immediately by inspection of binomial coefficients mod 2. We note that $\binom{b-j-1}{a-2j}=0$ if $b-j-1<a-2j$ for (iii), and also if $b-j-1$ is even and $a-2j$ is odd, for (v). □

We next give an alternative description of $\mathsf{Ad}_{\geq n}/\mathsf{Ad}_{\geq n+1}$.

Definition 22.3.2 The monomial $Sq^C=Sq^{c_1}\cdots Sq^{c_n}$ in A_2 is a **cone monomial** of length n if $c_i\geq c_{i+1}+\cdots+c_n$ for $1\leq i\leq n-1$.

Clearly, admissible monomials are cone monomials.

Proposition 22.3.3 *The subspace $\mathsf{Ad}_{\geq n}$ of A_2 is spanned by cone monomials of length $\geq n$, and $\mathsf{Ad}_{\geq n}/\mathsf{Ad}_{\geq n+1}$ is spanned by equivalence classes of cone monomials of length n.*

Proof We show that all cone monomials of length n belong to $\mathsf{Ad}_{\geq n}$. Suppose that adjacent factors $Sq^{c_r}Sq^{c_{r+1}}$ are changed to $Sq^{c_r+c_{r+1}-j}Sq^j$, where $c_r-c_{r+1}<j\leq c_r/2$, using an Adem relation. Note that $j>c_r-c_{r+1}\geq c_{r+2}+c_{r+3}+\cdots+c_n$ and $c_r+c_{r+1}-j\geq j+c_{r+2}+c_{r+3}+\cdots+c_n$ because $c_r\geq 2k$ and $c_{r+1}\geq c_{r+2}+c_{r+3}+\cdots+c_n$. This shows that the cone condition is preserved by applying an Adem relation. In particular, all terms in the admissible reduction of a cone monomial of length n have length n, and so the cone monomial is in $\mathsf{Ad}_{\geq n}$. □

The next result shows that $\mathsf{Ad}_{\geq n}$ is an A_2-module, where A_2 acts on $\mathsf{Ad}_{\geq n}$ by left multiplication.

Proposition 22.3.4 *Let* Sq^C *be a cone monomial of length n, and let* $k \geq 0$*. Then all terms in the admissible reduction of* $Sq^k Sq^C$ *have length n or* $n+1$.

Proof Let $Sq^C = Sq^{c_1} \cdots Sq^{c_n}$. By Proposition 22.3.1, an application of the Adem relations cannot increase the length of a monomial in A_2. It remains to show that the length of $Sq^k Sq^C$ cannot decrease by more than 1 during the repeated applications of Adem relations required to bring $Sq^k Sq^C$ to admissible form.

If $k \geq 2c_1$, then $Sq^k Sq^C$ is already a cone monomial of length $n+1$, and the result follows by Proposition 22.3.3. If $i < 2c_1$, then we can apply an Adem relation to the first two factors $Sq^k Sq^{c_1}$ of $Sq^k Sq^C$, to produce pairs of factors of the form $Sq^{c_1+k-k_1} Sq^{k_1}$, where $k_1 \leq k/2$. We continue to apply Adem relations to successive pairs of factors from left to right, to obtain a sum of terms of the form $Sq^{c_1+k-k_1} Sq^{c_2+k_1-k_2} \ldots Sq^{c_n+k_{n-1}-k_n} Sq^{k_n}$, noting, in particular, that $k_r \geq 2k_{r+1}$ for every term produced. This 'cascade' process makes the first factor Sq^k of $Sq^k Sq^C$ pass down the line of successive squares, using the Adem relations at each stage to form an increasing list of monomials.

Since the second index in an adjacent pair is at most $k/2^t$ at the tth step, the cascade process stops after about $\log_2(i)$ steps. In any particular term, it is possible for the process to stop at an earlier step, when the Adem relation allows a contraction. In such a term, all $Sq^{k_t} = Sq^0 = 1$ from some t onwards. As there can be at most one contraction in each term during the cascade process, the length of the original monomial $Sq^k Sq^C$ is reduced by at most 1.

To show that every term produced from $Sq^k Sq^C$ by the cascade process is a cone monomial, we require $c_{r+1} + k_r - k_{r+1} \geq \sum_{i=r+2}^{n} (c_i + k_{i-1} - k_i) + k_n$ for $1 \leq r \leq n-1$. This reduces to $c_{r+1} + k_r - k_{r+1} \geq k_{r+1} + c_{r+2} + \cdots + c_n$, and is true because $k_r \geq 2k_{r+1}$ and $c_{r+1} \geq c_{r+2} + \cdots + c_n$. Hence $Sq^k Sq^C$ reduces to cone monomials of length n or $n+1$, and the proof is completed by Proposition 22.3.3. □

Theorem 22.3.5 $\mathsf{MP}(n)$ *is isomorphic as an* A_2*-module to the Steinberg summand* $\mathsf{P}(\mathsf{St}(n))$ *of* $\mathsf{P}(n)$.

Proof The map $Sq^A \mapsto Sq^A(\mathsf{c}(n)^{-1})$, where A is an admissible sequence, is a map of A_2-modules which preserves the shifted grading on $\mathsf{MP}(n)$. By the results of Section 22.1, this map is a bijection from $\mathsf{MP}(n)$ to $\mathsf{P}(n)e(n)$. □

The Mitchell–Priddy model $\mathsf{MP}(n)$ for the Steinberg summand $\mathsf{P}(\mathsf{St}(n))$ is helpful in studying its structure as an A_2-module.

Proposition 22.3.6 *For $n \geq 2$, the A_2-module $\mathsf{P}(\mathsf{St}(n))$ is free over the subalgebra $\mathsf{A}_2(n-2)$ of A_2.*

Proof Recall from Definition 3.4.1 that if the sequence $A = (a_1,\dots,a_n)$ is admissible, the corresponding Milnor sequence $R = (r_1,\dots,r_n)$ is given by $r_j = a_j - 2a_{j+1}$ for $1 \leq j \leq n$, where $a_{n+1} = 0$. Thus $a_j = r_j + 2r_{j+1} + \cdots + 2^{n-j}r_n$ for $1 \leq j \leq n$.

The A_2-module $\mathsf{Ad}_{\geq n}/\mathsf{Ad}_{\geq n+1}$ has $\mathbb{F}_2$-basis given by the (equivalence classes of) admissible monomials of length n, with the action of A_2 given by left multiplication. The module $\mathsf{MP}(n)$ isomorphic with the Steinberg summand of $\mathsf{P}(n)$ is obtained by regrading this module, but as this regrading does not affect the result, we work with $\mathsf{Ad}_{\geq n}/\mathsf{Ad}_{\geq n+1}$. We use the Milnor basis to prove that this module is the free module over $\mathsf{A}_2(n-2)$ with basis given by admissible monomials Sq^A such that 2^{n-j} divides a_j for $1 \leq j \leq n$.

Since the correspondence $A \leftrightarrow R$ preserves length, $\mathsf{Ad}_{\geq n}/\mathsf{Ad}_{\geq n+1}$ has $\mathbb{F}_2$-basis given by the (equivalence classes of) Milnor basis of length n, with the action of A_2 given by left multiplication. The equations above show that the condition on A that 2^{n-j} divides a_j for $1 \leq j \leq n$ is equivalent to the same condition on R, i.e. 2^{n-j} divides r_j for $1 \leq j \leq n$. Thus it is sufficient to prove that $\mathsf{Ad}_{\geq n}/\mathsf{Ad}_{\geq n+1}$ is the free module over $\mathsf{A}_2(n-2)$ with basis given by Milnor basis elements $Sq(T)$, where $T = (t_1,\dots,t_n)$ and 2^{n-j} divides t_j for $1 \leq j \leq n$.

Recall from Definition 12.4.1 that $\mathsf{A}_2(n-2)$ has $\mathbb{F}_2$-basis given by the Milnor basis elements $Sq(S)$, where $S = (s_1,\dots,s_{n-1})$ is a sequence such that $0 \leq s_j \leq 2^{n-j}-1$ for $1 \leq j \leq n-1$. Using the Milnor product formula, we consider the products $Sq(S)Sq(T)$ with S and T as above. Since $s_j \leq 2^{n-j}-1$ and 2^{n-j} divides t_j, $\mathrm{bin}(s_j)$ and $\mathrm{bin}(t_j)$ are disjoint for all j, and hence the initial Milnor matrix contributes a term $Sq(S+T)$ to the product.

The correspondence $A \leftrightarrow R$ also preserves the right order $<_r$ on sequences. Since the initial Milnor matrix produces the right maximal term in the product if it is nonzero, we have $Sq(S)Sq(T) = Sq(S+T) +$ right lower terms. Given a Milnor basis element $Sq(R)$ where $R = (r_1,\dots,r_n)$, there are unique sequences S and T such that $R = S+T$ and $0 \leq s_j \leq 2^{n-j}-1$, $2^{n-j}|t_j$ for all j, obtained by taking s_j as the remainder of r_j on division by 2^{n-j} and

$t_j = r_j - s_j$. Thus the elements $Sq(S)Sq(T)$ with S and T as above give a $\mathbb{F}_2$-basis for $\mathsf{Ad}_{\geq n}/\mathsf{Ad}_{\geq n+1}$. □

22.4 The hit problem for $\mathsf{MP}(n)$

The module $\mathsf{MP}(n)$ is obtained by shifting the grading on $\mathsf{Ad}_{\geq n}/\mathsf{Ad}_{\geq n+1}$ by $-n$, so that the element $[Sq^A]$ represented by the admissible monomial A of length n has grading $|A| - n$ in $\mathsf{MP}(n)$. The object of this section is to prove that the equivalence classes of admissible monomials of the form $Sq^{2^{r_1}} \cdots Sq^{2^{r_n}}$, where $r_1 > \cdots > r_n \geq 0$, form a minimal basis for $\mathsf{MP}(n)$ as an A_2-module.

Definition 22.4.1 An element of A_2 is ***n*-indecomposable** if it cannot be written as a sum of monomials of length $\geq n+1$.

For example, an element is 1-indecomposable if it cannot be written as a sum of products of elements of positive degree. By Proposition 3.2.4, the 1-indecomposables in A_2 are the elements Sq^{2^i} for $i \geq 0$.

Proposition 22.4.2 *The admissible monomials $Sq^{2^{r_1}} \cdots Sq^{2^{r_n}}$ in A_2 are n-indecomposable for $r_1 > \cdots > r_n \geq 0$. Hence the corresponding elements $[Sq^{2^{r_1}} \cdots Sq^{2^{r_n}}]$ of $\mathsf{MP}(n)$ are not hit.*

Proof We argue by induction on n, using the stripping technique of Section 13.2. The base case $n = 1$ is true by Proposition 3.2.4. Assume that the result is true for $n-1$ and that $Sq^R = Sq^{2^{r_1}} \cdots Sq^{2^{r_n}}$ is a monomial of length n which is not n-indecomposable. Then Sq^R can be written as a sum of monomials in A_2 of length $\geq n+1$. By stripping all terms by (2^r), where $r = \max(r_1, \ldots, r_n)$, we obtain an expression for a string of $n-1$ distinct factors of the form $Sq^{2^{r_i}}$, in which every nonzero term has length $\geq n$. This contradicts the inductive step. □

We wish to show that any element $[Sq^A] = [Sq^{a_1} \cdots Sq^{a_n}]$ is hit in $\mathsf{MP}(n)$, where Sq^A is an admissible monomial of length n, if at least one index a_i is not a 2-power. This is clear if a_1 is odd, because the Adem relation $Sq^{a_1} = Sq^1 Sq^{a_1-1}$ gives $Sq^{a_1} \cdots Sq^{a_n} = Sq^1 Sq^{a_1-1} Sq^{a_2} \cdots Sq^{a_n}$, and $Sq^{a_1-1} Sq^{a_2} \cdots Sq^{a_n}$ is still admissible of length n. We show in Proposition 22.4.4 that this is true if any index a_i is odd. For this we need a preliminary result.

Proposition 22.4.3 *Given $a_i > 0$ for $1 \leq i \leq k$,*

$$(Sq^{2^{k-1}} Sq^{2^{k-2}} \cdots Sq^2 Sq^1) Sq^{2a_1} \cdots Sq^{2a_k} = Sq^{2a_1 + 2^{k-1}} \cdots Sq^{2a_k+1} + E,$$

where E is a sum of monomials with an odd index before position k.

Proof The proof is by induction on k. The base case $k = 1$ follows from the Adem relation $Sq^1 Sq^{2a} = Sq^{2a+1}$. Assuming the case $k-1$, we can write

$$(Sq^{2^{k-2}} \cdots Sq^2 Sq^1) Sq^{2a_1} \cdots Sq^{2a_{k-1}} = Sq^{2a_1+2^{k-2}} \cdots Sq^{2a_{k-1}+1} + F,$$

where every term in F has an odd index before position $k-1$. We next pre-multiply $Sq^{2a_1+2^{k-2}} \cdots Sq^{2a_{k-1}+1} Sq^{2a_k}$ by $Sq^{2^{k-1}}$, and apply the Adem relations successively to adjacent pairs of factors by the cascade process of Proposition 22.3.4. At step t of the cascade process, the monomial

$$Sq^{2a_1+2^{k-1}} \cdots Sq^{2a_t+2^{k-t}} Sq^{2^{k-t}} Sq^{2a_{t+1}+2^{k-t-1}} \cdots Sq^{2a_{k-1}+1} Sq^{2a_k}$$

appears, arising from the last term in each application of an Adem relation (see Proposition 22.3.1). At the last step $t = k$, the monomial $Sq^{2a_1+2^{k-1}} \cdots Sq^{2a_k+1}$ appears. This monomial occurs once only, since all other terms in the cascade process stop before reaching the step $t = k$, and therefore have an odd index before position k. Finally, by Proposition 22.3.1(v) an odd index in a position $< k-1$ in any monomial of F will occur in a position $< k$ after premultiplication of F by $Sq^{2^{k-1}}$. Hence every monomial in E has an odd square in a position $< k$. □

Proposition 22.4.4 *Let $Sq^A = Sq^{a_1} \cdots Sq^{a_n}$ be an admissible monomial of length n. If any a_i is odd, then $[Sq^A]$ is hit in* MP(n).

Proof The proof is by induction on k, where a_k is the first odd index in Sq^A. As observed above, the result is true when a_1 is odd. Assume that the result is true for $k-1$, and let $Sq^A = Sq^{A'} Sq^{A''}$, where $Sq^{A'} = Sq^{a_1} \cdots Sq^{a_k}$ and $Sq^{A''} = Sq^{a_{k+1}} \cdots Sq^{a_n}$. Let $Sq^B = Sq^{b_1} \cdots Sq^{b_k}$, where $b_i = a_i - 2^{k-i}$ for $1 \le i \le k$. We check that $Sq^B Sq^{A''}$ is an admissible monomial of length n. Multiplying on the right by $Sq^{A''}$, Proposition 22.4.3 gives

$$(Sq^{2^{k-1}} Sq^{2^{k-2}} \cdots Sq^2 Sq^1) Sq^B Sq^{A''} = Sq^A + E Sq^{A''},$$

where E is a sum of monomials in A_2 with an odd index $< k$. Hence Sq^A is hit modulo monomials with an odd index $< k$. Applying the induction hypothesis, $[Sq^A]$ is hit. □

The results of Section 13.3 give the following duplication property of relations in A_2. Notice that formal duplication preserves length and admissibility.

Proposition 22.4.5 *Any relation in A_2 remains a relation if the indices of all monomials are formally doubled, modulo error terms consisting of monomials Sq^A with at least one odd index.* □

We can now show that the n-indecomposables generate the module MP(n).

Theorem 22.4.6 *The admissible monomials of the form* $[Sq^{2^{r_1}} \cdots Sq^{2^{r_n}}]$, *where* $r_1 > \cdots > r_n \geq 0$, *form a minimal basis for* MP(n) *as an* A$_2$-*module. The Steinberg summand* P(St(n)) *of* P(n) *has a minimal generating set with one generator in each degree* d *such that* $\alpha(d+n) = n$.

Proof Consider an admissible monomial $Sq^{2A} = Sq^{2a_1} \cdots Sq^{2a_n}$ of length n such that every index is even, and some index is not a 2-power. By induction on the grading, we may assume that the element $[Sq^A] = [Sq^{a_1} \cdots Sq^{a_n}]$ of MP(n) is hit. Formally doubling every index in the corresponding hit equation for Sq^A produces a hit equation for Sq^{2A}, modulo terms with an odd index. The first statement then follows from Proposition 22.4.4.

The second statement follows from this and Theorem 22.3.5. There is a unique admissible monomial $Sq^{2^{r_1}} \cdots Sq^{2^{r_n}}$ in A^d when $\alpha(d) = n$. With the shift of grading, this corresponds to $\alpha(d+n) = n$ in P(St(n)). □

22.5 The symmetric Steinberg summand

In this section we give a minimal generating set for the symmetric Steinberg summand P$(n)e(n)$ as an A$_2$-module.

Definition 22.5.1 Let $r_1 > r_2 > \cdots > r_n \geq 0$. The **symmetrized spike** $\Sigma(r_1, \ldots, r_n)$ is the monomial symmetric function

$$\Sigma(r_1, \ldots, r_n) = \sum_{\pi \in \Sigma(n)} x_{\pi(1)}^{2^{r_1}-1} \cdots x_{\pi(n)}^{2^{r_n}-1}.$$

Note that the exponents are distinct, so that the monomial symmetric function is the same as the sum over all permutations of $1, 2, \ldots, n$.

Theorem 22.5.2 *The symmetrized spikes* $\Sigma(r_1, \ldots, r_n)$, $r_1 > r_2 > \cdots > r_n \geq 0$ *form a minimal generating set for* P$(n)e(n)$.

Proof Working in P$(n)[x_1^{-1}, \ldots, x_n^{-1}]$, $\Sigma(r_1, \ldots, r_n) = \mathrm{c}(n)^{-1}\Delta(r_1, \ldots, r_n)$, where

$$\Delta(r_1, \ldots, r_n) = \sum_{\pi \in \Sigma(n)} x_{\pi(1)}^{2^{r_1}} \cdots x_{\pi(n)}^{2^{r_n}}.$$

The numerator $\Delta(r_1, \ldots, r_n)$ is an element of the Dickson algebra D(n), as it is the element $c_n(d)$ of Proposition 15.1.7 where $d = 2^{r_1} + \cdots + 2^{r_n}$ with $\alpha(d) = n$. Hence $(\Delta(r_1, \ldots, r_n)\mathrm{c}(n)^{-1}) \cdot e(n) = \Delta(r_1, \ldots, r_n)((\mathrm{c}(n)^{-1}) \cdot e(n)) = \Delta(r_1, \ldots, r_n)\mathrm{c}(n)^{-1}$ by Proposition 22.2.2. It follows that the polynomials

$\Sigma(r_1,\ldots,r_n)$ are fixed by $e(n)$, and so they are in the symmetric Steinberg summand $\mathsf{P}(n)e(n)$. Since they involve linearly independent spikes, these elements must form part of any minimal generating set for $\mathsf{P}(n)e(n)$ as an A_2-module. There is a bijection $\Sigma(r_1,\ldots,r_n) \leftrightarrow Sq^{2^{r_1}} \cdots Sq^{2^{r_n}}$, where $r_1 > r_2 > \cdots > r_n \geq 0$, between the symmetrized spikes and the minimal generating set of Theorem 22.4.6 for $\mathsf{MP}(n)$. Hence the symmetrized spikes form a complete minimal generating set for $\mathsf{P}(n)e(n)$. □

Since the top Dickson invariant $\Delta_n = \Delta(n-1,n-2,\ldots,1,0)$, it follows that the first occurrence of $\mathsf{St}(n)$ in $\mathsf{P}(n)$ is generated by the symmetrized spike $\Sigma(n-1,n-2,\ldots,1,0) = \mathsf{c}(n)^{-1}\Delta_n$. Since Δ_n is the product of all nonzero elements $x \in \mathsf{P}^1(n)$, $\Sigma(n-1,n-2,\ldots,1,0)$ is the product of all nonzero linear polynomials $x \neq x_i$, $1 \leq i \leq n$. For example, in the case $n=3$, $\Sigma(2,1,0) = \sum_{i\neq j} x_i^3 x_j = (x_1+x_2)(x_1+x_3)(x_2+x_3)(x_1+x_2+x_3)$.

It is also possible to obtain a set of A_2-generators of $\mathsf{P}(n)e(n)$ by applying the admissible monomials $Sq^{2^{r_1}} \cdots Sq^{2^{r_n}}$ to $\mathsf{c}(n)^{-1}$. In general, these polynomials do not coincide with the symmetrized spikes: for example, in $\mathsf{P}^8(2)$ we have $Sq^8 Sq^2(x_1^{-1}x_2^{-1}) = Sq^8(x_1x_2^{-1}+1+x_1^{-1}x_2) = x_1^7x_2 + x_1x_2^7 + x_1^2x_2^6 + x_1^6x_2^2$.

22.6 The $\mathsf{B}(n)$-invariant Steinberg summand

In this section we give a minimal set of A_2-module generators for the Steinberg summand $\mathsf{P}(n)e'(n)$ which is invariant under the lower triangular subgroup $\mathsf{B}(n)$. Recall from Section 19.4 that the polynomial $\Delta_I = \Delta_{i_1} \cdots \Delta_{i_k} \in \mathsf{P}(n)$, where Δ_i is the top Dickson invariant for $\mathsf{P}(i)$, $1 \leq i \leq n$, generates the minimal degree occurrence of the irreducible $\mathbb{F}_2\mathsf{GL}(n)$-module corresponding to $I \subseteq Z[n-1]$ as a submodule in $\mathsf{P}(n)$. The polynomial Δ_I is also the unique $\mathsf{B}(n)$-invariant element of this submodule. In the case of the module $\mathsf{St}(n)$, $I = Z[n-1]$, and we denote Δ_I by $\Delta_{Z[n-1]}$. For example $\Delta_{Z[2]} = x \begin{vmatrix} x & y \\ x^2 & y^2 \end{vmatrix} = x^3y + x^2y^2$.

Proposition 22.6.1 $\Delta_{Z[n-1]} \cdot e'(n) = \Delta_{Z[n-1]}$.

Proof Since $e'(n)$ is a product of the idempotents e'_i, $1 \leq i \leq n-1$, it suffices to show that $\Delta_{Z[n-1]} \cdot e'_i = \Delta_{Z[n-1]}$ for $1 \leq i \leq n$. Since e'_i affects only the variables x_i and x_{i+1}, Δ_j is invariant under all four matrices in the sum e'_i when $j < i$ (when Δ_j involves neither x_i nor x_{i+1}) and when $j > i$ (when Δ_j involves both x_i and x_{i+1}). Thus it remains to show that $\Delta_i \cdot e'_i = \Delta_i$. Since $e'_i = (I_n + S_i)$

$(I_n + L_i)$ we obtain $\Delta_i \cdot e_i' = \Delta_i(\ldots,x_i,x_{i+1}\ldots) + \Delta_i(\ldots,x_{i+1},x_i,\ldots) + \Delta_i(\ldots,x_i,x_i+x_{i+1},\ldots)+\Delta_i(\ldots,x_i+x_{i+1},x_i,\ldots)$. Now Δ_i involves only the first i variables, so the first and third terms in the sum are equal, and $\Delta_i(\ldots,x_i+x_{i+1},x_i,\ldots) = \Delta_i(\ldots,x_i,x_i\ldots) + \Delta_i(\ldots,x_{i+1},x_i,\ldots)$ since $(x_i+x_{i+1})^{2^k} = x_i^{2^k} + x_{i+1}^{2^k}$ for $k \geq 0$. Thus the sum reduces to Δ_i, and the result follows. □

By the results of Section 19.1, all the A_2-module summands of $\mathsf{P}(n)$ corresponding to the same irreducible $\mathbb{F}_2\mathsf{GL}(n)$-module $L(\lambda)$ are isomorphic. We can use the formulae $e(n) = \overline{\mathsf{B}}(n)\overline{\mathsf{W}}(n)$ and $e'(n) = \overline{\mathsf{W}}(n)\overline{\mathsf{B}}(n)$ to make this isomorphism explicit in the case of the summands $\mathsf{P}(n)e(n)$ and $\mathsf{P}(n)e'(n)$.

Proposition 22.6.2 *Let $f \in \mathsf{P}(n)$. Then the map $f \mapsto f \cdot \overline{\mathsf{B}}(n)$ is an A_2-module isomorphism $\mathsf{P}(n)e(n) \to \mathsf{P}(n)e'(n)$, with inverse $f \mapsto f \cdot \overline{\mathsf{B}}(n)$. In particular, the symmetrized spike $\Sigma(n-1,n-2,\ldots,1,0) \in \mathsf{P}(n)e(n)$ corresponds to the element $\Delta_{Z[n-1]} \in \mathsf{P}(n)e'(n)$ under these bijections, and is the unique nonzero symmetric polynomial in the submodule which is the first occurrence of $\mathsf{St}(n)$ in $\mathsf{P}(n)$.*

Proof The first statement follows at once from the fact that $e(n)$ and $e'(n)$ are idempotents. Let $S \subseteq \mathsf{P}^{w(n-1)}(n)$ be the first occurrence of $\mathsf{St}(n)$ as a submodule of $\mathsf{P}(n)$. The element $\Delta_{Z[n-1]}$ is the unique nonzero element of S which is invariant under $\mathsf{B}(n)$. Since elements of S invariant under $\mathsf{B}(n)$ correspond to elements of S invariant under $\mathsf{W}(n)$ under these isomorphisms, and since $\Sigma(n-1,n-2,\ldots,1,0) \in S$ is a symmetric polynomial, the remaining statements follow. □

The next result gives an explicit set of generators for $\mathsf{P}(n)e'(n)$. Since any summand $\mathsf{P}(n,\lambda)$ of $\mathsf{P}(n)$ is a module over the Dickson algebra $\mathsf{D}(n)$, we look for generators of the form $\Delta_{Z[n-1]}f$ where $f \in \mathsf{D}(n)$. For $0 \leq i \leq n-1$ we denote the generator $\mathsf{d}_{n,i}$ of $\mathsf{D}(n)$ in degree $2^n - 2^i$ by d_i.

Proposition 22.6.3 *Let $\mu = (\mu_1,\ldots,\mu_r)$ be a decreasing sequence of size r with $\mu_1 \leq n-1$, and let d_μ be the monomial*

$$\mathsf{d}_\mu = \prod_{j=1}^{r} \mathsf{d}_{\mu_j}^{2^{r-j}}$$

in the Dickson invariants $\mathsf{d}_i = \mathsf{d}_{n,i} \in \mathsf{D}(n)$. Then the polynomials $\Delta_{Z[n-1]}\mathsf{d}_\mu$ are a minimal set of A_2-module generators for $\mathsf{P}(n)e'(n)$.

Proof By the remarks above, the polynomials $\Delta_{Z[n-1]}\mathsf{d}_\mu$ are in $\mathsf{P}(n)e'(n)$. Using the left lexicographic order on monomials in $\mathsf{P}(n)$, we denote the leading term of $f \in \mathsf{P}(n)$ by $\mathrm{LT}(f)$. We will show that the leading term of

$\Delta_{Z[n-1]}\mathsf{d}_\mu$ is a spike with n distinct exponents. In this way we obtain a one-one correspondence between the polynomials $\Delta_{Z[n-1]}\mathsf{d}_\mu$ and the generators of the symmetric Steinberg summand $\mathsf{P}(n)e(n)$.

The leading term of $\mathsf{d}_{n,i}$ is $x_1^{2^{n-1}}x_2^{2^{n-2}}\cdots x_{n-i}^{2^i}$. This follows either from the standard formula for $\mathsf{d}_{n,i}$ as a quotient of determinants, or from the Stong–Tamagawa formula

$$\mathsf{d}_{n,i} = \sum_{W\subseteq \mathsf{V}(n), \dim W=i}\ \prod_{v\notin W} v, \tag{22.5}$$

using the term corresponding to the subspace spanned by the last i variables. We calculate the exponent of x_i in $\mathrm{LT}(\mathsf{d}_\mu) = \prod_{j=1}^r (\mathrm{LT}(\mathsf{d}_{\mu_j}))^{2^{r-j}}$. The variable x_i divides the jth factor $\mathrm{LT}(\mathsf{d}_{\mu_j})$ of d_μ when $n-\mu_j \geq i$, or $\mu_j \leq n-i$, and its exponent in this factor is $2^{n-i}\cdot 2^{r-j}$. Hence the exponent of x_i in d_μ is $2^{n-i}(2^{r-j}+2^{r-j-1}+\cdots+1) = 2^{n-i}(2^{r-j+1}-1)$, where j is the least integer such that $\mu_j \leq n-i$. In terms of the conjugate partition $\lambda = \mu^{\mathrm{tr}}, j = 1+\mu^{\mathrm{tr}}_{n-i+1}$. It follows that we can write the exponent of x_i in d_μ as $2^{n-i}(2^{\lambda_i}-1)$, where λ is defined by

$$\lambda_i = r - \mu^{\mathrm{tr}}_{n-i+1}. \tag{22.6}$$

Since the exponent of x_i in $\mathrm{LT}(\Delta_{Z[n-1]})$ is $2^{n-i}-1$, we finally obtain

$$\mathrm{LT}(\Delta_{Z[n-1]}\mathsf{d}_\mu)) = \prod_{i=1}^{r} x_i^{2^{n-i+\lambda_i}-1},$$

a spike with n distinct exponents. Conversely, every spike with distinct exponents can be expressed in this form for some partition λ, and so is the leading term of $\Delta_{Z[n-1]}\mathsf{d}_\mu$, where μ is given in terms of λ by (22.6). □

The same method provides a set of generators for an arbitrary Steinberg summand of $\mathsf{P}(n)$. These summands each contain a single nonzero polynomial f in the $\mathbb{F}_2\mathsf{GL}(n)$-submodule generated by $\Delta_{Z[n-1]}$. For example, in the case $n=2$ there are three Steinberg summands, with $f = x, y, x+y$. In each case, the elements $f\mathsf{d}_\mu$ form a basis for the cohit module of the summand, where d_μ is the Dickson monomial indexed by the partition μ, as in Proposition 22.6.3. Note that, in the case of the symmetric summand $\mathsf{P}(n)e(n)$, this basis is not the same as that given by Theorem 22.5.2. For example, with $n=2$ and $\mu=(1,1)$, $\mathsf{d}_\mu = \mathsf{d}_{2,1}^3 = (x^2+xy+y^2)^3$ and $(x+y)\mathsf{d}_\mu \neq x^7+y^7$.

In the case $n=2$, we can relate the solution of the 2-variable hit problem given in Chapter 1 to the A_2-module decomposition $\mathsf{P}(2) = \mathsf{P}(2,(0)) \oplus 2\mathsf{P}(2,(1))$ obtained from the action of $\mathsf{GL}(2)$. Recall that $\mathsf{P}(2,(0)) = \mathsf{P}(2)e_0$, where

$$e_0 = \begin{pmatrix} 1 & 0 \\ 0 & 1 \end{pmatrix} + \begin{pmatrix} 0 & 1 \\ 1 & 1 \end{pmatrix} + \begin{pmatrix} 1 & 1 \\ 1 & 0 \end{pmatrix}$$

is a central idempotent in $\mathbb{F}_2\mathsf{GL}(2)$. Recall from Section 15.2 that the Dickson algebra $\mathsf{D}(2) = \mathbb{F}_2[\mathsf{d}_1, \mathsf{d}_0]$ is an A_2-submodule of $\mathsf{P}(2,(0))$, and $\mathsf{P}(2,(0))$ is generated as a module over $\mathsf{D}(2)$ by 1 and $g = x^3 + x^2y + y^3$. Further $g^2 = \mathsf{d}_1^3 + \mathsf{d}_0^2 + g\mathsf{d}_0$, where $\mathsf{d}_1 = x^2 + xy + y^2$ and $\mathsf{d}_0 = x^2y + xy^2$.

Using the table of Section 1.6 as a guide, we can find a minimal set of A_2-module generators of $\mathsf{P}(2,(0))$ in the form of monomials in g, d_1 and d_0 which contain spikes when expanded as polynomials in x and y. In particular, since $\mathsf{d}_1^{2^k} = x^{2^k} + x^{2^{k-1}}y^{2^{k-1}} + y^{2^k}$, it follows by induction on k that $\mathsf{d}_1^{2^k-1}$ contains the spike $x^{2^k-1}y^{2^k-1}$. Thus $\mathsf{D}(2)$ provides the cohit module in degrees $2^{k+1} - 2$, $k \geq 0$, corresponding to the first column of the diagram following Theorem 1.8.2.

From this diagram we also require one generator of $\mathsf{P}(2,(0))$ in each degree $d = 2^{i+1} + 2^j - 2$ where $i > j \geq 0$, corresponding to the third and subsequent columns. As in Proposition 15.4.3, rewriting d as $d = (2^{s+t} - 1) + (2^t - 1)$ where $s \geq 2$ and $t \geq 0$, the polynomial $g\mathsf{d}_1^{2^{s+t-1}-2^t-1}\mathsf{d}_0^{2^t-1}$ has leading term $x^{2^{s+t}-1}y^{2^t-1}$, and so provides a suitable choice of generator.

To complete the solution of the hit problem related to the decomposition $\mathsf{P}(2) = \mathsf{P}(2)e_0 \oplus \mathsf{P}(2)e(2) \oplus \mathsf{P}(2)e'(2)$, we include the polynomials $f\mathsf{d}_1^{2^i-2^j}\mathsf{d}_0^{2^j-1}$, where $i \geq j \geq 0$ and $f = x$ or $x+y$. In the case $f = x+y$ we can use the symmetrized spikes instead. Of course, there is a third summand $\mathsf{P}(2,(1))$, with generators as above with $f = y$. The choice $f = x$ or y for two linearly independent $\mathsf{P}(2,(1))$ summands corresponds to the choice of the orthogonal idempotents $g_1 = \overline{\mathsf{U}}(2)\overline{\mathsf{B}}(2)$ and $g_1' = \overline{\mathsf{B}}(2)\overline{\mathsf{U}}(2)$ of Example 18.1.3, which contain both upper and lower triangular matrices.

22.7 Remarks

The key results linking the Steinberg module to topology were obtained in the fundamental paper of S. A. Mitchell and S. B. Priddy [141], who highlighted the role played by the Steinberg idempotent in splitting off certain summands of the suspended classifying space $\Sigma(BV)$, where V is an n-dimensional vector space over $\mathbb{F}_2$, regarded as an elementary abelian 2-group. Theorem 22.3.5 is one of their main results. For all n and d, they determined the multiplicity of $\mathsf{St}(n)$ as a composition factor in $\mathsf{P}^d(n) = H^d(BV;\mathbb{F}_2)$, using topological methods. An algebraic proof of this was given by Mitchell and N. J. Kuhn in [118]. Mitchell and Priddy defined the cyclic A_2-module $\mathcal{D}[n]$ with basis $Sq^A(u_n)$ where $\text{len}(A) \leq n$. This arises topologically as the cohomology of the cofibre of the diagonal map $Sp^{2^{n-1}} \to Sp^{2^n}$, where Sp^k is the kth symmetric power of the sphere spectrum S^0. Then $\mathcal{D}[1] = \Sigma \mathsf{P}(1)$ and $\mathcal{D}[1] \otimes \cdots \otimes \mathcal{D}[1] = \Sigma^n \mathsf{P}(n)$.

The contribution of the Steinberg summand $\mathsf{P}(\mathsf{St}(n))$ to the cohit module $\mathsf{Q}^d(n)$ for all n and d was determined by M. Inoue in [96], who found the minimal generating set of Theorem 22.4.6 for the A_2-module $\mathsf{MP}(n)$. (Cone monomials are called 'semi-admissible monomials' in [96].) Another proof has been given by Hai [72]. For the Stong–Tamagawa formula (22.5) see [190, p.240].

It is shown in [219] that $\mathsf{Q}^d(n) \cong \mathsf{St}(n)$ for the first occurrence degree $d = w(n-1) = 2^n - n - 1$, so that the lower bound given by Theorem 22.4.6 is exact.

Proposition 22.4.3 may be described as a 'superposition lemma' because a product of monomials of Steenrod squares is obtained by adding corresponding indices, modulo certain error terms. For further superposition results, see [233, Section 5].

23

The d-spike module $\mathsf{J}(n)$

23.0 Introduction

In this chapter we extend the work of Chapter 10 by calculating the d-spike module $\mathsf{J}^d(n)$ for all n and degrees d containing spikes with strictly decreasing exponents. Recall that $\mathsf{J}^d(n)$ is the submodule of $\mathsf{K}^d(n)$ generated by d-spikes of degree d. Up to permutation of the variables, a d-spike $f = v_1^{(2^{\lambda_1}-1)} \cdots v_n^{(2^{\lambda_n}-1)}$ corresponds to a partition $\lambda = (\lambda_1, \ldots, \lambda_n)$ of length $\leq n$. By Proposition 5.1.13, $d = \sum_{i=1}^{n}(2^{\lambda_i} - 1)$ where $\lambda_1 > \lambda_2 > \cdots > \lambda_n \geq 0$ if and only if $\alpha(d+n) = n$, and in this case $(2^{\lambda_1} - 1, \ldots, 2^{\lambda_n} - 1)$ is the unique element of $\mathsf{Spike}^d(n)$, $\omega = \lambda^{\mathrm{tr}}$ is the unique element of $\mathsf{Dec}_d(n)$ and the set of terms $I(\omega)$ of ω is $Z[n-1]$ or $Z[n]$. Hence $\mathsf{Q}^d(n) = \mathsf{Q}^\omega(n)$ and $\mathsf{K}^d(n) = \mathsf{K}^\omega(n)$. We call such sequences ω, or degrees d, **1-dominant**, and if $\lambda_i - \lambda_{i+1} \geq 2$ for $1 \leq i < n$, so that the set of repeated terms $I_2(\omega)$ of ω is $Z[n-1]$ or $Z[n]$, we call ω, or d, **2-dominant**.

In the 2-dominant case we show (Theorem 23.4.1) that $\mathsf{J}^d(n) \cong \mathsf{FL}(n)$, and in the 1-dominant case (Theorem 23.4.4) that $\mathsf{J}^d(n)$ is isomorphic to a direct summand of $\mathsf{FL}(n)$ determined by $I_2(\omega)$. It follows that

$$\dim \mathsf{J}^d(n) = \binom{n}{c_1, \ldots, c_{r+1}}_2 \cdot \prod_{k=1}^{r+1} 2^{c_k(c_k-1)/2}$$

where $(c_1, \ldots, c_{r+1})$, $r = |I_2(\omega)|$, is the composition of n associated to $I_2(\omega)$. Since $\mathsf{J}^d(n) \subseteq \mathsf{K}^d(n)$ and $\dim \mathsf{K}^d(n) = \dim \mathsf{Q}^d(n)$, this is a lower bound for $\dim \mathsf{Q}^d(n)$ in 1-dominant degrees.

The minimal 1-dominant degree is $w(n-1) = 2^n - 1 - n = \sum_{k=1}^{n-1}(2^k - 1)$, when the lower bound on $\dim \mathsf{Q}^d(n)$ is $2^{n(n-1)/2}$. It follows from Theorem 22.4.6 that in this case $\mathsf{Q}^d(n)$ has a summand isomorphic to the Steinberg module $\mathsf{St}(n)$. In the 2-dominant case, the lower bound on $\dim \mathsf{Q}^d(n)$

is $\dim \mathsf{FL}(n) = \prod_{i=1}^{n}(2^i - 1)$. The minimal 2-dominant degree is given by $\sum_{k=1}^{n-1}(4^k - 1) = \frac{1}{3}(4^n - 1) - n$.

In particular, using Theorem 8.1.13 it follows that $\mathsf{K}^d(n) = \mathsf{J}^d(n)$ and that $\mathsf{Q}^d(n)$ and $\mathsf{K}^d(n)$ are isomorphic to $\mathsf{FL}(n)$ as $\mathbb{F}_2\mathsf{GL}(n)$-modules if $\lambda_i - \lambda_{i+1} \geq i + 1$, $1 \leq i < n$, and using Theorem 8.2.10 that the same is true when $\lambda_i - \lambda_{i+1} \geq n - i + 1$, $1 \leq i < n$. In these cases Theorem 23.4.1 provides an alternative method to Propositions 8.1.10 and 8.2.6 for obtaining the lower bound on $\dim \mathsf{Q}^d(n)$.

In Section 23.1 we define the Crabb–Hubbuck map $\phi^\omega : \mathsf{FL}(n) \longrightarrow \mathsf{DP}^d(n)$, which associates to a complete flag X in $\mathsf{V}(n)$ a **flag d-polynomial** $\phi^\omega(X)$ of degree $d = \deg_2 \omega$ in $\mathsf{J}^d(n)$. When ω is 1-dominant, we can identify $\mathsf{J}^d(n)$ with the image of ϕ^ω, and so we are concerned with linear relations between flag polynomials. We give examples of such relations in Proposition 23.1.7.

Given a decreasing sequence $\omega \in \mathsf{Dec}(n)$ and a permutation $\rho \in \Sigma(n)$, we introduce the **zip d-monomial** $z_\omega(\rho)$ in Section 23.2. These d-monomials are useful in proving linear independence of flag d-polynomials. The **zip monomial** $z^\omega(\rho)$ in $\mathsf{P}^d(n)$ dual to $z_\omega(\rho)$ is also of interest. In many cases the $\mathbb{F}_2\mathsf{GL}(n)$-module $\mathsf{Q}^d(n)$ is cyclic, but is not generated by spikes. For example x^3 and y^3 lie in a 2-dimensional submodule of $\mathsf{Q}^3(2)$, while the monomials x^2y and xy^2 are equivalent modulo hits and generate $\mathsf{Q}^3(2)$. In contrast, $\mathsf{K}^3(2)$ is generated by any d-spike, while the sum $u^{(2)}v + uv^{(2)}$ of the d-monomials dual to x^2y and xy^2 generates a 1-dimensional submodule. Propositions 10.7.7 and 10.7.8 give similar examples for $n = 3$. We return to $\mathsf{Q}^d(n)$ in Section 23.5.

The zip d-monomial $z_\omega(\rho)$ depends on a permutation $\rho \in \Sigma(n)$. We show in Proposition 23.2.5 that, when ω is 1-dominant, $z_\omega(\rho)$ is a term in the sum of the d-polynomials $\phi^\omega(X)$ for flags X in the Schubert cell $\mathsf{Sch}(\rho)$. Another important property of the zip d-monomials is proved in Proposition 23.3.6: if ω is 2-dominant and if $z_\omega(\rho)$ is a term in the flag d-polynomial $\phi^\omega(X)$, then $X \in \mathsf{Sch}(\sigma)$ where $\sigma \geq \rho$ in the Bruhat order. Taken together, these properties of zip d-monomials imply linear independence of suitable sets of flag d-polynomials by downward induction on any linear order on $\Sigma(n)$ which extends the Bruhat order.

In Section 23.3, we interpret the Bruhat order in terms of tableaux with integer entries whose shape depends on a permutation ρ and a 1-dominant sequence ω. This provides a combinatorial tool for working with zip d-monomials, which we use to prove Proposition 23.3.6.

In Section 23.4 we complete the linear independence arguments for flag d-polynomials in the 2-dominant and 1-dominant cases. In Theorem 23.4.4

we use the relations of Proposition 23.1.7 to show that the lower bound on $\dim \mathsf{J}^d(n)$ obtained in this way is exact.

In Section 23.5 we obtain parallel results for $\mathsf{P}(n)$, resulting in a lower bound for $\dim \mathsf{Q}^d(n)$ when d is 1-dominant. We associate the monomial $z^\omega(\rho)$ to the flag $\mathsf{B}(n)W$, where $W \in \mathsf{W}(n)$ is the permutation matrix associated to ρ. In general, we associate the **flag polynomial** $z^\omega(\rho)\cdot(B')^{\mathrm{tr}}$ to the flag $\mathsf{B}(n)WB' \in \mathsf{Sch}(\rho)$, where $B' \in \mathsf{B}(W)$, and show that these flag polynomials are linearly independent modulo hit polynomials when ω is 2-dominant, with appropriate restrictions in the 1-dominant case. Later (see Section 30.2) we identify the dual of the d-spike module $\mathsf{J}^d(n)$ as the quotient of $\mathsf{Q}^d(n)$ by the submodule $\mathsf{SF}^d(n)$ of 'strongly spike-free' polynomials.

23.1 Flag d-polynomials

Let $\omega \in \mathsf{Dec}_d(n)$ be a decreasing sequence of 2-degree d, regarded as a partition, and let $\omega^{\mathrm{tr}} = \lambda = (\lambda_1,\ldots,\lambda_n)$. The Ferrers block F associated with ω represents the spike $f = x_1^{2^{\lambda_1}-1}\cdots x_n^{2^{\lambda_n}-1}$, and $\omega(F) = \omega(f) = \omega$, $\alpha(F) = \alpha(f) = \lambda$. Given a permutation $\rho \in \Sigma(n)$, there is a corresponding spike $s^\omega(\rho) = x_1^{r_1}\cdots x_n^{r_n} \in \mathsf{P}^\omega(n)$ and a d-spike $s_\omega(\rho) = v_1^{(r_1)}\cdots v_n^{(r_n)} \in \mathsf{DP}^\omega(n)$, where $r_{\rho(i)} = 2^{\lambda_i}-1$, both represented by the block $S^\omega(\rho)$ obtained by applying the permutation ρ to the rows of F. Thus $F = S^\omega(\iota_n)$ where ι_n is the identity permutation. The spikes $s^\omega(\rho)$, $\rho \in \Sigma(n)$, are all distinct if and only if $\lambda_1 > \lambda_2 > \cdots > \lambda_n \geq 0$.

Example 23.1.1 Let $n = 4$ and $\omega = (3,2,2,1)$. Then $\lambda = (4,3,1)$ and for the identity permutation $\iota_4 = (1,2,3,4)$, $s^\omega(\iota_4) = x_1^{15}x_2^7x_3$ and $s_\omega(\iota_4) = v_1^{(15)}v_2^{(7)}v_3$. For the permutation $\rho = (2,4,3,1)$, $s^\omega(\rho) = x_2^{15}x_3x_4^7$ and $s_\omega(\rho_0) = v_2^{(15)}v_3v_4^{(7)}$. The corresponding blocks are

$$S^\omega(\iota_4) = \begin{matrix} 1111 \\ 1110 \\ 1000 \\ 0000 \end{matrix}, \quad S^\omega(\rho) = \begin{matrix} 0000 \\ 1111 \\ 1000 \\ 1110 \end{matrix}.$$

Recall that a complete flag X in $\mathsf{V}(n)$ corresponds to a right coset $\mathsf{B}(n)A$ in $\mathsf{GL}(n)$, where the ith subspace X^i of X is spanned by the first i rows $u_1,\ldots,u_i$ of the representative matrix A.

Definition 23.1.2 Let $\omega \in \mathsf{Dec}_d(n)$ and let $\omega^{\mathrm{tr}} = \lambda = (\lambda_1,\ldots,\lambda_n)$. The **Crabb–Hubbuck map** $\phi^\omega : \mathsf{FL}(n) \longrightarrow \mathsf{DP}^d(n)$ is the $\mathbb{F}_2\mathsf{GL}(n)$-module map

$$\phi^\omega(X) = s_\omega(\iota_n)\cdot A = \prod_{i=1}^{n} u_i^{(2^{\lambda_i}-1)},$$

where the right coset $\mathsf{B}(n)A$ of $\mathsf{GL}(n)$ corresponds to the flag X, and $u_1,\ldots,u_n$ are the rows of A, each regarded as an element of $\mathsf{V}(n) = \mathsf{DP}^1(n)$ by taking the entry 0 or 1 in the jth column as the coefficient of v_j, for $1 \le j \le n$. The d-polynomial $\phi^\omega(X)$ is the **flag d-polynomial** of X for ω.

Proposition 9.4.7 implies that $\phi^\omega(X)$ does not depend on the choice of A. Since λ is decreasing, the exponents of $s_\omega(\iota_n)$ are in decreasing order, so that $s_\omega(\iota_n)\cdot B = s_\omega(\iota_n)$ when $B \in \mathsf{B}(n)$, and hence $s_\omega(\iota_n)\cdot BA = s_\omega(\iota_n)\cdot A$. To see that $\phi^\omega(X)$ is a $\mathbb{F}_2\mathsf{GL}(n)$-module map, let $A' \in \mathsf{GL}(n)$, so that $\mathsf{B}(n)AA'$ is the coset corresponding to the flag $X\cdot A'$. Then $\phi^\omega(X)\cdot A' = (s_\omega(\iota_n)\cdot A)\cdot A' = s_\omega(\iota_n)\cdot AA' = \phi^\omega(X\cdot A')$.

Example 23.1.3 For $n=4$, $\omega=(3,2,2,1)$ and $\rho=(2,4,3,1)$, let $X \in \mathsf{FL}(4)$ be the flag in $\mathsf{Sch}(\rho)$ corresponding to the coset $\mathsf{B}(4)A$ of $\mathsf{GL}(4)$ where

$$A = \begin{pmatrix} a & 1 & 0 & 0 \\ b & 0 & c & 1 \\ d & 0 & 1 & 0 \\ 1 & 0 & 0 & 0 \end{pmatrix}. \tag{23.1}$$

Then $\lambda = (4,3,1)$ and $\phi^\omega(X) = (av_1+v_2)^{(15)}(bv_1+cv_3+v_4)^{(7)}(dv_1+v_3)$.

In general, $\phi^\omega(X)$ has terms with ω-sequence $>_l \omega$. For example, if $a=b=c=1$ then $\phi^\omega(X)$ has a term $v_1^{(15)}v_2^{(2)}v_3^{(3)}v_4^{(3)}$ with ω-sequence $(3,4,1,1)$.

Proposition 23.1.4 $\mathrm{Im}(\phi^\omega) \subseteq \mathsf{K}^{\ge_l \omega}(n)$, *and* $\mathrm{Im}(\phi^\omega)$ *is a submodule of the d-spike module* $\mathsf{J}^d(n)$. *If* ω *is the unique element of* $\mathsf{Dec}_d(n)$, *then* $\mathrm{Im}(\phi^\omega) = \mathsf{J}^d(n)$.

Proof Since the sequence $(1,\ldots,1)$ of length k is the minimal ω-sequence of 2-degree 2^k-1, the first statement follows from the superposition rule for products in $\mathsf{DP}(n)$. By definition, $\mathsf{J}^d(n)$ is the $\mathbb{F}_2\mathsf{GL}(n)$-module generated by the d-spikes of degree d. If all of these have ω-sequence ω, then they are of the form $s_\omega(\rho) = \phi^\omega(X)$ where $X \in \mathsf{FL}(n)$ corresponds to $\mathsf{B}(n)W$, and $W \in \mathsf{W}(n)$ is associated to the permutation $\rho \in \Sigma(n)$. □

Flag d-polynomials are most conveniently described by choosing row-reduced representatives A for the cosets. For flags in the Schubert cell $\mathsf{Sch}(\rho)$, these matrices A are the elements WB of the left coset $W\mathsf{B}(W)$ of the Bruhat subgroup $\mathsf{B}(W)$ by the Bruhat decomposition 16.2.7, where W is associated to ρ. Thus the flag d-polynomials for flags in $\mathsf{Sch}(W)$ are the d-polynomials $s_\omega(\rho)\cdot B$, $B\in\mathsf{B}(W)$. Example 23.1.3 shows A for $\mathsf{Sch}(2,4,3,1)$.

If ω is not 2-dominant, then there are linear dependence relations between the flag d-polynomials $\phi^\omega(X)$. More precisely, if $W_2 = S_iW_1$ for some $i\in Z[n-1]$, then every flag d-polynomial for $\mathsf{Sch}(W_1)$ is the sum of two flag d-polynomials for $\mathsf{Sch}(W_2)$. To prove this, we need a calculation in $\mathsf{DP}(n)$.

Proposition 23.1.5 *For $u,v\in\mathsf{V}(n)$ and $k\geq 1$,*

$$u^{(2^{k+1}-1)}v^{(2^k-1)} = v^{(2^{k+1}-1)}u^{(2^k-1)} + (u+v)^{(2^{k+1}-1)}u^{(2^k-1)}.$$

Proof By Proposition 9.1.3(iii), $(u+v)^{(2^{k+1}-1)} = \sum_{i+j=2^{k+1}-1}u^{(i)}v^{(j)}$. We have $u^{(i)}u^{(2^k-1)}=0$ unless 2^k divides i, so that $i=0$ or $i=2^k$. The term $i=0$ gives $v^{(2^{k+1}-1)}u^{(2^k-1)}$, and as $\binom{2^{k+1}-1}{2^k-1}$ is odd the term $i=2^k$ gives $u^{(2^{k+1}-1)}v^{(2^k-1)}$. □

This identity implies the following relations between flag polynomials when $\lambda_i-\lambda_{i+1}=1$, $i\in Z[n-1]$.

Proposition 23.1.6 *Let λ be a partition with $\lambda_i-\lambda_{i+1}=1$ for some i such that $1\leq i\leq n-1$, and let $\omega=\lambda^{tr}$. Let $u_1,\dots,u_i,u_{i+1},\dots,u_n$ be a basis for $\mathsf{V}(n)$ such that $u_1,\dots,u_n$ are the rows of a matrix in the coset corresponding to X, i.e. the ith subspace X^i of X is spanned by $u_1,\dots,u_i$ for all i. Let Y and Z be the corresponding flags for the bases $u_1,\dots,u_{i+1},u_i,\dots,u_n$ and $u_1,\dots,u_{i+1}+u_i,u_i,\dots,u_n$. Then $\phi^\omega(X)=\phi^\omega(Y)+\phi^\omega(Z)$.* □

Proposition 23.1.7 *Let λ be a partition with $\lambda_i-\lambda_{i+1}=1$ for some i such that $1\leq i\leq n-1$, let $\omega=\lambda^{tr}$. Let $W\in\mathsf{W}(n)$ with $i\notin des(W)$, and let $W'=S_iW$. Then for $X\in\mathsf{Sch}(W)$, $\phi^\omega(X)=\phi^\omega(Y)+\phi^\omega(Z)$ where $Y,Z\in\mathsf{Sch}(W')$.*

Proof Let $\mathsf{B}(n)A$ be the coset corresponding to X, so that $A\in W\mathsf{B}(n)$, and let $u_1,\dots,u_n$ be the rows of A. Then by Proposition 23.1.6 $\phi^\omega(X)=\phi^\omega(Y)+\phi^\omega(Z)$, where Y and Z are the flags corresponding to $\mathsf{B}(n)S_iA$ and $\mathsf{B}(n)U_iS_iA$ respectively, S_i switches v_i and v_{i+1}, and $U_i=T_{i,i+1}$ is the transvection which adds v_{i+1} to v_i. Then $\mathsf{B}(n)S_iA=\mathsf{B}(n)W'$, so $Y\in\mathsf{Sch}(W')$. We will show that $Z\in\mathsf{Sch}(W')$ also. We have $U_iS_i=S_iL_i$, where $L_i=T_{i+1,i}$ is the transvection which adds v_i to v_{i+1}. Thus $\mathsf{B}(n)U_iS_iA=\mathsf{B}(n)S_iL_iA$ is in the Bruhat cell $\mathsf{B}(n)S_iL_iW\mathsf{B}(n)$. Since $i\notin$ des(W), this is $\mathsf{B}(n)W'\mathsf{B}(n)$ by Proposition 16.2.8. □

Example 23.1.8 Let $W, W' \in \mathsf{W}(4)$ be associated to $(2,4,3,1)$ and $(4,2,3,1)$ respectively, so that $W' = S_1 W$. Then if $\lambda_1 - \lambda_2 = 1$, $\omega = \lambda^{\mathrm{tr}}$ and X is represented by the matrix A of (23.1), $\phi^\omega(X) = \phi^\omega(Y) + \phi^\omega(Z)$, where the flags Y and Z are represented by the matrices

$$\begin{pmatrix} b & 0 & c & 1 \\ a & 1 & 0 & 0 \\ d & 0 & 1 & 0 \\ 1 & 0 & 0 & 0 \end{pmatrix}, \quad \begin{pmatrix} a+b & 1 & c & 1 \\ a & 1 & 0 & 0 \\ d & 0 & 1 & 0 \\ 1 & 0 & 0 & 0 \end{pmatrix}.$$

Example 23.1.9 (Compare Theorem 10.6.2.) For each line L in $\mathsf{V}(3)$, there are three complete flags containing L, namely a flag $X \in \mathsf{Sch}(W)$ where $2 \notin \mathrm{des}(W)$, and flags $Y, Z \in \mathsf{Sch}(W')$ where $W' = S_2 W$. For $\omega = (2,1,1)$, the corresponding flag d-polynomials satisfy the relations $\phi^\omega(X) = \phi^\omega(Y) + \phi^\omega(Z)$. Hence the submodule $\mathsf{FL}^1(3)$ is in the kernel of $\phi^{(2,1,1)} : \mathsf{FL}(3) \to \mathsf{DP}^8(3)$. By Proposition 10.3.1 there are no other relations between the flag d-polynomials, so $\mathsf{J}^8(3) \cong \mathsf{FL}(3)/\mathsf{FL}^1(3)$.

Similarly, for each plane P in $\mathsf{V}(3)$ there are three complete flags containing P, namely a flag $X \in \mathsf{Sch}(W)$ where $1 \notin \mathrm{des}(W)$, and flags $Y, Z \in \mathsf{Sch}(W')$ where $W' = S_1 W$. For $\omega = (2,2,1)$, the corresponding flag d-polynomials satisfy the relations $\phi^\omega(X) = \phi^\omega(Y) + \phi^\omega(Z)$. Hence the submodule $\mathsf{FL}^2(3)$ is in the kernel of $\phi^{(2,2,1)} : \mathsf{FL}(3) \to \mathsf{DP}^{10}(3)$. By Proposition 10.4.1 there are no other relations between the flag d-polynomials, so $\mathsf{J}^{10}(3) \cong \mathsf{FL}(3)/\mathsf{FL}^2(3)$.

23.2 Zip d-monomials

Definition 23.2.1 A sequence $\omega \in \mathsf{Dec}_d(n)$ is **1-dominant** for n if $I(\omega) = Z[n-1]$ or $Z[n]$, i.e. i is a term of ω for $1 \le i \le n-1$, and **2-dominant** for n if $I_2(\omega) = Z[n-1]$ or $Z[n]$, i.e. i has multiplicity ≥ 2 in ω for $1 \le i \le n-1$. The corresponding 2-degrees $d = \deg_2 \omega$ are 1- (2-) **dominant** degrees.

Thus d is 1-dominant for n if and only if $\alpha(d+n) = n$ (cf. Proposition 5.1.13). For 1-dominant degrees we define auxiliary d-monomials which we use to prove that suitable sets of flag d-polynomials are linearly independent.

Definition 23.2.2 Regard a block B with 1-dominant ω-sequence ω as the concatenation $B_n|B_{n-1}|\cdots|B_1$ where $\omega(B_r) = (r, \ldots, r)$. For $1 \le r \le n-1$, the last column of B_r is **moveable** and all other columns of B_r are **fixed**. For $\rho \in \Sigma(n)$, the **zip block** $Z^\omega(\rho)$ is obtained by raising the entries 1 in the rth moveable column of the spike block $S^\omega(\rho)$ to the first r rows for $1 \le r \le n-1$. The corresponding d-monomial is the **zip d-monomial** $z_\omega(\rho)$.

Thus $z_\omega(\rho) \in \mathsf{DP}^d(n)$, where $d = \deg_2 \omega$. Since the ω-sequence of its fixed part is 1-dominant if and only if ω is 2-dominant, the zip d-monomials $z^\omega(\rho)$, $\rho \in \Sigma(n)$, are all distinct if and only if ω is 2-dominant.

Example 23.2.3 Let $n = 4$, $\omega = (3,3,2,2,1,1)$ and $\rho = (4,3,2,1)$. Then

$$S^\omega(\rho) = \begin{matrix} 0\,0\,0\,0\,0\,0 \\ 1\,1\,0\,0\,0\,0 \\ 1\,1\,1\,1\,0\,0 \\ 1\,1\,1\,1\,1\,1 \end{matrix}, \quad Z^\omega(\rho) = \begin{matrix} 0\,1\,0\,1\,0\,1 \\ 1\,1\,0\,1\,0\,0 \\ 1\,1\,1\,0\,0\,0 \\ 1\,0\,1\,0\,1\,0 \end{matrix},$$

and so $z_\omega(\rho) = v_1^{(42)} v_2^{(11)} v_3^{(7)} v_4^{(21)}$.

The flag d-polynomial $\phi^\omega(X) = s_\omega(\rho) \cdot A$, $A \in \mathsf{B}(\rho)$, is the product $\phi^\omega(X) = \prod_{i=1}^n u_i^{(2^{\lambda_i}-1)}$ of divided powers of linear forms u_i in $v_1, \ldots, v_n$. Each occurrence of $z_\omega(\rho)$ in this product corresponds to a factorization $z_\omega(\rho) = f_1 \cdots f_n$, where f_i is a d-monomial of degree $2^{\lambda_i} - 1$ in the variables occurring in u_i, for $1 \le i \le n$. By the d-binomial theorem 9.1.3(ii), f_i can be any d-monomial of this form, but the possibilities are reduced by the superposition rule: the product of two d-monomials is 0 if they are represented by blocks both of which have an entry 1 in the same position. Thus the 1s in the blocks $F_1, \ldots, F_n$ representing $f_1, \ldots, f_n$ form a set partition of the 1s in the block representing $z_\omega(\rho)$.

We shall prove that all but one of the possible factorizations $f_1 \cdots f_n$ of $z_\omega(\rho)$ occur in an even number of flag d-polynomials $\phi^\omega(X)$, $X \in \mathsf{Sch}(\rho)$, while the remaining factorization occurs in $\phi^\omega(X)$ only when X is the 'maximal' flag in $\mathsf{Sch}(\rho)$, defined by choosing all the arbitrary elements of $\mathbb{F}_2$ in the row-reduced representative matrix (as in (16.5)) to be 1s.

Example 23.2.4 In the situation of Example 23.2.3, we wish to determine all factorizations $z_\omega(\rho) = f_1 f_2 f_3$, (since $\omega_1 = 3, f_4 = 1$), where $\deg f_1 = 63$ and f_1 involves v_1, v_2, v_3, v_4, $\deg f_2 = 15$ and f_2 involves v_1, v_2, v_3, $\deg f_3 = 3$ and f_3 involves v_1, v_2, and where the 1s in the blocks for the monomials f_1, f_2 and f_3 form a set partition of the 1s in the block for $z_\omega(\rho)$. There is a unique solution

$$F_1 = \begin{matrix} 0\,0\,0\,0\,0\,1 \\ 0\,0\,0\,1\,0\,0 \\ 0\,1\,0\,0\,0\,0 \\ 1\,0\,1\,0\,1\,0 \end{matrix}, \quad F_2 = \begin{matrix} 0\,0\,0\,1\,0\,0 \\ 0\,1\,0\,0\,0\,0 \\ 1\,0\,1\,0\,0\,0 \\ 0\,0\,0\,0\,0\,0 \end{matrix}, \quad F_3 = \begin{matrix} 0\,1\,0\,0\,0\,0 \\ 1\,0\,0\,0\,0\,0 \\ 0\,0\,0\,0\,0\,0 \\ 0\,0\,0\,0\,0\,0 \end{matrix},$$

corresponding to the factorization $f_1 = v_1^{(32)} v_2^{(8)} v_3^{(2)} v_4^{(21)}$, $f_2 = v_1^{(8)} v_2^{(2)} v_3^{(5)}$, $f_3 = v_1^{(2)} v_2$ of $z_\omega(\rho)$. We show this factorization by replacing the entries 1 of the block $Z^\omega(\rho)$ by entries 1,2,3 corresponding to the factors f_1, f_2, f_3, and the entries 0 of $Z^\omega(\rho)$ by dots, giving the array

$$F^\omega(\rho) = \begin{array}{cccccc} \cdot & 3 & \cdot & 2 & \cdot & 1 \\ 3 & 2 & \cdot & 1 & \cdot & \cdot \\ 2 & 1 & 2 & \cdot & \cdot & \cdot \\ 1 & \cdot & 1 & \cdot & 1 & \cdot \end{array}.$$

Proposition 23.2.5 *Let ω be 1-dominant, let $\rho \in \Sigma(n)$ be a permutation and let $f(\rho) = \sum_{A \in \mathsf{B}(\rho)} s_\omega(\rho) \cdot A$ be the sum of the flag d-polynomials for* $\mathsf{Sch}(\rho)$. *Then $z_\omega(\rho)$ is a term in $f(\rho)$. In particular, $f(\rho) \neq 0$.*

Proof For $1 \le i \le n$, $v_{\rho(i)}$ is a term in u_i for all $X \in \mathsf{Sch}(\rho)$, and v_k can be a term in u_i for some $X \in \mathsf{Sch}(\rho)$ if and only if $k = \rho(j)$, where $i \le j$ and $\rho(i) \ge \rho(j)$. Let $z_\omega(\rho) = f_1 \cdots f_n$ be a factorization such that some f_i does not involve every variable v_k satisfying this condition. Then we may assign either coefficient 0 or 1 to v_k, and so the factorization $f_1 \cdots f_n$ contributes to the coefficient of $z_\omega(\rho)$ in an even number of flag d-polynomials. In this way, we can pair off all occurrences of $z_\omega(\rho)$ which appear when a flag d-polynomial is multiplied out in full, except those in which each f_i involves all the variables v_k which can be terms of u_i. These are a subset of the occurrences of $z_\omega(\rho)$ which appear when the 'maximal' flag d-polynomial is multiplied out.

To complete the proof, we shall prove that there is a unique set partition of the 1s in the block representing $z_\omega(\rho)$ which satisfies the required conditions. The variable $v_{\rho(1)}$ can appear only in f_1, and so all 1s in row $\rho(1)$ of $Z^\omega(\rho)$ must be assigned to f_1. By definition of $Z^\omega(\rho)$, there is a 1 in each of the fixed columns of this row, and in each of the first $n - \rho(1)$ moveable columns. Since f_1 must also involve all the variables v_k for $1 \le k < \rho(1)$, we must assign at least one 1 to f_1 from each row above row $\rho(1)$. As there remain only $\rho(1) - 1$ 1s to be assigned to f_1, we must choose exactly one of these from each row. Again by construction of $Z^\omega(\rho)$, by working along the block from right to left we see that these 1s can only be chosen as (part of) a NE to SW diagonal in the moveable columns of $Z^\omega(\rho)$. Thus there is a unique choice for f_1. The result follows by induction on n, since ρ determines a bijection from $\{2,3,\ldots,n\}$ to $\{1,2,\ldots,n\} \setminus \{\rho(1)\}$, and this can be regarded as a permutation of $\{1,2,\ldots,n-1\}$ by reindexing the variables while preserving their order. □

Example 23.2.6 Let $n = 4$, $\omega = (3,3,2,2,1,1)$ and $\rho = (2,4,3,1)$. Then the block $Z^\omega(\rho)$ is shown below, together with the corresponding array $F^\omega(\rho)$ which shows the factorization $z_\omega(\rho) = f_1 f_2 f_3$, constructed as in Example 23.2.4.

$$Z^\omega(\rho) = \begin{matrix} 0 & 1 & 0 & 1 & 0 & 1 \\ 1 & 1 & 1 & 1 & 1 & 0 \\ 1 & 1 & 0 & 0 & 0 & 0 \\ 1 & 0 & 1 & 0 & 0 & 0 \end{matrix}, \quad F^\omega(\rho) = \begin{matrix} \cdot & 3 & \cdot & 2 & \cdot & 1 \\ 1 & 1 & 1 & 1 & 1 & \cdot \\ 3 & 2 & \cdot & \cdot & \cdot & \cdot \\ 2 & \cdot & 2 & \cdot & \cdot & \cdot \end{matrix}$$

We relabel the entries i of $F^\omega(\rho)$ by $\rho(i)$ and split the resulting array $\rho(F^\omega(\rho))$ into its fixed and moveable columns, i.e.

$$\rho(F^\omega(\rho)) = \begin{matrix} \cdot & 3 & \cdot & 4 & \cdot & 2 \\ 2 & 2 & 2 & 2 & 2 & \cdot \\ 3 & 4 & \cdot & \cdot & \cdot & \cdot \\ 4 & \cdot & 4 & \cdot & \cdot & \cdot \end{matrix} \longrightarrow \begin{matrix} \cdot & \cdot & \cdot \\ 2 & 2 & 2 \\ 3 & \cdot & \cdot \\ 4 & 4 & \cdot \end{matrix}, \begin{matrix} 3 & 4 & 2 \\ 2 & 2 & \cdot \\ 4 & \cdot & \cdot \\ \cdot & \cdot & \cdot \end{matrix}.$$

In this example, the moveable columns play an active role and the fixed columns a passive one in the factorization of $z_\omega(\rho)$. Example 23.3.7 illustrates the use of the fixed columns.

Factorizations of d-monomials as in Example 23.2.6 will be an essential tool in Chapter 24, where we consider cases where ω is not 1-dominant. We formalize their properties as follows (see Definitions 24.2.1 and 24.3.1).

Definition 23.2.7 Let f be a d-monomial in $\mathsf{DP}^\omega(n)$, where $\omega = (\omega_1, \omega_2, \ldots)$ is decreasing, and let $\rho \in \Sigma(n)$ be a permutation. A **ρ-factorization** of f is one which is shown by a **zip array** $F^\omega(\rho)$ with integer entries corresponding to the entries 1 in the $(0,1)$-block representing f, and dots corresponding to the entries 0, so that

(i) for all j, the integer entries in column j are $1, 2, \ldots, \omega_j$,
(ii) (raised entries) if $k < \ell$ and $\rho(k) > \rho(\ell)$, there is an entry k in row $\rho(\ell)$,
(iii) (normal entries) all other entries k are in row $\rho(k)$.

Thus raised entries correspond to pairs (k, ℓ) which are reversed by ρ. In Example 23.2.6, the numbers in row 1 of $F^\omega(\rho)$ and the entry 2 in row 3 are raised and the remaining numbers are normal entries. The following result will be useful in considering sums of flag d-polynomials over Schubert cells in $\mathsf{FL}(n)$.

Proposition 23.2.8 *Let f be a d-monomial in $\mathsf{DP}^{\omega}(n)$, where ω is decreasing, and let $\rho \in \Sigma(n)$ be a permutation. Then f is a term in $f(\rho) = \sum_{A \in \mathsf{B}(\rho)} s_{\omega}(\rho) \cdot A$, the sum of the flag d-polynomials for $\mathsf{Sch}(\rho)$, if and only if the number of ρ-factorizations of f is odd. In particular, if some f has a unique ρ-factorization, then $f(\rho) \neq 0$.*

Proof By replacing $z_{\omega}(\rho)$ by f in the first paragraph of the proof of Proposition 23.2.8, we see that all factorizations of f as a term of the flag d-polynomial for $X \in \mathsf{Sch}(\rho)$ cancel in pairs unless X is the maximal flag in $\mathsf{Sch}(\rho)$ and the factorization satisfies the conditions of Definition 23.2.7. □

23.3 Permutation tableaux

The spike block $S^{\omega}(\rho)$ is obtained by applying the permutation ρ to the rows of the Ferrers block associated to a decreasing sequence ω. We may similarly permute the rows of Young tableaux. We begin by permuting the rows of the diagram associated to $\lambda = \omega^{\mathrm{tr}}$.

Definition 23.3.1 Given a partition $\lambda = (\lambda_1, \ldots, \lambda_n)$ and a permutation $\rho \in \Sigma(n)$, the **spike diagram** for $\omega = \lambda^{\mathrm{tr}}$ associated to ρ is obtained by replacing each entry 1 in the spike block $S^{\omega}(\rho)$ by a 'box', so that row $\rho(i)$ has λ_i boxes for $1 \leq i \leq n$.

For example, the spike diagram of $\rho = (2,4,3,1)$ for $\omega = (4,3,2,1)$ is

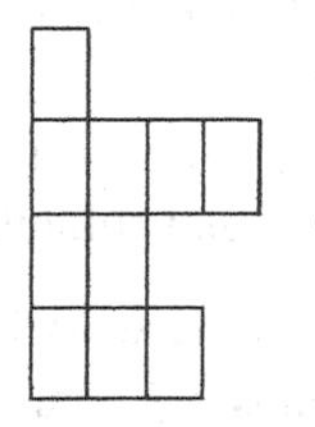

Definition 23.3.2 For ω 1-dominant and $\rho \in \Sigma(n)$, a **ρ-tableau** for ω is a filling of each box in the spike diagram of ρ for ω by an integer i with $1 \leq i \leq n$, so that the entries in each column are distinct. In particular, the **spike ρ-tableau** for ω is obtained by entering i in all the boxes in the ith row for each i. A ρ-tableau is **upward** if each i appears only in rows $\leq i$, and is **downward** if each i appears only in rows $\geq i$. For $\rho_1, \rho_2 \in \Sigma(n)$, a ρ_1-tableau for ω has **content** ρ_2 if it is filled by the entries of the spike ρ_2-tableau for ω.

Example 23.3.3 For $\omega = (4,3,2,1)$, let $\rho_1 = (2,3,1,4)$ and $\rho_2 = (2,4,3,1) \in \Sigma(4)$. Then the spike ρ_2-tableau S and an upward ρ_1-tableau T of content ρ_2

are shown below. We can regard T as the result of an upward shift of some of the entries of S.

$$S = \begin{array}{|c|c|c|c|}\hline 1 \\ \hline 2 & 2 & 2 & 2 \\ \hline 3 & 3 \\ \hline 4 & 4 & 4 \\ \hline \end{array}\ , \quad T = \begin{array}{|c|c|c|c|}\hline 1 & 4 \\ \hline 2 & 2 & 2 & 2 \\ \hline 3 & 3 & 4 \\ \hline 4 \\ \hline \end{array}\ .$$

In the same way, the spike ρ_1-tableau S and a downward ρ_2-tableau T of content ρ_1 are shown below. We can regard T as the result of a downward shift of some of the entries of S.

$$S = \begin{array}{|c|c|c|c|}\hline 1 & 1 \\ \hline 2 & 2 & 2 & 2 \\ \hline 3 & 3 & 3 \\ \hline 4 \\ \hline \end{array}\ , \quad T = \begin{array}{|c|c|c|c|}\hline 1 \\ \hline 2 & 2 & 2 & 2 \\ \hline 3 & 1 \\ \hline 4 & 3 & 3 \\ \hline \end{array}\ . \tag{23.2}$$

The Bruhat order $\geq$ on $\mathsf{W}(n)$ was defined in Section 16.1, where it was shown that for $W_1, W_2 \in \mathsf{W}(n)$ each of the following conditions is equivalent to $W_1 \geq W_2$.

(i) $\text{len}(W_1) - \text{len}(W_2) = k > 0$ and W_1 and W_2 can be exchanged by applying a sequence of k exchanges of two rows or a sequence of k exchanges of two columns.

(ii) $s_{i,j}(W_1) \geq s_{i,j}(W_2)$ for $1 \leq i,j \leq n$, where $s_{i,j}(W)$ for $W \in \mathsf{W}(n)$ in the number of entries 1 in rows $\leq i$ and columns $\geq j$.

In the proof of Proposition 23.3.4, we work instead with $r_{i,j}(W)$, the number of entries 1 in rows $\leq i$ and columns $\leq j$. Thus $r_{i,j}(W) + s_{i,j+1}(W) = i$, and (ii) is equivalent to

(ii') $W_1 \geq W_2$ if and only if $r_{i,j}(W_1) \leq r_{i,j}(W_2)$ for $1 \leq i,j \leq n$.

Proposition 23.3.4 *Let ω be 1-dominant for n and let $\rho_1, \rho_2 \in \Sigma(n)$. Then there exists an upward ρ_1-tableau for ω of content ρ_2 if and only if $\rho_1 \leq \rho_2$. Similarly, there is a downward ρ_2-tableau for ω of content ρ_1 if and only if $\rho_1 \leq \rho_2$.*

Proof If ρ is the permutation associated to $W \in \mathsf{W}(n)$, we write $r_{i,j}(W) = r_{i,j}(\rho)$. Since $\lambda = \omega^{\text{tr}}$ is strictly decreasing, $r_{i,j}(\rho_1)$ is the number of rows of length $\geq \lambda_i$ among the first j rows of the spike ρ_1-tableau for ω. The entries in the spike ρ_2-tableau S which are $\leq j$ appear in its first j rows. Assume that an upward tableau T exists, where the entries of S are used to fill the boxes in the spike diagram of ρ_1 for ω. Then the entries just described have to appear in T in the same columns, since the entries in each column are distinct, and in the same or higher rows. Hence $r_{i,j}(\rho_1) \geq r_{i,j}(\rho_2)$, and so $\rho_1 \leq \rho_2$.

For the converse, we use a chain in the Bruhat order from ρ_1 to ρ_2 to convert S to a ρ_1-tableau T of content ρ_2. Consider first the case where ρ_1 and ρ_2 are associated to W_1 and W_2 in $\mathsf{W}(n)$ and W_2 is obtained from W_1 by replacing the submatrix I_2 in columns i,j with $i < j$ and rows k,ℓ with $k < \ell$ by J_2, where

$$I_2 = \begin{pmatrix} 1\,0 \\ 0\,1 \end{pmatrix}, \ J_2 = \begin{pmatrix} 0\,1 \\ 1\,0 \end{pmatrix}.$$

Then $\rho_1(k) = i = \rho_2(\ell)$ and $\rho_1(\ell) = j = \rho_2(k)$. Since $k < \ell$ and λ is strictly decreasing, $\lambda_k > \lambda_\ell$. In the spike diagram of ρ_1 for ω, row $i = \rho_1(k)$ has length λ_k and row $j = \rho_1(\ell)$ has length λ_ℓ, and so row i is longer than row j. Similarly, row j of the spike diagram of ρ_2 for ω is longer than row i. Thus we may move entries j from row j of S up to row i so as to obtain a ρ_1-tableau T with content ρ_2.

For the general case, ρ_2 can be obtained from ρ_1 by a finite sequence of inversions of such pairs i,j. Corresponding to such a sequence, we can carry out a sequence of moves in which certain entries are moved up from the jth row to the ith row to give a tableau for the next lower permutation in the chain, with content ρ_2.

The second statement follows from this argument by using the reversal $\rho_0 = (n, n-1, \ldots, 1)$. The diagrams of $\rho_0 \circ \rho_1$ and $\rho_0 \circ \rho_2$ are obtained by reversing the rows of the diagrams of ρ_1 and ρ_2. These row reversals interchange upward and downward tableaux, while composition with ρ_0 reverses the Bruhat order. □

Example 23.3.5 Let $n = 4$, $\rho_1 = (2,3,1,4)$ and $\rho_2 = (2,4,3,1)$. Then there are two chains $(2,4,3,1) > (2,3,4,1) > (2,3,1,4)$ and $(2,4,3,1) > (2,4,1,3) > (2,3,1,4)$ linking ρ_1 and ρ_2, giving upward tableau moves as shown below.

$$\begin{array}{|c|c|c|c|} \hline 1 \\ \hline 2&2&2&2 \\ \hline 3&3 \\ \hline 4&4&4 \\ \hline \end{array} \longrightarrow \begin{array}{|c|c|c|c|} \hline 1 \\ \hline 2&2&2&2 \\ \hline 3&3&4 \\ \hline 4&4 \\ \hline \end{array} \longrightarrow \begin{array}{|c|c|c|c|} \hline 1&4 \\ \hline 2&2&2&2 \\ \hline 3&3&4 \\ \hline 4 \\ \hline \end{array}$$

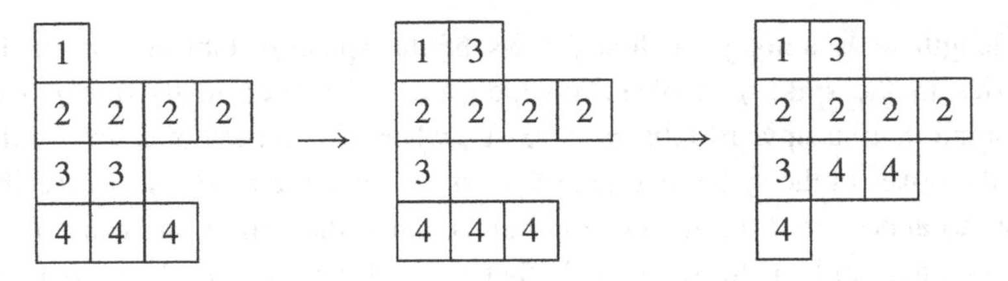

Proposition 23.3.6 *Let ω be 2-dominant. If the zip d-monomial $z_\omega(\rho_1)$ is a term in the flag d-polynomial $\phi^\omega(X)$, then $X \in \mathsf{Sch}(\rho_2)$, where $\rho_2 \succeq \rho_1$. More generally, if $z_\omega(\rho_1)$ is a term of a d-polynomial in the $\mathsf{B}(n)$-orbit of $s_\omega(\rho_2)$, then $\rho_2 \succeq \rho_1$.*

Proof Let B be the subblock of the zip block $Z^\omega(\rho_1)$ formed by the fixed columns. Since ω is 2-dominant, $\omega(B)$ is 1-dominant. By replacing the 1s in B by boxes, we obtain the spike diagram of ρ_1 for $\omega(B)$. By Proposition 23.3.4, there exists an upward ρ_1-tableau T for $\omega(B)$ of content ρ_2 if and only if $\rho_2 \succeq \rho_1$.

If we apply any linear substitution $A \in \mathsf{B}(n)$ to the d-spike $s_\omega(\rho_2)$, and show the factorizations of the d-monomials in the resulting product by integer arrays relabelled as in Example 23.2.6, the fixed columns form a collection of tableaux in which the entries in each column are distinct, and in which the entries of the spike tableau for ρ_2 appear in the same or higher rows. Thus, on restriction to the fixed columns of the block $Z^\omega(\rho_1)$, an occurrence of $z_\omega(\rho_1)$ will yield an upward ρ_1-tableau of content ρ_2. □

Example 23.3.7 As in Example 23.3.5, let $n = 4$, $\rho_1 = (2,3,1,4)$ and $\rho_2 = (2,4,3,1)$, and let $\omega = (3,3,2,2,1,1)$. Then $z_\omega(\rho_1)$ and $s_\omega(\rho_2)$ are represented by the blocks

$$Z^\omega(\rho_1) = \begin{matrix} 1\,1\,0\,1\,0\,1 \\ 1\,1\,1\,1\,1\,0 \\ 1\,1\,1\,0\,0\,0 \\ 0\,0\,0\,0\,0\,0 \end{matrix}, \quad S^\omega(\rho_2) = \begin{matrix} 0\,0\,0\,0\,0\,0 \\ 1\,1\,1\,1\,1\,1 \\ 1\,1\,0\,0\,0\,0 \\ 1\,1\,1\,1\,0\,0 \end{matrix}$$

and the flag d-polynomials for $X \in \mathsf{Sch}(\rho_2)$ are given by $\phi^\omega(X) = s_\omega(\rho_2) \cdot A$, where we may take A in the form (16.6). Taking all the stars in (16.6) as 1s, the d-polynomial for the maximal flag in $\mathsf{Sch}(\rho_2)$ is $(v_1 + v_2)^{(63)}(v_1 + v_3 + v_4)^{(15)}(v_1 + v_3)^{(3)}$. Several factorizations $f_1 f_2 f_3$ of the zip monomial $z_\omega(\rho_1)$ occur when this product is expanded, for example we may take $f_1 = v_1^{(32)} v_2^{(31)}$,

$f_2 = v_1^{(10)} v_3^{(5)}, f_3 = v_1 v_3^{(2)}$. We have

$$F^\omega(\rho_1) = \begin{matrix} 3 & 2 & \cdot & 2 & \cdot & 1 \\ 1 & 1 & 1 & 1 & 1 & \cdot \\ 2 & 3 & 2 & \cdot & \cdot & \cdot \\ \cdot & \cdot & \cdot & \cdot & \cdot & \cdot \end{matrix}, \quad \rho_2(F^\omega(\rho_1)) = \begin{matrix} 3 & 4 & \cdot & 4 & \cdot & 2 \\ 2 & 2 & 2 & 2 & 2 & \cdot \\ 4 & 3 & 4 & \cdot & \cdot & \cdot \\ \cdot & \cdot & \cdot & \cdot & \cdot & \cdot \end{matrix},$$

and the fixed columns of $\rho_2(F^\omega(\rho_1))$ give an upward ρ_1-tableau for $\omega' = (3,2,1)$ of content ρ_2 corresponding to the last tableau of Example 23.3.5.

23.4 $\mathsf{J}^d(n)$ for 1-dominant degrees

We begin by proving linear independence of the flag d-polynomials in the 2-dominant case. We argue by induction over the Schubert cells, using any total order which refines the Bruhat order, beginning with the largest Schubert cell, $\mathsf{Sch}(\rho_0)$, and ending with the smallest, $\mathsf{Sch}(\iota_n)$. Since $\rho > \sigma$ in the Bruhat order implies that $\text{len}(\rho) > \text{len}(\sigma)$, we may use any total order on $\Sigma(n)$ which is compatible with the length function. At each step we remove all terms which arise from one Schubert cell from a possible dependence relation. The method is illustrated in the case $n = 3$ in Sections 10.3 and 10.4.

Theorem 23.4.1 *Let ω be 2-dominant, with $\deg_2 \omega = d$. Then $\phi^\omega : \mathsf{FL}(n) \to \mathsf{DP}^d(n)$ is injective, so that $\mathsf{J}^d(n) \cong \mathsf{FL}(n)$. Equivalently, the flag d-polynomials $s_\omega(\rho) \cdot A$ are linearly independent for $\rho \in \Sigma(n)$ and $A \in \mathsf{B}(\rho)$.*

Proof The flag d-polynomials for $\mathsf{Sch}(\rho)$ form the $\mathsf{B}(\rho)$-orbit of the d-spike $s_\omega(\rho)$, and so they span the cyclic $\mathbb{F}_2\mathsf{B}(\rho)$-module $\langle s_\omega(\rho)\rangle$. We apply Proposition 10.1.2 with $G = \mathsf{B}(\rho)$, $V = \mathsf{DP}(n)$ and $x = s_\omega(\rho)$. Then $x \cdot \overline{G} = \sum_{A \in \mathsf{B}(\rho)} s_\omega(\rho) \cdot A = f(\rho)$, the sum of the flag d-polynomials for $\mathsf{Sch}(\rho)$. By Proposition 23.2.5 $f(\rho) \neq 0$. It follows from Proposition 10.1.2 that $\theta : \mathbb{F}_2\mathsf{B}(\rho) \xrightarrow{\cong} \langle s_\omega(\rho)\rangle$, where θ is the $\mathbb{F}_2\mathsf{B}(\rho)$-map defined by $\theta(I_n) = s_\omega(\rho)$, and $f(\rho)$ 'cogenerates' $\langle s_\omega(\rho)\rangle$, in the sense that given $g \neq 0 \in \langle s_\omega(\rho)\rangle$, there is an element $u \in \mathbb{F}_2\mathsf{B}(\rho)$ such that $g \cdot u = f(\rho)$.

Given a relation $\sum_X a_X \phi^\omega(X) = 0$ in $\mathsf{DP}(n)$, where the sum is over all complete flags X in $\mathsf{V}(n)$ and where $a_X \in \mathbb{F}_2$, we may write it as $g(\rho_0) + g' = 0$, where $g(\rho_0)$ is the sum of the terms involving flags in the top Schubert cell $\mathsf{Sch}(\rho_0)$, and g' is the sum of the remaining terms. Assume, for a contradiction, that $g(\rho_0) \neq 0$. Then by applying a suitable element $u \in \mathbb{F}_2\mathsf{B}(n)$ to the relation, we obtain $f(\rho_0) + g' \cdot u = 0$, where $g' \cdot u$ is a sum of elements of the $\mathsf{B}(n)$-orbits of the flag d-polynomials for flags in lower Schubert cells. By Proposition 23.2.5 the zip d-monomial $z_\omega(\rho_0)$ is a term in $f(\rho_0)$, and by Proposition 23.3.6, $z_\omega(\rho_0)$ is not a term in $g' \cdot u$. Hence $z_\omega(\rho_0)$ is a term of $f(\rho_0) + g' \cdot u$, contradicting the relation $f(\rho_0) + g' \cdot u = 0$. It follows that $g(\rho_0) = 0$. We continue by writing $g' = g(\rho) + g''$, where $g(\rho)$ is the sum of the terms involving flags in a maximal Schubert cell $\mathsf{Sch}(\rho) \neq \mathsf{Sch}(\rho_0)$, and iterating the argument. □

The preceding argument can be carried out over any subset of the Schubert cells for which the spike and zip d-monomials associated to ω are distinct. This allows us to prove a corresponding result for the 1-dominant case, using the direct sum decomposition $\mathsf{FL}(n) \cong \bigoplus_J \mathsf{FL}_J(n)$ of Theorem 16.5.3. We recall from Definition 5.1.1 that $I_2(\omega) \subseteq Z[n]$ is the set of repeated terms of ω, i.e. the set of all i such that $\lambda_i - \lambda_{i+1} \geq 2$, where $\lambda = \omega^{\mathrm{tr}}$.

Theorem 23.4.2 *Let ω be 1-dominant, and let $R \subseteq \Sigma(n)$ be a set of permutations associated to coset representatives for $\mathsf{W}^I(n)$ in $\mathsf{W}(n)$, where $I = I_2(\omega)$. Then the flag d-polynomials $s_\omega(\rho) \cdot A$ for $\rho \in R$ and $A \in \mathsf{B}(\rho)$ are linearly independent.*

Proof Since ω is 1-dominant, the ω-sequence ω' of its fixed columns is decreasing. For ρ_1 and $\rho_2 \in \Sigma(n)$, the d-spikes $s_{\omega'}(\rho_1)$ and $s_{\omega'}(\rho_2)$ are equal if and only if the corresponding permutation matrices are in the same right coset of $\mathsf{W}^I(n)$, and so the zip d-monomials $z_\omega(\rho_1)$ and $z_\omega(\rho_2)$ are equal under the same condition. The result follows by induction on the Schubert cells, as for Theorem 23.4.1. □

Here X is a complete flag, but $\phi^\omega(X)$ depends only on the partial flag given by restriction to subspaces with dimensions in $I_2(\omega)$. As a special case of Theorem 23.4.2, the flag d-polynomials for flags in the Schubert cells $\mathsf{Sch}(\rho)$ such that $\mathrm{des}(\rho) \subseteq I_2(\omega)$ are linearly independent. If ω is not 2-dominant, however, then the $\mathbb{F}_2$-subspace spanned by these d-polynomials is not a $\mathbb{F}_2\mathsf{GL}(n)$-submodule of $\mathsf{FL}(n)$: for example, it contains $s_\omega(\iota_n)$ but not $s_\omega(\rho_0)$. We can obtain a better result by using a set of coset representatives for $\mathsf{W}^I(n)$ which is maximal, rather than minimal, for the Bruhat order.

Example 23.4.3 Let $n = 4$ and $\omega = (3,2,2,1)$, so that $I = \{2\}$ and $\lambda = (4,3,1)$. The diagram below shows the Bruhat order on permutations σ such that

$\{1,3\} \subseteq \operatorname{des}(\sigma)$.

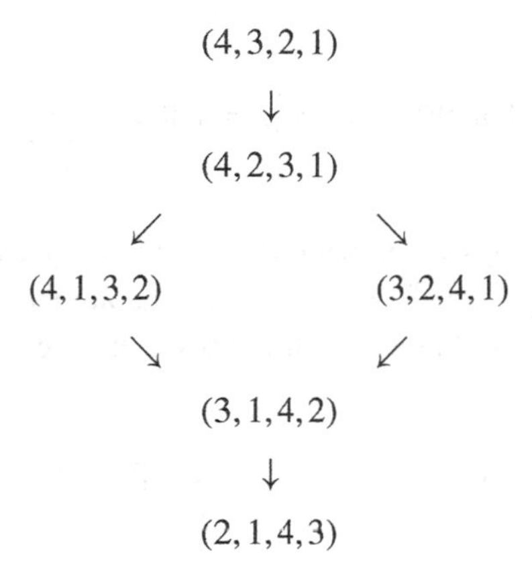

The matrices $W \in \mathsf{W}(n)$ corresponding to these permutations form a set of representatives for the cosets $\mathsf{W}^I W$ in $\mathsf{W}(4)$, and the Schubert cells $\mathsf{Sch}(\sigma)$ give 140 linearly independent flag d-polynomials of the form $u_1^{(15)} u_2^{(7)} u_3 \in \mathsf{DP}^{23}(4)$, where $u_1, u_2, u_3 \in \mathsf{DP}^1(4)$. For $1 \leq j \leq n-1$, as in Definition 17.3.3 we denote by $\mathsf{FL}^{-\{j\}}(n)$ the maximal partial flag module given by flags which omit only subspaces of dimension j. Since the associated composition is $(1,\ldots,1,2,1,\ldots,1)$, with jth component 2, the parabolic subgroup $\mathsf{P}^{-\{j\}}(n)$ is obtained from $\mathsf{B}(n)$ by replacing $\mathsf{B}(2)$ by $\mathsf{GL}(2)$ in the 2×2 submatrix in rows and columns $j, j+1$. Hence each maximal partial flag is contained in three complete flags.

Theorem 23.4.4 *Let ω be 1-dominant of 2-degree d and let $I = I_2(\omega)$. Then $\mathsf{J}^d(n) \cong \mathsf{FL}(n)/\sum_{j \notin I} \mathsf{FL}^{-\{j\}}(n) \cong \sum_{I \cup K = Z[n-1]} \mathsf{FL}_K(n)$, as $\mathbb{F}_2\mathsf{GL}(n)$-modules, and*

$$\dim \mathsf{J}^d(n) = \binom{n}{c_1, \ldots, c_{r+1}}_2 \cdot \prod_{k=1}^{r+1} 2^{c_k(c_k-1)/2},$$

where $r = |I|$ and $(c_1, \ldots, c_{r+1})$ is the composition of n associated to I.

Proof The upper bound for $\dim \mathsf{J}^d(n)$ is obtained as follows. By Proposition 23.1.7, the sum of the flag d-polynomials for the three complete flags containing a partial flag of type $-\{j\} = \{1, \ldots, j-1, j+1, \ldots, n-1\}$ is 0 when $j \notin I$. Hence the partial flag modules $\mathsf{FL}^{-\{j\}}(n)$ for $j \notin I$ are in $\operatorname{Ker}(\phi^\omega)$.

The lower bound for $\dim \mathsf{J}^d(n)$ follows from Theorem 23.4.2, by choosing the coset representatives $W_0^I W$ for $\mathsf{W}^I(n)$, where $\operatorname{des}(W) \subseteq I$. The corresponding permutations σ are decreasing on the intervals $[i_{j-1}, i_j]$, $1 \leq j \leq r+1$, i.e.

$\mathrm{des}(\sigma) \supseteq -I = Z[n-1] \setminus I$. We have

$$\mathrm{len}(W_0^I W) = \mathrm{len}(W) + \mathrm{len}(W_0^I) = \mathrm{len}(W) + \sum_{k=1}^{r+1} c_k(c_k - 1)/2.$$

Since the number of flags in the Schubert cells $\mathsf{Sch}(\rho)$ such that $\mathrm{des}(\rho) \subseteq I$ is $\dim \mathsf{FL}^I(n) = \sum_{\mathrm{des}(W)\subseteq I} 2^{\mathrm{len}(W)} = \binom{n}{c_1,\ldots,c_{r+1}}_2$, the number of linearly independent flag d-polynomials obtained by this choice of coset representatives is

$$\sum_{\mathrm{des}(\rho)\supseteq -I} 2^{\mathrm{len}(\rho)} = \binom{n}{c_1,\ldots,c_{r+1}}_2 \cdot \prod_{k=1}^{r+1} 2^{c_k(c_k-1)/2}. \qquad \square$$

Proposition 23.4.5 *If d is 1-dominant, $\mathsf{K}^d(n)$ has a direct summand isomorphic to the Steinberg module* $\mathsf{St}(n)$.

Proof By Theorem 17.6.8, $\mathsf{FL}_I(n) \cong \mathsf{St}(n)$ for $I = Z[n-1]$. By Theorem 23.4.4, this summand of $\mathsf{FL}(n)$ occurs in $\mathsf{J}^d(n)$ for all 1-dominant degrees d. $\square$

Example 23.4.6 The results of Theorem 23.4.4 are tabulated below for $n = 4$, using the minimum ω for each $I = I_2(\omega)$. In the table, $\mathsf{FL}_I(4)$ is written as FL_I.

d	ω	$I_2(\omega)$	$\mathsf{J}^d(4)$	$\dim \mathsf{J}^d(4)$
11	$(3,2,1)$	$\emptyset$	$\mathsf{FL}_{1,2,3}$	64
19	$(3,2,1,1)$	$\{1\}$	$\mathsf{FL}_{1,2,3} \oplus \mathsf{FL}_{2,3}$	120
23	$(3,2,2,1)$	$\{2\}$	$\mathsf{FL}_{1,2,3} \oplus \mathsf{FL}_{1,3}$	140
25	$(3,3,2,1)$	$\{3\}$	$\mathsf{FL}_{1,2,3} \oplus \mathsf{FL}_{1,2}$	120
39	$(3,2,2,1,1)$	$\{1,2\}$	$\mathsf{FL}_{1,2,3} \oplus \mathsf{FL}_{2,3} \oplus \mathsf{FL}_{1,3} \oplus \mathsf{FL}_3$	210
41	$(3,3,2,1,1)$	$\{1,3\}$	$\mathsf{FL}_{1,2,3} \oplus \mathsf{FL}_{2,3} \oplus \mathsf{FL}_{1,2} \oplus \mathsf{FL}_2$	210
49	$(3,3,2,2,1)$	$\{2,3\}$	$\mathsf{FL}_{1,2,3} \oplus \mathsf{FL}_{1,3} \oplus \mathsf{FL}_{1,2} \oplus \mathsf{FL}_1$	210
81	$(3,3,2,2,1,1)$	$\{1,2,3\}$	$\mathsf{FL}(4)$	315

By Example 7.4.6 $\mathsf{J}^{11}(4) = \mathsf{K}^{11}(4) \cong \mathsf{St}(4)$. Computer calculation of $\dim \mathsf{K}^d(4)$ shows that in the cases $d = 25, 49$ and 81 we also have $\mathsf{J}^d(4) = \mathsf{K}^d(4)$, while

$\dim \mathsf{K}^d(4)/\mathsf{J}^d(4)$ is 20 in degree $d = 19$, and 15 in degrees $d = 23, 39$ and 41. We discuss the cases where $\dim \mathsf{K}^d(4)/\mathsf{J}^d(4) \neq 0$ in Section 30.3.

23.5 Zip monomials and $\mathsf{Q}^d(n)$

The monomial in $\mathsf{P}^d(n)$ corresponding to the zip block $Z^\omega(\rho)$ is the **zip monomial** $z^\omega(\rho)$. When we look for an analogue of Proposition 23.2.5 for $\mathsf{P}(n)$, we find that the roles of the spike and zip blocks are exchanged. This motivates the following definition.

Definition 23.5.1 Let ω be 1-dominant. Given a permutation $\rho \in \Sigma(n)$ and an upper triangular matrix $A \in \mathsf{U}(\rho) = \mathsf{B}(\rho)^{\mathrm{tr}}$, $z^\omega(\rho) \cdot A$ is the **flag polynomial** for the flag $\mathsf{B}(n)A^{\mathrm{tr}} \in \mathsf{Sch}(\rho)$.

Example 23.5.2 Let $n = 4$, $\omega = (3,2,2,1)$ and $\rho = (2,4,3,1)$. Then

$$S^\omega(\rho) = \begin{matrix} 0\,0\,0\,0 \\ 1\,1\,1\,1 \\ 1\,0\,0\,0 \\ 1\,1\,1\,0 \end{matrix}\,, \quad Z^\omega(\rho) = \begin{matrix} 1\,0\,1\,1 \\ 1\,1\,1\,0 \\ 1\,0\,0\,0 \\ 0\,1\,0\,0 \end{matrix}\,, \quad A = \begin{pmatrix} 1 & a & b & c \\ 0 & 1 & 0 & 0 \\ 0 & 0 & 1 & d \\ 0 & 0 & 0 & 1 \end{pmatrix},$$

where $a,b,c,d \in \mathbb{F}_2$, referring to (16.6) for $\mathsf{B}(\rho)$. Hence $z^\omega(\rho) = x_1^{13}x_2^7x_3x_4^2$, and the flag polynomial $z^\omega(\rho) \cdot A = (x_1 + ax_2 + bx_3 + cx_4)^{13}x_2^7(x_3 + dx_4)x_4^2$.

We wish to show that the sum of the flag polynomials for $\mathsf{Sch}(\rho)$ contains the spike $s^\omega(\rho)$. Thus we wish to count all occurrences of $s^\omega(\rho)$ mod 2 in the flag polynomials $z^\omega(\rho) \cdot A$, $A \in \mathsf{U}(\rho)$. Writing $z^\omega(\rho) \cdot A = \prod_{i=1}^n u_i^{r_i}$, where $u_1, \ldots, u_n \in \mathsf{P}^1(n)$, each occurrence of $s^\omega(\rho)$ in this product corresponds to a factorization $s^\omega(\rho) = f_1 \cdots f_n$, where f_i is a monomial of degree r_i in those variables x_k which are terms in u_i. By Proposition 1.4.11, the entries 1 in the corresponding blocks $F_1, \ldots, F_n$ form a set partition of the entries 1 in $S^\omega(\rho)$.

We shall prove that all but one of the possible factorizations $f_1 \cdots f_n$ of $s_\omega(\rho)$ occur in an even number of flag polynomials, while the remaining factorization occurs in $\phi^\omega(X)$ only when X is the 'maximal' flag in $\mathsf{Sch}(\rho)$.

Example 23.5.3 Let $n = 4$, $\omega = (3,3,2,2,1,1)$ and $\rho = (4,3,2,1)$. Then $S^\omega(\rho)$ and $Z^\omega(\rho)$ are as in Example 23.2.3, $s^\omega(\rho) = x_2^3x_3^{15}x_4^{63}$, $z^\omega(\rho) = x_1^{42}x_2^{11}x_3^7x_4^{21}$ and $\mathsf{U}(\rho) = \mathsf{U}(4)$. We wish to determine all factorizations $s^\omega(\rho) = f_1f_2f_3f_4$, where $\deg f_1 = 42$ and f_1 involves x_2, x_3, x_4 and possibly x_1, $\deg f_2 = 11$ and f_2 involves x_3, x_4 and possibly x_2, $\deg f_3 = 7$ and f_3 involves x_4 and possibly x_3, and $f_4 = x_4^{21}$, and where the entries 1 in the corresponding blocks form a set partition of those in $S^{\omega}(\rho)$. There is a unique solution, shown (as in Example 23.2.6) by the integer array

$$F^{\omega}(\rho) = \begin{array}{cccccc} \cdot & \cdot & \cdot & \cdot & \cdot & \cdot \\ 2 & 1 & \cdot & \cdot & \cdot & \cdot \\ 3 & 2 & 3 & 1 & \cdot & \cdot \\ 4 & 3 & 4 & 2 & 4 & 1 \end{array}.$$

Proposition 23.5.4 *When ω is 1-dominant and $\rho \in \Sigma(n)$, the spike polynomial $s^{\omega}(\rho)$ is a term in $h(\rho) = \sum_{A \in \mathsf{U}(\rho)} z^{\omega}(\rho) \cdot A$, the sum of the flag polynomials for* $\mathsf{Sch}(\rho)$. *In particular, $h(\rho)$ is not hit.*

Proof The proof is similar to that of Proposition 23.2.5. Consider a factorization $s^{\omega}(\rho) = f_1 \cdots f_n$, where some monomial f_i does not involve a variable x_k which can occur in u_i for some flag in $\mathsf{Sch}(\rho)$. Since we may assign either coefficient 0 or 1 to x_k, this factorization must arise in an even number of flag polynomials. In this way, we can pair off all occurrences of $s^{\omega}(\rho)$ which arise when a flag polynomial $z^{\omega}(\rho) \cdot A$ is multiplied out in full, except those in which, for each i, all the variables x_k which can arise in u_i are involved in f_i. These are a subset of the occurrences of $s^{\omega}(\rho)$ which arise when the 'maximal' flag polynomial is multiplied out. We prove that there is a unique set partition of the 1s in the spike block $S^{\omega}(\rho)$ which satisfies the required conditions.

First consider $x_{\rho(1)}$. Since $f_{\rho(1)}$ is a power of $x_{\rho(1)}$ given by the 1s in row $\rho(1)$ of $Z^{\omega}(\rho)$, the corresponding 1s in row $\rho(1)$ of $S^{\omega}(\rho)$ must be assigned to $F_{\rho(1)}$. These occur in all the fixed columns and in $n - \rho(1)$ of the moveable columns. By definition of $S^{\omega}(\rho)$, there is a 1 in every column of this row. The variable $x_{\rho(1)}$ must also occur in f_k for $1 \leq k < \rho(1)$, so we must assign at least one 1 in row $\rho(1)$ of $S^{\omega}(\rho)$ to F_k. As there are only $\rho(1) - 1$ more 1s in row $\rho(1)$ of $S^{\omega}(\rho)$ to be assigned, we must assign exactly one of these to each F_k for $1 \leq k < \rho(1)$. Again by construction of $Z^{\omega}(\rho)$, by working along the block from right to left, we see that these 1s must correspond to (part of) a NW to SE diagonal in the moveable columns $S'(\rho) = Z'(\rho)$ of $Z^{\omega}(\rho)$. Thus there is a unique choice for the selection of powers of the variable $x_{\rho(1)}$. The result follows by induction on n, as for Proposition 23.2.5. □

Example 23.5.5 For $n = 4$, $\omega = (3,3,2,2,1,1)$ and $\rho = (2,4,3,1)$ the block $S^{\omega}(\rho)$ is shown below, together with the integer array $F^{\omega}(\rho)$ showing the special factorization of the spike $s^{\omega}(\rho)$.

$$S^{\omega}(\rho) = \begin{matrix} 0\,0\,0\,0\,0\,0 \\ 1\,1\,1\,1\,1\,1 \\ 1\,1\,0\,0\,0\,0 \\ 1\,1\,1\,1\,0\,0 \end{matrix}, \quad F^{\omega}(\rho) = \begin{matrix} \cdot\,\cdot\,\cdot\,\cdot\,\cdot\,\cdot \\ 2\,2\,2\,2\,2\,1 \\ 3\,1\,\cdot\,\cdot\,\cdot\,\cdot \\ 4\,3\,4\,1\,\cdot\,\cdot \end{matrix}.$$

Splitting $F^{\omega}(\rho)$ into its fixed and moveable columns, we obtain the arrays

$$\begin{matrix} \cdot\,\cdot\,\cdot \\ 2\,2\,2 \\ 3\,\cdot\,\cdot \\ 4\,4\,\cdot \end{matrix}, \quad \begin{matrix} \cdot\,\cdot\,\cdot \\ 2\,2\,1 \\ 1\,\cdot\,\cdot \\ 3\,1\,\cdot \end{matrix},$$

showing that the moveable columns of ω play an active role and the fixed columns a passive one in the factorization of $s^{\omega}(\rho)$. Example 23.5.7 illustrates the use of the fixed columns.

Proposition 23.5.6 *Let ω be 2-dominant. If the spike $s^{\omega}(\rho)$ is a term of a flag polynomial for* $\mathsf{Sch}(\sigma)$, *then $\sigma \le \rho$ in the Bruhat order. More generally, if $s^{\omega}(\rho)$ is a term of a polynomial in the* $\mathsf{U}(n)$*-orbit of the zip monomial $z^{\omega}(\sigma)$, then $\sigma \le \rho$.*

Proof The proof is similar to that of Proposition 23.3.6. Let B be the subblock of the zip block $Z^{\omega}(\rho)$ formed by the fixed columns. Since ω is 2-dominant, $\omega' = \omega(B)$ is 1-dominant. By replacing the entries 1 in B by boxes, we obtain the spike diagram of ρ for ω'. By Proposition 23.3.4 there exists a downward ρ-tableau for ω' of content σ if and only if $\sigma \le \rho$ in the Bruhat order.

If we apply any linear substitution $A \in \mathsf{U}(n)$ to the zip monomial $z^{\omega}(\sigma)$ and encode the factorizations of the monomials in the resulting product as above, we obtain a collection of tableaux in which the entries in each column are distinct and the entries of the spike tableau for σ appear in the same or lower rows. Thus, on restriction to the fixed columns of the corresponding block, an occurrence of $s^{\omega}(\rho)$ will yield a downward ρ-tableau of content σ. □

Example 23.5.7 As in Example 23.3.7, let $n = 4$, $\rho_1 = (2,3,1,4)$ and $\rho_2 = (2,4,3,1)$, and let $\omega = (3,3,2,2,1,1)$. Then $z_{\omega}(\rho_1) = x_1^{43}x_2^{31}x_3^{7}$ and $s_{\omega}(\rho_2) = x_2^{63}x_3^{3}x_4^{15}$ are represented by the blocks $Z^{\omega}(\rho_1)$ and $S^{\omega}(\rho_2)$ shown in Example 23.3.7. The flag polynomials $z_{\omega}(\rho_1) \cdot A$ for $A \in \mathsf{U}(\rho_1)$ do not involve x_4, and so they do not contain $s_{\omega}(\rho_2)$. However, taking A as in Example 23.5.2 with $a = b = c = d = 1$, the $\mathsf{U}(\rho_2)$-orbit of $z_{\omega}(\rho_1)$ contains

$(x_1+x_2+x_3+x_4)^{43}x_2^{31}(x_3+x_4)^7$. Several factorizations $f_1f_2f_3$ of the spike $s^\omega(\rho_2)$ occur when this product is expanded: for example, we may take $f_1 = x_2^{32}x_3^3x_4^8, f_2 = x_2^{31}, f_3 = x_4^7$, corresponding to the integer array

$$F^\omega(\rho_2) = \begin{array}{cccccc} \cdot & \cdot & \cdot & \cdot & \cdot & \cdot \\ 2 & 2 & 2 & 2 & 2 & 1 \\ 1 & 1 & \cdot & \cdot & \cdot & \cdot \\ 3 & 3 & 3 & 1 & \cdot & \cdot \end{array}$$

Selecting the fixed columns (with no replacement of the entries), we obtain the tableau obtained from the tableau T of (23.2) by removing the first column. This is a downward ρ_2-tableau of content ρ_1.

The following result is the analogue of Theorem 23.4.1 for $\mathsf{P}(n)$.

Theorem 23.5.8 *Let ω be 2-dominant, with $\deg_2 \omega = d$. Then the flag polynomials $z^\omega(\rho)\cdot A$, for $\rho \in \Sigma(n)$ and $A \in \mathsf{U}(\rho)$, are linearly independent in $\mathsf{Q}^d(n)$.*

Proof The argument is similar to that for Theorem 23.4.1, but at each stage we obtain a contradiction stating that a polynomial containing a spike is hit. The module $\langle s_\omega(\rho)\rangle$ is replaced by the module $\langle z^\omega(\rho)\rangle$ spanned by the orbit of the zip monomial $z^\omega(\rho)$ under the action of $\mathsf{U}(\rho)$, i.e. the cyclic right $\mathbb{F}_2\mathsf{U}(\rho)$-module generated by $z^\omega(\rho)$. For $\mathsf{P}(n)$ the induction over the Schubert cells is carried out in increasing Bruhat order using Propositions 23.5.4 and 23.5.6, starting with ι_n and ending with ρ_0. □

The following result is the analogue for $\mathsf{P}(n)$ of Theorem 23.4.2, and is proved in a similar way.

Theorem 23.5.9 *Let ω be 1-dominant and let $R \subseteq \Sigma(n)$ be a set of permutations associated to coset representatives for $\mathsf{W}^I(n)$ in $\mathsf{W}(n)$, where $I = I_2(\omega)$. Then the flag polynomials in $z^\omega(\rho)\cdot A$ for $\rho \in R$, $A \in \mathsf{U}(\rho)$, are linearly independent in $\mathsf{Q}^d(n)$.* □

By Proposition 16.5.6, the structure of the quotient of $\mathsf{Q}^d(n)$ dual to the submodule $\mathsf{J}^d(n)$ of $\mathsf{K}^d(n)$ is obtained by exchanging I and J, where $j \in J$ if and only if $n-j \in I$. Thus the structure of the dual module can be read from the table in Example 23.4.6 by exchanging the entries for $\mathsf{J}^d(4)$ in rows 2 and 4 and in rows 5 and 7. Dualizing Theorem 23.4.4, we have the following result, where $j \in \mathrm{tr}(I)$ if and only if $n-j \in I$.

Proposition 23.5.10 *Let ω be 1-dominant and let $(\phi^\omega)^* : \mathsf{P}^d(n) \to \mathsf{FL}(n)$ be the dual of the Crabb–Hubbuck map $\phi^\omega : \mathsf{FL}(n) \to \mathsf{DP}^d(n)$. Let $J = tr(I_2(\omega))$. Then $(\phi^\omega)^*\mathsf{P}^d(n) \cong \mathsf{FL}(n)/\sum_{i\notin J}\mathsf{FL}^{-\{i\}}(n) \cong \sum_{J\cup K=Z[n-1]}\mathsf{FL}_K(n)$ as $\mathbb{F}_2\mathsf{GL}(n)$-modules.* □

23.6 Remarks

Since $\mathsf{J}(n)$ is generated as a ring by the (2^k-1)th divided powers of elements of degree 1, it is called the 'ring of lines' in [41], where the map ϕ^ω is defined by M. C. Crabb and J. R. Hubbuck. This map is used implicitly in Chapter 10. The dual map $(\phi^\omega)^*$ is essentially the same as the map θ of J. Repka and P. S. Selick [173]. By duality, the hit polynomials are in $\mathrm{Ker}(\phi^\omega)^*$, so that $\mathrm{Im}(\phi^\omega)^*$ is a quotient module of $\mathsf{Q}^d(n)$. The first discrepancy between $\mathsf{J}^d(n)$ and $\mathsf{K}^d(n)$ occurs for $n=3$ and $d=8$ (see [7, 21, 173]), when $\dim \mathsf{K}^8(3)=15$ and $\dim \mathsf{J}^8(3)=14$. Dually, the polynomial $x^6yz+xy^6z+xyz^6$ of Singer [186] lies in $\ker(\phi^\omega)^*$ but is not hit: it is an invariant modulo hit polynomials, and so represents a 1-dimensional submodule of $\mathsf{Q}^8(3)$. J. M. Boardman [21] has shown that $\mathsf{Q}^d(3)$ and $\mathsf{K}^d(3)$ are cyclic $\mathbb{F}_2\mathsf{GL}(3)$-modules except in the cases $s\geq 5$, $t=0$ of the tables in Theorem 10.6.2.

The computed values of $\dim \mathsf{K}^d(4)$ referred to in Example 23.4.6 agree with the values of $\dim \mathsf{Q}^d(n)$ given by Kameko [108] or Sum [204, 207, 209]. A more detailed comparison shows the effect of replacing the minimal ω-sequence with a 1-dominant ω-sequence of higher 2-degree with the same I, so that $\mathsf{J}^d(4)$ is as in 23.4.6. We obtain $\mathsf{J}^d(4)=\mathsf{K}^d(4)$ except for $\omega = (3,2,2,1,\ldots,1)$ (with at least three 1s) or $(3,\ldots,3,2,1,1)$ (with at least three 3s), when the discrepancy remains as in lines 5 and 6 of 23.4.6.

Tran Ngoc Nam [155] has extended the results of [41] in another direction, and in particular he has given a different proof that $\mathsf{J}^d(n)\cong\mathsf{FL}(n)$ when d is 2-dominant. We discuss these results in the next chapter.

This chapter is based on [221]. In the final sentences of Propositions 2.10 and 2.12 of [221], the 'if' part of the 'if and only if' statement is not proved, but it is not needed for the rest of the argument. We wish to apologize for the inaccurate statements in [221].

24

Partial flags and $\mathsf{J}(n)$

24.0 Introduction

In this chapter we find sufficient conditions for the local d-spike module $\mathsf{J}^{\omega}(n)$ to have a submodule isomorphic to the indecomposable summand $\mathsf{FL}_J(n)$ of the flag module $\mathsf{FL}(n)$, where ω is decreasing and $J \subseteq I(\omega) \subseteq Z[n-1]$. In place of the recursive argument of Chapter 23 using Schubert cells, we use the direct sum decomposition $\mathsf{FL}^I(n) \cong \bigoplus_{J \subseteq I} \mathsf{FL}_J(n)$ of Theorem 16.5.3.

Theorem 17.6.2 identifies an element s_J of the 0-Hecke algebra $\mathsf{H}_0(n)$ such that $s_J(\mathsf{B}(n)) \in \mathsf{FL}(n)$ generates the socle of the summand $\mathsf{FL}_J(n)$. We consider the image of $s_J(\mathsf{B}(n))$ under the Crabb–Hubbuck map $\phi^{\omega} : \mathsf{FL}(n) \to \mathsf{DP}^d(n)$. If $\phi^{\omega}(s_J(\mathsf{B}(n))) \neq 0$, then it follows that the restriction of ϕ^{ω} to $\mathsf{FL}_J(n)$ is injective. Since the image of ϕ^{ω} is in $\mathsf{DP}^{\geq\omega}(n)$, we obtain a map $\mathsf{FL}(n) \to \mathsf{DP}^{\omega}(n)$ by composition of ϕ^{ω} with the projection on $\mathsf{DP}^{\omega}(n)$, and we denote this map also by ϕ^{ω}. Thus the condition $\phi^{\omega}(s_J(\mathsf{B}(n))) \neq 0$ will be taken to mean that this d-polynomial has a term with ω-sequence ω.

The element s_J is the sum of the basis elements f_W of $\mathsf{H}_0(n)$ over permutation matrices W in the coset $\mathsf{W}^J(n)W_0$ of block anti-diagonal matrices in $\mathsf{W}(n)$, where the size of the blocks is given by the composition of n associated to J. As $f_W(\mathsf{B}(n))$ is the sum of the flags in the Schubert cell $\mathsf{Sch}(W)$, our task is to show that a particular large sum of flag d-polynomials is nonzero, and not, as in Chapter 23, that a set of smaller sums is linearly independent. To help comparison of the two methods, the case $I(\omega) = \{2\}$ is treated in both ways in Section 24.2.

If $\rho \in \Sigma(n)$ is the permutation corresponding to W, we use the notion of ρ-factorization (Definition 23.2.7) to detect d-monomials in $\phi^{\omega}(f_W(\mathsf{B}(n)))$, under suitable conditions on ω. In particular, the lower bound on $\mathsf{J}^d(n)$ of Theorem 23.4.4 can be obtained in this way by taking ω 1-dominant and $I_2(\omega) \cup J = Z[n-1]$.

In the 1-dominant case $I(\omega) = Z[n-1]$, the zip d-monomial $z_\omega(\rho)$ is defined in Section 23.2. In Section 24.3 we construct $z_\omega(\rho) \in \mathsf{DP}^\omega(n)$, and a zip array $F^\omega(\rho)$ giving a ρ-factorization of $z_\omega(\rho)$, for all $I(\omega)$, provided that the descent set $\mathrm{des}(\rho) \subseteq I(\omega)$. The entries of $F^\omega(\rho)$ which correspond to 1s in the $(0,1)$-block $Z_\omega(\rho)$ representing $z_\omega(\rho)$ are positive integers indicating the factors, while the 0s in $Z_\omega(\rho)$ are replaced by dots in $F^\omega(\rho)$. When $I(\omega) = Z[n-1]$, $z_\omega(\rho)$ coincides with the zip d-monomial of Definition 23.2.2.

In Section 24.2 we discuss the 'Grassmannian' case $I(\omega) = \{k\}$, when the partial flags are subspaces of dimension k in $\mathsf{V}(n)$. The general case follows in Section 24.3, the main result being Theorem 24.3.9. In all cases, our aim is to show that the sum $\phi^\omega(s_J(\mathsf{B}(n)))$ of flag d-polynomials contains a term $z_\omega(\rho)$, where ρ is the permutation corresponding to the top Schubert cell of $\mathsf{FL}^{I(\omega)}(n)$. This is the unique permutation of maximal length such that $\mathrm{des}(\rho) = I(\omega)$.

In Section 24.4 we apply Theorem 24.3.9 to the case $n = 4$ of the hit problem, so as to obtain a lower bound for $\dim \mathsf{J}^\omega(4)$ when the terms of ω have sufficient multiplicities, together with the structure of $\mathsf{J}^\omega(4)$ as a direct sum of the modules $\mathsf{FL}_J(4)$. We also tabulate similar results for smaller values of ω, when quotient modules of $\mathsf{FL}_J(4)$ appear, as in the 'tail' and 'head' cases of Sections 6.7 and 6.8.

When $\omega = (n-1, n-2, \ldots, 1)$ and $J = I(\omega) = Z[n-1]$, $\mathsf{FL}_J(n) \cong \mathsf{St}(n)$, the Steinberg module. Since $s_J = f_{W_0}$, the generator of $\mathsf{FL}_J(n)$ is the sum of the flags in the top Schubert cell $\mathsf{Sch}(W_0)$. The image of the top flag in $\mathsf{Sch}(W_0)$ under ϕ^ω is the d-polynomial

$$(v_1 + \cdots + v_n)^{(2^{n-1}-1)}(v_1 + \cdots + v_{n-1})^{(2^{n-2}-1)} \cdots (v_1 + v_2 + v_3)^{(3)}(v_1 + v_2),$$

whose formal expansion contains the d-spike $s = v_1^{(2^{n-1}-1)} v_2^{(2^{n-2}-1)} \cdots v_{n-1}$ as the product

$$s = (v_1^{(2^{n-2})} v_2^{(2^{n-3})} \cdots v_{n-1})(v_1^{(2^{n-3})} v_2^{(2^{n-4})} \cdots v_{n-2}) \cdots (v_1^{(2)} v_2) v_1.$$

This is the unique ρ_0-factorization of s, where $\rho_0 = (n, n-1, \ldots, 1)$, and all other occurrences of s in $\phi^\omega(s_J(\mathsf{B}(n)))$ cancel in pairs. For example, in the case $n = 3$, $s = v_1^{(3)} v_2$ occurs as $v_1^{(3)} \cdot v_2$ in the four flag d-polynomials $(v_1 + av_2 + v_3)^{(3)}(bv_1 + v_2)$ for $a, b \in \mathbb{F}_2$, but as $v_1^{(2)} v_2 \cdot v_1$ only in $(v_1 + v_2 + v_3)^{(3)}(v_1 + v_2)$, and in the case $n = 4$, $s = v_1^{(7)} v_2^{(3)} v_3$ with the ρ_0-factorization $v_1^{(4)} v_2^{(2)} v_3 \cdot v_1^{(2)} v_2 \cdot v_1$. It follows that $\phi^\omega(s_J(\mathsf{B}(n))) \neq 0$. This gives another proof that the Steinberg module $\mathsf{St}(n)$ embeds in $\mathsf{J}^\omega(n)$. This case is exceptional, as the zip d-monomial is a d-spike.

Section 24.1 establishes a preliminary result which allows us to fix attention on the minimal ω in the main argument. For example, using Proposition 24.1.1 the argument above implies that $\mathsf{St}(n) \subseteq \mathsf{J}^{\omega}(n)$ when ω is 1-dominant. More generally, we relate $\mathsf{J}^{\omega}(n)$ and $\mathsf{J}^{\omega'}(n)$, where ω' is obtained from ω by deleting some repeated entry, by showing that the 'd-duplication' map $\varepsilon_* : \mathsf{J}^{\omega}(n) \to \mathsf{J}^{\omega'}(n)$, which removes a repeated column, is a surjection of cyclic $\mathbb{F}_2\mathsf{GL}(n)$-modules.

In Section 24.5 we apply Theorem 24.3.10 to find counterexamples for $n \geq 5$ to the conjecture of Kameko that $\dim \mathsf{Q}^d(n) \leq \dim \mathsf{FL}(n)$ for all d. The image of $\phi^{\omega} : \mathsf{FL}^I(n) \to \mathsf{DP}^d(n)$ lies in the submodule $\mathsf{DP}^{\geq \omega}(n)$, and as $z_{\omega}(\rho) \in \mathsf{DP}^{\omega}(n)$, the composition of ϕ^{ω} with the projection on to $\mathsf{DP}^{\omega}(n)$ is also injective. Thus for suitable values of d, we can combine the lower bounds for $\mathsf{Q}^{\omega}(n)$ for several sequences ω with 2-degree d to show that $\dim \mathsf{Q}^d(n) > \dim \mathsf{FL}(n)$. We use the permutation modules on maximal partial flags which omit subspaces of only one dimension. There are $n-1$ such modules, each of dimension $\dim \mathsf{FL}(n)/3$. By choosing d so that there are sequences ω of 2-degree d corresponding to each of these, we obtain $\dim \mathsf{Q}^d(n) \geq (n-1)\dim \mathsf{FL}(n)/3$.

24.1 d-spikes and d-duplication

By Definition 9.4.9, for $\omega \in \mathsf{Dec}(n)$ the local d-spike module $\mathsf{J}^{\omega}(n)$ is the submodule of $\mathsf{K}^{\omega}(n)$ generated by d-spikes. All d-spikes in $\mathsf{K}^{\omega}(n)$ are obtained by permuting exponents of the d-spike $s_{\omega}(\iota_n)$ whose exponents are in decreasing order. Hence $\mathsf{J}^{\omega}(n)$ is the cyclic $\mathbb{F}_2\mathsf{GL}(n)$-module generated by $s_{\omega}(\iota_n)$. Since the image of $\phi^{\omega} : \mathsf{FL}(n) \to \mathsf{K}(n)$ is a submodule of $\mathsf{K}^{\geq \iota\omega}(n)$ by Proposition 23.1.4, $\mathsf{J}^{\omega}(n)$ is the image of $\mathrm{Im}(\phi^{\omega})$ under the projection of $\mathsf{K}^{\geq \iota\omega}(n)$ on to $\mathsf{K}^{\omega}(n)$.

The d-duplication map δ_* (Definition 9.5.7) removes the first column of a block if its first two columns are identical, and maps it to zero otherwise. When $\mu(d) = n-1$ and $\omega_1 = \omega_2 = n-1$, δ_* restricts to a $\mathbb{F}_2\mathsf{GL}(n)$-module map $\mathsf{K}^{\omega}(n) \to \mathsf{K}^{\omega'}(n)$, where $\omega = (n-1)|\omega'$. Since δ_* maps d-spikes to d-spikes, it further restricts to a $\mathbb{F}_2\mathsf{GL}(n)$-module map $\mathsf{J}^{\omega}(n) \to \mathsf{J}^{\omega'}(n)$. We shall show that such a map, and indeed a surjection, is defined more generally by removing any duplicated column from a d-spike. In Section 24.4 we apply Proposition 24.1.1 to study small ω-sequences when $n = 4$.

Proposition 24.1.1 *Let* $\omega \in \mathsf{Dec}(n)$ *have a repeated entry* $\omega_r = \omega_{r+1}$ *for some* $r \geq 1$, *and let* ω' *be the sequence obtained by deleting* ω_r. *Then there is a surjection of cyclic* $\mathbb{F}_2\mathsf{GL}(n)$*-modules* $\varepsilon_* : \mathsf{J}^{\omega}(n) \longrightarrow \mathsf{J}^{\omega'}(n)$ *which maps the* *generator $s_\omega(\iota_n)$ of $\mathsf{J}^\omega(n)$ to the generator $s_{\omega'}(\iota_n)$ of $\mathsf{J}^{\omega'}(n)$. For $f \in \mathsf{J}^\omega(n)$, $\varepsilon_*(f) = f'$, where the blocks B' for f' are obtained by deleting the rth column of a block B for f if column $r+1$ repeats column r, and by deleting the block B otherwise.*

Proof If the block B' represents a d-monomial $m' \in \mathsf{DP}^{\omega'}(n)$, then the block B obtained by repeating the rth column of B' represents a d-monomial $m \in \mathsf{DP}^\omega(n)$, and all blocks B in which column $r+1$ repeats column r are obtained in this way. Let $\lambda = \omega^{\mathrm{tr}} = (\lambda_1, \dots, \lambda_n)$ and let $\lambda' = (\omega')^{\mathrm{tr}}$. Then $\lambda_i' = \lambda_i - 1$ for $1 \le i \le \omega_r$ and $\lambda_i' = \lambda_i$ for $\omega_r + 1 \le i \le n$. We have $s_\omega(\iota_n) = \prod_{i=1}^n v_i^{(2^{\lambda_i}-1)}$ and $s_{\omega'}(\iota_n) = \prod_{i=1}^n v_i^{(2^{\lambda_i'}-1)}$. A general flag d-polynomial $\phi^\omega(X)$ is of the form $\prod_{i=1}^n u_i^{(2^{\lambda_i}-1)}$, where $u_1, \dots, u_n$ are the rows of a matrix in the right coset of $\mathsf{B}(n)$ in $\mathsf{GL}(n)$ corresponding to the flag X.

A relation between the flag d-polynomials for flags X in $\mathsf{V}(n)$ can be written as $\sum_{X \in \mathcal{F}} \phi^\omega(X) = 0$, where $\mathcal{F}$ is a set of flags. Using the d-multinomial theorem $(u_1 + \cdots + u_k)^{(r)} = \sum_{s_1+\cdots+s_k=r} u_1^{(s_1)} \cdots u_k^{(s_k)}$, we may expand each flag d-polynomial $\phi^\omega(X)$ as a sum of d-monomials $m \in \mathsf{DP}^{\ge\omega}(n)$. Let S denote the sum of blocks obtained in this way from all $X \in \mathcal{F}$. Since S represents a relation in $\mathsf{DP}^{\ge\omega}(n)$, each block occurs in S an even number of times. In particular, if $\omega(m) = \omega$ and m corresponds to a block B with identical rth and $(r+1)$st columns, then m occurs in S an even number of times. By deleting the rth columns of all blocks B corresponding to such d-monomials, we obtain a sum S' of blocks B' which have ω-sequence ω', each occurring an even number of times in S'.

The blocks B' in S' correspond to d-monomials m' obtained by expanding the d-polynomials $\phi^{\omega'}(X)$ for flags $X \in \mathcal{F}$ using the d-multinomial theorem. Hence $\sum_{X \in \mathcal{F}} \phi^{\omega'}(X) = 0$. Thus each relation in the cyclic $\mathbb{F}_2\mathsf{GL}(n)$-module $\mathsf{J}^\omega(n)$ corresponds to a relation in the cyclic $\mathbb{F}_2\mathsf{GL}(n)$-module $\mathsf{J}^{\omega'}(n)$, and so ε_* is a surjection of $\mathbb{F}_2\mathsf{GL}(n)$-modules. □

Proposition 24.1.2 *Let $\omega' \in \mathsf{Dec}(n)$ and let $I = I(\omega')$. Then the $\mathbb{F}_2\mathsf{GL}(n)$-module $\mathsf{J}^{\omega'}(n)$ is a quotient of the partial flag module $\mathsf{FL}^I(n)$. In particular, when λ is a column 2-regular partition and $\omega' = \lambda^{tr}$, the restricted Weyl module $\Delta(\lambda, n)$ is a quotient of the partial flag module $\mathsf{FL}^I(n)$.*

Proof By Proposition 24.3.10, $\phi^\omega : \mathsf{FL}^I(n) \to \mathsf{DP}^d(n)$ is injective when the terms of $\omega \in \mathsf{Dec}_d(n)$ are the elements of I, each with sufficiently large multiplicity. The generator $s_\omega(\iota_n)$ is a minimal spike in degree d and so $\mathsf{J}^\omega(n) \cong \mathrm{Im}(\phi^\omega) \cong \mathsf{FL}^I(n)$. The result follows from Proposition 24.1.1 by iteration. The last statement follows from Theorem 20.5.2, since $\mathsf{J}^{\omega'}(n) = \mathsf{K}^{\omega'}(n) \cong \Delta(\lambda, n)$ when $\omega' = \lambda^{\mathrm{tr}}$ is strictly decreasing. □

24.2 The Grassmannian case

In this section we give examples which illustrate Theorem 24.3.9 in the 'Grassmannian' case $I = I(\omega) = \{k\}$, so that $\omega = (k, \dots, k)$ of length ℓ, and the partial flags are subspaces of $\mathsf{V}(n)$ of fixed dimension k. Thus $J = \{k\}$ or $\emptyset$, and $\mathsf{FL}^I(n) \cong \mathsf{FL}_I(n) \oplus \mathsf{FL}_\emptyset(n)$, where $\mathsf{FL}_\emptyset(n)$ is the 1-dimensional submodule generated by the sum of all k-planes and the elements of $\mathsf{FL}_I(n)$ are sums of an even number of k-planes. In this case Theorem 24.3.9 states that $\mathsf{FL}_J(n) \subseteq J^\omega(n)$ for $\ell \geq n-1$ if $J = \{k\}$ and for $\ell \geq n$ if $J = \emptyset$, so that $\mathsf{FL}^I(n) \subseteq J^\omega(n)$ for $\ell \geq n$.

We begin by defining the zip d-monomials $z_\omega(\rho)$ and their associated factorizations. Since $\mathrm{des}(\rho) \subseteq I = \{k\}$, $\rho(1) < \cdots < \rho(k)$ and $\rho(k+1) < \cdots < \rho(n)$. The zip d-monomial $z_\omega(\rho)$ is represented by a $n \times \ell$ block $Z^\omega(\rho)$ with k entries 1 in each column. As in Section 23.2, we represent the factorization $f_1 f_2 \cdots f_k$ of $z_\omega(\rho)$ by an array $F^\omega(\rho)$, replacing the entries 1 in the block $Z^\omega(\rho)$ by integers $1 \leq i \leq k$ corresponding to the factors f_i, and the entries 0 in $Z^\omega(\rho)$ by dots. Thus every column of $F^\omega(\rho)$ contains each integer $1, 2, \dots, k$ once, and $n-k$ dots.

Definition 24.2.1 Given $1 \leq k \leq n$, $\omega = (k, \dots, k)$ of length $\ell \geq n-1$ and $\rho \in \Sigma(n)$ with $\mathrm{des}(\rho) \subseteq \{k\}$, we define the $n \times \ell$ **zip array** $F^\omega(\rho)$ as follows. For $j = 1, 2, \dots, k$, enter j in rows $< \rho(j)$, omitting those rows among $\rho(1), \dots, \rho(j-1)$ which precede row $\rho(j)$, diagonally from the right, in columns $\ell - j + 1, \ell - j, \dots$, and enter j in row $\rho(j)$ in all other columns. Finally enter dots in all other positions.

The **zip block** $Z^\omega(\rho)$ is obtained by replacing the integer entries of $F^\omega(\rho)$ by 1s and the dots by 0s, and represents the **zip d-monomial** $z_\omega(\rho)$. The last $n-1$ columns of $F^\omega(\rho)$ are the 'moveable' columns, and the first $\ell - n + 1$ columns are the 'fixed' columns. The definition implies that, in the fixed columns, j appears in row $\rho(j)$ for $1 \leq j \leq k$, and dots appear in the remaining rows. The entries j in row $\rho(j)$ are 'normal' entries, and the entries j in rows $< \rho(j)$ are 'raised' entries. Thus the fixed columns contain only normal entries, while the moveable columns can contain entries of both kinds. Since raised entries correspond to pairs (i,j) with $i < j$ and $\rho(i) > \rho(j)$, the total number of raised entries is $\mathrm{len}(\rho)$.

Example 24.2.2 Let $n = 4$, $I(\omega) = \{2\}$, $\omega = (2,2,2,2)$. Then $\rho = (3,4,1,2)$ is the maximal permutation in $\Sigma(4)$ with descent set $I(\omega)$, and the zip array is

$$F^{\omega}(\rho) = \begin{array}{cccc} \cdot & \cdot & 2 & 1 \\ \cdot & 2 & 1 & \cdot \\ 1 & 1 & \cdot & \cdot \\ 2 & \cdot & \cdot & 2 \end{array}. \tag{24.1}$$

The numbering of the array indicates the ρ-factorization $f_1 f_2$ of the zip monomial $z_\omega(\rho)$, which occurs in the flag d-polynomial $(v_1+v_2+v_3)^{(15)}(v_1+v_2+v_4)^{(15)}$ for the maximal flag in $\mathsf{Sch}(\rho)$, namely $f_1 = v_1^{(8)}v_2^{(4)}v_3^{(3)}$, $f_2 = v_1^{(4)}v_2^{(2)}v_4^{(9)}$. We show that $z_\omega(\rho) = f_1 f_2$ does not occur in the sum of the flag d-polynomials $\phi^\omega(X)$ for the flags X in any other Schubert cell $\mathsf{Sch}(\sigma)$. Since the $(1,1)$ and $(2,1)$ entries of the array are dots, $z_\omega(\rho)$ does not occur in the formal expansion of $\phi^\omega(X)$ if $\sigma(1) = 1$ or 2. If $\sigma(1) = 4$, then there is no σ-factorization of $F^\omega(\rho)$, because normal entries 4 must be placed in positions $(4,1)$ and $(4,4)$, leaving only columns 2 and 3 to accommodate three raised entries 4. Hence $\sigma(1) = 3$, and, since row 4 of the array contains 4s, $\sigma(2) = 4$. However $F^\omega(\rho)$ does not have a σ-factorization for $\sigma = (3,4,2,1)$, because the star in row 3 of the representative matrix

$$\begin{pmatrix} * & * & 1 & 0 \\ * & * & 0 & 1 \\ * & 1 & 0 & 0 \\ 1 & 0 & 0 & 0 \end{pmatrix}$$

can be 0 or 1, so that the flag d-polynomials for $\mathsf{Sch}(\sigma)$ cancel in pairs in the sum. We conclude that the image of the sum of all the flag d-polynomials under $\phi^\omega : \mathsf{FL}(4) \to \mathsf{J}^\omega(4)$ is nonzero, and so the 1-dimensional summand $\mathsf{FL}_\emptyset(4)$ of $\mathsf{FL}(4)$ appears in $\mathsf{J}^\omega(4)$.

A similar argument applies to $\mathsf{FL}_2(4)$. The socle of $\mathsf{FL}_2(4)$ is generated by $s_2(\mathsf{B}(4))$, the sum of the flag d-polynomials for the Schubert cells $\mathsf{Sch}(\sigma)$, where $\sigma = (3,4,1,2)$, $(4,3,1,2)$, $(3,4,2,1)$ or $(4,3,2,1)$ (see (17.2)). Hence $\mathsf{FL}_2(4) \subseteq \mathsf{J}^\omega(4)$, and so the partial flag module $\mathsf{FL}^2(4) \subseteq \mathsf{J}^\omega(4)$.

In the next example, we outline an alternative argument that $\mathsf{FL}^2(4) \subseteq \mathsf{J}^\omega(4)$ for $\omega = (2,2,2,2)$ by induction over the Schubert cells, as in Chapter 23.

Example 24.2.3 The following diagram shows the zip arrays $F^\omega(\rho)$ for the Schubert cells $\mathsf{Sch}(\sigma)$ of $\mathsf{FL}^2(4)$ for $\omega = (2,2,2,2)$ and $\mathrm{des}(\sigma) \subseteq I(\omega) = \{2\}$ (see Example 16.4.12). The top array corresponds to the factorization

$z_\omega(\rho) = f_1 f_2$ of the zip d-monomial for the maximal Schubert cell $\mathsf{Sch}(\rho)$, where $\rho = (3,4,1,2)$. Note that a factor $v_3^{(s)}$ with s odd can appear only in f_1, and a factor $v_4^{(t)}$ with t odd only in f_2, because the $(3,1)$ and $(4,1)$ entries of $Z^\omega(\rho)$ are 1s. No flag d-polynomial $\phi^\omega(X)$ for $X \in \mathsf{Sch}(\sigma)$, $\sigma \neq \rho$, involves an odd exponent of v_3 in one factor and of v_4 in the other. For example, the next array corresponds to $\sigma = (2,4,1,3)$, for which the top flag d-polynomial is $(v_1+v_2)^{(15)}(v_1+v_3+v_4)^{(15)}$.

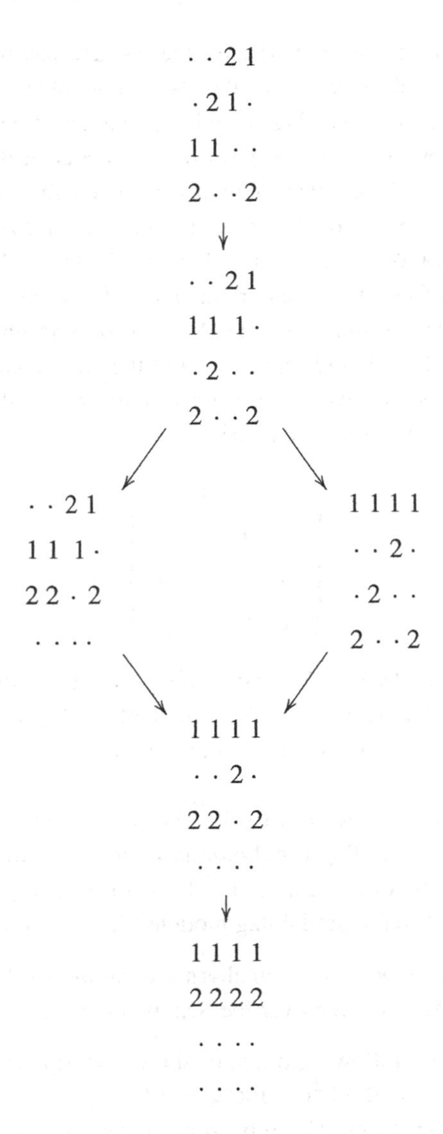

More generally, the inductive argument over the Schubert cells requires that, as in Proposition 23.3.6, the zip d-monomial $z_\omega(\rho)$ does not appear in $\phi^\omega(X)$ for a flag $X \in \mathsf{Sch}(\sigma)$ if $\sigma < \rho$. A similar argument using the first column of the zip array shows that this condition is satisfied for all pairs of permutations giving Schubert cells in $\mathsf{FL}^2(4)$. We first observe that the first column of $Z^\omega(\rho)$ has entries 1 in rows $\rho(j)$ for $1 \le j \le k$, and entries 0 in all other rows, and that in the zip array $F^\omega(\rho)$ the entry 1 in row $\rho(j)$ is replaced by j. This argument can be extended to the general Grassmannian case $\mathsf{FL}^k(n)$, by taking $\omega = (k,\ldots,k)$ of length $\ell \ge n$ and using any fixed column in the zip array for the top Schubert cell $\mathsf{Sch}(\rho)$. For the variables $v_{\rho(1)}, v_{\rho(2)}, \ldots, v_{\rho(k)}$ to appear in different factors of the flag d-polynomial $\phi^\omega(\sigma)$, where $\mathrm{des}(\sigma) \subseteq \{k\}$, σ must satisfy $\sigma(j) \ge \rho(j)$ for $1 \le j \le k$, and so $\sigma \ge \rho$ in the Bruhat order. If $\ell = n-1$, then this argument fails, as there is no fixed column. In this case, the method of Example 24.2.2 can be used to show that $\mathsf{FL}_k(n) \subseteq \mathsf{J}^\omega(n)$.

24.3 Zip d-monomials for $\mathsf{FL}_J(n)$

In this section we define the zip array $F^\omega(\rho)$ for a general nonempty subset $I = I(\omega)$ of $Z[n-1]$ and a permutation $\rho \in \Sigma(n)$ whose descents are in I, and show that $F^\omega(\rho)$ represents the unique ρ-factorization of the zip monomial $z_\omega(\rho)$. For $I = \{i_1, i_2, \ldots, i_r\} \subseteq Z[n-1]$, where $i_1 < i_2 < \cdots < i_r$, we regard $F^\omega(\rho)$ as the concatenation of subblocks F_j for $1 \le j \le r$, where $\omega(F_j) = (i_j, \ldots, i_j)$.

As in Definition 24.2.1, the d-monomial $z_\omega(\rho)$ is represented by the $(0,1)$-block $Z^\omega(\rho)$ obtained from $F^\omega(\rho)$ by replacing integer entries by 1s and dots by 0s. We denote the associated composition of n by $C = (c_1, \ldots, c_{r+1})$, so that $c_j = i_j - i_{j-1}$ for $1 \le j \le r+1$, where $i_0 = 0$ and $i_{r+1} = n$, and define $A = (a_1, \ldots, a_r)$, where $a_j = i_{j+1} - i_{j-1} = c_{j+1} + c_j$.

An integer entry of $F^\omega(\rho)$ is either 'raised' or 'normal'. All normal entries k are in row $\rho(k)$, and one raised entry appears in row $\rho(\ell)$ when $\rho(\ell) < \rho(k)$ and $\ell > k$. Thus the total number of raised entries is $\mathrm{len}(\rho)$, the number of pairs which are reversed by ρ. In Example 24.2.3, the raised entries are the entries in rows $\rho(3) = 1$ and $\rho(4) = 2$, and the normal entries are the entries in rows $\rho(1) = 3$ and $\rho(2) = 4$, where $\rho = (3,4,1,2)$. A row of $F^\omega(\rho)$ can contain both raised and normal entries.

In practice, it is convenient to insert the raised entries k from top to bottom and from right to left. For this reason we number the columns of F_j from right to left. As before, columns of F_j are either 'fixed' or 'moveable', and all raised entries are in the moveable columns.

Definition 24.3.1 With notation as above, let $\omega \in \mathsf{Dec}(n)$ have terms i_j of multiplicity $m_j \geq a_j - 1$ for $1 \leq j \leq r$, and let $\rho \in \Sigma(n)$ be a permutation with descent set $\mathrm{des}(\rho) \subseteq I$. The **zip array** $F^{\omega}(\rho) = F_r|\cdots|F_1$ is the concatenation of subarrays F_j, where the columns of F_j are numbered from 1 to m_j from the *right* and contain one entry k for $1 \leq k \leq i_j$ and $n - i_j$ dots. Columns $\leq a_j - 1$ of F_j are 'moveable' and all others are 'fixed'. For $i_{j-1} < k \leq i_j$, k is entered once in each column of $F_{j'}$ for $j' \geq j$, so that:

(Raised entries) k is entered once in row $\rho(\ell) < \rho(k)$ for $\ell > k$, in increasing order of rows, starting in column $k - i_{j-1}$ of F_j, and continuing in consecutive columns from the right, omitting columns $< c_{j'}$ of $F_{j'}$ for $j' > j$ and all fixed columns;

(Normal entries) k is entered in row $\rho(k)$ in all other columns of $F_{j'}, j' \geq j$.

Example 24.3.2 Let $n = 8$ and $I = I(\omega) = \{2,3,6\}$, so that $C = (2,1,3,2)$ and $A = (3,4,5)$. Let $\omega = \{6^5, 3^4, 2^3\}$ and $\rho = (7,8,6,3,4,5,1,2)$. Thus $\mathrm{des}(\rho) = I$ and $\mathsf{Sch}(\rho)$ is the maximal Schubert cell in $\mathsf{FL}^I(n)$. Then

$$F^{\omega}(\rho) = \begin{array}{ccccc|cccc|ccc}
\cdot & \cdot & 6 & 5 & 4 & \cdot & \cdot & \cdot & 3 & \cdot & 2 & 1 \\
\cdot & 6 & 5 & 4 & \cdot & \cdot & \cdot & 3 & 2 & \cdot & 1 & \cdot \\
4 & 4 & 4 & \cdot & \cdot & \cdot & 3 & 2 & 1 & \cdot & \cdot & \cdot \\
5 & 5 & 3 & \cdot & 5 & \cdot & 2 & 1 & \cdot & \cdot & \cdot & \cdot \\
6 & 3 & 2 & 6 & 6 & \cdot & 1 & \cdot & \cdot & \cdot & \cdot & \cdot \\
3 & 2 & 1 & 3 & 3 & 3 & \cdot & \cdot & \cdot & \cdot & \cdot & \cdot \\
1 & 1 & \cdot & 1 & 1 & 1 & \cdot & \cdot & \cdot & 1 & \cdot & \cdot \\
2 & \cdot & \cdot & 2 & 2 & 2 & \cdot & \cdot & \cdot & 2 & \cdot & 2
\end{array}.$$

In this example $m_j = a_j$ for $j = 1,2,3$, so that each subarray F_j has one fixed column, but in general F_j can have several identical fixed columns, or none.

Proposition 24.3.3 *The zip array $F^{\omega}(\rho)$ is well defined, since*

(1) *there are sufficient columns available for the raised entries,*
(2) *no position is occupied by raised entries with different values,*
(3) *no position is occupied by both a raised entry and a normal entry.*

Proof For (1), if there is a raised entry k in row $\rho(\ell)$ then $\ell > k$, so the number of raised entries k is $\leq n - k$. We show that the number of columns available for raised entries k is $n - k$. For $i_{j-1} < k \leq i_j$, entries k appear in $F_{j'}$ for $j' \geq j$. The subarray F_j has $a_j - 1$ moveable columns, and raised entries k do not appear in the first $k - i_{j-1} - 1$ columns (nor in the fixed columns). Hence the number of columns of F_j available for raised entries k is $(a_j - 1) - (k - i_{j-1} - 1) = i_{j+1} - k$. For $j' > j$, raised entries k do not appear in the first $c_{j'} - 1$ columns nor in the fixed columns, so the number of available columns is $a_{j'} - c_{j'} = c_{j'+1}$. Thus the total number of available columns is $(i_{j+1} - k) + \sum_{j'>j} c_{j'+1} = (i_{j+1} - k) + (n - i_{j+1}) = n - k$.

For (2), we observe that successive raised entries of the same integer are inserted in earlier columns, the raised entries in each row and in each column form a decreasing sequence.

For (3), we show that if a raised entry k appears in a row $\rho(\ell) < \rho(k)$ in column c of the array, and if ℓ also appears in column c, then this entry ℓ is also a raised entry, and $\ell > k$ by definition of the raised entries. Suppose that the raised entry k in column c is the $(s+1)$th raised entry k, working from right to left. Then there are s integers i such that $k < i$, $\rho(k) > \rho(i)$ and $\rho(i) < \rho(\ell)$. Since the number of integers i such that $k < i \leq \ell$ is $\ell - k$, there are $\geq s - \ell + k$ integers i such that $k < \ell < i$ and $\rho(i) < \rho(\ell) < \rho(k)$. In these cases, there is a raised entry ℓ in row $\rho(i)$.

We prove that one of these raised entries ℓ is in column c. Since $k < \ell$ and $\rho(k) > \rho(\ell)$, there is at least one descent of ρ between k and ℓ. Thus, working from right to left, raised entries k begin in a subarray earlier than raised entries ℓ. In this situation, if there is a raised entry k in a column c' where ℓ also appears, then column c' is also available for a raised entry ℓ. In particular, this applies when $c' = c$ or when column c' is to the left of column c. Since the number of columns available for raised entries k is $n - k$, and the corresponding number for raised entries ℓ is $n - \ell$, and since column c is used for the sth raised entry k, column c is available for the $(s - \ell + k)$th raised entry ℓ. Since the array contains $\geq s - \ell + k$ raised entries ℓ, a raised entry ℓ occurs in column c. □

By construction, the array $F^{\omega}(\rho)$ satisfies the conditions of Definition 23.2.7. Hence $F^{\omega}(\rho)$ represents a ρ-factorization of $z_{\omega}(\rho)$. We wish to prove that this ρ-factorization is unique.

Example 24.3.4 In Example 24.3.2, X is the maximal flag in $\mathsf{Sch}(\rho)$, where $\rho = (7,8,6,3,4,5,1,2)$. Then $\phi^{\omega}(X) = \prod_{i=1}^{6} u_i^{(2^{\lambda_i}-1)}$ where $u_1 = v_1 + v_2 + v_3 + v_4 + v_5 + v_6 + v_7$, $u_2 = v_1 + v_2 + v_3 + v_4 + v_5 + v_6 + v_8$, $u_3 = v_1 + v_2 + v_3 + v_4 + v_5 + v_6$, $u_4 = v_1 + v_2 + v_3$, $u_5 = v_1 + v_2 + v_4$ and $u_6 = v_1 + v_2 + v_5$ and $\lambda = \omega^{\mathrm{tr}} = \{12^2, 9, 5^3\}$. We wish to show that there is a unique factorization $z_{\omega}(\rho) =$

$\prod_{j=1}^{6} f_j$, where f_j is a d-monomial which appears in u_j and which involves all the variables v_i occurring in u_j. We show by induction on j that f_j must be represented by the entries numbered j in $F^{\omega}(\rho)$. For $j = 1$, the variable v_7 occurs only in u_1, and so all entries in row 7 must be assigned to f_1. Each of the remaining variables $v_1, \ldots, v_6$ occurring in u_1 are to be involved in f_1, and only six columns of the array remain. It follows by working from right to left that we must assign the diagonal entries marked 1 to f_1. A similar argument is repeated for $j = 2, 3, \ldots 6$.

Proposition 24.3.5 *Let* $I(\omega) = \{i_1, \ldots, i_r\} \subseteq Z[n-1]$, $\omega \in \mathsf{Dec}(n)$ *with terms* i_j *of multiplicity* $m_j \geq a_j = i_{j+1} - i_{j-1}$ *for* $1 \leq i \leq r$, *and let* $\rho \in \Sigma(n)$ *with* $des(\rho) \subseteq I$. *Then the zip array* $F^{\omega}(\rho)$ *gives the unique* ρ*-factorization of* $z_{\omega}(\rho)$.

Proof We argue as in Proposition 23.2.5. Starting with the (0,1)-block $Z^{\omega}(\rho)$, we wish to replace each entry 1 by an integer k such that $1 \leq k \leq n$ in accordance with Definition 23.2.7. We first consider row $\rho(1)$, and observe that all integer entries in this row must be normal entries 1. We also have to place at least one raised entry 1 in each of rows $1, 2, \ldots, \rho(1) - 1$. As there remain only $\rho(1) - 1$ columns in which these can be placed, we must choose exactly one of these from each row. Again by construction of $Z^{\omega}(\rho)$, by working along the block from right to left we see that these 1s can only be chosen as (part of) a NE to SW 'broken' diagonal in the moveable columns of $Z^{\omega}(\rho)$. Thus there is a unique way to place the entries 1. Next we consider row $\rho(2)$. All integer entries in this row must be 2s, except for one raised entry 1 in the case that $\rho(1) > \rho(2)$. Again, by counting the number of raised 2s required and the number of columns available for them, we see that there is a unique way to place the entries 1, again diagonally from upper right to lower left. The result follows by iterating the argument. □

Proposition 24.3.6 *Let* $\omega \in \mathsf{Dec}(n)$ *and* $\rho \in \Sigma(n)$ *be as in Proposition 24.3.5, and let* $W \in \mathsf{W}(n)$ *be the permutation matrix associated to* ρ. *Then* $s \neq 0$, *where* $s = \phi^{\omega}(f_W(\mathsf{B}(n)))$ *is the sum of the flag d-polynomials for flags in* $\mathsf{Sch}(\rho)$.

Proof It follows from Proposition 24.3.5 that the zip monomial $z_{\omega}(\rho)$ is a term in s, and so $s \neq 0$. □

Example 24.3.7 In Examples 24.3.2 and 24.3.4, $\mathsf{Sch}(\rho)$ is the top Schubert cell for the partial flag module $\mathsf{FL}^{(2,3,6)}(8)$, but any permutation σ such that $\mathrm{des}(\sigma) \subseteq I$ can be used. For example, for $\sigma = (5, 8, 7, 1, 4, 6, 2, 3)$,

$$F^{\omega}(\sigma) = \left.\begin{array}{ccccc|cccc|ccc}
4 & 4 & 4 & 4 & 4 & \cdot & \cdot & \cdot & 3 & \cdot & 2 & 1 \\
\cdot & \cdot & 6 & 5 & \cdot & \cdot & \cdot & 3 & 2 & \cdot & 1 & \cdot \\
\cdot & 6 & 5 & \cdot & \cdot & \cdot & 3 & 2 & 1 & \cdot & \cdot & \cdot \\
5 & 5 & 3 & \cdot & 5 & \cdot & 2 & 1 & \cdot & \cdot & \cdot & \cdot \\
1 & 1 & 1 & 1 & 1 & 1 & 1 & \cdot & \cdot & 1 & \cdot & \cdot \\
6 & 3 & 2 & 6 & 6 & \cdot & \cdot & \cdot & \cdot & \cdot & \cdot & \cdot \\
3 & 2 & \cdot & 3 & 3 & 3 & \cdot & \cdot & \cdot & \cdot & \cdot & \cdot \\
2 & \cdot & \cdot & 2 & 2 & 2 & \cdot & \cdot & \cdot & 2 & \cdot & 2
\end{array}\right. .$$

The flag d-polynomial for the maximal flag $X \in \mathsf{Sch}(\sigma)$ is $\phi^{\omega}(X) = \prod_{i=1}^{6} u_i^{(2^{\lambda_i}-1)}$, where $u_1 = v_1 + v_2 + v_3 + v_4 + v_5$, $u_2 = v_1 + v_2 + v_3 + v_4 + v_6 + v_7 + v_8$, $u_3 = v_1 + v_2 + v_3 + v_4 + v_6 + v_7$, $u_4 = v_1$, $u_5 = v_2 + v_3 + v_4$ and $u_6 = v_2 + v_3 + v_6$. The unique factorization argument for $F^{\omega}(\sigma)$ works in the same way as before, starting by identifying f_1, then f_2 and so on, according to the numbering of the entries in the array above.

The next example illustrates the proof of Theorem 24.3.9.

Example 24.3.8 As in Example 24.3.2, let $n = 8$ and $I(\omega) = \{2,3,6\}$, so that $\rho = (7,8,6,3,4,5,1,2)$. Let $J = \{2,3\}$, so that $\sigma = (\sigma(1),\ldots,\sigma(8))$ where $\{\sigma(1),\sigma(2)\}$ permutes $\{7,8\}$ and $\{\sigma(4),\ldots,\sigma(8)\}$ permutes $\{1,2,3,4,5\}$. For $\omega = (6^5,3^3,2^2)$, the zip array $F^{\omega}(\rho)$ is obtained by removing the fixed columns in the subblocks F_1, F_2 from the array shown in Example 24.3.2.

We show that if $\sigma \neq \rho$ then there is no σ-factorization of the block $Z^{\omega}(\rho)$. First consider the case $\sigma(1) = 8$. For a σ-factorization we require all integer entries in row 8 to be 1s, and this leaves only 6 columns to accommodate raised entries in each of the first 7 rows. Hence $\sigma(1) < 8$ and so $\sigma(1) = 7$. Thus $\sigma = (7,8,6,*,*,*,*,*)$ where the stars are a permutation of $\{1,2,3,4,5\}$. At this point, we may assume that the placement of all the entries $1,2,3$ in the proposed σ-factorization coincide with that in $F^{\omega}(\rho)$.

Next suppose that $\sigma(4) = 5$. Then all remaining entries in row 5 must be 4s, and this leaves only 2 columns to accommodate 4 raised entries 5. Similarly, if $\sigma(4) = 4$, then all remaining entries in row 4 must be 4s, and this leaves only 2 columns to accommodate 3 raised entries 4. Hence $\sigma(4) \leq 3$. Continuing in this way, we see that $\sigma \leq_{\ell} \rho$.

Finally we observe that as neither v_1 nor v_2 is a factor of the zip monomial, $\sigma(7)$ and $\sigma(8)$ must be 1 and 2 in some order, and in fact $\sigma(7) = 1$ and $\sigma(8) = 2$, as otherwise a raised entry 7 would be required in the array. Hence $\sigma = \rho$.

Theorem 24.3.9 gives sufficient conditions on ω for the summand $\mathsf{FL}_J(n)$ of the flag module to be embedded as a submodule of $\mathsf{J}^\omega(n)$. Given $J \subseteq I(\omega)$, we obtain a lower bound on the multiplicity of each element of $I(\omega)$ as a term of ω. Although it suffices to prove the result for the minimal case, as Proposition 24.1.1 will then imply the result for greater multiplicities, we continue to work with the general case, because this makes no real difference to the argument.

Theorem 24.3.9 *Let $J \subseteq I = \{i_1, \ldots, i_r\} \subseteq Z[n-1]$ and let $a_k = i_{k+1} - i_{k-1}$ for $1 \leq k \leq r$, where $i_0 = 0$ and $i_{r+1} = n$. Let $\omega \in \mathsf{Dec}(n)$ with terms i_k of multiplicity m_k, where $m_k \geq a_k - 1$ if $i_k \in J$ and $m_k \geq a_k$ if $i_k \notin J$. Then $\phi^\omega : \mathsf{FL}_J(n) \to \mathsf{DP}^\omega(n)$ is injective, where $d = \deg_2 \omega$.*

Proof It is sufficient to prove that the projection on $\mathsf{DP}^\omega(n)$ of the image of the generator $s_J(\mathsf{B}(n))$ of the socle of $\mathsf{FL}_J(n)$ under ϕ^ω is nonzero.

By Definition 17.5.1, $s_J \in \mathsf{H}_0(n)$ is the sum of the elements f_W, where $W \in \mathsf{W}^J(n)W_0$. The permutations σ associated to these permutation matrices W are obtained by splitting the sequence $(n, n-1, \ldots, 2, 1)$ into subsequences corresponding to the composition of n corresponding to J, and then permuting the elements of each subsequence. For example, if $n = 4$, $J = \{2\}$ then $\sigma \in \{(3,4 \mid 1,2), (3,4 \mid 2,1), (4,3 \mid 1,2), (4,3 \mid 2,1)\}$. Since $J \subseteq I$, this set of permutations includes the unique permutation ρ of maximal length such that $\mathrm{des}(\rho) = I$, as ρ refines the splitting of $(n, n-1, \ldots, 2, 1)$ into subsequences and preserves the order of each smaller subsequence. For example, if $I(\omega) = \{1,2\}$ and $n = 4$ then $\rho = (4 \mid 3 \mid 1,2)$. By Proposition 24.3.5 the zip array $F^\omega(\rho)$ defines the unique ρ-factorization of the corresponding zip d-monomial $z_\omega(\rho)$. (Note that ρ depends only on $I(\omega)$ and not on J.) We shall prove that $z_\omega(\rho)$ does not have a σ-factorization for any $\sigma \neq \rho$ associated to a matrix $W \in \mathsf{W}^J(n)W_0$. It follows as in Proposition 24.3.6 that $z_\omega(\rho) \in \mathsf{DP}^\omega(n)$ appears as a term in $\phi^\omega(s_J(\mathsf{B}(n)))$, which is therefore nonzero.

We assume that $z_\omega(\rho)$ has a σ-factorization, where $\sigma \in \Sigma(n)$ is associated to a matrix $W \in \mathsf{W}^J(n)W_0$. We shall show that (1) $\sigma \leq_\ell \rho$, and (2) $\{\rho(1), \ldots, \rho(\ell)\} = \{\sigma(1), \ldots, \sigma(\ell)\}$ when $\ell \in I \setminus J$, and it follows that $\sigma = \rho$.

To prove (1), we consider the minimal integer k such that $\sigma(k) \neq \rho(k)$. If $\sigma(k) > \rho(k)$, we shall show that it is impossible to place the entries k in the zip array with underlying $(0,1)$-block $Z^\omega(\rho)$. For this we observe that, as in the proof of Proposition 24.3.3, the k's must be placed in all the remaining positions available in row $\sigma(k)$, and that this leaves too few positions in rows $< k$ to contain the raised entries k.

To prove (2), we use the fixed columns of $F^\omega(\rho)$. For each such ℓ there is a 'fixed' column in $F^\omega(\rho)$ with dots in the first $n-\ell$ rows and integer entries in the last ℓ rows. These positions must be occupied by the integers $1,2,\ldots,\ell$ in any σ-factorization. □

Since the partial flag module $\mathsf{FL}^I(n)$ is the direct sum of the modules $\mathsf{FL}_J(n)$ for $J \subseteq I$, Theorem 24.3.9 implies the following result.

Theorem 24.3.10 *Let $I(\omega) = \{i_1,\ldots,i_r\} \subseteq Z[n-1]$, $\omega \in \mathsf{Dec}(n)$ with terms i_j of multiplicity $m_j \geq a_j = i_{j+1} - i_{j-1}$ for $1 \leq j \leq r$, and let $d = \deg_2 \omega$. Then $\phi^\omega : \mathsf{FL}^I(n) \to \mathsf{DP}^\omega(n)$ is injective.* □

The minimum degree $d = \sum_{j=1}^{r} c_j(2^{b_j} - 1)$ of the embedding arises when i_j has multiplicity $m_j = a_j$ in ω for all j, where $b_j = a_j + \cdots + a_r = (n+i_r) - (i_j + i_{j-1})$ for $1 \leq j \leq r$ and $c_j = i_j - i_{j-1}$ for $1 \leq j \leq r+1$.

Example 24.3.11 We apply Theorem 24.3.9 to the 3-variable case, giving the minimal ω-sequences $(2,1)$, $(2,1,1)$, $(2,2,1)$ or $(2,2,1,1)$ for which the summand $\mathsf{FL}_J(3)$ occurs in $\mathsf{J}^\omega(3)$,

1. $\mathsf{FL}_{1,2}(3)$ for $\omega = (2,1)$. The socle generator of $\mathsf{FL}_2(3)$ is $s_1(\mathsf{B}(3))$ where $s_2 = f_{(3,2,1)}$. The zip d-monomial $v_1^{(3)}v_2$ has the ρ_0-factorization $v_1^{(2)}v_2 \cdot v_1$ in the top flag d-polynomial for $\mathsf{Sch}(\rho_0)$, where $\rho_0 = (3,2,1)$. Hence $\mathsf{FL}_{1,2}(3)$ embeds in $\mathsf{J}^\omega(3)$.

2. $\mathsf{FL}_2(3)$ for $\omega = (2,1,1)$. The socle generator of $\mathsf{FL}_2(3)$ is $s_2(\mathsf{B}(3))$ where $s_2 = f_{(2,3,1)} + f_{(3,2,1)}$. The zip d-monomial $v_1^{(5)}v_2v_3^{(2)}$ does not occur in the flag d-polynomials for $\mathsf{Sch}(2,3,1)$, and has the ρ_0-factorization $v_1^{(4)}v_2v_3^{(2)} \cdot v_1$ in the top flag d-polynomial for $\mathsf{Sch}(\rho_0)$. Hence $\mathsf{FL}_2(3)$ embeds in $\mathsf{J}^\omega(3)$.

3. $\mathsf{FL}_1(3)$ for $\omega = (2,2,1)$. The socle generator of $\mathsf{FL}_1(3)$ is $s_3(\mathsf{B}(3))$ where $s_3 = f_{(3,1,2)} + f_{(3,2,1)}$. The zip d-monomial $v_1^{(6)}v_2^{(3)}v_3$ does not occur in the flag d-polynomials for $\mathsf{Sch}(3,1,2)$, and has the ρ_0-factorization $v_1^{(4)}v_2^{(2)}v_3 \cdot v_1^{(2)}v_2$ in the top flag d-polynomial for $\mathsf{Sch}(\rho_0)$. Hence $\mathsf{FL}_1(3)$ embeds in $\mathsf{J}^\omega(3)$.

4. $\mathsf{FL}_\emptyset(3)$ for $\omega = (2,2,1,1)$. The generator of $\mathsf{FL}_\emptyset(3)$ is $s_\emptyset(\mathsf{B}(3))$ where $s_\emptyset$ is the sum of all 21 flag d-polynomials. The zip d-monomial $v_1^{(10)}v_2^{(3)}v_3^{(5)}$ does not occur in the flag d-polynomials for $\mathsf{Sch}(\rho)$ when $\rho \neq (3,2,1)$, and has the ρ_0-factorization $v_1^{(8)}v_2^{(2)}v_3^{(5)} \cdot v_1^{(2)}v_2$ in the top flag d-polynomial for $\mathsf{Sch}(\rho_0)$. Hence $\mathsf{FL}_\emptyset(3)$ embeds in $\mathsf{J}^\omega(3)$.

In the following tables, we give the zip arrays corresponding to these special factorizations. The integers s and t are the numbers of 1s and 2s respectively in the minimal ω-sequence. The dimensions are the lower bounds on $\dim \mathsf{J}^{\omega}(3)$ given by Theorem 24.3.9, and are cumulative, e.g. for $s=2$, $t=1$, $\mathsf{J}^{\omega}(3)$ contains both $\mathsf{FL}_{1,2}(3)$ and $\mathsf{FL}_{2}(3)$, and so has dimension ≥ 14.

$t=1$	$s=1$	$s=2$
module	$\mathsf{FL}_{1,2}(3)$	$\mathsf{FL}_{2}(3)$
permutation ρ	$(3,2,1)$	$(3,2,1)$
zip array $Z^{\omega}(\rho)$	$2\,1$ $1\,\cdot$ $\cdot\,\cdot$	$2\,\cdot\,1$ $1\,\cdot\,\cdot$ $\cdot\,1\,\cdot$
dimension	8	14

$t=2$	$s=1$	$s=2$
module	$\mathsf{FL}_{1}(3)$	$\mathsf{FL}_{\emptyset}(3)$
permutation ρ	$(3,2,1)$	$(3,2,1)$
zip array $Z^{\omega}(\rho)$	$\cdot\,2\,1$ $2\,1\,\cdot$ $1\,\cdot\,\cdot$	$\cdot\,2\,\cdot\,1$ $2\,1\,\cdot\,\cdot$ $1\,\cdot\,1\,\cdot$
dimension	14	21

24.4 Lower bounds for $\mathsf{J}^{\omega}(4)$

In this section we apply Theorem 24.3.9 to the 4-variable case, and tabulate similar results for small ω-sequences, where only a quotient of the relevant module $\mathsf{FL}_J(n)$ is embedded in $\mathsf{DP}^d(n)$ by ϕ^{ω}. We write such a quotient as $\mathsf{FL}_J(n) \downarrow L(\lambda)$, where $L(\lambda)$ is the isomorphism type of its socle. For example $\mathsf{FL}_{1,2}(4) \downarrow L(2,1,1)$ is the 35-dimensional quotient of $\mathsf{FL}_{1,2}(4)$ with composition factors $L(2,1)$, $L(0)$ and $L(2,1,1)$ shown in the diagram of Chapter 17. These results are obtained by considering the image under ϕ^{ω} of the generator of the socle $L(\lambda)$, in the same way as for the socle of $\mathsf{FL}_J(4)$ itself given by Theorem 24.3.9. In each case, we list the choice of zip array used in the argument.

This section is divided into subsections depending on $I(\omega) \subseteq \{1,2,3\}$. In the tables, $\omega = (3,\ldots,3,2,\ldots,2,1,\ldots,1)$ where there are s 1s, t 2s and u 3s. We regard a block in $\mathsf{DP}^{\omega}(4)$, or a corresponding array, as the concatenation of a 'head' with ω-sequence $(3,\ldots,3)$ of length u, a 'body' with ω-sequence $(2,\ldots,2)$ of length t and a 'tail' with ω-sequence $(1,\ldots,1)$ of length s.

In each table, we list

(i) a quotient module of $\mathsf{FL}(4)$ which is mapped injectively into $\mathsf{J}^{\omega}(4)$ by ϕ^{ω},
(ii) a permutation $\rho \in \Sigma(4)$,
(iii) a zip array showing the unique ρ-factorization of $z_{\omega}(\rho)$,
(iv) the lower bound on $\dim \mathsf{J}^{\omega}(4)$ given by Theorem 24.3.9.

The dimension listed in the tables is the cumulative dimension obtained by applying Proposition 24.1.1. For example, the dimension 15 for $s = 4$, $t = u = 0$ is obtained by adding the dimension of FL_1 to that of $\mathsf{FL}_\emptyset$. Note that in the case $\omega = (3,1)$ this method fails to detect the 1-dimensional submodule of $\mathsf{J}^{(3,1)}(4) \cong \Delta((2,1,1),4)$, the restricted Weyl module, as given by Theorem 20.5.1, so that the lower bound tabulated is 14, and not the exact value $\dim \mathsf{J}^{(3,1)}(4) = 15$.

We begin with the 'tail' case $I(\omega) = \{1\}$, the 'body' case $I(\omega) = \{2\}$ and the 'head' case $I(\omega) = \{3\}$,

Tail case: $s > 0, t = u = 0$.

s	1	2	3	4
module	$L(1)$	$\mathsf{FL}_1 \downarrow L(1,1)$	FL_1	$\mathsf{FL}_\emptyset$
permutation ρ	$(2,1,3,4)$	$(3,1,2,4)$	$(4,1,2,3)$	$(4,1,2,3)$
zip array $Z^{\omega}(\rho)$	1 $\cdot$ $\cdot$ $\cdot$	$\cdot\,1$ $1\,\cdot$ $\cdot\,\cdot$ $\cdot\,\cdot$	$\cdot\,\cdot\,1$ $\cdot\,1\,\cdot$ $1\,\cdot\,\cdot$ $\cdot\,\cdot\,\cdot$	$\cdot\,\cdot\,\cdot\,1$ $\cdot\,\cdot\,1\,\cdot$ $\cdot\,1\,\cdot\,\cdot$ $1\,\cdot\,\cdot\,\cdot$
dimension	4	10	14	15

Body case: $t > 0, s = u = 0$.

t	1	2	3	4
module	$L(1,1)$	$\mathsf{FL}_2 \downarrow L(2,1,1)$	FL_2	$\mathsf{FL}_\emptyset$
permutation ρ	$(1,3,2,4)$	$(2,4,1,3)$	$(3,4,1,2)$	$(3,4,1,2)$
zip array $Z^{\omega}(\rho)$	1 2 $\cdot$ $\cdot$	$2\,1$ $1\,\cdot$ $\cdot\,2$ $\cdot\,\cdot$	$\cdot\,2\,1$ $2\,1\,\cdot$ $1\,\cdot\,\cdot$ $\cdot\,\cdot\,2$	$\cdot\,\cdot\,2\,1$ $\cdot\,2\,1\,\cdot$ $1\,1\,\cdot\,\cdot$ $2\,\cdot\,\cdot\,2$
dimension	6	20	34	35

Head case: $u > 0$, $s = t = 0$.

u	1	2	3	4
module	$L(1,1,1)$	$\mathsf{FL}_3 \downarrow L(1,1)$	FL_3	$\mathsf{FL}_\emptyset$
permutation ρ	$(1,2,4,3)$	$(1,3,4,2)$	$(2,3,4,1)$	$(2,3,4,1)$
zip array $Z^\omega(\rho)$	1 2 3 ·	1 1 3 2 2 · · 3	3 2 1 1 1 · 2 · 2 · 3 3	· 3 2 1 1 1 1 · 2 2 · 2 3 · 3 3
dimension	4	10	14	15

In the tail case, the socle generators of $\mathsf{FL}_1(4) \downarrow L(1)$, $\mathsf{FL}_1(4) \downarrow L(1,1)$ and $\mathsf{FL}_1(4) = \mathsf{FL}_1(4) \downarrow L(1,1,1)$ are the images of $\mathsf{B}(4)$ under the elements g_1, $g_1 b$ and $g_1 bc$ of $\mathsf{H}_0(4)$ respectively. (See the diagram at the end of Chapter 17.) Here a, b, c are the elements $f_{(2,1,3,4)}, f_{(1,3,2,4)}$ and $f_{(1,2,4,3)}$ and $g_1 = (1+b)(1+c)(1+b)a$. Thus, in the case $\omega = (1)$, $\phi^\omega(\mathsf{B}(4))$ is the sum of the flag d-polynomials in the Schubert cells corresponding to the 6 permutations $\rho \in \Sigma(4)$ such that $\rho(1) = 2$. For $\rho = (2,1,3,4)$ this sum is $(v_1 + v_2) + v_2 = v_1$, while for the other 5 permutations the corresponding sum is 0. Thus the quotient $\mathsf{FL}_1(4) \downarrow L(1) \cong L(1)$ embeds in $\mathsf{J}^{(1)}(4)$. The corresponding calculation for $\omega = (1,1)$ gives $g_1 b(\mathsf{B}(4)) = v_1^{(2)} v_2 + v_1 v_2^{(2)} + v_1 v_2 v_3$, and for $\omega = (1,1,1)$ it gives $g_1 bc(\mathsf{B}(4))$ as the symmetrization of $v_1^{(4)} v_2^{(2)} v_3$.

The body-tail case: $I(\omega) = \{1,2\}$

The minimal cases for the occurrence of a particular quotient of $\mathsf{FL}^{1,2}(4)$ in $\mathsf{J}^\omega(4)$ are shown in the table below. The table is cumulative, e.g. for $t = 1$, $s \geq 3$, $\mathsf{J}^\omega(4) \cong \mathsf{FL}_{1,2} \downarrow L(2,1,1) \oplus \mathsf{FL}_2 \downarrow L(1)$; for $t = 2$, $s \geq 2$ $\mathsf{J}^\omega(4) \cong \mathsf{FL}_{1,2} \oplus \mathsf{FL}_2$, etc.

Body and tail case: $s, t > 0$, $u = 0$.

$t\backslash s$	1	2	3
1	$L(2,1)$	$\mathsf{FL}_{1,2} \downarrow L(2,1,1) \oplus L(1,1)$	$\mathsf{FL}_2 \downarrow L(1)$
	$(3,2,1,4)$	$(4,2,1,3)$ $(3,1,2,4)$	$(4,2,1,3)$
	2 1	2 · 1 2 · 1	2 · · 1
	1 ·	· 1 · · 1 ·	· · 1 ·
	· ·	1 · · 1 · ·	· 1 · ·
	· ·	· · · · · ·	1 · · ·
	20	41	45
2	$\mathsf{FL}_{1,2}$	FL_2	
	$(4,3,1,2)$	$(4,3,1,2)$	
	· 2 1	· 2 · 1	
	2 1 ·	2 1 · ·	
	1 · ·	1 · · ·	
	· · ·	· · 1 ·	
	56	90	90
3	FL_1	$\mathsf{FL}_\emptyset$	
	$(4,3,1,2)$	$(4,3,1,2)$	
	· · 2 1	· · 2 · 1	
	· 2 1 ·	· 2 1 · ·	
	2 1 · ·	2 1 · · ·	
	1 · · ·	1 · · 1 ·	
	70	105	105

Example 24.4.1 Let $\omega = (2,2,1)$ and $J = I = \{1,2\}$. The socle generator $s_J(\mathsf{B}(4))$ is the sum of the flags for $\mathsf{Sch}(\rho)$ and $\mathsf{Sch}(\rho_0)$, where $\rho = (4,3,1,2)$ and $\rho_0 = (4,3,2,1)$. The zip monomial $z_\omega(\rho) = v_1^{(6)} v_2^{(3)} v_3$ has the unique ρ-factorization given by the zip array shown. All integer entries are raised entries in this case. The sum of the flag d-polynomials for $\mathsf{Sch}(\rho_0)$ is 0 since they cancel in pairs, because of the star in row 3 of the representative matrix. Hence $z_\omega(\rho)$ is a term in $\phi^\omega(s_J(\mathsf{B}(4)))$, which is therefore nonzero, and so ϕ^ω embeds $\mathsf{FL}_{1,2}(4)$ in $\mathsf{J}^\omega(4)$.

The head-tail case: $I(\omega) = \{1,3\}$

The minimal cases for the occurrence of a particular quotient of $\mathsf{FL}^{1,3}(4)$ in $\mathsf{J}^{\omega}(4)$ are tabulated. The table is again cumulative, e.g. for $u=2$, $s \geq 3$, $\mathsf{J}^{\omega}(4) \cong \mathsf{FL}_{1,3} \oplus \mathsf{FL}_3$.

Head and tail case: $s, u > 0$, $t = 0$.

$u \setminus s$	1	2	3
1	$L(2,1,1)$	$\mathsf{FL}_{1,3} \downarrow L(2,2,1)$	$\mathsf{FL}_3 \downarrow L(1,1)$
	$(2,1,4,3)$	$(4,1,3,2)$	$(4,1,3,2)$
	2 1	2 ·1	2 ··1
	1 ·	3 1·	3 ·1·
	3 ·	1 ··	· 1··
	· ·	· ··	1 ···
	14	35	45
2	$\mathsf{FL}_{1,3} \downarrow L(2,1)$	$\mathsf{FL}_{1,3}$	FL_3
	$(3,2,4,1)$	$(4,2,3,1)$	$(4,2,3,1)$
	3 2 1	3 2 ·1	3 2 ··1
	2 1 ·	2· 1·	2· ·1·
	1· ·	1 3 ··	1 3 ···
	·3 ·	·1 ··	·1 1··
	35	76	90
3	$\mathsf{FL}_1 \downarrow L(1,1)$	FL_1	$\mathsf{FL}_\emptyset$
	$(3,2,4,1)$	$(4,2,3,1)$	$(4,2,3,1)$
	·3 2 1	·3 2 ·1	·3 2 ··1
	2 2 1 ·	2 2· 1·	2 2· ·1·
	1 1· ·	3 1 3 ··	3 1 3 ···
	3 · 3 ·	1 · 1 ··	1 · 1 1··
	45	90	105

Example 24.4.2 Let $\omega = (3,3,3,1,1)$ and $J = \{1\}$, $I(\omega) = \{1,3\}$. The socle generator $s_J(\mathsf{B}(4))$ is the sum of the flags for the 6 Schubert cells $\mathsf{Sch}(\sigma)$ where $\sigma(1) = 4$. Thus the first factor of the flag d-polynomial for the maximal flag of $\mathsf{Sch}(\sigma)$ is $(v_1 + v_2 + v_3 + v_4)^{(31)}$. The zip block is

$$Z^\omega(\rho) = \begin{array}{cc} 011 & 01 \\ 110 & 10 \\ 111 & 00 \\ 101 & 00 \end{array}$$

and so the entries 1 in a σ-factorization of $Z^\omega(\rho)$ must be placed as in the zip array. For $\sigma = \rho = (4,2,3,1)$, we check that this represents the unique ρ-factorization of $Z^\omega(\rho)$. If $\sigma^{-1}(1) = 2$ or 3, then $v_1^{(7)}$ divides all flag d-polynomials for $\mathsf{Sch}(\sigma)$, and so $z_\omega(\rho)$ is not a term of these. If $\sigma = (4,3,2,1)$, the flag d-polynomial for the maximal flag is $(v_1 + v_2 + v_3 + v_4)^{(31)}(v_1 + v_2 + v_3)^{(7)}(v_1 + v_2)^{(7)}$. After placing the 1s as above, the $(3,1)$ and $(3,3)$ entries of a possible σ-factorization must be 2s, leaving only column 2 for two required raised entries. Hence no σ-factorization of $Z^\omega(\rho)$ exists. Thus $z_\omega(\rho)$ is a term in $\phi^\omega(s_J(\mathsf{B}(4)))$, which is therefore nonzero, and so ϕ^ω embeds $\mathsf{FL}_1(4)$ in $\mathsf{J}^\omega(4)$.

The head-body case: $I(\omega) = \{2,3\}$

The minimal cases for the occurrence of a particular quotient of $\mathsf{FL}^{2,3}(4)$ in $\mathsf{J}^\omega(4)$ are tabulated. The table is again cumulative, e.g. for $u \geq 2$, $s = 2$, $\mathsf{J}^\omega(4) \cong \mathsf{FL}_{2,3} \oplus \mathsf{FL}_2$.

Head and body case: $t, u > 0$, $s = 0$.

$u \backslash t$	1	2	3
1	$L(2,2,1)$	$\mathsf{FL}_{2,3}$	FL_3
	$(1,4,3,2)$	$(3,4,2,1)$	$(3,4,2,1)$
	1 1 3 2 2 · · ·	3 2 1 2 1 · 1 · · · · 2	3 · 2 1 2 · 1 · 1 1 · · · 2 · 2
	20	56	70

	$\mathsf{FL}_{2,3} \downarrow L(2,1,1) \oplus L(1,1)$		FL_2	$\mathsf{FL}_\emptyset$
	$(2,4,3,1)$	$(1,3,4,2)$	$(3,4,2,1)$	$(3,4,2,1)$
	3 2 1	1 1 1	· 3 2 1	· 3 · 2 1
2	1 1 ·	3 · 2	3 2 1 ·	3 2 · 1 ·
	· 3 2	2 2 ·	1 1 · ·	1 1 1 · ·
	2 · ·	· 3 ·	2 · · 2	2 · 2 · 2
	41		90	105

	$\mathsf{FL}_2 \downarrow L(1,1,1)$		
	$(2,4,3,1)$		
	· 3 2 1		
3	1 1 1 ·		
	3 · 3 2		
	2 2 · ·		
	45	90	105

Example 24.4.3 Let $\omega = (3,3,2)$ and $J = I = \{2,3\}$. From the data on $\mathsf{FL}(4)$ at the end of Chapter 17, the socle generator of the quotient module of $\mathsf{FL}_{2,3}(4)$ with socle $L(2,1,1)$ is $g_{2,3}a(\mathsf{B}(4))$ where $g_{2,3}a = (1+a)bcba \in \mathsf{H}_0(4)$ maps $\mathsf{B}(4)$ to the sum of the flags for $\mathsf{Sch}(\rho)$ and $\mathsf{Sch}(\sigma)$ where $\rho = (2,4,3,1)$ and $\sigma = (4,2,3,1)$.

We check that the zip array shown in the table represents the unique ρ-factorization of the zip monomial $z_\omega(\rho) = v_1^{(7)} v_2^{(3)} v_3^{(6)} v_4$. The first factor of the flag d-polynomial for the maximal flag of $\mathsf{Sch}(\sigma)$ is $(v_1+v_2+v_3+v_4)^{(7)}$, and so the $(4,1)$ entry of a possible σ-factorization must be 1, leaving only two columns for three required raised entries. Hence no σ-factorization of $Z^\omega(\rho)$ exists. It follows that $z_\omega(\rho)$ is a term in $\phi^\omega(g_{2,3}a(\mathsf{B}(4)))$, which is therefore nonzero, and so ϕ^ω embeds $\mathsf{FL}_{2,3}(4) \downarrow L(2,1,1)$ in $\mathsf{J}^\omega(4)$.

Example 24.4.4 Again let $\omega = (3,3,2)$ and $J = I = \{2,3\}$. From the data on $\mathsf{FL}(4)$ at the end of Chapter 17, the generator of the head $L(1,1)$ of $\mathsf{FL}_{2,3}(4)$ is $g_2(\mathsf{B}(4))$, where $g_2 = (1+a)(1+c)b \in \mathsf{H}_0(4)$ maps $\mathsf{B}(4)$ to the sum of the flags for $\mathsf{Sch}(\sigma)$, where $\sigma = (1,3,2,4)$, $(3,1,2,4)$, $(1,3,4,2)$ or $(3,1,4,2)$.

For $\rho = (1,3,4,2)$, we check that the zip array shown in the table represents the unique ρ-factorization of the zip monomial $z_\omega(\rho) = v_1^{(7)} v_2^{(5)} v_3^{(3)} v_4^{(2)}$. For

$\sigma = (1,3,2,4)$ and $\sigma = (3,1,2,4)$, all flag d-polynomials have a factor $v_2^{(3)}$, and so there is no σ-factorization of $Z^{\omega}(\rho)$. For $\sigma = (3,1,4,2)$, the flag d-polynomial for the maximal flag of $\mathsf{Sch}(\sigma)$ is $(v_1+v_2+v_3)^{(7)}v_1^{(7)}(v_2+v_4)^{(3)}$, so all entries in row 1 of a possible σ-factorization must be 2s, leaving no position for a raised entry 1 in row 1. Hence no σ-factorization of $Z^{\omega}(\rho)$ exists. It follows that $z_{\omega}(\rho)$ is a term in $\phi^{\omega}(g_2(\mathsf{B}(4)))$, which is therefore nonzero, and so ϕ^{ω} embeds the quotient $\mathsf{FL}_2(4) \downarrow L(1,1) \cong L(1,1)$ in $\mathsf{J}^{\omega}(4)$.

The 1*-dominant case:* $I(\omega) = \{1,2,3\}$

The 8 minimal cases for embeddings of summands of $\mathsf{FL}(4)$ in $\mathsf{J}^{\omega}(4)$ are tabulated below, for comparison with the table in Example 23.4.6. The tables are again cumulative, e.g. for $u = 1$, $t \geq 2$, $s \geq 2$, $\mathsf{J}^{\omega}(4) \cong \mathsf{FL}_{1,2,3} \oplus \mathsf{FL}_{1,3} \oplus \mathsf{FL}_{2,3} \oplus \mathsf{FL}_3$, and for $u,t,s \geq 2$, $\mathsf{J}^{\omega}(4) \cong \mathsf{FL}(4)$.

1-dominant case: $s,t > 0$, $u = 1$.

$t \backslash s$	1	2
1	$\mathsf{FL}_{1,2,3}$	$\mathsf{FL}_{2,3}$
	$(4,3,2,1)$	$(4,3,2,1)$
	3 2 1 2 1 · 1 · · · · ·	3 2 ·1 2 1 ·· 1 · ·· · · 1·
	64	120
2	$\mathsf{FL}_{1,3}$	FL_{3}
	$(4,3,2,1)$	$(4,3,2,1)$
	3 ·2 1 2 ·1 · 1 2· · · 1· ·	3 ·2 ·1 2 ·1 ·· 1 2· ·· · 1· 1·
	140	210

1-dominant case: $s,t > 0$, $u = 2$.

$t \backslash s$	1	2
1	$\mathsf{FL}_{1,2}$	FL_{2}
	$(4,3,2,1)$	$(4,3,2,1)$
	·3 2 1 32 1 · 21 · · 1· · ·	·3 2 ·1 32 1 ·· 21 · ·· 1· · 1·
	120	210
2	FL_{1}	$\mathsf{FL}_{\emptyset}$
	$(4,3,2,1)$	$(4,3,2,1)$
	·3 ·2 1 32 ·1 · 21 2· · 1· 1· ·	·3 ·2 ·1 32 ·1 ·· 21 2· ·· 1· 1· 1·
	210	315

24.5 Counterexamples to Kameko's conjecture

In this section we obtain examples which disprove the conjecture of Kameko that $\dim \mathsf{Q}^d(n) \leq \dim \mathsf{FL}(n) = \prod_{i=1}^n (2^i - 1)$ for all d and n. The following result gives such examples for all $n \geq 5$. The method is to construct degrees containing decreasing sequences $\omega(k)$ for $1 \leq k \leq n-1$ to which Theorem 24.3.10 applies with $I(\omega) = -\{k\} = Z[n-1] \setminus \{k\}$, so that $\phi^{\omega(k)}$: $\mathsf{FL}^{-\{k\}}(n) \to \mathsf{DP}^{\omega(k)}(n)$ is an embedding of the maximal partial flag module which omits only subspaces of dimension k. Since each of these $n-1$ modules has dimension $\dim \mathsf{FL}(n)/3$ and $\mathsf{DP}^d(n)$ has a filtration whose quotients include all $\mathsf{DP}^{\omega(k)}(n)$, $\dim \mathsf{DP}^d(n) \geq (n-1)\dim \mathsf{FL}(n)/3$. In the case $n = 5$, the minimum degree $d = (2^{13}-1)+(2^9-1)+(2^5-1) = 8733$.

Proposition 24.5.1 *Let $n \geq 3$ and $d = \sum_{i=1}^{n-2}(2^{\lambda_i}-1)$, where $\lambda_i - \lambda_{i+1} \geq 4$ for $1 \leq i \leq n-3$ and $\lambda_{n-2} \geq 5$. Then* $\dim \mathsf{Q}^d(n) \geq \frac{1}{3}(n-1)\prod_{i=1}^n (2^i - 1)$.

Proof For $1 \leq k \leq n-2$, we have $2^{\lambda_k} - 1 = (2^{\lambda_k - 1} - 1) + (2^{\lambda_k - 1} - 1) + 1$, and so $d = \sum_{i=1}^{k-1}(2^{\lambda_i}-1) + (2^{\lambda_k-1}-1) + (2^{\lambda_k-1}-1) + \sum_{i=k+1}^{n-2}(2^{\lambda_i}-1) + 1$. Including the given partition, this gives $n-1$ spike partitions of d, the kth partition having equal kth and $(k+1)$st parts.

For example, in the case $n = 4$ with $\lambda = (9,5)$ and $d = (2^9-1)+(2^5-1) = 542$, the corresponding Ferrers blocks $F(k)$ for $k = 3,2,1$ respectively are

$$\begin{matrix} 1\,1\,1\,1\,1\,1\,1\,1\,1 \\ 1\,1\,1\,1\,1 \\ 0 \\ 0 \end{matrix}\;,\quad \begin{matrix} 1\,1\,1\,1\,1\,1\,1\,1\,1 \\ 1\,1\,1\,1 \\ 1\,1\,1\,1 \\ 1 \end{matrix}\;,\quad \begin{matrix} 1\,1\,1\,1\,1\,1\,1\,1 \\ 1\,1\,1\,1\,1\,1\,1\,1 \\ 1\,1\,1\,1\,1 \\ 1 \end{matrix}\;.$$

Hence $\mathsf{K}^d(3)$ has composition factors which include $\mathsf{FL}^{1,2}(3)$, $\mathsf{FL}^{1,3}(3)$ and $\mathsf{FL}^{2,3}(3)$ each of dimension 105 and so $\dim \mathsf{K}^d(3) \geq 315$. Since there are no more spike types $\mathsf{Dec}_d(3)$, it follows using Theorem 29.2.1 that $\dim \mathsf{K}^d(4) = 315$.

Let $\omega(k) = \omega(F(k))$, $1 \leq k \leq n-1$. For $k < n-1$, $\omega(k) = (n)|\omega(k)'$, and we denote by $F(k)'$ the Ferrers block obtained by deleting the first column of $F(k)$. The down Kameko map $\kappa : \mathsf{P}^d(n) \to \mathsf{P}^{d'}(n)$, where $d' = (d-n)/2$, induces a surjection $\mathsf{Q}^d(n) \to \mathsf{Q}^{d'}(n)$ of $\mathbb{F}_2\mathsf{GL}(n)$-modules. In terms of the $\leq_l$-filtrations of $\mathsf{Q}^d(n)$ and $\mathsf{Q}^{d'}(n)$, κ induces a surjection $\mathsf{Q}^{\omega}(n) \to \mathsf{Q}^{\omega'}(n)$ when $\omega = (n)|\omega' \in \mathsf{Dec}_d(n)$. Hence for $k < n-1$, $\dim \mathsf{Q}^{\omega(k)}(n) \geq \dim \mathsf{Q}^{\omega(k)'}(n)$.

In forming the Ferrers blocks $F(k)'$ for $k < n-1$ from $F(n-1)$, the multiplicity of the longest column is reduced by 1 or 2, and the multiplicities of all other columns are not reduced by more than 1. Thus, under the hypotheses of Proposition 24.5.1, all the column multiplicities of $\omega(n-1)$ and of $\omega(k)'$, for $k < n-1$, are ≥ 3. Since $\omega(n-1)$ has terms $1,2,\ldots,n-2$, it follows from Theorem 24.3.10 that $\phi^{\omega(n-1)} : \mathsf{FL}^{-\{n-1\}}(n) \to \mathsf{DP}^{\omega(n-1)}(n)$ is injective, where $-\{n-1\} = \{1,2,\ldots,n-2\}$, since for $I(\omega) = -\{n-1\}$ we have $a_j = i_{j+1} - i_{j-1} = 2$ or 3 for $1 \leq j \leq n-2$. In the same way, it follows from Theorem 24.3.10 that for $k < n-1$ the map $\phi^{\omega(k)'} : \mathsf{FL}^{-\{k\}}(n) \to \mathsf{DP}^{\omega(k)'}(n)$ is injective, where $-\{k\} = \{1,2,\ldots,k-1,k+1,\ldots,n-1\}$, since all $a_j \leq 3$ when $I(\omega) = -\{k\}$. Hence $\phi^{\omega(k)} : \mathsf{FL}^{-\{k\}}(n) \to \mathsf{DP}^{\omega(k)}(n)$ is also injective for $2 \leq k \leq n-1$.

By Proposition 16.4.4, each of the maximal partial flag modules $\mathsf{FL}^{-\{k\}}(n)$ has dimension $\dim \mathsf{FL}(n)/3$. By duality between $\mathsf{P}(n)$ and $\mathsf{DP}(n)$ it follows that $\dim \mathsf{Q}^{\omega(k)}(n) \geq \dim \mathsf{FL}(n)/3$ for $1 \leq k \leq n-1$. Since $\mathsf{Q}^d(n)$ has a filtration in which each of the $n-1$ modules $\mathsf{Q}^{\omega(k)}(n)$ appears as a quotient, $\dim \mathsf{Q}^d(n) \geq (n-1)\dim \mathsf{FL}(n)/3$. □

24.6 Remarks

The main result of this chapter, Theorem 24.3.9, was motivated by the work of Tran Ngoc Nam [155], who proved Theorem 24.3.10, the embedding of the partial flag module $\mathsf{FL}^I(n)$ for $\mathsf{GL}(n)$ in the d-spike module $\mathsf{J}^d(n)$.

The map ϕ^ω was introduced by Michael Crabb and John Hubbuck in [41], who used it to prove the following result (CH): if ω is decreasing with $I(\omega) = \{1,2,\ldots,r\}$ and if $\lambda = \omega^{\mathrm{tr}}$ satisfies $2^{\lambda_i - \lambda_{i+1}} > n-i+1$ for $1 \leq i \leq r$, then ϕ^ω restricts to an embedding of $\mathsf{FL}^{I(\omega)}(n)$ in $\mathsf{DP}^d(n)$.

In [155], it is stated that Theorem 24.3.10 is a generalization of (CH). This is true in the sense that it applies to a general $I(\omega)$ rather than the special cases $I(\omega) = \{1,2,\ldots,r\}$, $1 \leq r \leq n$, but in these special cases (CH) is stronger than Theorem 24.3.10, since the 2-powers involved in the degree d grow linearly as n increases for a fixed $I(\omega)$, rather than logarithmically. In Theorem 24.3.10 the embedding is in $\mathsf{DP}^\omega(n)$, i.e. Theorem 24.3.10 is 'local' and (CH) 'global'. For example, when $n=4$ and $d=34$, there are 3 sequences $\omega_1 = (2,2,1,1,1)$, $\omega_2 = (4,1,1,1,1)$ and $\omega_3 = (4,3,2,2)$. Theorem 24.3.9 shows that the summands $\mathsf{FL}_{1,2}(4)$ and $\mathsf{FL}_2(4)$ of $\mathsf{FL}^{1,2}(4)$ are in $\mathsf{J}^{\omega_1}(4)$, and, using Kameko iteration, that the remaining summands $\mathsf{FL}_1(4)$ and $\mathsf{FL}_\emptyset(4)$ are in $\mathsf{J}^{\omega_2}(4)$, proving this case of (CH). We also have an embedding of $\mathsf{FL}_{2,3}(4)$ in $\mathsf{J}^{\omega_3}(4)$. Kameko's computations [108] give $\dim \mathsf{Q}^{\omega_1}(4) = 90$, $\dim \mathsf{Q}^{\omega_2}(4) = 15$ and $\dim \mathsf{Q}^{\omega_3}(4) = 60$, so that $\dim \mathsf{Q}^{34}(4) = 165$, in agreement with the results of Nguyen Sum [204, 207].

Kameko's conjecture that $\dim \mathsf{Q}^d(n) \leq \prod_{i=1}^n (2^i - 1)$ for all n and d appears in [106]. In [205, 206] Sum gave counterexamples for all $n \geq 5$. The examples given in Section 24.5 are similar to those of Sum. The results of [108, 204, 207] show that Kameko's conjecture is true for $n \leq 4$. The question of whether the local version $\dim \mathsf{Q}^\omega(n) \leq \prod_{i=1}^n (2^i - 1)$ of Kameko's conjecture is true remains an open problem.

25
The symmetric hit problem

25.0 Introduction

For any subgroup G of $\mathsf{GL}(n)$, the algebra $\mathsf{P}(n)^G$ of G-invariant polynomials is an A_2-submodule of $\mathsf{P}(n)$, since the action of G on $\mathsf{P}(n)$ commutes with that of A_2. In particular, when G is the group of permutation matrices $\mathsf{W}(n)$, $\mathsf{P}(n)^{\mathsf{W}(n)} = \mathsf{S}(n)$, the algebra of **symmetric polynomials** in $x_1,\ldots,x_n$. The action of A_2 on $\mathsf{S}(n)$ has topological significance, as $\mathsf{S}(n)$ can be identified with the mod 2 cohomology of the Grassmannian of n-dimensional subspaces of $\mathbb{R}^\infty$ or $\mathbb{C}^\infty$. The symmetric hit problem concerns the quotient $\mathsf{Q}(\mathsf{S}(n)) = \mathsf{S}(n)/\mathsf{A}_2^+(\mathsf{S}(n))$ of the A_2-module $\mathsf{S}(n)$ by the hit elements.

Given a monomial $f \in \mathsf{P}(n)$ and a group of permutation matrices $G \subseteq \mathsf{W}(n)$, the sum $\sigma_G(f)$ of the distinct monomials $f \cdot A$ in the G-orbit of f is G-invariant. Conversely, if f is a term of a G-invariant polynomial h, then any monomial $f \cdot A$ which is a term of $\sigma_G(f)$ must also be a term of h. Hence the set of orbit sums of monomials spans $\mathsf{P}(n)^G$ as a vector space over $\mathbb{F}_2$. Since the G-orbits are the equivalence classes of monomials in $\mathsf{P}(n)$, orbit sums have no terms in common. Hence the nonzero orbit sums $\sigma_G(f)$ form a $\mathbb{F}_2$-basis for $\mathsf{P}(n)^G$. For example, when $n = 3$ and $G \cong \mathbb{Z}/3$ permutes the variables x, y, z cyclically, the four polynomials $x^3+y^3+z^3$, $x^2y+y^2z+z^2x$, $xy^2+yz^2+zx^2$, xyz form a basis for the invariants of degree 3.

In the case $G = \mathsf{W}(n)$, $\sigma_G(f)$ is the usual monomial symmetric function $m(\lambda)$ corresponding to the partition $\lambda = (\lambda_1,\ldots,\lambda_n)$ obtained by writing the exponents of f in decreasing order. In this case we write $\sigma_G(f)$ as $\sigma_n(f)$ or $\sigma_n(F)$, where F is the n-block corresponding to f. The fundamental theorem of symmetric functions states that the ring of $\mathsf{W}(n)$-invariants of $\mathbb{Z}[x_1,\ldots,x_n]$ is the polynomial ring $\mathbb{Z}[e_1,\ldots,e_n]$, where e_i is the ith elementary symmetric function in $x_1,\ldots,x_n$. Hence $\mathsf{S}(n) = \mathbb{F}_2[e_1,\ldots,e_n]$. The monomial symmetric functions $m(\lambda)$ corresponding to partitions of length $\leq n$ form a $\mathbb{F}_2$-basis for $\mathsf{S}(n)$.

A spike monomial f cannot be a term of $Sq^k(h)$ for $h \in \mathsf{P}(n)$ and $k > 0$. It follows that if λ is a spike partition then $m(\lambda)$ cannot appear in the reduced expression for $Sq^k(h)$ in the basis of monomial symmetric functions for $\mathsf{S}(n)$, for $h \in \mathsf{S}(n)$ and $k > 0$. Hence the symmetrized spikes $\sigma_n(f)$ represent linearly independent elements of the quotient $\mathsf{Q}(\mathsf{S}(n)) = \mathsf{S}(n)/\mathsf{A}_2^+\mathsf{S}(n)$ of $\mathsf{S}(n)$ by the hit elements. More generally, for any permutation group G, the orbit sums $\sigma_G(f)$ of spikes f represent linearly independent elements of the cohits $\mathsf{Q}(\mathsf{P}(n)^G) = \mathsf{P}(n)^G/\mathsf{A}_2^+\mathsf{P}(n)^G$.

In Section 25.1 we show that several features of the hit problem for $\mathsf{P}(n)$ can be generalized to the corresponding problem for $\mathsf{P}(n)^G$, where $G \subseteq \mathsf{W}(n)$. These include the analogue of the Peterson conjecture, Theorem 25.1.6, which states that generators for $\mathsf{S}(n)$ as an A_2-module can be found in degrees d such that $\mu(d) \leq n$. In Section 25.2 we show that the symmetrized spikes generate $\mathsf{S}(n)$ as an A_2-module for $n = 2$, but not for $n \geq 3$. The main result of this chapter, Theorem 25.3.1, states that $\mathsf{S}(3)$ is generated by the symmetrized spikes together with symmetrizations of three sequences of monomials, those of type C_1 occurring in degrees d such that $\mu(d) = 2$, and those of types C_2 and C_3 in degrees d such that $\mu(d) = 1$. The proof for the case $\mu(d) = 2$ is given in Section 25.3, and that for the case $\mu(d) = 1$ in Section 25.4.

In Section 25.5 we give a proof of the classical Wu formula for evaluation of Steenrod squares on elementary symmetric functions. This formula is included because of its historic importance in algebraic topology, but we make no use of it in studying the symmetric hit problem, as we work throughout with the basis of monomial symmetric functions for $\mathsf{S}(n)$.

25.1 The hit problem for permutation groups

As for the Dickson algebra $\mathsf{D}(n)$ in Chapter 15, we can consider both the 'absolute' problem of determining a minimal set of A_2-generators for $\mathsf{P}(n)^G$ and the 'relative' problem of determining which elements of $\mathsf{P}(n)^G$ are hit in $\mathsf{P}(n)$. When $|G|$ is even, there is no general process for converting a hit equation in $\mathsf{P}(n)$ into a hit equation in $\mathsf{P}(n)^G$. For example, $Sq^1(x^2y) = x^2y^2$ in $\mathsf{P}(2)$, but x^2y^2 is not in the image of Sq^1 acting on $\mathsf{S}(2)$, although $x^2y^2 = Sq^2(xy)$ is hit in $\mathsf{S}(2)$. However, this can be done when $|G|$ is odd.

Proposition 25.1.1 *Let $G \subset \mathsf{GL}(n)$ be a group of odd order, and let $f \in \mathsf{P}(n)^G$ be hit in $\mathsf{P}(n)$. Then f is hit in $\mathsf{P}(n)^G$.*

Proof Let $f = \sum_{i\geq 1} Sq^i(h_i)$, where $h_i \in \mathsf{P}(n)$. Then

$$\sum_{A\in G} f\cdot A = \sum_{i\geq 1}\sum_{A\in G} Sq^i(h_i)\cdot A = \sum_{i\geq 1} Sq^i(s_i), \tag{25.1}$$

where $s_i = \sum_{A\in G} h_i \cdot A \in \mathsf{P}(n)^G$. Since $f\cdot A = f$ for all $A \in G$ and $|G|$ is odd, $\sum_{A\in G} f\cdot A = |G|f = f$. Hence (25.1) is a hit equation for f in $\mathsf{P}(n)^G$. □

A useful case when a hit equation in $\mathsf{P}(n)$ can be symmetrized over a permutation group $G \subseteq \mathsf{W}(n)$ occurs when the stabilizer of the monomial f is the identity matrix I_n, so that the orbit sum $\sigma_n(f) = \sum_{A\in G} f\cdot A$.

Proposition 25.1.2 *Let $G \subseteq \mathsf{W}(n)$ be a permutation group. Let f be a monomial in $\mathsf{P}^d(n)$ whose stabilizer in G is the identity element. Then any hit equation $f = f' + \sum_{i>0} Sq^i(h_i)$, where $f' \in \mathsf{P}^d(n)$ and $h_i \in \mathsf{P}^{d-i}(n)$ for $i > 0$, can be symmetrized to give a hit equation*

$$\sigma_G(f) = \sum_{A\in G} f'\cdot A + \sum_{i>0} Sq^i\left(\sum_{A\in G} h_i\cdot A\right)$$

in $\mathsf{P}(n)^G$. In particular, if f has distinct exponents and is hit in $\mathsf{P}(n)$, then $\sigma_G(f)$ is hit in $\mathsf{P}(n)^G$. □

Recall that $\omega_1(f)$ is the number of odd exponents of a monomial $f \in \mathsf{P}(n)$.

Proposition 25.1.3 *Let $G \subseteq \mathsf{W}(n)$ be a permutation group, and let f be a monomial in $\mathsf{P}^d(n)$. If $\omega_1(f) < \mu(d)$, then $\sigma_G(f)$ is hit in $\mathsf{P}(n)^G$.*

Proof Let S be the stabilizer of f in G, and let $SA_1, \ldots, SA_t$ be the right cosets of S in G, so that $\sigma_G(f) = \sum_{i=1}^t f\cdot A_i$. Let $f = f_1 f_2^2$ where $f_1 = x_{i_1}\cdots x_{i_r}$ is a product of $r = \omega_1(f)$ distinct variables, and let $k = \deg f_2$, so that $d = 2k + r$ and $f = f_1 Sq^k(f_2)$. If $A \in S$, then $f_1\cdot A = f_1$ and $f_2\cdot A = f_2$, since A acts by permuting the variables. For any element $\theta \in \mathsf{A}_2$, $(\theta(f_1)f_2)\cdot A = (\theta(f_1)\cdot A)(f_2\cdot A) = (\theta(f_1\cdot A))(f_2\cdot A) = \theta(f_1)f_2$. Hence the stabilizer of $\theta(f_1)f_2$ contains S, and it follows that $\sum_{i=1}^t(\theta(f_1)f_2)\cdot A_i$ is a scalar multiple of the coset sum of $\theta(f_1)f_2$. In particular, $\sum_{i=1}^t(\theta(f_1)f_2)\cdot A_i \in \mathsf{P}(n)^G$. By Proposition 2.5.1

$$f_1 Sq^k(f_2) + Xq^k(f_1)f_2 = \sum_{j=1}^{k} Sq^j(Xq^{k-j}(f_1)f_2) \tag{25.2}$$

for all k. Since $r < \mu(d)$, $r < \mu(k)$ by Proposition 2.4.5, and so $Xq^k(f_1) = 0$ since Xq^k has excess $\mu(k)$. Hence (25.2) is a hit equation for f in $\mathsf{P}(n)$ such that every term on the right hand side is fixed by A for all $A \in S$. Summing (25.2) over the coset representatives for S in G, we obtain

$$\sigma_G(f) = \sum_{i=1}^{t} \left(\sum_{j=1}^{k} Sq^j (Xq^{k-j}(f_1)f_2 \right) \cdot A_i = \sum_{j=1}^{k} Sq^j \left(\sum_{i=1}^{t} (Xq^{k-j}(f_1)f_2) \cdot A_i \right).$$

Since $\sum_{i=1}^{t}(Xq^{k-j}(f_1)f_2) \cdot A_i$ is G-invariant for all j, this is a hit equation for $\sigma_G(f)$ in $\mathsf{P}(n)^G$. □

By iterating the standard hit equation (25.2) on compositions of operations Sq^k and using linearity, we obtain the following more general formula, which we use to solve the hit problem for $\mathsf{S}(3)$ in Theorem 25.3.1.

Proposition 25.1.4 *For homogeneous polynomials $f_1, f_2 \in \mathsf{P}(n)$ and $\theta \in \mathsf{A}_2$, there is a hit equation of the form*

$$f_1\theta(f_2) + (\chi(\theta)f_1)f_2 = \sum_{j>0} Sq^j \left(\sum_{\ell} \theta_{j,\ell}(f_1)\phi_{j,\ell}(f_2) \right),$$

where the operations $\theta_{j,\ell}, \phi_{j,\ell} \in \mathsf{A}_2$ depend on θ but not on f_1 and f_2. In particular $f_1\theta(f_2) \sim (\chi(\theta)f_1)f_2$. □

We next generalize Proposition 25.1.3, so as to obtain the symmetric analogue of Theorem 6.3.12.

Proposition 25.1.5 *Let $G \subseteq \mathsf{W}(n)$ be a permutation group. Let f be a monomial in $\mathsf{P}^d(n)$ where $\mu(d) \leq n$. If $\omega(f) <_l \omega^{\min}(d)$, then $\sigma_G(f)$ is hit in $\mathsf{P}(n)^G$.*

Proof Let $s \in \mathsf{P}^d(n)$ be a minimal spike, and let $k \geq 1$ be minimal such that $\omega_k(f) < \omega_k(s)$. We argue by induction on k. In the base case $k = 1$, $\omega_1(s) = \mu(d)$, and the result follows from Proposition 25.1.3. For the inductive step, let $f = f_1 f_2^{2^k}$ and $s = s_1 s_2^{2^k}$, where all exponents of f_1 are $< 2^k$. Then s_2 is a minimal spike in degree $\deg f_2$, and $\omega_1(f_2) < \omega_1(s_2)$. Applying the case $k = 1$ in degree $\deg f_2$, we obtain a hit equation of the form $f_2 = \sum_{i>0} Sq^i(h_i)$, where the stabilizer S of f in G also stabilizes h_i for each i. Since S also stabilizes f_1 and f_2, $\sigma_G(f)$ is symmetrically reducible to a sum of symmetric polynomials whose ω-sequences are $\leq_l \omega(f)$ at a position $< k$. This completes the induction. □

The next result is the analogue of the Peterson conjecture (Theorem 2.5.5) for invariants of permutation groups.

Theorem 25.1.6 *If $G \subseteq \mathsf{W}(n)$ is a permutation group, then $\mathsf{Q}^d(\mathsf{P}(n)^G) \neq 0$ if and only if $\mu(d) \leq n$.*

Proof If $\mu(d) \leq n$, then $\mathsf{P}^d(n)$ contains a spike s, and the symmetrized spike $\sigma_G(s) \in \mathsf{P}^d(n)^G$ is not hit in $\mathsf{P}(n)$. Hence $\sigma_G(s)$ is not hit in $\mathsf{P}(n)^G$. Conversely, if $\mu(d) > n$ then the orbit sums $\sigma_G(f)$ of monomials $f \in \mathsf{P}^d(n)$ form a basis of $\mathsf{P}^d(n)^G$, so it suffices to show that every orbit sum of degree d is hit in $\mathsf{P}(n)^G$. This follows from Proposition 25.1.3, since $\omega_1(f) \leq n < \mu(d)$. □

If $G \subseteq \mathsf{GL}(n)$ has odd order, it follows from Proposition 25.1.1 and Theorem 2.5.5 that $\mathsf{P}(n)^G$ has no generators in degree d when $\mu(d) > n$. But if G is not a permutation group, there may also be no generators in certain degrees d such that $\mu(d) \leq n$, as we have seen in the cases of $\mathsf{D}(2)$, $\mathsf{P}(2)^{H_2}$ and $\mathsf{P}(3)^{H_3}$ in Chapter 15. In general, the orbit sum $\sigma_G(s)$ of a spike s may contain no spikes. For example, when $n = 2$, $G = H_2 \cong \mathbb{Z}/3$ and $s = x^3y$, $\sigma_G(s) = x^4 + x^2y^2 + y^4 = \mathsf{d}_1^2$ is hit, and it is the only nonzero G-invariant of degree 4.

The filtration of $\mathsf{P}(n)$ by the left order on ω-sequences of monomials f gives a filtration on invariant polynomials $\mathsf{P}(n)^G$ for all subgroups $G \subseteq \mathsf{GL}(n)$, where the ω-sequence of a polynomial $f \in \mathsf{P}(n)^G$ is the maximum of the ω-sequences of its terms (Definition 6.2.1). When $G \subseteq \mathsf{W}(n)$, all terms of f have the same ω-sequence and $\omega(\sigma_G(f)) = \omega(f)$, and we have the following sufficient condition for left reducibility of $\sigma_G(f)$.

Proposition 25.1.7 *Let $G \subseteq \mathsf{W}(n)$ be a permutation group. Let $f = f_1 f_2^{2^k}$ be a monomial in $\mathsf{P}(n)$, where all exponents of f_1 are $< 2^k$, and let S be the stabilizer of f in G. If $f_2 = \sum_{i \geq 1} Sq^i(h_i)$, where $h_i \in \mathsf{P}(n)^S$ for all i, then $\sigma_G(f)$ is left reducible in $\mathsf{P}(n)^G$.*

Proof The 2^kth power of the given hit equation gives $f_2^{2^k} = \sum_{i \geq 1} Sq^{2^k i}(h_i^{2^k})$. Since G is a permutation group, $f_1 \cdot A = f_1$ and $f_2 \cdot A = f_2$ for all $A \in S$. The χ-trick gives $f_1(Sq^i(h_i^{2^k})) = Xq^i(f_1)h_i^{2^k} + \sum_{j=1}^{i} Sq^j(Xq^{j-i}(f_1)h_i^{2^k})$ for each $i > 0$. All the terms in this equation are stabilized by S. Summing over coset representatives $A_1, \ldots, A_t$ for S in G, we obtain a hit equation

$$\sigma_G(f) = \sum_{\ell=1}^{t} Xq^i(f_1 \cdot A_\ell)(h_i \cdot A_\ell)^{2^k} + \sum_{j=1}^{i} Sq^j \left(\sum_{\ell=1}^{t} Xq^{j-i}(f_1 \cdot A_\ell)(h_i \cdot A_\ell)^{2^k} \right).$$

This is a hit equation for $\sigma_G(f) + \sum_{\ell=1}^{t} Xq^i(f_1 \cdot A_\ell)(h_i \cdot A_\ell)^{2^k}$ in $\mathsf{P}(n)^G$. Since the action of G preserves ω-sequences, and the action of A_2^+ lowers ω-sequences in the left order, the result follows. □

Example 25.1.8 Let $f = x_1^t \cdots x_n^t$ be a monomial in $\mathsf{P}(n)$ with all exponents equal, so that $\sigma_G(f) = f$. If f is not a spike, then the block representing f has a zero column, and we can write $f = f_1 f_2^{2^k}$ for some k such that all exponents of

f_1 are $< 2^k$ and $f_2 = g^2 = Sq^r(g)$ where $r = \deg g$. By Proposition 25.1.7, f is left reducible in $\mathsf{P}(n)^G$.

Recall that the up Kameko map $\upsilon : \mathsf{P}^d(n) \to \mathsf{P}^{2d+n}(n)$ is the linear map defined by $\upsilon(f) = \mathsf{c}(n)f^2$ for $f \in \mathsf{P}^d(n)$, where $\mathsf{c}(n) = x_1 \cdots x_n$, and that if $\mu(2d+n) = n$ then υ induces a linear map $\mathsf{Q}^d(n) \to \mathsf{Q}^{2d+n}(n)$, which we again denote by υ. Clearly $\upsilon(f)$ is symmetric if f is. The following result is the symmetric analogue of Proposition 6.5.3.

Proposition 25.1.9 *Let $G \subseteq \mathsf{W}(n)$ be a permutation group. If $\mu(2d+n) = n$, then υ induces an isomorphism $\upsilon : \mathsf{Q}^d(\mathsf{P}(n)^G) \longrightarrow \mathsf{Q}^{2d+n}(\mathsf{P}(n)^G)$.*

Proof If $f \in \mathsf{P}^d(n)$ is a monomial, then $\upsilon(\sigma_G(f)) = \sigma_G(\upsilon(f))$, so υ restricts to a map $\upsilon : \mathsf{P}^d(n)^G \to \mathsf{P}^{2d+n}(n)^G$. Let $h \in \mathsf{P}^d(n)^G$ satisfy the symmetric hit equation $h = \sum_{i \geq 1} Sq^i(h_i)$, where $h_i \in \mathsf{P}(n)^G$ for all i. Then $\upsilon(h) = \mathsf{c}(n)h^2 = \sum_{i \geq 1} \mathsf{c}(n)Sq^{2i}(h_i^2)$. By Proposition 25.1.7, $\upsilon(h)$ is left reducible in $\mathsf{P}^{2d+n}(n)^G$ to a polynomial e with $\omega_1(e) < n$. Since $\mu(2d+n) = n$, e is hit in $\mathsf{P}^{2d+n}(n)^G$ by Proposition 25.1.3, and hence $\upsilon(h)$ is hit in $\mathsf{P}^{2d+n}(n)^G$. Hence $\upsilon : \mathsf{Q}^d(\mathsf{P}(n)^G) \to \mathsf{Q}^{2d+n}(\mathsf{P}(n)^G)$ is well defined.

We shall prove that υ is surjective. By Proposition 25.1.5, $\sigma_G(f') \in \mathsf{P}^{2d+n}(n)^G$ is hit in $\mathsf{P}(n)^G$ if $\omega_1(f) < n$, since the condition $\mu(2d+n) = n$ implies that $\omega_1(s) = n$, where $s \in \mathsf{P}^{2d+n}(n)$ is a minimal spike. Hence every element of $\mathsf{Q}^{2d+n}(\mathsf{P}(n)^G)$ is represented by a polynomial in $\mathsf{P}^{2d+n}(n)^G$ of the form $\upsilon(h)$ where $h \in \mathsf{P}^d(n)^G$.

Finally we show that υ is injective. Let $\upsilon(h) = \mathsf{c}(n)h^2$ be hit in $\mathsf{P}^{2d+n}(n)^G$. In any symmetric hit equation $\mathsf{c}(n)h^2 = \sum_{i \geq 1} Sq^i(h_i)$, a monomial with $\omega_1 = n$ can appear in $Sq^i(h_i)$ only if $h_i = \mathsf{c}(n)f_i^2$ for some f_i, and then only as the term $\mathsf{c}(n)Sq^i(f_i^2)$ in the expansion of $Sq^i(\mathsf{c}(n)f_i^2)$ by the Cartan formula. In particular, $i = 2j$ is even. Hence by equating terms with $\omega_1 = n$ in the symmetric hit equation and cancelling $\mathsf{c}(n)$, we obtain the symmetric hit equation $h^2 = \sum_{j \geq 1} Sq^{2j}(f_{2j}^2)$. It follows by taking square roots that h is hit in $\mathsf{P}^d(n)^G$. □

25.2 The hit problem for $\mathsf{S}(2)$

We begin by showing that the symmetrized spikes form a minimal generating set for $\mathsf{S}(2)$. We write $\mathsf{P}(2)$ as $\mathbb{F}_2[x, y]$.

Theorem 25.2.1 *The symmetric cohit space $\mathsf{Q}^d(\mathsf{S}(2))$ has dimension 1 if $\mu(d) = 1$ or 2, and 0 otherwise. If $d = (2^{s+t} - 1) + (2^t - 1)$ where $t \geq 0$, then $\mathsf{Q}^d(\mathsf{S}(2))$ is generated by $x^{2^{s+t}-1}y^{2^t-1} + x^{2^t-1}y^{2^{s+t}-1}$ if $s > 0$, and by $x^{2^t-1}y^{2^t-1}$ if $s = 0$.*

Proof Since the symmetrized spikes cannot be hit, the result states that $f \in \mathsf{S}^d(2)$ is hit if it does not contain a spike. By Theorem 25.1.6 we may assume that $\mu(d) = 1$ or 2, and Proposition 25.1.9 reduces the case $\mu(d) = 2$ to the case $\mu(d) = 1$.

It follows from Theorem 1.8.2 that two monomials f_1 and f_2 of degree $2^s - 1$ in $\mathsf{P}(2)$ which are not spikes are equivalent modulo hits. Given a hit equation $f_1 + f_2 = \sum_{i \geq 1} Sq^i(h_i)$, we obtain a symmetric hit equation $(f_1 + f_1 \cdot S) + (f_2 + f_2 \cdot S) = \sum_{i \geq 1} Sq^i(h_i + h_i \cdot S)$, where $S = \left(\begin{smallmatrix} 0 & 1 \\ 1 & 0 \end{smallmatrix}\right)$ exchanges x and y. Hence $f_1 + f_1 \cdot S$ and $f_2 + f_2 \cdot S$ represent the same element of $\mathsf{Q}^d(\mathsf{S}(2))$. In particular, this element is represented by $h = x^{2^{s-1}} y^{2^{s-1}-1} + x^{2^{s-1}-1} y^{2^{s-1}}$. But $h = Sq^1(x^{2^{s-1}-1} y^{2^{s-1}-1})$ is a symmetric hit equation for h, so this element is 0. $\square$

The following example shows that the symmetrized spikes are not sufficient to span $\mathsf{Q}(\mathsf{S}(3))$.

Example 25.2.2 Let $c_1 = x^8 y^7 z$. Then $\sigma_3(c_1)$ is hit in $\mathsf{P}(3)$ but not in $\mathsf{S}(3)$.

We first show that $\sigma_3(c_1)$ is hit in $\mathsf{P}(3)$. By 1- and 2-back splicing operations in $\mathsf{P}(3)$ on the corresponding block C_1,

$$\begin{array}{llllll} & 0001 & & 011 & & 0101 \\ C_1 = & 111 & \sim & 1001 & \sim & 101 \\ & 1 & & 1 & & 1 \end{array} .$$

We also have

$$\begin{array}{lllllllll} 101 & & 101 & & 101 & & 011 & & 101 \\ 101 & \sim & 11 & \sim & 1001 & \sim & 1001 & + & 0101 \\ 011 & & 0001 & & 01 & & 1 & & 1 \end{array} .$$

Hence

$$Sq^1 \begin{pmatrix} 101 \\ 101 \\ 101 \end{pmatrix} = \sigma_3 \begin{pmatrix} 101 \\ 101 \\ 011 \end{pmatrix} \sim \sigma_3 \begin{pmatrix} 0001 \\ 111 \\ 1 \end{pmatrix} = \sigma_3(C_1).$$

A direct calculation shows that $\sigma_3(c_1)$ is not symmetrically hit. Suppose that $\sigma_3(c_1) = Sq^1(f_1) + Sq^2(f_2) + Sq^4(f_3)$ where f_1, f_2, f_3 are symmetric. (We can ignore Sq^8.) Then $Sq^5 Sq^1(\sigma_3(c_1)) = Sq^5 Sq^1(Sq^4(f_3)) = Sq^5 Sq^5(f_3) = Sq^9 Sq^1(f_3)$. In fact $Sq^5 Sq^1(\sigma_3(c_1)) = \sigma_3(x^{12} y^8 z^2) + \sigma_3(x^{10} y^8 z^4) = m(12,8,2) + m(10,8,4)$. We need only check $Sq^9 Sq^1(m(\lambda))$ for partitions λ of degree 12 and length 3. The result is $m(16,4,2) + m(12,8,2) + m(10,8,4)$ for $\lambda = (7,3,2)$ and $(7,4,1)$, $m(12,6,4) + m(10,8,4)$ for $\lambda = (6,5,1)$, $(6,3,3)$, $(5,5,2)$ and $(5,4,3)$, and 0 otherwise. Since $m(12,8,2)+m(10,8,4)$ is not a linear combination of these, we have a contradiction.

25.3 The hit problem for S(3)

The following theorem gives a solution of the hit problem for S(3).

Theorem 25.3.1 *The dimension of the symmetric cohit space $Q^d(S(3))$ is 0 unless $d=(2^{u+t+s}-1)+(2^{u+t}-1)+(2^u-1)$, where $u,s,t\geq 0$. In this case, the dimension is independent of u when $s>0$ and depends on t and s as follows.*

	$u=0,s=0$	$s=1$	$s=2$	$s=3$	$s\geq 4$
$t=0$	1	1	2	3	4
$t\geq 1$	1	1	2	2	2

When $u=0$, a minimal spanning set for $Q^d(S(3))$ is given by symmetrized spikes together with $\sigma_3(C_1)=m(2^{s+t-1},2^{s+t-1}-1,2^t-1)$ for $t\geq 1$, $s\geq 2$, $\sigma_3(C_2)=m(3,2^{s-1}-2,2^{s-1}-2)$ for $t=0$, $s\geq 4$ and $\sigma_3(C_3)=m(2^{s-1}-1,2^{s-2},2^{s-2})$ for $t=0$, $s\geq 3$, where

$$C_1=\begin{matrix}0\cdots0\,0\cdots0\,1\\1\cdots1\,1\cdots1\\1\cdots1\end{matrix}\;,\quad C_2=\begin{matrix}1\,1\\0\,1\,1\cdots1\\0\,1\,1\cdots1\end{matrix}\;,\quad C_3=\begin{matrix}1\cdots1\,1\\0\cdots0\,1\\0\cdots0\,1\end{matrix}\;.$$

In this chapter we prove that these polynomials span $Q(S(3))$. The proof of Theorem 25.3.1 is completed in Chapter 26, where we show that this spanning set is minimal by considering the dual problem.

Since the result for $\mu(d)>3$ follows from Theorem 25.1.6, we may assume that $d=(2^{s+t+u}-1)+(2^{t+u}-1)+(2^u-1)$ where $u,t,s\geq 0$. If $u>t>s>0$, this is the only spike partition of d, and the corresponding symmetrized spike is $m(\lambda)$ where $\lambda=(2^{s+t+u}-1,2^{t+u}-1,2^u-1)$. By Proposition 25.1.9, we can reduce the problem to the cases $\mu(d)=2$ and $\mu(d)=1$ by using the Kameko maps. We show in this section that the stated polynomials span $Q^d(S(3))$ in the case $\mu(d)=2$, and prove a similar result for the case $\mu(d)=1$ in Section 25.4.

Since the 2-variable case gives only symmetrized spikes as generators, we may start with the symmetrization $\sigma_3(F)$ of a 3-block $F=\begin{pmatrix}F_1\\F_2\\F_3\end{pmatrix}$ with nonzero rows, so that F involves all three variables. If $F_1=F_2=F_3$ then F

is symmetrically reducible in the left order by Example 25.1.8, so we may assume that F_1 is not equal to F_2 or F_3. Thus $\sigma_3(F) = m_3(F)$ is the sum of 3 or 6 blocks obtained by inserting F_1 into a 2-block $\widehat{F}$ whose rows are F_2 and F_3 in some order, to obtain a 3-block $F \cdot A$, $A \in \mathsf{W}(3)$.

Proposition 25.3.2 *Let $\mu(d) = 2$, so that $d = (2^{s+t} - 1) + (2^t - 1)$ where $s \geq 0$ and $t > 0$, and let $\lambda = (2^{s+t} - 1, 2^t - 1, 0)$. Then $\mathsf{Q}(\mathsf{S}^d(3))$ is spanned by the symmetrized spike $m(\lambda)$ if $s = 0$ or 1, and by $m(\lambda)$ and $\sigma_3(C_1)$ if $s \geq 2$.*

Proof The proof is in three steps.

Step 1. We show that $\sigma_3(F)$ is symmetrically equivalent to a sum of symmetrized blocks $\sigma_3(B)$ such that $\alpha_1(B) = 1$.

By Theorem 25.2.1, $\sigma_2(\widehat{F}) = \sum_i \theta_i(S_i)$ where $\theta_i \in \mathsf{A}_2$ and each S_i is a symmetrized spike in $\mathsf{S}(2)$. We next use the χ-trick 25.1.4 on each term $\theta_i(S_i)$ to transfer the action of $\theta_i \in \mathsf{A}_2$ on the spike S_i in two variables to the action of $\chi(\theta_i)$ on the monomial in the third variable represented by F_1. In the case where $f_1 = F_1$ is row 1 of $F \cdot A$ (i.e. $A = I_3$ or S_2) and f_2 is a term in S_i represented by a 2-block R, the hit equation 25.1.4 takes the form

$$\begin{matrix} F_1 \\ \theta_i(R) \end{matrix} = \begin{matrix} \chi(\theta_i)F_1 \\ R \end{matrix} + \sum_{j>0} Sq^j \left(\sum_{\ell} \begin{matrix} \theta_{j,\ell}(F_1) \\ \phi_{j,\ell}(R) \end{matrix} \right). \tag{25.3}$$

If the two rows of R are different, then a second hit equation is obtained by exchanging them (25.3), and we symmetrize on the second and third variables by adding these equations. By summing over i, we therefore obtain a hit equation involving $F_1\sigma_2(\widehat{F})$. Finally we symmetrize over all three variables by adding the corresponding equations in which F_1 appears as the second or third row, to obtain a hit equation $\sigma_3(F) = \sum_i \sigma_3(G_i) + \sum_{j>0} Sq^j(H_j)$, where each block G_i has rows 2 and 3 of spike form, and where $H_j \in \mathsf{S}(3)$.

Since each block G_i involves all three variables and $\deg G_i = d$ is even, row 1 of G_i has an initial section of the form $0 \cdots 0\,1 = Sq^1(1 \cdots 1)$. Thus, in a typical case, we have a hit equation of the form

$$Sq^1 \begin{pmatrix} 1\,1\,0\,1 \\ 1\,1 \\ 1\,1\,1 \end{pmatrix} = \begin{matrix} 0\,0\,1\,1 \\ 1\,1 \\ 1\,1\,1 \end{matrix} + \begin{matrix} 1\,1\,0\,1 \\ 0\,0\,1 \\ 1\,1\,1 \end{matrix} + \begin{matrix} 1\,1\,0\,1 \\ 1\,1 \\ 0\,0\,0\,1 \end{matrix},$$

where the first term on the right is G_i. Rows 2 and 3 of G_i are of spike type, but may have equal length; however, in all cases row 1 of G_i is different from row 2 or row 3. By symmetrizing this hit equation with respect to permutations of the rows, we therefore obtain a hit equation in $\mathsf{S}(3)$ of the form $\sigma_3(G_i) =$

$\sigma_3(H_i)+\sigma_3(H_i')$, where H_i and H_i' are blocks in which either row 2 or row 3 has α-count 1 and initial entry 0. This completes Step 1.

Step 2. We show that $\sigma_3(F)$ is symmetrically equivalent to a sum of symmetrized blocks $\sigma_3(B)$, where B has at least two rows in spike form and at least one row with α-count 1. This is done by repeating the first part of Step 1, beginning with a symmetrized block $\sigma_3(F)$, where $\alpha_1(F)=1$ and row 1 of F is different from row 2 or row 3. By applying (25.3), we can bring rows 2 and 3 to spike form, while α_1 cannot be further reduced.

Step 3. Since $\mu(d)=2$, $d=(2^{t+s}-1)+(2^s-1)$ where $s,t\geq 0$. We show that $\sigma_3(B)$ is symmetrically hit if $s=0$ or $s=1$, and that, for $s>1$, B can be reduced to a row permutation of the block C_1 of Theorem 25.3.1.

Let F be a block in $\mathsf{P}^d(3)$ with $\alpha_1(F)=1$ and rows 2 and 3 in spike form, corresponding to the monomial $x^{2^a}y^{2^b-1}z^{2^c-1}$, where $b,c\geq 1$ and $a\geq 1$ since $d=0 \bmod 2$. Then $d=(2^a-1)+(2^b-1)+(2^c-1)+1$, and since $\mu(d)=2$ at least two of a,b,c must be equal. If $b\neq c$, then the cases $a=b$ and $a=c$ give the following forms for F:

$$B_1=\begin{array}{lll} 0\cdots 0 & 0\cdots 0 & 1\\ 1\cdots 1 & 1\cdots 1 & \\ 1\cdots 1 & & \end{array}\ ,\qquad B_2=\begin{array}{lll} 0\cdots 0 & 1 & \\ 1\cdots 1 & 1 & 1\cdots 1\\ 1\cdots 1 & & \end{array}\ ,$$

while if $b=c$ then a is arbitrary. However $\omega^{\min}(d)=(2,\ldots,2,1,\ldots,1)$ with t 2s and s 1s, and if $\omega(F)<\omega^{\min}(d)$ then $\sigma_3(F)$ is symmetrically hit, by Proposition 25.1.5. Hence we may assume that $a\leq b$, so that F has the form

$$B_3=\begin{array}{lll} 0\cdots 0 & 1 & \\ 1\cdots 1 & 1 & 1\cdots 1\\ 1\cdots 1 & 1 & 1\cdots 1 \end{array}\ .$$

Clearly $B_1=C_1$, and in this case $s\geq 2$. We shall show that $\sigma_3(B_2)$ is symmetrically hit and that B_3 can be reduced to C_1. For B_2, consider the blocks

$$H_2=\begin{array}{ll} 1\cdots 1 & \\ 1\cdots 1 & 1\cdots 1\\ 1\cdots 1 & \end{array}\ ,\qquad E_2=\begin{array}{lll} 1\cdots 1 & & \\ 0\cdots 0 & 0\cdots 0 & 1\\ 1\cdots 1 & & \end{array}\ ,$$

where $\deg H_2=d-1$ and $\deg E_2=d$. We have $Sq^1(\sigma_3(H_2))=\sigma_3(B_2)+\sigma_3(E_2)$. Since $\omega(E_2)<\omega^{\min}(d)$, $\sigma_3(E_2)$ is hit in $\mathsf{S}(3)$ by Proposition 25.1.5. Hence

$\sigma_3(B_2)$ is hit in $\mathsf{S}(3)$. For B_3, consider the blocks

$$H_3 = \begin{matrix} 1\cdots 1 & 0 & \\ 1\cdots 1 & 1 & 1\cdots 1 \\ 1\cdots 1 & 1 & 1\cdots 1 \end{matrix}, \quad E_3 = \begin{matrix} 1\cdots 1 & 0 & & \\ 1\cdots 1 & 1 & 1\cdots 1 & \\ 0\cdots 0 & 0 & 0\cdots 0 & 1 \end{matrix},$$

where $\deg H_3 = d-1$ and $\deg E_3 = d$. We have $Sq^1(\sigma_3(H_3)) = \sigma_3(B_3) + \sigma_3(E_3)$. Since E_3 is obtained by permuting the rows of C_1, it follows that $\sigma_3(B_3)$ is equivalent to $\sigma_3(C_1)$. □

25.4 The case $n = 3$, $\mu(d) = 1$

In this section, we show that the stated polynomials in Theorem 25.3.1 span $\mathsf{Q}^d(\mathsf{S}(3))$ in the case $\mu(d) = 1$.

Proposition 25.4.1 *Let $\mu(d) = 1$, so that $d = 2^s - 1$ where $s > 0$. Then $\mathsf{Q}(\mathsf{S}^d(3))$ is spanned by $x+y+z$ if $s = 1$, by $x^3+y^3+z^3$ and xyz if $s = 2$, and by $x^7+y^7+z^7$, $x^3y^3z + x^3yz^3 + xy^3z^3$ and $x^3y^2z^2 + x^2y^3z^2 + x^2y^2z^3$ if $s = 3$. For $s \geq 4$, $\mathsf{Q}(\mathsf{S}^d(3))$ is spanned by the symmetrized spikes $m(2^s-1,0,0)$, $m(2^{s-1}-1, 2^{s-1}-1, 1)$ together with $\sigma_3(C_2)$ and $\sigma_3(C_3)$.*

Proof The proof is again in three steps. We consider symmetrizations $\sigma_3(F)$ of blocks F of degree d having three nonzero rows. We may assume that F is not a spike, and in particular that $d = 2^s - 1$ where $s > 2$.

Step 1. We reduce to symmetrizations $\sigma_3(F)$ of blocks F with $\omega_1(F) = 1$.

If $\omega_1(F) = 3$ then $F = \begin{smallmatrix}1\\1\\1\end{smallmatrix}|B$ where $\deg B = d' = 2^{s-1} - 2$. By the case $\mu(d') = 2$ proved above, $\sigma_3(B)$ is hit in $\mathsf{S}(3)$ modulo the symmetrized spike $\sigma_3((xy)^{2^{s-2}-1})$. Thus, modulo the symmetrized spike $\sigma_3((xy)^{2^{s-1}-1}z)$, the monomial corresponding to F has the form $xyzb^2$ where b is represented by the subblock B of F, and $\sigma_3(b) = \sum_{k>0} Sq^k(\sigma_3(b_k))$ for some $b_k \in \mathsf{P}^{d'-k}(3)$. Thus $\sigma_3(xyzb^2) = xyz(\sigma_3(b))^2 = xyz\left(\sum_{k>0} Sq^k(\sigma_3(b_k))\right)^2 = xyz\sum_{k>0} Sq^{2k}(\sigma_3(b_k^2))$. Hence by the χ-trick 2.5.2,

$$\sigma_3(xyzb^2) = \sum_{k>0}(Xq^{2k}(xyz))\sigma_3(b_k^2) + \sum_{i,k>0} Sq^i(Xq^{2k-i}(xyz)\sigma_3(b_k^2)).$$

Since $\omega_1 = 1$ for the first term on the right, and the second term on the right is hit in $\mathsf{S}(3)$, we have reduced the problem to the case $\omega_1(F) = 1$. This completes Step 1.

Step 2. We show that $\sigma_3(F)$ is hit in $\mathsf{S}(3)$ if $\omega_1(F) = 1$ and F has three distinct rows. We may assume that row 1 of F has leading entry 1 and rows 2 and 3 have leading entry 0. Since F has three distinct rows, by Proposition 25.1.2 any hit equation for F can be symmetrized to give a hit equation for $\sigma_3(F)$.

By 1-back splicing we can reduce F modulo hit polynomials to a sum of blocks in which row 1 has spike form, while maintaining the initial 0 entries of rows 2 and 3. We can then splice the initial section of 0s in row 2 of such a block to make the α-count of row 1 equal to 1. By further 1-back splicing we can reduce to a sum of blocks of the form

$$C = \begin{array}{c|c} 1 & 0\cdots 0 \\ 0 & \multirow{2}{*}{B} \\ 0 & \end{array}$$

where the subblock B represents a monomial in $\mathsf{P}^{d'}(2)$, $d' = 2^{s-1}-1$. By Theorem 25.2.1, $\mathsf{Q}^{d'}(\mathsf{S}(2))$ is generated by the symmetrized spike $y^{d'}+z^{d'}$. Since we are assuming that the monomial in $\mathsf{P}(3)$ corresponding to F involves all three variables, both rows of B are nonzero, and hence $\sigma_2(B)$ is hit in $\mathsf{S}^{d'}(2)$. By symmetrizing over the three 2-variable subsets of x, y, z it follows that $\sigma_3(B)$ is hit in $\mathsf{S}^{d'}(3)$. We next apply the χ-trick to the block C as in Step 1, to reduce C to hit elements of $\mathsf{S}(3)$ modulo elements of the form $Xq^{2k}(x)\sigma_3(b_k^2)$. Since such elements are 0, this completes Step 2.

Step 3. We show that if $\omega_1(F)=1$ and F has two equal rows, then $\sigma_3(F)$ is equivalent to $\sigma_3(C_3)$ when $s=3$, and $\sigma_3(F)$ is equivalent to a linear combination of $\sigma_3(C_2)$ and $\sigma_3(C_3)$ when $s>3$.

The block F has the form

$$F = \begin{array}{c|c} 1 & F_1 \\ 0 & \multirow{2}{*}{B} \\ 0 & \end{array}$$

where the subblock B has two equal rows. We next apply the argument of Step 1 of the case $\mu(d)=2$ to the subblock $\frac{F_1}{B}$. By Theorem 25.2.1, $\sigma_2(B) = \sum_i \theta_i(S_i)$ where $\theta_i \in \mathsf{A}_2$ and each S_i is a symmetrized spike in $\mathsf{S}(2)$. Using the χ-trick we obtain a hit equation of the form (25.3), where C_1 and C_2 are the rows of S_i. If $C_1 \neq C_2$ then a second hit equation is obtained by exchanging C_1 and C_2 in (25.3), and we symmetrize on the second and third variables y and z by adding these equations. We therefore obtain a hit equation involving F_1B by summing over i. By prefixing the first column of F, we obtain an equation in $\mathsf{P}^d(3)$ of the form $F = \sum_i B_i$, where

$$B_i = \begin{array}{c|c} 1 & \chi(\theta_i)F_1 \\ 0 & \\ 0 & S_i \end{array} + \sum_{j>0} \begin{array}{c|c} 1 & \\ 0 & Sq^j \left(\sum_\ell \begin{array}{c} \theta_{j,\ell}(F_1) \\ \phi_{j,\ell}(S_i) \end{array} \right) \\ 0 & \end{array}.$$

By applying the χ-trick again to each term in the sum on the right, it can be written in the form $\sum_{k>0} Sq^k(E_k)$, where E_k represents a polynomial in $\mathsf{P}^{d-k}(3)$ which is symmetric in y and z. Finally we symmetrize over all three variables by adding the corresponding equations in which F_1 appears as the second or third row, to obtain a hit equation $\sigma_3(F) = \sum_i \sigma_3(G_i) + \sum_{j>0} Sq^j(H_j)$, where each block G_i has rows 2 and 3 of spike form, and where $H_j \in \mathsf{S}(3)$.

The argument so far has not used the assumption that B has two equal rows. However we can now use Step 2 to reduce to blocks $G = G_i$ with two equal rows 2 and 3 of spike form. Such a block must have the form

$$G = \begin{array}{l} 1\,1\,0 \cdots 0 \\ 0\,1\,1 \cdots 1 \\ 0\,1\,1 \cdots 1 \end{array} \quad \text{or} \quad \begin{array}{l} 1\,1\,0 \cdots 0\,0\,1 \cdots 1 \\ 0\,1\,1 \cdots 1 \\ 0\,1\,1 \cdots 1 \end{array}.$$

In the first case, $G = C_2$. If $d = 7$, then $G = C_2 = C_3$. In the second case, let

$$H_1 = \begin{array}{l} 1 \cdots 1\,1\,0\,1 \cdots 1 \\ 0 \cdots 0\,1 \\ 0 \cdots 0\,1 \end{array}, \quad H_2 = \begin{array}{l} 1\,1\,1 \cdots 1\,1\,0\,1 \cdots 1 \\ 0\,1\,1 \cdots 1 \\ 0\,1\,1 \cdots 1 \end{array},$$

where H_2 is formed by splicing the 0-section in row 1 of G. Then we have the symmetric hit equation

$$Sq^4(\sigma_3(H_2)) = \sigma_3(G) + \sigma_3(H_1) + E,$$

where $E \in \mathsf{S}^d(3)$ is a sum of blocks with three distinct rows and $\omega_1 = 1$. In a typical case, this is the sum of three equations of the form

$$Sq^4 \left(\begin{array}{l} 1\,1\,1\,1\,0\,1 \\ 0\,1\,1 \\ 0\,1\,1 \end{array} \right) = \begin{array}{l} 1\,1\,0\,0\,1\,1 \\ 0\,1\,1 \\ 0\,1\,1 \end{array} + \begin{array}{l} 1\,1\,1\,1\,0\,1 \\ 0\,0\,0\,1 \\ 0\,0\,0\,1 \end{array} + E_1,$$

where

$$E_1 = \begin{array}{l} 111101 \\ 0101 \\ 011 \end{array} + \begin{array}{l} 111101 \\ 011 \\ 0101 \end{array} + \begin{array}{l} 100011 \\ 0001 \\ 011 \end{array} + \begin{array}{l} 100011 \\ 0001 \\ 011 \end{array}.$$

By Step 2, E is hit in $\mathsf{S}(3)$, so $\sigma_3(G) = \sigma_3(H_1)$ in $\mathsf{Q}^d(\mathsf{S}(3))$. Finally, iterated 1-back splicing of the first row of H_1 shows that $\sigma_3(H_1) = \sigma_3(C_3) + E'$ in $\mathsf{Q}^d(\mathsf{S}(3))$, where $E' \in \mathsf{S}^d(3)$ is again a sum of blocks with three distinct rows and $\omega_1 = 1$. In a typical case, this equation in $\mathsf{S}(3)$ is the sum of three equations of the form

$$Sq^4 \left(\begin{array}{l} 111 \\ 01 \\ 01 \end{array} \right) = \begin{array}{l} 1101 \\ 01 \\ 01 \end{array} + \begin{array}{l} 111 \\ 001 \\ 001 \end{array} + \begin{array}{l} 1001 \\ 001 \\ 01 \end{array} + \begin{array}{l} 1001 \\ 01 \\ 001 \end{array}.$$

Again E' is hit in $\mathsf{S}(3)$ by Step 2, and so we conclude that $\sigma_3(F) = \sigma_3(C_3)$. Hence $\sigma_3(C_2)$ and $\sigma_3(C_3)$ generate $\mathsf{Q}^d(\mathsf{S}(3))$ when $\mu(d) = 1$. □

25.5 The Wu formula

The following expression for the action of Sq^k on the elementary symmetric functions e_i is known as the Wu formula. Note that by (2.2) $\binom{k-r}{i} = (-1)^i\binom{r-k+i-1}{i}$, and all binomial coefficients are taken mod 2.

Proposition 25.5.1 *For $1 \leq i \leq n$, let $e_r \in \mathsf{S}(n)$ be the rth elementary symmetric function, and let $0 \leq k \leq r$. Then*

$$Sq^k(e_r) = \sum_{i=0}^{k} \binom{k-r}{i} e_{k-i}e_{r+i}.$$

Proof By the Cartan formula,

$$Sq^k(x_1 \cdots x_r) = \sigma_r(x_1^2 \cdots x_k^2 x_{k+1} \cdots x_r).$$

By summing over all $\binom{n}{r}$ subsets $I \subseteq Z[n]$ with $|I| = r$, it follows that $Sq^k(e_r) = m(2^k, 1^{r-k})$, the sum of all monomials in $\mathsf{P}(n)$ with k exponents 2, $r-k$ exponents 1 and $n-r$ exponents 0.

Since all terms in the symmetric polynomial $e_s e_t$ have exponents 2, 1 or 0, $e_s e_t$ can be expressed as a sum of monomial symmetric functions $m(2^a, 1^b)$, where $a+b \leq n$ and $s+t = 2a+b$. For example, $e_2e_3 = m(2^2, 1) + 3m(2, 1^3) +$

$10m(1^5)$ when $n \geq 5$, and so $e_2e_3 = m(2^2,1) + m(2,1^3) \bmod 2$. The coefficients here are binomial coefficients which arise from the assignment of the variables with exponent 1 in a monomial f in $m(2^a,1^b)$ to either e_s or e_t. Hence

$$e_se_t = \sum_{a=0}^{s} \binom{b}{s-a} m(2^a1^b) = \sum_{a=0}^{s} \binom{b}{s-a} Sq^a(e_{a+b}). \tag{25.4}$$

We shall obtain the Wu formula by inverting (25.4). Putting $s = k-i$, $t = r+i$, where $1 \leq k \leq r$, (25.4) becomes

$$e_{k-i}e_{r+i} = \sum_{a=0}^{k-i} \binom{k+r-2a}{k-i-a} Sq^a(e_{k+r-a}),$$

and, setting $\ell = k-a$, we can rewrite this as

$$e_{k-i}e_{r+i} = \sum_{\ell=i}^{k} \binom{r-k+2\ell}{\ell-i} Sq^{k-\ell}(e_{r+\ell}).$$

We weight this equation by $\binom{k-r}{i}$ and sum over $0 \leq i \leq k$. After interchanging the order of summation, this gives

$$\sum_{i=0}^{k} \binom{k-r}{i} e_{k-i}e_{r+i} = \sum_{\ell=i}^{k} \left(\sum_{i=0}^{\ell} \binom{k-r}{i} \binom{r-k+2\ell}{\ell-i} \right) Sq^{k-\ell}(e_{r+\ell}).$$

The coefficient of $Sq^{k-\ell}(e_{r+\ell})$ is $\sum_{i=0}^{\ell} \binom{-s}{i}\binom{s+2\ell}{\ell-i}$, where $s = r-k$. This is the coefficient of x^ℓ in the expansion of $(1+x)^{-s}(1+x)^{s+2\ell}$ by the binomial theorem, and this is $\binom{2\ell}{\ell} = 0 \bmod 2$ if $\ell > 0$. Hence $\sum_{i=0}^{k} e_{k-i}e_{r+i} = Sq^k(e_r)$. □

25.6 Remarks

The symmetric polynomial algebra $\mathsf{S}(n)$ arises in algebraic topology as the cohomology algebra $H^*(BO(n);\mathbb{F}_2)$ of the Grassmannian of n-dimensional vector subspaces of $\mathbb{R}^\infty$. The space $BO(n)$ is the classifying space of the orthogonal group $O(n)$. The symmetric polynomials in n variables divisible by all of them can be identified with $H^*(MO(n))$, where $MO(n)$ is the Thom space of the standard n-plane bundle over $BO(n)$ associated with the bordism theory of closed smooth manifolds. Thom's classic work [212] shows that a smooth closed manifold M^d of dimension d is bordant to zero, i.e. it is the boundary of some smooth manifold, if and only if all its Stiefel–Whitney numbers vanish. As a consequence of the Peterson conjecture [168], as proved in Chapter 2, we have the following result proved by Peterson in [169]: if all products of

Stiefel–Whitney classes of the normal bundle of M^d of length greater than n vanish, then $\alpha(d) \leq n$ or M^d is bordant to zero.

In algebraic combinatorics, the algebra $\mathsf{S}(n)$ arises as the mod 2 reduction of the algebra $\Lambda(n)$ of symmetric polynomials in n variables, as in [128, Chapter 1]. In [233, Section 3] it is explained how the action of A_2 on $\mathsf{S}(n)$ can be lifted to an action of integral Steenrod operations on $\Lambda(n)$ using differential operators. The Wu formula [136, 237] for evaluating Steenrod squares on Stiefel–Whitney classes, and its generalization to odd primes, are discussed in [122], alongside further formulae for the action of Steenrod operations on symmetric polynomials. Work on the hit problem for $\mathsf{S}(n)$ appears in [99, 102, 103, 104, 105, 236]; the treatment here follows [236]. Details needed to complete some proofs in [105] and Theorem 25.3.1 can be found in [103].

No example is known (to the authors) of a permutation group G and an element of $\mathsf{P}(n)^G$ which is hit in $\mathsf{P}(n)$ but not in $\mathsf{P}(n)^G$. Proposition 25.1.1 shows that no such example can exist when G has odd order.

26
The dual of the symmetric hit problem

26.0 Introduction

In this chapter we discuss the dual form of the hit problem for $\mathsf{S}(n)$. We complete the solution in the 3-variable case by proving that the symmetrized monomials $\sigma_3(C_1)$, $\sigma_3(C_2)$, $\sigma_3(C_3)$ of Theorem 25.3.1 are not hit in $\mathsf{S}(3)$, and further that, in degrees $d = 2^s - 1$ for $s \geq 4$, $\sigma_3(C_2)$ and $\sigma_3(C_3)$ are linearly independent modulo symmetrically hit polynomials and symmetrized spikes, or in effect that $\sigma_3(C_2) + \sigma_3(C_3)$ is also not hit in $\mathsf{S}(3)$. We do this indirectly by computing elements in the kernel of down Steenrod squares acting on a coalgebra $\mathsf{DS}(3)$ dual to $\mathsf{S}(3)$. The result follows since $\dim \mathsf{Q}^d(\mathsf{S}(3)) = \dim \mathsf{K}^d(\mathsf{DS}(3))$.

In Section 26.1 we construct the coalgebra $\mathsf{DS}(n)$ and the right action of A_2 on $\mathsf{DS}(n)$. The algebra S of symmetric functions in a sequence of variables $x_1, x_2, \ldots$ is the inverse limit of the symmetric polynomial algebras $\mathsf{S}(n)$ with respect to the projection maps $j_n : \mathsf{S}(n+1) \to \mathsf{S}(n)$ defined by $j_n(e_i) = e_i$ for $1 \leq i \leq n$ and $j_n(e_{n+1}) = 0$. As an algebra, $\mathsf{S} = \mathbb{F}_2[e_1, e_2, \ldots]$ is a polynomial algebra on generators e_k, $k \geq 1$, which project to the elementary symmetric functions in $\mathsf{S}(n)$ for $k \leq n$. The coproduct $\phi : \mathsf{S} \to \mathsf{S} \otimes \mathsf{S}$ defined by $\phi(e_k) = \sum_{i+j=k} e_i \otimes e_j$ makes S into a Hopf algebra. The antipode of S exchanges the elementary symmetric function e_k with the complete symmetric function h_k for $k \geq 1$.

The dual Hopf algebra DS of S is isomorphic to S itself. In particular, it is a polynomial algebra $\mathbb{F}_2[b_1, b_2, \ldots]$ with one generator b_k in each degree $k \geq 1$. We also define $b_0 = 1$. The generator b_k is dual to the power sum function $p_k = \sum_{i \geq 1} x_i^k = m(k)$ with respect to the basis $\{m(\lambda)\}$ of S, and the product $b_{i_1} \cdots b_{i_n}$ is dual to the symmetrization $\sigma_n(f) \in \mathsf{S}$ of $f = x_1^{i_1} \cdots x_n^{i_n} \in \mathsf{P}(n)$. We define $\mathsf{DS}(n)$ for $n \geq 1$ as the sub-coalgebra of DS spanned by monomials $b_{i_1} \cdots b_{i_n}$ of length $\leq n$. The Steenrod algebra A_2 acts on the right of DS by

$b_k \cdot Sq^r = \binom{k-r}{r} b_{k-r}$ and the Cartan formula. The dual of the hit problem for S is to determine the elements which are in the kernel $\mathsf{K}(\mathsf{DS})$ of all Steenrod operations of positive degree. The dual up Kameko map κ_* maps products of length r in $\mathsf{K}^d(\mathsf{DS})$ to products of length r in $\mathsf{K}^{2d+r}(\mathsf{DS})$. In Sections 26.2 and 26.3 we find elements in $\mathsf{K}(\mathsf{DS}(2))$ and $\mathsf{K}(\mathsf{DS}(3))$ dual to the symmetrized spikes and to $\sigma_3(C_1)$, $\sigma_3(C_2)$ and $\sigma_3(C_3)$, so completing the solution of the symmetric hit problem for $n = 3$.

In Section 26.4 we introduce an algebra $\widetilde{\mathsf{A}}_2$ of operations on the left of DS which commutes with the down action of A_2 on the right. The algebra $\widetilde{\mathsf{A}}_2$ is a bigraded analogue of A_2, defined by generators $\widetilde{Sq}^a$ for $a \geq 0$ subject to the Adem relations. It differs from A_2 in that the relation $Sq^0 = 1$ in A_2 has no counterpart in $\widetilde{\mathsf{A}}_2$. The operation of $\widetilde{Sq}^0$ on DS is given by identifying $\widetilde{Sq}^0$ with the dual up Kameko map κ_*. In general, the operation of $\widetilde{Sq}^a$ on DS increases the number of variables by a. We refer to $\widetilde{\mathsf{A}}_2$ as the 'bigraded' or 'big' Steenrod algebra.

26.1 The dual algebra DS

Recall that the symmetrization $\sigma_n(f)$ of the monomial $f = x_1^{a_1} \cdots x_n^{a_n}$ in $\mathsf{P}(n)$ is the monomial symmetric function $m(\lambda)$, where $\lambda = (\lambda_1, \ldots, \lambda_n)$ is the partition obtained by writing the exponents $a_1, \ldots, a_n$ of f in decreasing order. If f involves r of the variables, where $0 \leq r \leq n$, then λ has length $\ell(\lambda) = r$.

Since the action of A_2 on f preserves the set $Y \subseteq Z[n]$ of the variables $x_1, \ldots, x_n$ which divide f, the action of A_2 on a monomial symmetric function $m(\lambda)$ in the algebra of symmetric polynomials $\mathsf{S}(n)$ preserves the length $\ell(\lambda)$. Hence $\mathsf{S}(n)$ is the direct sum of the A_2-submodules $\mathsf{MS}(r)$ spanned by the elements $m(\lambda)$ such that $\ell(\lambda) = r$, $0 \leq r \leq n$.

Let $j_n : \mathsf{P}(n+1) \to \mathsf{P}(n)$ be the linear projection which maps $x_i \in \mathsf{P}(n+1)$ to $x_i \in \mathsf{P}(n)$ for $1 \leq i \leq n$ and maps x_{n+1} to 0. The map j_n is an algebra map and a map of A_2-modules, and j_n maps $\mathsf{S}(n+1)$ to $\mathsf{S}(n)$. Its restriction $\mathsf{S}(n+1) \to \mathsf{S}(n)$ is again an algebra map and a map of A_2-modules, and we denote it also by j_n.

Since $\mathsf{S}(n) = \mathbb{F}_2[e_1, \ldots, e_n]$ where e_k is the kth elementary symmetric function in $x_1, \ldots, x_n$, we can define $j_n : \mathsf{S}(n+1) \to \mathsf{S}(n)$ directly in terms of symmetric functions by $j_n(e_k) = e_k$ for $1 \leq k \leq n$ and $j_n(e_{n+1}) = 0$. In terms of the $\mathbb{F}_2$-bases of $\mathsf{S}(n+1)$ and $\mathsf{S}(n)$ given by the monomial symmetric functions, $j_n(m(\lambda)) = m(\lambda)$ if $\ell(\lambda) \leq n$, and $j_n(m(\lambda)) = 0$ if $\ell(\lambda) = n+1$. Hence, as a map of A_2-modules, j_n maps the summand $\mathsf{MS}(r)$ of $\mathsf{P}(n+1)$ isomorphically to the summand $\mathsf{MS}(r)$ in $\mathsf{P}(n)$ for $r \leq n$, and maps the summand $\mathsf{MS}(n+1)$ to 0. We use j_n to identify the A_2-submodules $\mathsf{MS}(r)$ in $\mathsf{P}(n)$ and in $\mathsf{P}(n+1)$. This point of view allows us to combine the algebras $\mathsf{S}(n)$ for $n \geq 1$ into a large algebra S which has each of the algebras $\mathsf{S}(n)$ as a quotient, and which splits as the direct sum $\bigoplus_{r\geq 0} \mathsf{MS}(r)$ as an A_2-module.

Definition 26.1.1 The algebra S of symmetric functions in a sequence of variables $x_1, x_2, \ldots$ is the inverse limit of the algebras $\mathsf{S}(n)$ with respect to the projection maps j_n.

The description of the elements of S as 'symmetric functions in $x_1, x_2, \ldots$' is informal. We can think of expressions such as $\sum_i x_i = x_1 + x_2 + \cdots$ and $\sum_{i<j} x_i x_j = x_1x_2 + x_1x_3 + \cdots + x_2x_3 + \cdots$ as elements of S, but since these expressions have an infinite number of 'terms' they are not polynomials: they are referred to as 'functions' for lack of a better word. By definition, the elements of S are sequences $f = (f_0, f_1, f_2, \ldots)$ such that $f_n \in \mathsf{S}(n)$ and $j_n(f_{n+1}) = f_n$ for all n. For example, the informal expressions above are correctly described as the elementary symmetric functions e_1 and e_2 in S. More generally, if we extend the definition of $e_k \in \mathsf{S}(n)$ so that $e_k = 0$ if $k > n$, and similarly extend the definition of $m(\lambda)$ so that $m(\lambda) = 0$ in $\mathsf{S}(n)$ if $\ell(\lambda) > n$, then we have naturally defined elements e_k of S for $k \geq 0$, where $e_0 = 1$ is the identity element of S, and elements $m(\lambda)$ of S for all partitions λ. The complete symmetric function $h_k = \sum_{|\lambda|=k} m(\lambda)$. The A_2-module structure of S may be defined by the Wu formula 25.5.1 and the Cartan formula, but as in Chapter 25 we prefer to work with the basis $\{m(\lambda)\}$.

Proposition 26.1.2 (i) *The algebra* S *is bigraded as* $\mathsf{S} = \sum_{d,r\geq 0} \mathsf{S}^{d,r}$, *where* $\mathsf{S}^{d,r}$ *has an* $\mathbb{F}_2$*-basis given by the monomial symmetric functions* $m(\lambda)$ *indexed by partitions* λ *with modulus* $|\lambda| = d$ *and length* $\ell(\lambda) = r$.

(ii) *The algebra* S *is the polynomial algebra* $\mathbb{F}_2[e_1, e_2, \ldots]$ *generated by the elementary symmetric functions* $e_k = m(1^k) \in \mathsf{S}^{k,k}$, $k \geq 1$.

(iii) *With the coproduct* $\phi : \mathsf{S} \to \mathsf{S} \otimes \mathsf{S}$ *defined by* $\phi(e_k) = \sum_{i+j=k} e_i \otimes e_j$ *for* $k \geq 1$, *and the counit* $\varepsilon : \mathsf{S} \to \mathbb{F}_2$ *defined by* $\varepsilon(e_k) = 0$ *for* $k \geq 1$, S *is a Hopf algebra over* $\mathbb{F}_2$ *with antipode* χ *defined by* $\chi(e_k) = h_k$ *for* $k \geq 1$.

(iv) *As an* A_2*-module,* $\mathsf{S} \cong \sum_{r\geq 0} \mathsf{MS}(r)$, *where* $\mathsf{MS}(r) = \sum_{d\geq 0} \mathsf{S}^{d,r}$ *and* $Sq^k : \mathsf{S}^{d,r} \to \mathsf{S}^{d+k,r}$.

Proof Statements (i) and (ii) follow from the description of the maps j_n and the $\mathbb{F}_2$-basis and algebra structure of $\mathsf{S}(n)$. For (iii), the coproduct ϕ and the counit ε are algebra maps since by (ii) we may define such a map on S by assigning arbitrary values to the generators e_k, $k \geq 1$. The statement about χ follows from the standard identity $\sum_{i+j=k} (-1)^i e_i h_j = 0$ for $k \geq 1$ for the corresponding symmetric functions over the integers $\mathbb{Z}$. Finally (iv) follows from our previous discussion of the maps j_n as projections of A_2-modules. □

The coproduct ϕ on S can be viewed in another way. For $n \geq 0$, let $\phi_n : \mathsf{P}(2n) \to \mathsf{P}(n) \otimes \mathsf{P}(n)$ be the algebra map defined on the generators by $\phi_n(x_i) = x_i \otimes 1$ if $1 \leq i \leq n$ and $\phi_n(x_i) = 1 \otimes x_{i-n}$, if $n+1 \leq i \leq 2n$. Then ϕ_n is an isomorphism which restricts to an injective map $\mathsf{S}(2n) \to \mathsf{S}(n) \otimes \mathsf{S}(n)$, which we again denote by ϕ_n. A monomial f in $x_1,\dots,x_{2n}$ has a unique factorization as the product of monomials f_1 in $x_1,\dots,x_n$ and f_2 in $x_{n+1},\dots,x_{2n}$, and so $\phi_n(f) = f_1 \otimes f_2$. Hence ϕ_n is given on the generators $e_1,\dots,e_{2n}$ of $\mathsf{S}(2n)$ by $\phi_n(e_k) = \sum_{i+j=k} e_i \otimes e_j$, where $e_i = 0$ in $\mathsf{S}(n)$ for $i > n$. Hence we have a sequence of commutative diagrams

$$\begin{array}{ccc} \mathsf{S}(2n+2) & \xrightarrow{j_{2n} \circ j_{2n+1}} & \mathsf{S}(2n) \\ \downarrow{\scriptstyle \phi_{n+1}} & & \downarrow{\scriptstyle \phi_n} \\ \mathsf{S}(n+1) \otimes \mathsf{S}(n+1) & \xrightarrow{j_n \otimes j_n} & \mathsf{S}(n) \otimes \mathsf{S}(n) \end{array} .$$

It follows that the maps ϕ_n fit together to give the coproduct $\phi : \mathsf{S} \to \mathsf{S} \otimes \mathsf{S}$.

A similar calculation shows that the coproduct of a monomial symmetric function is given by

$$\phi(m(\lambda)) = \sum_{\lambda = \mu \cup \nu} m(\mu) \otimes m(\nu) \qquad (26.1)$$

where $\mu \cup \nu$ is obtained by putting the parts of μ and ν in decreasing order, so that $\ell(\lambda) = \ell(\mu) + \ell(\nu)$. The defining formula $\phi(e_k) = \sum_{i+j=k} e_i \otimes e_j$ is a special case. In the case $\lambda = (k)$, $m(\lambda) = p_k$ is the kth power sum, and we obtain $\phi(p_k) = p_k \otimes 1 + 1 \otimes p_k$. Thus S has a primitive element p_k in each degree $k \geq 1$. Since the elements $m(\lambda)$ are a $\mathbb{F}_2$-basis of S, and the term $m(\mu) \otimes m(\nu)$ occurs only in $\phi(\mu \cup \nu)$, it follows from (26.1) that the power sums p_k form a $\mathbb{F}_2$-basis for the primitive elements of S.

The Hopf algebra point of view allows us to dualize the symmetric hit problem. Let DS denote the graded dual of S, so that $\mathsf{DS}^d = \mathrm{Hom}(\mathsf{S}^d, \mathbb{F}_2)$ for $d \geq 0$. Then DS is a graded Hopf algebra with product $\mu_* : \mathsf{DS} \otimes \mathsf{DS} \to \mathsf{DS}$ dual to ϕ and coproduct $\phi_* : \mathsf{DS} \to \mathsf{DS} \otimes \mathsf{DS}$ dual to the product $\mu : \mathsf{S} \otimes \mathsf{S} \to \mathsf{S}$, and for $m \in \mathsf{S}^d$ and $b \in \mathsf{DS}^d$ we write $b(m)$ as a Kronecker product $\langle b, m \rangle$. For $k \geq 1$, we denote the element of DS^k element dual to the power sum p_k with respect to the $\mathbb{F}_2$-basis $\{m(\lambda)\}$ of S^k by b_k, i.e. $\langle b_k, m(\lambda) \rangle = 1$, if $\lambda = (k)$ and $\langle b_k, m(\lambda) \rangle = 0$ otherwise.

Proposition 26.1.3 (i) *The dual Hopf algebra* DS *is a polynomial algebra with one generator* b_k *in each degree* $k \geq 1$.

(ii) *The monomial basis in* $b_1, b_2, \dots$ *is dual to the basis* $\{m(\lambda)\}$ *of* S, *i.e.* $\langle b_\lambda, m(\mu) \rangle = 1$ *if* $\lambda = \mu$ *and* $\langle b_\lambda, m(\mu) \rangle = 0$ *otherwise.*

(iii) *The algebra* DS *is bigraded where* $\mathsf{DS}^{d,r}$ *has* $\mathbb{F}_2$*-basis given by monomials* b_λ *where* $|\lambda| = d$ *and* $\ell(\lambda) = r$.

(iv) *The Hopf algebras* S *and* DS *are isomorphic.*

Proof Statements (i) and (iii) follow immediately from (ii). For (ii), we argue by induction on r. Given $\lambda = (\lambda_1, \ldots, \lambda_r)$ of length $r \geq 1$, let $b_\lambda = b_{\lambda_1} \cdots b_{\lambda_r}$ be the corresponding monomial in $b_1, b_2, \ldots$. If $r = 1$, then this is the definition of b_k.

Let $r > 1$ and assume as induction hypothesis that (ii) is true for partitions ν of length $< r$. Then $b_\lambda = b_\nu b_{\lambda_r}$ where $\nu = (\lambda_1, \ldots, \lambda_{r-1})$, and so for any partition μ we have $\langle b_\lambda, m(\mu)\rangle = \langle \mu_*(b_\nu \otimes b_{\lambda_r}), m(\mu)\rangle = \langle b_\nu \otimes b_{\lambda_r}, \phi(m(\mu))\rangle = \langle b_\nu \otimes b_{\lambda_r}, \sum_{\mu=\rho\cup\sigma} m(\rho) \otimes m(\sigma)\rangle = \langle b_\lambda, m(\mu)\rangle = \sum_{\mu=\rho\cup\sigma} \langle b_\nu, m(\rho)\rangle\langle b_{\lambda_r}, m(\sigma)\rangle$ by (26.1). By the induction hypothesis, $\langle b_\nu, m(\rho)\rangle = 1$ if $\nu \neq \rho$ and is 0 otherwise, and $\langle b_{\lambda_r}, m(\sigma)\rangle = 1$ if $(\lambda_r) = \sigma$ and is 0 otherwise. Hence $\langle b_\lambda, m(\mu)\rangle = 1$ if $\mu = \rho \cup \sigma = \nu \cup (\lambda_r) = \lambda$, and is 0 otherwise. This completes the induction and the proof of (ii).

It follows from (i) that the map $\mathsf{S} \to \mathsf{DS}$ defined by $e_k \mapsto b_k$ for $k \geq 1$ is an isomorphism of algebras. To show that it is an isomorphism of Hopf algebras, we show that $\phi_*(b_k) = \sum_{i+j=k} b_i \otimes b_j$ for $k \geq 1$. Since ϕ_* is dual to the product μ in S, this follows from the fact that if the power sum p_k is a term in $m(\mu)m(\nu)$ when this is written in the basis $\{m(\lambda)\}$, then $m(\mu) = p_i$ and $m(\nu) = p_j$ for some i and j such that $i + j = k$. □

Definition 26.1.4 For $n \geq 0$, we denote by $\mathsf{DS}(n)$ the linear dual of $\mathsf{S}(n)$, i.e. the sub-coalgebra of DS spanned by monomials in $b_1, b_2, \ldots$ of length $\leq n$.

By duality, $\mathsf{DS}^{d,r}$ is a right A_2-module, where Sq^k maps $\mathsf{DS}^{d,r}$ to $\mathsf{DS}^{d-k,r}$. In contrast to our convention for $\mathsf{DP}(n)$, we write $b \cdot Sq^k$ for the element obtained by acting on $b \in \mathsf{DS}$ with $Sq^k \in \mathsf{A}_2$. The action of A_2 satisfies the Cartan formula

$$(bb') \cdot Sq^r = \sum_{i+j=r} (b \cdot Sq^i)(b' \cdot Sq^j) \tag{26.2}$$

with respect to the product in DS. Since b_k is dual to the power sum symmetric function $p_k = m(k)$, $b_k \in \mathsf{DS}^{k,1}$. Since $\mathsf{S}(1) = \mathsf{P}(1)$, the A_2-action on elements of length grading 1 is the down action of A_2 on $\mathsf{DP}(1)$. Thus by Proposition 9.3.6

$$b_k \cdot Sq^r = \binom{k-r}{r} b_{k-r}. \tag{26.3}$$

The dual of the hit problem for S is to determine the elements of which are in the kernel of all Steenrod operations of positive degree. It follows from (26.3) that in the case $n = 1$, $b_k \cdot Sq^r = 0$ for all $r > 0$ if and only if $k = 2^j - 1$.

Definition 26.1.5 The **symmetric Steenrod kernel** $\mathsf{K}(\mathsf{DS})$ is the set of all elements $b \in \mathsf{DS}$ such that $b \cdot Sq^r = 0$ for all $r > 0$.

The Cartan formula 26.2 implies that $\mathsf{K}(\mathsf{DS})$ is a subalgebra of DS. It is bigraded by d and r, where $\mathsf{K}^{d,r}(\mathsf{DS}) = \mathsf{K}(\mathsf{DS}) \cap \mathsf{S}^{d,r}$. We begin by identifying the monomials in $\mathsf{K}(\mathsf{DS})$.

Definition 26.1.6 A monomial $b_\lambda \in \mathsf{DS}$ is a ***b*-spike** if λ is a spike partition.

Proposition 26.1.7 *A monomial in* DS *is in* $\mathsf{K}(\mathsf{DS})$ *if and only if it is a b-spike.*

Proof This follows directly from (26.3) and (26.2). □

This result is analogous to Proposition 9.4.4 for $\mathsf{DP}(n)$. Thus the dual hit problem is effectively to find polynomials in $\mathsf{K}(\mathsf{DS})$ which are not in the subalgebra $\mathsf{J}(\mathsf{DS})$ generated by the b-spikes. It follows by duality from Theorem 25.2.1 that $\mathsf{K}(\mathsf{DS}(2)) = \mathsf{J}(\mathsf{DS}(2))$, and from the work of Sections 25.3 and 25.4 that $\mathsf{K}(\mathsf{DS}(3))/\mathsf{J}(\mathsf{DS}(3))$ is spanned by elements dual to the generators C_1, C_2 and C_3 of $\mathsf{Q}(\mathsf{S}(3))$.

Example 26.1.8 We show that $c = b_3b_2^2 + b_5b_1^2 \in \mathsf{K}^{7,3}(\mathsf{DS})$. It suffices to check that $c \cdot Sq^1 = 0$ and $c \cdot Sq^2 = 0$, and we have $b_3b_2^2 \cdot Sq^1 = 0$, $b_5b_1^2 \cdot Sq^1 = 0$ and $b_3b_2^2 \cdot Sq^2 = b_3b_1^2 = b_5b_1^2 \cdot Sq^2$.

We next consider the analogues of the Kameko maps for S and DS. The down Kameko map $\kappa : \mathsf{P}^{2d+n}(n) \to \mathsf{P}^d(n)$ is defined on monomials by $\kappa(g) = f$ if $g = \mathsf{c}(n)f^2$, and $\kappa(g) = 0$ otherwise. Hence κ maps symmetric polynomials to symmetric polynomials. Given a partition λ of length $\ell(\lambda) = r$, let $\lambda_i' = 2\lambda_i + 1$ for $1 \le i \le r$, so that λ' is a partition of length r with all parts odd. Then $\kappa : \mathsf{S}^{2d+n}(n) \to \mathsf{S}^d(n)$ is given on monomial symmetric functions by $\kappa(m(\mu)) = m(\lambda)$ if $\mu = \lambda'$ and $r = n$, i.e. μ has length n and all parts of μ are odd, and $\kappa(m(\mu)) = 0$ otherwise.

By Proposition 6.5.2, $\kappa \circ Sq^{2k+1} = 0$ and $\kappa \circ Sq^{2k} = Sq^k \circ \kappa$. It follows by considering symmetric hit equations that κ induces a linear map $\kappa : \mathsf{Q}^{2d+n}(\mathsf{S}(n)) \to \mathsf{Q}^d(\mathsf{S}(n))$. By Proposition 25.1.9, this is an isomorphism if $\mu(2d+n) = n$ with inverse $\upsilon : \mathsf{Q}^d(\mathsf{S}(n)) \to \mathsf{Q}^{2d+n}(\mathsf{S}(n))$.

Recall that S is a bigraded algebra and that $\mathsf{S}^{d,r}$ has $\mathbb{F}_2$-basis given by the monomial symmetric functions $m(\lambda)$ where $|\lambda| = d$ and $\ell(\lambda) = r$. Since S splits as the direct sum of graded A_2-modules $\mathsf{MS}(r) = \sum_{d \ge 0} \mathsf{S}^{d,r}$, we define the down Kameko map on S so that it agrees with the down Kameko map κ on $\mathsf{S}(n)$ on the summand $\mathsf{MS}(n)$ for all $n \ge 1$.

Definition 26.1.9 $\kappa^S : \mathsf{S} \to \mathsf{S}$ is the linear map $\mathsf{S}^{2d+r,r} \to \mathsf{S}^{d,r}$ which is given on the $\mathbb{F}_2$-basis of monomial symmetric functions by $\kappa^S(m(\mu)) = m(\lambda)$ if $\mu = \lambda'$, so that all parts of μ are odd, and $\kappa(m(\mu)) = 0$ otherwise.

We show that the down Kameko map κ^S commutes with the action of A_2 on S in the sense of Proposition 6.5.2.

Proposition 26.1.10 *For all* $s \geq 0$,

(i) $\kappa^S \circ Sq^{2s+1} = 0$, *and* (ii) $\kappa^S \circ Sq^{2s} = Sq^s \circ \kappa^S$.

Proof Let $\mu = \lambda'$ be a partition of length r with all parts odd. For (i), we note that if the block C arises from the action of an odd squaring operation on a given block B then $\omega_1(C) < \omega_1(B)$. Hence it follows from the Cartan formula that $Sq^{2s+1}(m(\mu))$ is a sum of monomial symmetric functions $\sum_\rho m(\rho)$, where $\ell(\rho) = r$ and ρ has at least one even part, so that $\kappa^S(m(\rho)) = 0$. For (ii), we note that if the block C arises from the action of an even squaring operation Sq^{2k} on a given block B and if $\omega_1(C) = \omega_1(B)$, then C arises from the action of Sq^k on the block obtained by deleting the first column of B. Hence the Cartan formula together with the formula $Sq^{2k}(f^2) = (Sq^k(f))^2$ shows that $Sq^{2s}(m(\mu))$ is a sum $\sum_\rho m(\rho)$, and that, if all parts of ρ are odd, then $\rho = \sigma'$, where $m(\sigma)$ is a term in $Sq^k(m(\lambda)) = \sum_\sigma m(\sigma)$. □

It follows from Proposition 26.1.10 that κ^S maps hit elements of $\mathsf{S}^{2d+r,r}$ to hit elements of $\mathsf{S}^{d,r}$ for all $d, r \geq 0$, and so induces a map $\kappa^S : \mathsf{Q}(\mathsf{S}^{2d+r}(r)) \to \mathsf{Q}(\mathsf{S}^d(r))$. In the case $\mu(2d+r) = r$, κ^S is the inverse of the isomorphism $\upsilon : \mathsf{Q}^d(\mathsf{S}(r)) \to \mathsf{Q}^{2d+r}(\mathsf{S}(r))$ of Proposition 25.1.9.

As the basis $\{b_\lambda\}$ of DS is dual to the basis $\{m(\lambda)\}$ of S, the linear dual $\kappa^S_* : \mathsf{DS}^{d,r} \to \mathsf{DS}^{2d+r,r}$ of κ^S is given by $\kappa^S_*(b_\lambda) = b_{\lambda'}$. In particular, $\kappa^S_*(b_k) = b_{2k+1}$ for $k \geq 1$. We shall show that κ^S_* restricts to a map of the symmetric Steenrod kernel. When $\mu(2d+r) = r$, κ^S_* is an isomorphism on the symmetric Steenrod kernel whose inverse is the dual υ^S_* of the up Kameko map υ^S, defined by $\upsilon^S_*(b_\mu) = b_\lambda$ if $\mu = \lambda'$ and $\ell(\mu) = r$, and $\upsilon^S_*(b_\mu) = 0$ otherwise.

Proposition 26.1.11 *The map* $\kappa^S_* : \mathsf{DS} \to \mathsf{DS}$ *is the map of bigraded algebras defined on the generators* b_k, $k \geq 1$ *of* DS *by* $\kappa^S_*(b_k) = b_{2k+1}$ *for* $k \geq 1$. *For* $s \geq 0$, (i) $\kappa^S_*(b_\lambda) \cdot Sq^{2s+1} = 0$ *and* (ii) $\kappa^S_*(b_\lambda) \cdot Sq^{2s} = \kappa^S_*(b_\lambda) \cdot Sq^s$.

Proof The preceding discussion shows that $\kappa^S_*(b_k) = b_{2k+1}$ for $k \geq 1$. For all partitions ρ and σ, $b_\rho b_\sigma = b_{\rho \cup \sigma}$ and $(\rho \cup \sigma)' = \rho' \cup \sigma'$. It follows that $\kappa^S_*(b_\rho)\kappa^S_*(b_\sigma) = \kappa^S_*(b_\rho b_\sigma)$.

We prove (i) and (ii) by induction on $r = \ell(\lambda)$. For $r = 1$, $b_\lambda = b_k$ where $k \geq 1$. Then $\kappa^S_*(b_k) \cdot Sq^{2s+1} = b_{2k+1} \cdot Sq^{2s+1} = \binom{2k-2s}{2s+1} b_{2k-2s} = 0$, proving (i).

Further $\kappa_*^S(b_k) \cdot Sq^{2s} = b_{2k+1} \cdot Sq^{2s} = \binom{2k+1-2s}{2s} b_{2k+1-2s}$, while $\kappa_*^S(b_k \cdot Sq^s) = \binom{k-s}{s}\kappa_*^S(b_{k-s}) = \binom{k-s}{s} b_{2k-2s+1}$, which proves (ii) since the mod 2 binomial coefficients are equal.

For $r > 1$, let $\lambda = \rho \cup \sigma$ where ρ and σ have length $< r$, so that $b_\lambda = b_\rho b_\sigma$. Then for (i) $\kappa_*^S(b_\lambda) \cdot Sq^{2s+1} = (\kappa_*^S(b_\rho)\kappa_*^S(b_\sigma)) \cdot Sq^{2s+1} = \sum_{i+j=2s+1}(\kappa_*^S(b_\rho) \cdot Sq^i)(\kappa_*^S(b_\sigma) \cdot Sq^j)$ by the Cartan formula (26.2). Since in each term either i or j is odd, $\kappa_*^S(b_\lambda) \cdot Sq^{2s+1} = 0$ by the induction hypothesis. This establishes the inductive step for (i).

For (ii), $\kappa_*^S(b_\lambda) \cdot Sq^{2s} = \sum_{i+j=2s}(\kappa_*^S(b_\rho) \cdot Sq^i)(\kappa_*^S(b_\sigma) \cdot Sq^j)$ by (26.2). By (i) the terms with i and j odd are 0, and by the induction hypothesis the term with $i = 2i'$ and $j = 2j'$ is $\kappa_*^S(b_\rho \cdot Sq^{i'})\kappa_*^S(b_\sigma \cdot Sq^{j'})$. On the other hand, $\kappa_*^S(b_\lambda \cdot Sq^s) = \sum_{i'+j'=s}\kappa_*^S(b_\rho \cdot Sq^{i'})\kappa_*^S(b_\sigma \cdot Sq^{j'})$ by (26.2). This establishes the inductive step for (ii). □

Proposition 26.1.12 *κ_*^S maps $\mathsf{K}^{d,r}(\mathsf{DS})$ to $\mathsf{K}^{2d+r,r}(\mathsf{DS})$.*

Proof For $k \geq 0$ and $b \in \mathsf{DS}^{d,r}$, $\kappa_*^S(b) \cdot Sq^{2k+1} = 0$ by (i) of Proposition 26.1.11, and if $b \in \mathsf{K}(\mathsf{DS})$, then $\kappa_*^S(b) \cdot Sq^{2k} = \kappa_*^S(b \cdot Sq^s) = 0$ by (ii). □

26.2 The symmetric Steenrod kernel K(DS(3))

In this section and the next we determine generators of $\mathsf{K}(\mathsf{DS}(3))$. These comprise the b-spikes $b_\lambda = b_{\lambda_1}b_{\lambda_2}b_{\lambda_3}$ where $\lambda = (\lambda_1, \lambda_2, \lambda_3)$ is a spike partition of length ≤ 3, together with polynomials in $b_1, b_2, \ldots$ dual to the monomials C_1, C_2, C_3 of Theorem 25.3.1. These polynomials are sums of monomials b_λ dual to elements $m(\lambda)$ in $\mathsf{Q}(\mathsf{S}(3))$. We represent both the monomial b_λ in $\mathsf{DS}(3)$ and the monomial symmetric function $m(\lambda)$ in $\mathsf{S}(3)$ by the block whose rows are the reversed binary expansions of $\lambda_1, \lambda_2, \lambda_3$. Thus in $\mathsf{DS}(3)$, as in $\mathsf{S}(3)$, blocks which differ by a permutation of rows represent the same element. As for $\mathsf{S}(3)$, it is sufficient to consider degree d such that $\mu(d) \leq 2$. We treat the case $\mu(d) = 2$ in this section and the case $\mu(d) = 1$ in Section 26.3.

Proposition 26.2.1 *Let $d = (2^{s+t} - 1) + (2^s - 1)$ where $s \geq 2$ and $t \geq 1$. Let c_1 be the sum of all monomials b_λ in $\mathsf{DS}^d(3)$, where $\lambda_1 > \lambda_2 \geq 2^t > \lambda_3 > 0$, represented by blocks of the form $V = A|B$, where $\omega(A) = (2, \ldots, 2)$ of length t, $\omega(B) = (1, \ldots, 1)$ of length s, together with the two blocks W_1, W_2 given by*

$$W_1 = \begin{matrix} 0\cdots0 & 1\cdots1 \\ 1\cdots1 & \\ 1\cdots1 & \end{matrix}\quad, \qquad W_2 = \begin{matrix} 1\cdots1 & 1 & 1\cdots1 \\ 1\cdots1 & 1 & 1\cdots1 \\ 0\cdots0 & 1 & \end{matrix}\ .$$

Then $c_1 \in \mathsf{K}^d(\mathsf{DS}(3))$.

The condition $\lambda_1 > \lambda_2$ ensures that the entry 1 in the last column of B is in the first row, the conditions $\lambda_2 \geq 2^t > \lambda_3$ ensure that the block B has exactly two nonzero rows, and the condition $\lambda_3 > 0$ ensures that the block A has three nonzero rows. Hence there are $3^t - 1$ choices for A and $2^{s-1} - 1$ choices for B, so that c_1 is the sum of $(3^t - 1)(2^{s-1} - 1) + 2$ monomials. The conditions on A and B ensure that b_λ is not a b-spike and that two blocks $A|B$ and $A'|B'$ are not row permutations of each other.

Proof of Proposition 26.2.1 We begin by observing that if $b_\lambda = b_{\lambda_1} b_{\lambda_2} b_{\lambda_3}$ is a monomial which appears in c_1, then $b_\lambda \cdot Sq^{2^k}$ is either 0 or is one of the monomials $b_{\lambda_1 - 2^k} b_{\lambda_2} b_{\lambda_3}$, $b_{\lambda_1} b_{\lambda_2 - 2^k} b_{\lambda_3}$, $b_{\lambda_1} b_{\lambda_2} b_{\lambda_3 - 2^k}$ obtained by subtracting 2^k from the index of one of the factors. This can be seen by applying the Cartan formula (26.2) to the block representation of b_λ, since the conditions on a block $V = A|B$ in c_1 imply that V has at most one nontrailing 0 in each column, and this is also true of the special blocks W_1 and W_2.

Since the elements Sq^{2^k} generate A_2, it is sufficient to prove that $c_1 \cdot Sq^{2^k} = 0$ for $k \geq 0$. By (26.3), the action of Sq^{2^k} on a block representing a monomial in c_1 is thus to replace a section of one of the rows of the form $0 \cdots 0\ 1$ in columns $k+1, \ldots, \ell$ by a section of the form $1 \cdots 1\ 0$. In the simplest case, entries 0 in position $(i, k+1)$ and 1 in position $(i, k+2)$ are exchanged. If the 0 entry in column $k+1$ is trailing, then Sq^{2^k} maps the block to 0. Our strategy is to show that if $b_\lambda \cdot Sq^{2^k} = b_\mu$ is nonzero, then there is exactly one other monomial b_ν in c_1 such that $b_\nu \cdot Sq^{2^k} = b_\mu$, so that these terms cancel (mod 2) in $c_1 \cdot Sq^{2^k}$.

We split all the blocks in c_1 as concatenations $A|B$, where the left subblock A has t columns and $\omega(A) = (2, \ldots, 2)$. Thus B has s columns and $\omega(B) = (1, \ldots, 1)$ except in the case of W_2, where the right subblock B has $s - 1$ columns and $\omega(B) = (3, 1, \ldots, 1)$. We begin with a block $V = A|B$ in c_1, $V \neq W_1$ or W_2, on which Sq^{2^k} acts as above, changing one of the rows by replacing entries $0 \cdots 0\ 1$ in columns $k+1, \ldots, \ell$ by $1 \cdots 1\ 0$. We separate the three cases $\ell \leq t$, $k+1 \leq t < \ell$, and $k+1 > t$. As explained above, our aim is to find a unique partner $V' = A'|B'$ for V in c_1 such that $V' \cdot Sq^{2^k} = V \cdot Sq^{2^k}$.

Case 1: $\ell \leq t$. In this case, up to a permutation of rows the subblock of A given by columns $k+1, \ldots, \ell$ has the form

$$\begin{matrix} 0 \cdots 0\ 1 \\ 1 \cdots 1\ 0 \\ 1 \cdots 1\ 1 \end{matrix} .$$

The partner V' of V is the block obtained from V by exchanging rows 1 and 2 of this subblock.

Case 2: $k+1 \leq t < \ell$. In this case, up to exchange of rows 1 and 2, the subblock of V comprising columns $k+1,\ldots,\ell$ has the form

$$C = \begin{array}{c|c} 0\cdots0 & 0\cdots0\,1 \\ 1\cdots1 & 1\cdots1\,0 \\ 1\cdots1 & 0\cdots0\,0 \end{array},$$

and as in Case 1, the partner V' of V in general is the block obtained from V by exchanging rows 1 and 2 of this subblock. However, there are two exceptional situations (a) and (b) where this fails.

The first exception (a) occurs when $\ell = t+s$, so that the last column of B is involved. The entries in this column cannot be exchanged, as the resulting block is not in c_1. However, $V = A'|C$ where A' is the subblock comprising the first k columns of V, and $V \cdot Sq^{2^k}$ can be seen as the block obtained by replacing row 1 of C by row 2, so that these two rows of the subblock become identical. In general we can obtain the partner block V' by exchanging rows 1 and 2 of the subblock A' of V. After applying Sq^{2^k}, V and V' give blocks which are interchanged by exchanging the first two rows, and so represent occurrences of the same monomial in $c_1 \cdot Sq^{2^k}$, as required. An exceptional case arises when rows 1 and 2 of A' are identical. In particular, this occurs when $k = 0$, so that the block A' is empty. In this case

$$V = \begin{array}{cc|c} 1\cdots1 & 0\cdots0 & 0\cdots0\,1 \\ 1\cdots1 & 1\cdots1 & 1\cdots1 \\ 0\cdots0 & 1\cdots1 & \end{array},$$

where the left subblock A coincides with the corresponding subblock of W_2, and $V' = W_2$ is the partner of V.

The second exception (b) occurs when $\ell = t+1$ and row 2 of B is $1\,0\cdots0$. In this case, exchanging rows 1 and 2 of the subblock

$$C = \begin{array}{c|c} 1\cdots1 & 0 \\ 0\cdots0 & 1 \\ 1\cdots1 & 0 \end{array}$$

produces a block which is not in c_1 because the right subblock has only one nonzero row. The block $V \cdot Sq^{2^k}$ can again be seen as the block obtained by replacing row 1 of C by row 2, but in this case rows 2 and 3 of the subblock obtained by removing the first k columns are identical. Thus in general we can obtain the partner block V' by exchanging rows 2 and 3 of the subblock A'

of V. After applying Sq^{2^k}, V and V' give blocks which are interchanged by exchanging rows 2 and 3, and so represent occurrences of the same monomial in $c_1 \cdot Sq^{2^k}$, as required.

Once again, an exceptional case arises when rows 2 and 3 of A' are identical. In particular, this occurs when $k=0$, so that the block A' is empty. In this case,

$$V = \begin{array}{c|c|c} 0\cdots 0 & 1\cdots 1 & 0\,1\cdots 1 \\ 1\cdots 1 & 0\cdots 0 & 1\,0\cdots 0 \\ 1\cdots 1 & 1\cdots 1 & \end{array}$$

where the left subblock A' coincides with the corresponding subblock of W_1, and $V' = W_1$ is the partner of V.

Case 3: $k+1 > t$. In this case, up to a permutation of rows the subblock of B given by columns $k+1, \ldots, \ell$ has the form

$$\begin{array}{c} 0\cdots 0\,1 \\ 1\cdots 1\,0 \\ 0\cdots 0\,0 \end{array}$$

up to exchange of rows 1 and 2. Again, the partner V' of V is generally obtained by exchanging rows 1 and 2 of this subblock. As in Case 2, this fails if $\ell = s+t$, and as before this subcase is resolved by exchanging the first k entries in row 1 with the first k entries of row 2. As in Cases 1 and 2, this would fail if these sections of rows 1 and 2 are identical. However, this situation does not arise in Case 3, as it would imply that row 3 of V is zero, contrary to hypothesis. Finally, we note that the exceptional blocks W_1 and W_2 are unpaired in Case 3, since $W_1 \cdot Sq^{2^k} = 0$ and $W_2 \cdot Sq^{2^k} = 0$. □

Example 26.2.2 The exceptional pairings of blocks in c_1 are shown below in the case $k=1$, $t=2$ and $s=3$.

$$\begin{array}{c|l} 1 & 111 \\ 1 & 111 \\ 0 & 01 \end{array} + \begin{array}{c|l} 1 & 0001 \\ 1 & 111 \\ 0 & 1 \end{array} \xrightarrow{Sq^2} 0, \qquad \begin{array}{c|l} 1 & 0001 \\ 0 & 111 \\ 1 & 1 \end{array} + \begin{array}{c|l} 0 & 0001 \\ 1 & 111 \\ 1 & 1 \end{array} \xrightarrow{Sq^2} 0,$$

$$\begin{array}{c|l} 0 & 0111 \\ 1 & 1 \\ 1 & 1 \end{array} + \begin{array}{c|l} 0 & 1011 \\ 1 & 01 \\ 1 & 1 \end{array} \xrightarrow{Sq^2} 0, \qquad \begin{array}{c|l} 1 & 1011 \\ 0 & 01 \\ 1 & 1 \end{array} + \begin{array}{c|l} 1 & 1011 \\ 1 & 01 \\ 0 & 1 \end{array} \xrightarrow{Sq^2} 0.$$

26.3 The case $n = 3$, $\mu(d) = 1$

The following result describes elements c_2 and c in $\mathsf{K}^d(\mathsf{DS}(3))$ in the case $\mu(d) = 1$.

Proposition 26.3.1 *Let $d = 2^s - 1$ where $s \geq 3$. Let c_2 be the sum of all monomials in $\mathsf{DS}^d(3)$ represented by blocks with ω-sequence $(3,2,\dots,2)$, together with all blocks with two equal rows representing $b_i b_j^2$, where $i = 2^{r+1}+1$ and $j = 2^{s-1} - 2^r - 1$, $0 \leq r \leq s-2$. Then $c_2 \in \mathsf{K}^d(\mathsf{DS}(3))$. Further, for $s \geq 4$ the sum $c = \sum_{i+2j=d} b_i b_j^2$ of all monomials in $\mathsf{DS}^d(3)$ represented by blocks with two equal rows is also in $\mathsf{K}^d(\mathsf{DS}(3))$.*

In the case $r = 0$, the monomial $b_3 b^2_{2^{s-1}-2}$ is represented by the block C_2 of Theorem 25.3.1, while the monomial $b_{2^{s-1}-1} b^2_{2^{s-2}}$ is represented by the block C_3 of Theorem 25.3.1, and is a term of c but is not a term of c_2.

We need two preliminary results for the proof of Proposition 26.3.1. These describe the action of the indecomposable operations Sq^{2^k} on the terms of c or c_2. The first result states that Sq^{2^k} acts on a monomial in $\mathsf{DS}(3)$ with ω-sequence $(2,\dots,2)$ by subtracting 2^k from the row of the representing block which has an entry 0 in column $k+1$.

Proposition 26.3.2 *Let $b_\lambda \in \mathsf{DS}^d(3)$ be a monomial of degree $d = 2^{r+1} - 2$ represented by a block with ω-sequence $(2,\dots,2)$ of length r. Then $b_\lambda \cdot Sq^{2^k} = 0$ for $k \geq r$, and $b_\lambda \cdot Sq^{2^k} = b_\mu$ for $k < r$, where μ is obtained by subtracting 2^k from the part λ_i of λ such that $2^k \notin \mathrm{bin}(\lambda_i)$.*

Proof The result for $k \geq r$ is true for degree reasons, so let $k < r$. By (26.2) and (26.3), a nonzero term in the expansion of $b_\lambda \cdot Sq^{2^k}$ by the Cartan formula corresponds to a choice of $a_1, a_2, a_3 \geq 0$ such that $a_1 + a_2 + a_3 = 2^k$ and $b_{\lambda_i} \cdot Sq^{a_i} = \binom{\lambda_i - a_i}{a_i} b_{\lambda_i - a_i}$ is odd for $i = 1,2,3$. Unless $a_1 = 2^k$ for some i, two of a_1, a_2, a_3 are odd multiples of 2^s for the same $s \geq 0$. Since b_λ has ω-sequence $(2,\dots,2)$, column $s+1$ of the block representing b_λ contains two 1s. Hence for at least one of the mod 2 binomial coefficients $\binom{\lambda_i - a_i}{a_i}$ we have $a_i = 2^s q$ and $\lambda_i = 2^s q' + r'$, where q and q' are odd and $0 \leq r' < 2^s$. But then $\binom{\lambda_i - a_i}{a_i} = \binom{2^s(q'-q)+r'}{2^s q} = \binom{q'-q}{q} = 0 \bmod 2$. □

The second result is similar and deals with blocks of 2-degree $2^s - 1$ with two equal rows.

Proposition 26.3.3 *Let $i + 2j = 2^s - 1$ and suppose that $b_i b_j^2 \cdot Sq^{2^k} \neq 0$. Then $k \leq s-2$ and $b_i b_j^2 \cdot Sq^{2^k} = b_{i-2^k} b_j^2$ or $b_i b^2_{j-2^{k-1}}$, the first case occurring when $i \geq 2^{k+1}$ and $2^k \in \mathrm{bin}(i - 2^k)$, and the second when $j \geq 2^k$ and $2^k \in \mathrm{bin}(2j - 2^k)$.*

Proof By the Cartan formula,

$$b_i b_j^2 \cdot Sq^{2^k} = \sum_{a,b} \binom{i-a}{a}\binom{2j-2b}{2b} b_{i-a} b_{j-b}^2,$$

where the sum is over all $a,b \geq 0$ such that $2a \leq i$, $2b \leq j$ and $a+2b = 2^k$. Thus if the sum is nonzero then $2^{k+1} = 2(a+2b) \leq i+2j = 2^s - 1$, and so $k \leq s-2$. We shall show by contradiction that the coefficient is 0 mod 2 if $a \neq 0$ and $b \neq 0$. Thus let $a = 2^{u_1} v_1$ and $2b = 2^{u_2} v_2$ where v_1, v_2 are odd and $u_1, u_2 \geq 0$. Then $2^k = a + 2b = 2^{u_1} v_1 + 2^{u_2} v_2$ where $u_1, u_2 < k$. Hence $u_1 = u_2 = u$, say, and $k > u$. Let $i' = i - a$ and $j' = j - b$. If $\binom{i'}{a}$ and $\binom{2j'}{2b}$ are odd, then $2^u \in \mathrm{bin}(a) \subseteq \mathrm{bin}(i')$ and $2^u \in \mathrm{bin}(2b) \subseteq \mathrm{bin}(2j')$. But $i' + 2j' = 2^s - 1 - 2^k$ and $u < k < s$, so $i' + 2j' = -1 \bmod 2^{u+1}$. Hence 2^u is in exactly one of $\mathrm{bin}(i')$ and $\mathrm{bin}(2j')$, giving a contradiction.

Hence $b_i b_j^2 \cdot Sq^{2^k} = \binom{i-2^k}{2^k} b_{i-2^k} b_j^2 + \binom{2j-2^k}{2^k} b_i b_{j-2^{k-1}}^2$, and we shall show that at most one of these terms is nonzero. Since $(i-2^k)+(2j-2^k) = 2^s - 1 - 2^{k+1}$ and $k < s$, $(i-2^k)+(2j-2^k) = -1 \bmod 2^{k+1}$, so when $i - 2^k$ and $2j - 2^k$ are both ≥ 0, 2^k is in exactly one of $\mathrm{bin}(i-2^k)$ and $\mathrm{bin}(2j-2^k)$. □

We outline the proof of Proposition 26.3.1 as follows. First we show that $c \cdot Sq^r = 0$ for $r > 0$. It follows from Proposition 26.3.3 that the image of each monomial in c under Sq^{2^k} is a monomial or 0, and that these monomials cancel in pairs, so that $c \cdot Sq^{2^k} = 0$.

Next we assume $s \geq 4$ and consider $c_2 = c_2' + c_2''$, where c_2' is the sum of all monomials in $\mathrm{DS}^d(3)$ represented by blocks with ω-sequence $(3,2,\ldots,2)$, and c_2'' is the sum of all monomials in $\mathrm{DS}^d(3)$ represented by blocks with two equal rows representing $b_i b_j^2$, where $j = 2^{s-1} - 2^r - 1$, $0 \leq r \leq s-2$. Clearly $c_2' \cdot Sq^1 = c_2'' \cdot Sq^1 = 0$.

Using Proposition 26.3.2 we show that $c_2' \cdot Sq^{2^k}$ is the sum of monomials $b_i b_j^2$ where $i = 2^a - 1$ and $3 \leq a \leq s-1$, e.g. $b_7 b_{27}^2 + b_{15} b_{23}^2 + b_{31} b_{15}^2$ for $s = 6$. For $c_2' \cdot Sq^4$ we obtain 0 for $s = 4$, $b_{13} b_7^2$ for $s = 5$ and $b_{13} b_{23}^2 + b_{29} b_{15}^2$ for $s = 6$. For $c_2' \cdot Sq^8$ we obtain 0 for $s = 4$ or 5 and $b_{25} b_{15}^2$ for $s = 6$.

The proof of Proposition 26.3.1 is completed by Proposition 26.3.4, which shows that $c_2'' \cdot Sq^{2^k} = c_2' \cdot Sq^{2^k}$. For example when $s = 4$, $c_2'' = b_9 b_3^2 + b_5^3 + b_3 b_6^2$, and when $s = 5$, $c_2'' = b_{17} b_7^2 + b_9 b_{11}^2 + b_5 b_{13}^2 + b_3 b_{14}^2$. For each k, Sq^{2^k} gives the same value on two of these blocks, and its value on the other blocks is either 0 or one of the terms in $c_2' \cdot Sq^{2^k}$.

Proposition 26.3.4 *With c_2', c_2'' defined as above in $\mathrm{DS}^d(3)$ for $d = 2^s - 1$, and $k \geq 1$, $c_2' \cdot Sq^{2^k} = c_2'' \cdot Sq^{2^k} = \sum b_i b_j^2$, where $i = 2^a - 2^k + 1$, $j = 2^{s-1} - 2^{a-1} - 1$ and the sum is over all such i and j for $k+2 \leq a \leq s-1$.*

Proof First we consider c_2'. Since a monomial b_λ in c_2' has ω-sequence $(3,2,\dots,2)$, we can apply Sq^{2^k} to the block representing b_λ by deleting the first column, applying $Sq^{2^{k-1}}$ to the rest of the block and then replacing the first column. By Proposition 26.3.2, the result of applying $Sq^{2^{k-1}}$ to any monomial f with ω-sequence $(3,2,\dots,2)$ is the monomial obtained by applying $Sq^{2^{k-1}}$ to the row of the corresponding block which has a 0 entry in column k.

We begin with $c_2' \cdot Sq^2$, or equivalently with $f \cdot Sq^1$, where f is the sum of all monomials in $\mathrm{DS}^{d'}(3)$ for $d' = 2^{s-1} - 2$ with ω-sequence $(2,\dots,2)$ of length $s-2$. One such monomial is the b-spike b_ℓ^2 where $\ell = 2^{s-2} - 1$, and is in the kernel of Sq^{2^k} for all $k \geq 0$. All other monomials in c_2' are represented by blocks with three distinct rows, so the number of such monomials is $(3^{s-2} - 3)/6 = (3^{s-3} - 1)/2$. We show that all but $s-3$ of these cancel in pairs when we apply Sq^1.

If columns 1 and 2 of the block are different, then we can assume that they have the form

$$\begin{matrix} 1\,1 \\ 1\,0 \\ 0\,1 \end{matrix} \quad \text{or} \quad \begin{matrix} 1\,1 \\ 0\,1 \\ 1\,0 \end{matrix}. \tag{26.4}$$

If the rest of the blocks coincide, then these two cancel under Sq^1 unless the whole of the rest of rows 2 and 3 coincide. Next suppose that columns 1 and 2 of the block are the same, but column 3 is different. We can assume that they have the form

$$\begin{matrix} 1\,1\,1 \\ 1\,1\,0 \\ 0\,0\,1 \end{matrix} \quad \text{or} \quad \begin{matrix} 1\,1\,1 \\ 0\,0\,1 \\ 1\,1\,0 \end{matrix}. \tag{26.5}$$

If the rest of the blocks coincide, then these two cancel under Sq^1, unless the whole of the rest of rows 2 and 3 coincide. Continuing in this way, we are reduced to considering Sq^1 on $s-3$ blocks of the form

$$\begin{matrix} 1\cdots1\,1\,0\cdots0 \\ 1\cdots1\,0\,1\cdots1 \\ 0\cdots0\,1\,1\cdots1 \end{matrix},$$

where the right subblock (but not the left) may be empty. Applying Sq^1 to these blocks, we obtain $s-3$ blocks with row 1 of spike type and rows 2 and 3 equal, representing the terms $b_i b_j^2$ stated.

We next consider $f \cdot Sq^2$. If columns 2 and 3 of the block are different, then we may assume that they have the form (26.4). If the rest of the blocks coincide, then these two cancel under Sq^2, unless the whole of the rest of rows 2 and 3 coincide. Next suppose that columns 2 and 3 of the block are the same, but column 3 is different. We can assume that they have the form (26.4). If the rest of the blocks coincide, then these two cancel under Sq^2, unless the whole of the rest of rows 2 and 3 coincide. Continuing this way, we are reduced to considering Sq^2 on $s-4$ blocks of the form

$$\begin{array}{c} 0\,1\cdots1\,1\,0\cdots0 \\ 1\,1\cdots1\,0\,1\cdots1 \\ 1\,0\cdots0\,1\,1\cdots1 \end{array},$$

together with blocks where columns $2,\ldots,s-2$ (i.e. all except column 1) coincide. There are just two blocks of this last type, the b-spike and the block

$$\begin{array}{c} 1\,1\cdots1 \\ 0\,1\cdots1 \\ 1\,0\cdots0 \end{array},$$

and both these map to 0 under Sq^2. Applying Sq^2 to these blocks, we obtain $s-4$ blocks with row 1 of the form $0\,1\cdots1$ and rows 2 and 3 equal, representing the stated terms $b_i b_j^2$.

The argument generalizes to Sq^{2^k}, starting with columns k and $k+1$ of the form (26.4) and reducing to $s-k-2$ blocks which give the stated terms $b_i b_j^2$, together with blocks where all columns from the kth onwards are equal, which map to 0 under Sq^{2^k}.

Next we consider $c_2'' \cdot Sq^2$, where as above $c_2'' = \sum_{0\le r\le s-2} f_r$, where f_r is the monomial $b_i b_j^2$ in $\mathrm{DS}^d(3)$ for $i = 2^{r+1}+1$ and $j = 2^{s-1} - 2^r - 1$. By Proposition 26.3.3, for all i and j such that $i+2j = 2^s - 1$

$$b_i b_j^2 \cdot Sq^{2^k} = \begin{cases} b_{i-2^k} b_j^2, & \text{if } i \ge 2^{k+1} \text{ and } 2^k \in \mathrm{bin}(i-2^k) \\ b_i b_{j-2^{k-1}}^2, & \text{if } j \ge 2^k \text{ and } 2^k \in \mathrm{bin}(2j-2^k) \\ 0, & \text{otherwise.} \end{cases} \qquad (26.6)$$

It follows that $f_r \cdot Sq^{2^k} = 0$ when $r < k-1$, $f_{k-1} \cdot Sq^{2^k} = f_k \cdot Sq^{2^k} = b_{i'} b_{j'}^2$ where $i' = 2^k + 1$ and $j' = 2^{s-1} - 2^k - 1$. Thus these two terms cancel, and so $c_2'' \cdot Sq^{2^k} = \sum_{k+1\le r\le s-2} f_r \cdot Sq^{2^k}$.

To complete the proof, we show that each of the remaining terms matches a term in $c_2' \cdot Sq^{2^k}$, i.e. for $k+1 \le r \le s-2$ we have $f_r \cdot Sq^{2^k} = b_{i'} b_j^2$ where

$i' = i - 2^k$. Since $i = 2^{r+1} - 1$ and $2^{k+1} \leq 2^r$, $i \geq 2^{k+1}$, and since $i - 2^k = i' = 2^{r+1} - 2^k + 1 = 2^r + 2^{r-1} + \cdots + 2^k + 1$, $2^k \in \mathrm{bin}(i - 2^k)$. Hence by (26.6) $f_r \cdot Sq^{2^k} = b_{i'} b_j^2$, where $i' = i - 2^k$. □

26.4 The bigraded Steenrod algebra $\widetilde{\mathsf{A}}_2$

In this section we introduce the bigraded Steenrod algebra $\widetilde{\mathsf{A}}_2$, and show that A_2 is an algebra of operations on the dual symmetric algebra DS which commutes with the right action of A_2 in the sense of Proposition 26.4.9, i.e. $\Theta(b \cdot \phi) = \Theta(b) \cdot V(\phi)$, for all $b \in \mathsf{DS}$, where $\Theta \in \widetilde{\mathsf{A}}_2$, $\phi \in \mathsf{A}_2$ and V is the halving map of A_2. It follows that $\widetilde{\mathsf{A}}_2$ provides a valuable tool for studying the symmetric Steenrod kernel $\mathsf{K}(\mathsf{DS})$.

The difference between A_2 and $\widetilde{\mathsf{A}}_2$ is that the relation $Sq^0 = 1$, the identity element, has no counterpart in $\widetilde{\mathsf{A}}_2$. Thus no contraction occurs in the Adem relations when a term involving $\widetilde{Sq}^0$ is present. For example, the Adem relation $Sq^1 Sq^2 = Sq^3$ in A_2 is replaced by the relation $\widetilde{Sq}^1 \widetilde{Sq}^2 = \widetilde{Sq}^3 \widetilde{Sq}^0$ in $\widetilde{\mathsf{A}}_2$. One effect of this change is that all relations in A_2 preserve the length ℓ of monomials $\widetilde{Sq}^{a_1} \cdots \widetilde{Sq}^{a_\ell}$. It follows that $\widetilde{\mathsf{A}}_2$ is a graded algebra with respect to $\ell \geq 0$, in addition to its usual grading by degree, so that it is a bigraded algebra, i.e. $\widetilde{\mathsf{A}}_2^{d,\ell}$ is the $\mathbb{F}_2$-subspace of $\widetilde{\mathsf{A}}_2$ spanned by monomials $\widetilde{Sq}^A = \widetilde{Sq}^{a_1} \cdots \widetilde{Sq}^{a_\ell}$, where $A = (a_1, \ldots, a_\ell)$ with $a_i \geq 0$ for $1 \leq i \leq \ell$ and $|A| = d$. Thus $\widetilde{Sq}^A$ has length ℓ even if $a_\ell = 0$, in which case $\mathrm{len}(A) < \ell$. The identity element 1 of $\widetilde{\mathsf{A}}_2$ is regarded as a monomial of length 0 and generates $\widetilde{\mathsf{A}}_2^{0,0}$, which we may identify with $\mathbb{F}_2$, whereas $\widetilde{Sq}^0$ generates $\widetilde{\mathsf{A}}_2^{0,1}$. The product respects both gradings, i.e. it is a map $\widetilde{\mathsf{A}}_2^{d_1,\ell_1} \otimes \widetilde{\mathsf{A}}_2^{d_2,\ell_2} \to \widetilde{\mathsf{A}}_2^{d_1+d_2,\ell_1+\ell_2}$.

Definition 26.4.1 The **bigraded** (or **big**) **Steenrod algebra** $\widetilde{\mathsf{A}}_2$ is the associative algebra over $\mathbb{F}_2$ generated by symbols $\widetilde{Sq}^k$, $k \geq 0$, subject to the Adem relations

$$\widetilde{Sq}^a \widetilde{Sq}^b = \sum_{j=0}^{[a/2]} \binom{b-j-1}{a-2j} \widetilde{Sq}^{a+b-j} \widetilde{Sq}^j, \quad \text{for } 0 < a < 2b, \tag{26.7}$$

where the binomial coefficient is reduced mod 2.

In particular, the Adem relations $\widetilde{Sq}^0 \widetilde{Sq}^b = \widetilde{Sq}^b \widetilde{Sq}^0$ for $b > 0$ show that $\widetilde{Sq}^0$ commutes with all elements of $\widetilde{\mathsf{A}}_2$.

We begin by defining the action of $\widetilde{\mathsf{A}}_2$ on DS. As in Chapter 1, we start by introducing the total squaring operation $\widetilde{\mathsf{Sq}}$. Since DS is a polynomial algebra over $\mathbb{F}_2$, an algebra endomorphism of DS is defined uniquely by assigning an arbitrary element of DS to each generator b_k, $k \geq 1$.

Definition 26.4.2 The **total big Steenrod square** $\widetilde{\mathrm{Sq}} : \mathrm{DS} \to \mathrm{DS}$ is the algebra map defined by

$$\widetilde{\mathrm{Sq}}(b_k) = b_{2k+1} + b_k^2$$

for all $k \geq 1$.

If we consider the bigraded algebra $\mathrm{DS} = \sum_{k,r\geq 0} \mathrm{DS}^{k,r}$ as singly graded by the total degree $k+r$, then $b_k \in \mathrm{DS}^{k,1}(k,1)$ has total degree $k+1$, while $b_{2k+1} \in \mathrm{DS}^{2k+1,1}$ and $b_k^2 \in \mathrm{DS}^{2k,2}$ have total degree $2k+2$. Thus the action of $\widetilde{\mathrm{Sq}}$ doubles the total degree.

Definition 26.4.3 For $a \geq 0$ and $k, r \geq 0$, the **big Steenrod square** $\widetilde{Sq}^a$: $\mathrm{DS}^{k,r} \to \mathrm{DS}^{2k+r-a,r+a}$ is the linear map defined by restricting $\widetilde{\mathrm{Sq}}$ to elements of bigrading (k,r) and projecting on to elements of bigrading $(2k+r-a, r+a)$. Thus $\widetilde{\mathrm{Sq}} = \sum_{a\geq 0} \widetilde{Sq}^a$ is the formal sum of its graded parts, and $\widetilde{Sq}^0$ is the term in this sum which fixes the length grading r in DS.

Thus $\widetilde{Sq}^0(b_k) = b_{2k+1}$, $\widetilde{Sq}^1(b_k) = b_k^2$ and $\widetilde{Sq}^a(b_k) = 0$ for $a \geq 2$. Since $\widetilde{\mathrm{Sq}}(c_1c_2) = \widetilde{\mathrm{Sq}}(c_1)\widetilde{\mathrm{Sq}}(c_2)$ for all $c_1, c_2 \in \mathrm{DS}$, and $\widetilde{Sq}^0$ is the part of $\widetilde{\mathrm{Sq}}$ which fixes the second grading, $\widetilde{Sq}^0$ is also multiplicative, i.e. $\widetilde{Sq}^0(b_\rho b_\sigma) = \widetilde{Sq}^0(b_\rho)\widetilde{Sq}^0(b_\sigma)$ for all $b_\rho, b_\sigma \in \mathrm{DS}$.

Proposition 26.4.4 *$\widetilde{Sq}^0 : \mathrm{DS}^{k,r} \to \mathrm{DS}^{2k+r,r}$ is the up Kameko map κ_*^S.*

Proof This follows on comparing $\widetilde{Sq}^0$ with κ_*^S and using Proposition 26.1.11. □

Example 26.4.5 As in Example 26.1.8, let $c = b_3b_2^2 + b_5b_1^2 \in \mathrm{DS}^{7,3}$. Then $\widetilde{\mathrm{Sq}}(b_3b_2^2) = \widetilde{\mathrm{Sq}}(b_3)(\widetilde{\mathrm{Sq}}(b_2))^2 = (b_7 + b_3^2)(b_5 + b_2^2)^2 = b_7b_5^2 + b_3^2b_5^2 + b_7b_2^4 + b_3^2b_2^4$, and $\widetilde{\mathrm{Sq}}(b_5b_1^2) = \widetilde{\mathrm{Sq}}(b_5)(\widetilde{\mathrm{Sq}}(b_1))^2 = (b_{11} + b_5^2)(b_3 + b_1^2)^2 = b_{11}b_3^2 + b_5^2b_3^2 + b_{11}b_1^4 + b_5^2b_1^4$.

Using the length grading to equate terms, we obtain $\widetilde{Sq}^0(c) = b_7b_5^2 + b_{11}b_3^2$, $\widetilde{Sq}^1(c) = b_3^2b_5^2 + b_5^2b_3^2 = 0$, $\widetilde{Sq}^2(c) = b_7b_2^4 + b_{11}b_1^4$, $\widetilde{Sq}^3(c) = b_3^2b_2^4 + b_5^2b_1^4$ and $\widetilde{Sq}^a(c) = 0$ for $a \geq 4$.

Since $\widetilde{\mathrm{Sq}}$ is an algebra map, the operations $\widetilde{Sq}^a$ satisfy the Cartan formula.

Proposition 26.4.6 *For all $k \geq 0$ and $b_\rho, b_\sigma \in \mathrm{DS}$,*

$$\widetilde{Sq}^k(b_\rho b_\sigma) = \sum_{i+j=k} \widetilde{Sq}^i(b_\rho)\widetilde{Sq}^j(b_\sigma). \quad \square$$

To show that Definition 26.4.2 defines an action of $\widetilde{\mathsf{A}}_2$ on DS, we must show that it respects the Adem relations. The following result corresponds to Proposition 3.2.1. We define $\widetilde{Sq}^a = 0$ for $a < 0$, so as to extend the Adem relations to all pairs of integers a and b.

Proposition 26.4.7 *For all integers a and b, let $\widetilde{R}^{a,b} \in \widetilde{\mathsf{A}}_2$ be the element*

$$\widetilde{R}^{a,b} = \widetilde{Sq}^a \widetilde{Sq}^b + \sum_{j=0}^{[a/2]} \binom{b-j-1}{a-2j} \widetilde{Sq}^{a+b-j} \widetilde{Sq}^j.$$

Then $\widetilde{R}^{a,b}(c) = 0$ for all $c \in \mathsf{DS}$.

Proof The proof is essentially the same as that of Proposition 3.2.1. We argue by double induction on $a+b$ and on the length $r = \ell(\lambda)$ of a monomial $c = b_\lambda$ in DS. The induction on r starts with $r = 0$, when $c = 1$. Since $\widetilde{\mathsf{Sq}}(1) = \widetilde{Sq}^0(1) = 1$, the case $r = 0$ is easily verified. For $r = 1$, $c = b_k$ where $k \geq 1$, and we find using Definition 26.4.2 that $\widetilde{Sq}^a \widetilde{Sq}^b(b_k) = b_{4k+3}$ if $(a,b) = (0,0)$, b_{2k+1}^2 if $(a,b) = (1,0)$ or $(0,1)$, b_k^4 if $(a,b) = (2,1)$ and 0 otherwise. Thus for the inductive step with $c = b_\lambda = b_k b_\nu$, equation (3.3) for the action of A_2 on $\mathsf{P}(n)$ is replaced by the equation

$$\widetilde{R}^{a,b}(b_k b_\nu) = b_{4k+3}\widetilde{R}^{a,b}(b_\nu) + b_{2k+1}^2 \widetilde{S}^{a,b}(b_\nu) + b_k^4 \widetilde{T}^{a,b}(b_\nu) \tag{26.8}$$

for the action of $\widetilde{\mathsf{A}}_2$ on DS. The proof is completed by making the corresponding changes in the proof of Proposition 3.2.1. □

Since the Adem relations are a set of defining relations for $\widetilde{\mathsf{A}}_2$, we have the following result.

Proposition 26.4.8 *The operations $\widetilde{Sq}^k$ on DS define a left $\widetilde{\mathsf{A}}_2$-module structure on DS.* □

The total operation $\widetilde{\mathsf{Sq}}$ commutes with the right action of A_2 on DS via the halving map V of A_2.

Proposition 26.4.9 *For all $s \geq 0$,*

(i) $\widetilde{\mathsf{Sq}}(b_\lambda) \cdot Sq^{2s+1} = 0$, *and* (ii) $\widetilde{\mathsf{Sq}}(b_\lambda) \cdot Sq^{2s} = \widetilde{\mathsf{Sq}}(b_\lambda \cdot Sq^s)$.

Proof By Proposition 26.1.11 we need only check the terms involving $\widetilde{Sq}^a$ where $a > 0$. Both (i) and (ii) are proved by induction on the length $r = \ell(\lambda)$ of b_λ, and in each case the inductive step is proved by replacing $\kappa_*^S = \widetilde{Sq}^0$ by $\widetilde{\mathsf{Sq}}$ in the corresponding argument for Proposition 26.1.11. The base cases are proved as follows.

(i) For $r = 1$, $b_\lambda = b_k$ for some $k \geq 1$, and $\widetilde{\mathrm{Sq}}(b_k) \cdot Sq^{2s+1} = (b_{2k+1} + b_k^2) \cdot Sq^{2s+1} = 0$, since the first term is 0 by Proposition 26.1.11 and the second term is 0 by the Cartan formula (26.2) for the right action of A_2.

(ii) For $r = 1$, we need only check that $\widetilde{Sq}^a(b_k) \cdot Sq^{2s} = \widetilde{Sq}^a(b_k \cdot Sq^s)$ when $a = 1$, since Proposition 26.1.11 gives the result for $a = 0$ and both sides are 0 for $a > 1$. When $a = 1$, $\widetilde{Sq}^1(b_k) \cdot Sq^{2s} = b_k^2 \cdot Sq^{2s} = (b_k \cdot Sq^s)^2 = (\binom{k-s}{s} b_{k-s})^2 = \binom{k-s}{s} b_{k-s}^2$, and $\widetilde{Sq}^1(b_k \cdot Sq^s) = \widetilde{Sq}^1(\binom{k-s}{s} b_{k-s}) = \binom{k-s}{s} \widetilde{Sq}^1(b_{k-s}) = \binom{k-s}{s} b_{k-s}^2$ also. □

The next result follows at once, as for Proposition 26.1.12.

Proposition 26.4.10 *For $a \geq 0$, $\widetilde{Sq}^a$ maps $\mathsf{K}^{d,r}(\mathsf{DS})$ to $\mathsf{K}^{2d+r-a,r+a}(\mathsf{DS})$.* □

Example 26.4.11 Let $c = b_3 b_2^2 + b_5 b_1^2$. By Example 26.1.8, $c \in \mathsf{K}^{7,3}(\mathsf{DS})$. Hence by Example 26.4.5 and Proposition 26.4.10, $b_7 b_5^2 + b_{11} b_3^2 \in \mathsf{K}^{17,3}(\mathsf{DS})$, $b_7 b_2^4 + b_{11} b_1^4 \in \mathsf{K}^{15,5}(\mathsf{DS})$ and $b_3^2 b_2^4 + b_5^2 b_1^4 \in \mathsf{K}^{14,6}(\mathsf{DS})$.

We conclude by finding a $\mathbb{F}_2$-basis of $\widetilde{\mathsf{A}}_2$. Since A_2 is the quotient algebra of $\widetilde{\mathsf{A}}_2$ by the ideal generated by the central element $\widetilde{Sq}^0$, the results of Chapter 3 have counterparts for $\widetilde{\mathsf{A}}_2$.

Definition 26.4.12 A finite sequence $A = (a_1, a_2, \ldots, a_r)$ of length r of integers ≥ 0, and the corresponding monomial $\widetilde{Sq}^A = \widetilde{Sq}^{a_1} \cdots \widetilde{Sq}^{a_r} \in \widetilde{\mathsf{A}}_2$, are **admissible** if $a_i \geq 2a_{i+1}$ for $1 \leq i < r$.

Thus a finite admissible sequence may end in a string of 0s. The following example is the counterpart of Example 3.1.8.

Example 26.4.13

$$\widetilde{Sq}^4 \widetilde{Sq}^2 \widetilde{Sq}^3 = \widetilde{Sq}^9 \widetilde{Sq}^0 \widetilde{Sq}^0 + \widetilde{Sq}^8 \widetilde{Sq}^1 \widetilde{Sq}^0 + \widetilde{Sq}^7 \widetilde{Sq}^2 \widetilde{Sq}^0 + \widetilde{Sq}^6 \widetilde{Sq}^2 \widetilde{Sq}^1.$$

Proposition 26.4.14 *The admissible monomials form a $\mathbb{F}_2$-basis for $\widetilde{\mathsf{A}}_2$.*

Proof The argument of Proposition 3.1.7 can be adapted to show that the admissible monomials span $\widetilde{\mathsf{A}}_2$. To prove that they are linearly independent, we argue by contradiction. Let $\widetilde{Sq}^{A_1} + \ldots + \widetilde{Sq}^{A_t} = 0$ be a dependence relation in $\widetilde{\mathsf{A}}_2$, where $t > 0$ and $A_1, \ldots, A_t$ are distinct admissible sequences. Since the Adem relations preserve length, we may assume that the finite sequences A_i, $1 \leq i \leq t$ all have the same length r. For each i let $A_i = A_i' | 0_i$, where A_i' contains the positive terms and 0_i contains the zero terms of A_i. Applying the projection map $\widetilde{\mathsf{A}}_2 \to \mathsf{A}_2$ to the given relation, we obtain the relation $Sq^{A_1'} + \ldots + Sq^{A_t'} = 0$ between admissible monomials in A_2. Hence the sequences $A_1', \ldots, A_t'$ must be equal in pairs, and in particular such pairs of sequences have the same length. Hence, if $A'_i = A'_j$, then $A_i = A_j$, a contradiction. Hence the admissible monomials are linearly independent. □

26.5 Remarks

The standard references for symmetric functions are the books by I. G. Macdonald [128, Chapter 1] and R. P. Stanley [192, Chapter 7]. It is usual in algebraic combinatorics to identify the Hopf algebra $\Lambda = \mathbb{Z}[e_1, e_2, \ldots]$ with its dual Λ_* using the isomorphism which maps the complete symmetric function h_k to b_k. The Kronecker product then becomes the classical inner product on Λ defined by $\langle h_\lambda, m_\mu \rangle = \delta_{\lambda,\mu}$. Since the antipode χ of Λ exchanges e_k and h_k, the map $\Lambda \to \Lambda_*$ which sends e_k to b_k is also an isomorphism. The self-duality of Λ as a Hopf algebra is proved in [62] from an algebraic viewpoint, while a treatment which includes applications to cohomology of Grassmannians can be found in the notes by Arunas Liulevicius [125, Chapter IV]. In this chapter we prefer not to identify the Hopf algebra S of symmetric functions over $\mathbb{F}_2$ with its dual DS.

The hit problem for $\mathsf{S}(3)$ was solved by A. S. Janfada [99] by giving criteria on a monomial $f \in \mathsf{P}(3)$ for its symmetrization $\sigma_3(f)$ to be hit in $\mathsf{S}(3)$ [103]. The treatment given in this chapter using the dual is essentially equivalent to that of Janfada. Section 26.4 is based on the work of Singer [188]. Earlier work on the bigraded Steenrod algebra $\widetilde{\mathsf{A}}_2$ can be found in [124] and [130]. The work of David Pengelley and Frank Williams [163, 164] on the A_2-module structure of $H_*(BO)$ uses the algebras of Kudo–Araki–May operations and of Dyer–Lashof operations in place of $\widetilde{\mathsf{A}}_2$. Hadi Zare [239] has used the Dyer–Lashof algebra to compute elements of the Steenrod kernel for $H_*(BO)$.

27
The cyclic splitting of $\mathsf{P}(n)$

27.0 Introduction

A solution of the hit problem for $\mathsf{P}(n)$ can be sought by decomposing $\mathsf{P}(n)$ as the direct sum of A_2-submodules given by a complete set of primitive orthogonal idempotents $e_1, \ldots, e_r$ for $\mathbb{F}_2\mathsf{GL}(n)$, and solving the hit problem separately for each submodule $\mathsf{P}(n)e_i$, $1 \leq i \leq r$. Except in the case of the Steinberg idempotents $e(n)$ and $e'(n)$, this approach is frustrated by the difficulty of finding sufficiently explicit forms for the idempotents e_i.

A less ambitious procedure is to find a coarser decomposition of $\mathsf{P}(n)$ arising from a suitably chosen subgroup of $\mathsf{GL}(n)$, where the idempotents can be more easily managed. But then another problem arises. In the case of $\mathsf{GL}(n)$, a full set of irreducible representations and corresponding idempotents can be constructed over $\mathbb{F}_2$, which is a splitting field (Proposition 18.2.1). For subgroups of $\mathsf{GL}(n)$ this need not be the case.

For example, the matrix $C = \left(\begin{smallmatrix} 0 & 1 \\ 1 & 1 \end{smallmatrix}\right)$ generates a cyclic group $\mathsf{C}(2)$ of order 3 in $\mathsf{GL}(2)$ and $e_0 = I + C + C^2$, $e_0' = C + C^2$ form a pair of orthogonal idempotents in the group algebra $\mathbb{F}_2\mathsf{C}(2)$, decomposing the identity matrix I. Since $\mathsf{C}(2)$ has odd order, applying e_0 to $\mathsf{P}(2)$ is the classical transfer map, and the image space of the idempotent $\mathsf{P}(2,0) = \mathsf{P}(2) \cdot e_0$ may also be described as the algebra of invariants of C acting on $\mathsf{P}(2)$.

As in Example 18.0.1, to obtain the full set of three 1-dimensional representations of $\mathsf{C}(2)$ and corresponding idempotents, we need to extend $\mathbb{F}_2$ to the Galois field $\mathbb{F}_4$ by adjoining a cube root ζ of 1 satisfying $\zeta^2 + \zeta + 1 = 0$. Then $e_0' = e_1 + e_2$, where $e_1 = I + \zeta C + \zeta^2 C^2$ and $e_2 = I + \zeta^2 C + \zeta C^2$ are orthogonal idempotents decomposing e_0'. The idempotents e_j, $j = 0, 1, 2$ form a complete set for $\mathsf{C}(2)$ corresponding to the 1-dimensional representations of $\mathsf{C}(2)$ defined by $1, \zeta, \zeta^2$. Let $\mathsf{P}(2,j) = \mathsf{P}(2) \cdot e_j$ denote the image of e_j for $j = 0, 1, 2$.

Working in the polynomial algebra $\mathbb{F}_4[x_1,x_2]$, it is convenient to make a change of variables in order to diagonalize the matrix action of C over $\mathbb{F}_4$. Let $t_1 = x_1 + \zeta x_2$, $t_2 = x_1 + \zeta^2 x_2$. Then $t_1 = x_1 \cdot B$ and $t_2 = x_2 \cdot B$, where $B = \begin{pmatrix} 1 & \zeta \\ 1 & \zeta^2 \end{pmatrix}$. Since $\det(B) = 1$, the transformation of variables is invertible and $\mathbb{F}_4[t_1,t_2] = \mathbb{F}_4[x_1,x_2]$. Now BCB^{-1} is the diagonal matrix $D = \begin{pmatrix} \zeta & 0 \\ 0 & \zeta^2 \end{pmatrix}$. The action of C on the t-variables is given by $t_1 \cdot C = (x_1 \cdot B) \cdot C = x_1 \cdot BC = x_1 \cdot DB = \zeta t_1$, and similarly $t_2 \cdot C = \zeta^2 t_2$.

Let $\widetilde{\mathsf{P}}(2) = \mathbb{F}_2[t_1,t_2]$ and let $\widetilde{\mathsf{P}}(2,j) = \widetilde{\mathsf{P}}(2) \cdot e_j$ denote the image of e_j for $j = 0,1,2$. The image of e_0 is determined on monomials in t_1 and t_2 by $t_1^{a_1} t_2^{a_2} \cdot e_0 = (1 + \zeta^{a_1+2a_2} + \zeta^{2a_1+a_2}) t_1^{a_1} t_2^{a_2}$. This expression is 0 unless $a_1 + 2a_2 = 0 \bmod 3$, in which case it is $t_1^{a_1} t_2^{a_2}$. Hence $\widetilde{\mathsf{P}}(2,0)$ is identified as the vector subspace of $\widetilde{\mathsf{P}}(2)$ spanned by monomials $t_1^{a_1} t_2^{a_2}$ with $a_1 + 2a_2 = 0 \bmod 3$. Similar calculations show that $\widetilde{\mathsf{P}}(2,1)$ and $\widetilde{\mathsf{P}}(2,2)$ are vector subspaces of $\widetilde{\mathsf{P}}(2)$ spanned by monomials $t_1^{a_1} t_2^{a_2}$ with $a_1 + 2a_2 = 2 \bmod 3$ and $a_1 + 2a_2 = 1 \bmod 3$ respectively. The vector spaces $\widetilde{\mathsf{P}}(2,j)$ are defined over $\mathbb{F}_2$, and have monomial bases. The original idempotent images $\mathsf{P}(2,j)$ have been transformed by conjugation with B into the eigenspaces $\widetilde{\mathsf{P}}(2,j)$ of the three irreducible representations of the cyclic group generated by D. Multiplication of monomials induces a linear pairing $\widetilde{\mathsf{P}}(2,i) \otimes \widetilde{\mathsf{P}}(2,j) \to \widetilde{\mathsf{P}}(2,i+j)$, where $i+j$ is taken mod 3. In particular, $\widetilde{\mathsf{P}}(2,0)$ can be identified with the algebra of invariants of $\mathsf{C}(2)$, and each $\widetilde{\mathsf{P}}(2,j)$ is a module over $\widetilde{\mathsf{P}}(2,0)$.

For any extension field F of $\mathbb{F}_2$, the Steenrod squares act on $F[x_1,\ldots,x_n]$ by $Sq^k(\alpha f) = \alpha Sq^k(f)$, where f is a monomial in $x_1,\cdots,x_n$ and $\alpha \in F$. Working in the extension field does not change the hit problem as far as the dimensions of the cohits are concerned, but the Steenrod squares do not in general commute with the action of $\mathsf{GL}(n,F)$. In particular, we can no longer write $Sq^d(f) = f^2$ if f is a homogeneous polynomial of degree d. For example, $Sq^1(\alpha x_1) = \alpha x_1^2 \neq \alpha^2 x_1^2$ unless $\alpha \in \mathbb{F}_2$. However f^2 is still hit, and the stability condition $Sq^k(f) = 0$ for $k > d$ is still valid.

Working over $\mathbb{F}_4$, we have $Sq^1(t_1) = Sq^1(x_1 + \zeta x_2) = x_1^2 + \zeta x_2^2 = t_2^2$, and similarly $Sq^1(t_2) = t_1^2$. As in Chapter 1, we can define a *twisted* total squaring operation on $\widetilde{\mathsf{P}}(2)$ by $\mathsf{Sq}(t_1) = t_1 + t_2^2$, $\mathsf{Sq}(t_2) = t_2 + t_1^2$. Note that $\widetilde{\mathsf{P}}(2)$ and $\mathsf{P}(2)$ become the same A_2-algebra after extension of the ground field to $\mathbb{F}_4$. This shows that $\widetilde{\mathsf{P}}(2)$ is an A_2-module under the twisted action, in fact an A_2-algebra, extending the action of Sq^1 as above. For all j, $\widetilde{\mathsf{P}}(2,j)$ is invariant under the twisted action of A_2, since $\mathsf{Sq}(t_1^{a_1} t_2^{a_2})$ is a sum of monomials $t_1^{b_1} t_2^{b_2}$ such that $b_1 + 2b_2 = a_1 + 2a_2 \bmod 3$. We observe that switching t_1 and t_2 induces an A_2-module isomorphism between $\widetilde{\mathsf{P}}(2,1)$ and $\widetilde{\mathsf{P}}(2,2)$.

Thus $\widetilde{\mathsf{P}}(2) = \widetilde{\mathsf{P}}(2,0) \oplus \widetilde{\mathsf{P}}(2,1) \oplus \widetilde{\mathsf{P}}(2,2)$, where each cyclic summand $\widetilde{\mathsf{P}}(2,j)$ is an A_2-module with corresponding hit elements $\widetilde{\mathsf{H}}(2,j)$ and cohit quotient $\widetilde{\mathsf{Q}}(2,j)$. Since $\widetilde{\mathsf{P}}(2)$ and $\mathsf{P}(2)$ become equal after extension to $\mathbb{F}_4$, $\dim \mathsf{Q}^d(2) = \dim \widetilde{\mathsf{Q}}^d(2) = \dim \widetilde{\mathsf{Q}}^d(2,0) + \dim \widetilde{\mathsf{Q}}^d(2,1) + \dim \widetilde{\mathsf{Q}}^d(2,2)$, where the last two terms are equal.

In Section 27.1 we set up the cyclic decomposition for the general case. In Section 27.2 we adapt block technology and associated techniques to apply to the cyclic splitting. In Section 27.3 we introduce 'prefixing' and 'postfixing' maps which relate the spaces $\widetilde{\mathsf{Q}}^\omega(n,j)$ and the twisted analogues of the Kameko maps, and discuss the 2-variable case. Section 27.4 contains results about tail and head blocks for the twisted action of A_2, analogous to those of Sections 6.7 and 6.8. The 3-variable case is examined in Section 27.5.

27.1 The twisted A_2-module $\widetilde{\mathsf{P}}(n)$

Let ζ denote a generator of the cyclic group $\mathbb{F}_{2^n}^*$ of nonzero elements in the Galois field $\mathbb{F}_{2^n}$. The Frobenius automorphism F of $\mathbb{F}_{2^n}$ is defined by $F(\alpha) = \alpha^2$ for $\alpha \in \mathbb{F}_{2^n}$, and we identify $\mathbb{F}_2$ with the fixed points of F. Let

$$Z = \begin{pmatrix} 1 & \zeta & \zeta^2 & \cdots & \zeta^{n-1} \\ 1 & \zeta^2 & \zeta^4 & \cdots & \zeta^{2(n-1)} \\ \vdots & \vdots & \vdots & \ddots & \vdots \\ 1 & \zeta^{2^{n-1}} & \zeta^{2^n} & \cdots & \zeta^{2^{n-1}(n-1)} \end{pmatrix} \tag{27.1}$$

be the $n \times n$ matrix with (i,j)th entry $\zeta^{(j-1)\cdot 2^{i-1}}$, for $1 \le i,j \le n$. Since F permutes the rows of Z cyclically, the Vandermonde determinant $\det(Z) = \prod_{i<j}(\zeta^{2^j} - \zeta^{2^i})$ is fixed by F. Since the elements ζ^{2^i}, $0 \le i \le n-1$, are distinct, $\det(Z) \neq 0$ and hence $\det(Z) = 1$. If $\mathbf{t}$ and $\mathbf{x}$ are the column vectors of the variables t_i and x_j, the change of variables $\mathbf{t} = Z\mathbf{x}$, namely

$$t_i = x_1 + \zeta^{2^{i-1}} x_2 + \zeta^{2\cdot 2^{i-1}} x_3 + \cdots + \zeta^{(n-1)\cdot 2^{i-1}} x_n \tag{27.2}$$

is invertible. Hence $\mathbb{F}_{2^n}[t_1,\ldots,t_n] = \mathbb{F}_{2^n}[x_1,\ldots,x_n]$. Also

$$Sq^1(t_i) = x_1^2 + \zeta^{2^{i-1}} x_2^2 + \zeta^{2\cdot 2^{i-1}} x_3^2 + \cdots + \zeta^{(n-1)\cdot 2^{i-1}} x_n^2 = t_{i-1}^2,$$

where $t_0 = t_n$. As in the 2-variable case (Section 27.0), we define the 'twisted' action of A_2 on a polynomial algebra over $\mathbb{F}_2$ in n variables as follows.

Definition 27.1.1 The twisted A_2-module $\widetilde{P}(n) = \mathbb{F}_2[t_1, \dots, t_n]$ for $n \geq 1$, with the total squaring operation $\mathsf{Sq} : \widetilde{P}(n) \to \widetilde{P}(n)$ defined on the generators by $\mathsf{Sq}(t_i) = t_i + t_{i-1}^2$ and extended to $\widetilde{P}(n)$ as an algebra map. The twisted cohits are the nonzero elements of the quotient $\mathbb{F}_2$-space $\widetilde{Q}(n) = \widetilde{P}(n)/A_2^+\widetilde{P}(n)$.

Proposition 27.1.2 *$\widetilde{P}(n)$ is an algebra over A_2.*

Proof As in Chapter 1, Sq defines linear maps $Sq^k \colon \widetilde{P}^d(n) \to \widetilde{P}^{d+k}(n)$ satisfying the Cartan formula $Sq^k(fg) = \sum_{i+j=k} Sq^i(f)Sq^j(g)$ for $f,g \in \widetilde{P}(n)$. Since the extension of coefficients to $\mathbb{F}_{2^n}$ makes $\widetilde{P}(n)$ and $P(n)$ equal, and the formulae for $\mathsf{Sq}(t_1), \dots, \mathsf{Sq}(t_n)$ are derived by substitution from the formulae for $\mathsf{Sq}(x_1), \dots, \mathsf{Sq}(x_n)$, it follows that the operations Sq^k with the twisted action on $P(n)$ satisfy the Adem relations, and so make $\widetilde{P}(n)$ into an algebra over A_2. □

Proposition 27.1.3 *For all n and d,* $\dim \widetilde{Q}^d(n) = \dim Q^d(n)$.

Proof The vector spaces $\mathbb{F}_{2^n} \otimes_{\mathbb{F}_2} \widetilde{Q}^d(n)$ and $\mathbb{F}_{2^n} \otimes_{\mathbb{F}_2} Q^d(n)$ over $\mathbb{F}_{2^n}$ are equal, so $\widetilde{Q}^d(n)$ and $Q^d(n)$ have the same dimension as vector spaces over $\mathbb{F}_2$. □

The twisted squaring operations are given on powers of the generators by the following formula.

Proposition 27.1.4

$$Sq^k(t_i^r) = \binom{r}{k} t_i^{r-k} t_{i-1}^{2k}.$$

Proof By definition, $Sq^k(t_i^r)$ is the term of degree $k+r$ in the expansion of $\mathsf{Sq}(t_i^r) = (t_i + t_{i-1}^2)^r = \sum_{a+b=r} \binom{r}{a} t_i^a t_{i-1}^{2b}$. Putting $k+r = a+2b$ and $a+b=r$ gives $b=k$ and $a=r-k$. □

Proposition 27.1.5 *If $f \in \widetilde{P}(n)$ is hit in $\mathbb{F}_{2^n}[x_1, \dots, x_n]$, then f is hit in $\widetilde{P}(n)$.*

Proof From Galois theory, ζ satisfies an irreducible equation of degree n, and the powers ζ^i for $0 \leq i < n$ form a basis for $\mathbb{F}_{2^n}$ as a vector space over $\mathbb{F}_2$. Let f be an element in $\widetilde{P}(n)$ satisfying a hit equation $f = \sum_{i>0} Sq^i(g_i)$ for g_i in $\mathbb{F}_{2^n}[x_1, \dots, x_n]$. Write $g_i = f_i + \sum_{j=1}^{n-1} \zeta^j g_{i,j}$, where f_i and $g_{i,j}$ are in $\widetilde{P}(n)$. Comparing coefficients of 1, we obtain the hit equation $f = \sum_{i>0} Sq^i(f_i)$ in $\widetilde{P}(n)$. □

Definition 27.1.6 The **cyclic weight** of a monomial $f = t_1^{a_1} \cdots t_n^{a_n}$ in $\widetilde{P}(n)$ is $\mathrm{cw}(f) = \sum_{i=1}^n 2^{i-1} a_i \bmod 2^n - 1$. For $0 \leq j \leq 2^n - 2$, we denote by $\widetilde{P}(n,j)$ the vector subspace of $\widetilde{P}(n)$ spanned by monomials f with $\mathrm{cw}(f) = j$.

In n-block notation, the cyclic weight is the sum of the 2-powers 2^{i+j-2} over all (i,j) where the entry is 1, reduced mod $2^n - 1$. Since these 2-powers depend only on $i+j$ mod n, this can be visualized using the array

$$\begin{array}{cccccccc} 1 & 2 & 4 & \cdots & 2^{n-1} & 1 & \cdots \\ 2 & 4 & 8 & \cdots & 1 & 2 & \cdots \\ \vdots & \vdots & \vdots & \cdots & \vdots & \vdots & \cdots \\ 2^{n-1} & 1 & 2 & \cdots & 2^{n-2} & 2^{n-1} & \cdots \end{array}.$$

The spaces $\widetilde{\mathsf{P}}(n,j)$ are the eigenspaces of the 1-dimensional representations of the cyclic group of order $2^n - 1$ generated by the diagonal matrix

$$D = \begin{pmatrix} \zeta & 0 & \cdots & 0 \\ 0 & \zeta^2 & \cdots & 0 \\ \vdots & \vdots & \ddots & \vdots \\ 0 & 0 & \cdots & \zeta^{2^{n-1}} \end{pmatrix}.$$

Since the cyclic weight satisfies $\mathrm{cw}(fg) = \mathrm{cw}(f) + \mathrm{cw}(g)$ for monomials f, g, multiplication of monomials induces a pairing $\widetilde{\mathsf{P}}(n,i) \otimes \widetilde{\mathsf{P}}(n,j) \to \widetilde{\mathsf{P}}(n,i+j)$. In particular, $\widetilde{\mathsf{P}}(n,0)$ is an algebra, and $\widetilde{\mathsf{P}}(n,j)$ is a module over $\widetilde{\mathsf{P}}(n,0)$ for all j.

Let $c_0 + c_1 t + c_2 t^2 + \cdots + c_{n-1} t^{n-1} + t^n \in \mathbb{F}_2[t]$ be the minimal polynomial of ζ. The rows of the matrix Z in (27.1) are the left eigenvectors of the companion matrix

$$C = \begin{pmatrix} 0 & \cdots & 0 & c_0 \\ 1 & \cdots & 0 & c_1 \\ \vdots & \ddots & \vdots & \vdots \\ 0 & \cdots & 1 & c_{n-1} \end{pmatrix}$$

in $\mathsf{GL}(n, \mathbb{F}_2)$, corresponding to the eigenvalues $\zeta^{2^{i-1}}$ for $1 \le i \le n$. It follows that $ZCZ^{-1} = D$, as we saw for the 2-variable case in Section 27.0.

Example 27.1.7 For $n = 3$ we obtain the Galois field $\mathbb{F}_8$ by adjoining ζ to $\mathbb{F}_2$, satisfying the minimal polynomial equation $1 + \zeta^2 + \zeta^3 = 0$. The matrices C and Z are given by

$$C = \begin{pmatrix} 0 & 0 & 1 \\ 1 & 0 & 0 \\ 0 & 1 & 1 \end{pmatrix}, \quad Z = \begin{pmatrix} 1 & \zeta & \zeta^2 \\ 1 & \zeta^2 & \zeta^4 \\ 1 & \zeta^4 & \zeta \end{pmatrix},$$

and the variables t_i by

$$t_1 = x_1 + \zeta x_2 + \zeta^2 x_3, \; t_2 = x_1 + \zeta^2 x_2 + \zeta^4 x_3, \; t_3 = x_1 + \zeta^4 x_2 + \zeta x_3.$$

Theorem 27.1.8 *The spaces $\widetilde{\mathsf{P}}(n,j)$ are A_2-submodules of $\widetilde{\mathsf{P}}(n)$. There is a direct sum splitting of A_2-modules*

$$\widetilde{\mathsf{P}}(n) = \sum_{j=0}^{2^n-2} \widetilde{\mathsf{P}}(n,j).$$

Proof Let $f = t_1^{a_1} \cdots t_n^{a_n}$ be a monomial in $\widetilde{\mathsf{P}}(n)$. A typical monomial appearing in $\mathsf{Sq}(m) = (t_1 + t_n^2)^{a_1}(t_2 + t_1^2)^{a_2} \cdots (t_n + t_{n-1}^2)^{a_n}$ has the form

$$g = \prod_{k=1}^{n} t_k^{i_k}(t_{k-1}^2)^{a_k - i_k} = t_1^{i_1 + 2a_2 - 2i_2} t_2^{i_2 + 2a_3 - 2i_3} \cdots t_n^{i_n + 2a_1 - 2i_1}$$

where $t_0 = t_n$. The cyclic weight $\mathrm{cw}(g) = (i_1 + 2a_2 - 2i_2) + 2(i_2 + 2a_3 - 2i_3) + \cdots + 2^{n-1}(i_n + 2a_1 - 2i_1) = a_1 + 2a_2 + \cdots + 2^{n-1}a_n \bmod (2^n - 1) = \mathrm{cw}(f)$ is independent of $i_1, \ldots, i_n$. Hence the twisted action of the squaring operations preserves the cyclic summands $\widetilde{\mathsf{P}}(n,j)$. □

We next explain the effect of rotating the variables t_i.

Proposition 27.1.9 *Let ρ denote the automorphism of $\widetilde{\mathsf{P}}(n)$ induced by the rotation of variables defined by $\rho(t_i) = t_{i+1}$ for $i < n$ and $\rho(t_n) = t_1$. Then ρ induces an A_2-module isomorphism of $\widetilde{\mathsf{P}}(n,j)$ with $\widetilde{\mathsf{P}}(n,2j)$.*

Proof Since $\mathrm{cw}(\rho(f)) = 2\mathrm{cw}(f)$, ρ maps $\widetilde{\mathsf{P}}(n,j)$ to $\widetilde{\mathsf{P}}(n,2j)$, and since $\rho(\mathsf{Sq}(t_i)) = \rho(t_i + t_{i-1}^2) = t_{i+1} + t_i^2 = \mathsf{Sq}(t_{i+1}) = \mathsf{Sq}(\rho(t_i))$, ρ commutes with Sq. □

Definition 27.1.10 We define $\widetilde{\mathsf{H}}(n)$ for the hit elements in $\widetilde{\mathsf{P}}(n)$ and $\widetilde{\mathsf{Q}}(n) = \widetilde{\mathsf{P}}(n)/\widetilde{\mathsf{H}}(n)$. For $0 \le j \le 2^n - 2$, we define $\widetilde{\mathsf{H}}(n,j)$ for the hit elements in $\widetilde{\mathsf{P}}(n,j)$ and $\widetilde{\mathsf{Q}}(n,j) = \widetilde{\mathsf{P}}(n,j)/\widetilde{\mathsf{H}}(n,j)$.

It follows that $\dim \widetilde{\mathsf{Q}}^d(n) = \dim \mathsf{Q}^d(n)$ for all n and d by Proposition 27.1.2 and $\dim \widetilde{\mathsf{Q}}^d(n) = \sum_{j=0}^{2^n-2} \dim \widetilde{\mathsf{Q}}^d(n,j)$ by Theorem 27.1.8. In practice, the work in evaluating $\dim \widetilde{\mathsf{Q}}(n)$ is reduced by the use of Proposition 27.1.9. For example, if $n = 3$ we need only determine $\dim \widetilde{\mathsf{Q}}(n,j)$ for $j = 0, 1, 3$, since $\dim \widetilde{\mathsf{Q}}(3) = \dim \widetilde{\mathsf{Q}}(3,0) + 3\dim \widetilde{\mathsf{Q}}(3,1) + 3\dim \widetilde{\mathsf{Q}}(3,3)$.

The action of A_2 on $\widetilde{\mathsf{P}}(n)$ does not in general commute with matrix action on $t_1, \ldots, t_n$ over $\mathbb{F}_2$. Hence we cannot usually specialize variables as we did with the standard action in Chapter 1. However, there are exceptions. Since equating all variables in $\widetilde{\mathsf{P}}(n)$ to the same variable commutes with Sq, Theorem 1.4.12 yields the following result.

Proposition 27.1.11 *For $r \geq 0$, a monomial in $\widetilde{\mathsf{P}}^{2^r-1}(n)$ is not hit.* □

More generally, if $n = pq$, then there is an A_2-module map $\epsilon : \widetilde{\mathsf{P}}(n) \to \widetilde{\mathsf{P}}(q)$, obtained by equating variables in sets of p elements by putting $t_i = t_{i+kq}$ for $1 \leq i \leq q$, $0 \leq k < p$. We also have $\mathrm{cw}(\epsilon(f)) = \mathrm{cw}(f) \bmod 2^q - 1$ for a monomial f in $\widetilde{\mathsf{P}}(n)$.

27.2 The cyclic splitting and blocks

In this section we refine the splitting process by combining the cyclic splitting with the filtration by ω-sequences in $\widetilde{\mathsf{P}}(n)$. As usual, we identify a monomial $t_1^{a_1} \cdots t_n^{a_n}$ in $\widetilde{\mathsf{P}}(n)$ with an n-block, so that the rows of the block are the reverse binary expansions of the exponents $a_1, \ldots, a_n$, and we consider the twisted action of A_2 on a block.

We begin with Sq^1. For r odd, $Sq^1(t_i^r) = t_i^{r-1} t_{i-1}^2$ by Proposition 27.1.4.

Example 27.2.1 In the case $n = 3$, the Cartan formula gives $Sq^1(t_1^7 t_2 t_3) = Sq^1(t_1^7) t_2 t_3 + t_1^7 Sq^1(t_2) t_3 + t_1^7 t_2 Sq^1(t_3) = t_1^6 t_2 t_3^3 + t_1^9 t_3 + t_1^7 t_2^3$. Note that all four monomials have cyclic weight 6 mod 7. In terms of blocks, this becomes

$$Sq^1 \begin{pmatrix} 1\,1\,1 \\ 1\,0\,0 \\ 1\,0\,0 \end{pmatrix} = \begin{matrix} 0\,1\,1 \\ 1\,0\,0 \\ 1\,1\,0 \end{matrix} + \begin{matrix} 1\,0\,0\,1 \\ 0\,0\,0\,0 \\ 1\,0\,0\,0 \end{matrix} + \begin{matrix} 1\,1\,1 \\ 1\,1\,0 \\ 0\,0\,0 \end{matrix}. \tag{27.3}$$

Given a block B, $Sq^1(B) = \sum_r B_r$ where the sum is over the rows r for which B has an entry 1 in column 1, and B_r is formed by removing this entry and adding 1 arithmetically to row $r-1$ of B in column 2, where the rows are indexed mod n. We may view (27.3) as the 3-back splicing of the block B_2 at position $(2,1)$ by removing the digit 1 at position $(1,4)$ and inserting digits 1 at positions $(1,2)$, $(1,3)$ and $(2,1)$, to produce B. Then $B_2 \sim B_1 + B_3$.

We recall that the ω-sequence of a block is the sequence of its column sums. The effect of Sq^1 on a nonzero block is to move at least one digit 1 forward and no digits 1 backwards. By iteration of the Cartan formula, the evaluation of any Steenrod operation can be reduced to the evaluation of Sq^1. Hence the action of A_2^+ on $\widetilde{\mathsf{P}}(n)$ lowers the left and right orders of the ω-sequences of blocks. It follows that the filtration of $\widetilde{\mathsf{P}}^d(n)$ and $\widetilde{\mathsf{P}}^d(n,j)$ by ω-sequences, reducibility, and the spaces $\widetilde{\mathrm{Q}}^{\omega}(n)$ and $\widetilde{\mathrm{Q}}^{\omega}(n,j)$ for $\omega \in \mathsf{Seq}_d(n)$ go through as in Section 6.2, and that we obtain the following result.

Proposition 27.2.2 *For* $\omega \in \mathsf{Seq}(n)$, $\dim \widetilde{\mathrm{Q}}^{\omega}(n) = \dim \mathrm{Q}^{\omega}(n)$. □

For example, we deduce from Proposition 8.3.1 that $\widetilde{\mathrm{Q}}^{\omega}(3) = 0$ unless ω is decreasing. In Example 27.2.1, we see that B_1 and B_3 are equivalent modulo blocks which are hit or have left lower ω-sequences.

For the standard action of A_2, 1-back splicing involves the movement of digits 1 to right or left along rows of a block. For the twisted action, there is a corresponding movement along NE to SW diagonals.

Definition 27.2.3 The kth **diagonal** of an n-block B consists of the positions (i,j), where $i+j=k+1$ and i,j are taken mod n in the range $1,\dots,n$.

Thus the cyclic weight of an n-block is the sum mod $2^n - 1$ of the numbers obtained by replacing each digit 1 in the kth diagonal by 2^{k-1}, for $1 \le k \le n$.

We next describe the twisted version of 1-back splicing. By Theorem 27.1.8, this operation preserves the cyclic weight. Let B be an n-block with an entry 1 in position $(i-1,j)$ and an entry 0 in position $(i,j-1)$, and suppose that B has an entry 1 in position $(r,j-1)$ and an entry 0 in position $(r-1,j)$. Let B' be formed from B by moving the entry 1 in position $(i-1,j)$ one place to the left down its diagonal into position $(i,j-1)$, and form B_r from B' by moving the entry 1 in position $(r,j-1)$ one place to the right up its diagonal into position $(r-1,j)$. Using Proposition 27.1.4 to evaluate $Sq^{2^{j-2}}(B')$, we obtain the twisted version $B \sim \sum_r B_r$ of 1-back splicing of B at position $(i,j-1)$. More error blocks are involved in the application of $Sq^{2^{j-2}}$, but these have ω-sequences $<_l \omega(B)$.

Example 27.2.4 Consider the equivalences of tail blocks

$$\begin{matrix}1\,0\,1\\0\,1\,0\\0\,0\,0\end{matrix} \sim \begin{matrix}0\,0\,1\\0\,0\,0\\1\,1\,0\end{matrix} \sim \begin{matrix}0\,0\,0\\0\,1\,1\\1\,0\,0\end{matrix}$$

obtained by successive 1-back splicing at $(3,1)$ in the first block and at $(2,2)$ in the second. The digit 1 in the first column of the first block is moved along its diagonal to the last column of the third block. We may also interpret the result from right to left, as moving the digit 1 in position $(2,3)$ of the last block to position $(1,1)$ of the first block along the same diagonal.

Next consider an analogous result for head blocks

$$\begin{matrix} 0\,1\,0 & & 1\,1\,0 & & 1\,1\,1 \\ 1\,0\,1 & \sim & 1\,1\,1 & \sim & 1\,0\,0 \\ 1\,1\,1 & & 0\,0\,1 & & 0\,1\,1 \end{matrix}$$

obtained by successive 1-back splicing in position $(1,1)$ in the first block and $(3,2)$ in the second. The effect is to move the digit 0 from the first column of the first block along its diagonal to the last column of the third block.

The following general result is clear.

Proposition 27.2.5 *If a block has at least one digit* 1 *in some diagonal, then it is equivalent to a sum of blocks with a digit* 1 *in the same diagonal and in the first column. For a tail block it is also equivalent to a tail block with a digit* 1 *in the same diagonal and in the last column. A similar statement applies to head blocks, by interchanging the roles of* 0 *and* 1. □

We can visualize the diagonals of a block by writing it in diagonal form, so that the k-diagonal of a (n,c)-block occupies the positions $(k,1),(k-1,2),\ldots,(k-c+1,c)$, where row positions can be ≤ 0. The following illustration shows Example 27.2.4 in diagonal form.

Example 27.2.6

$$\begin{array}{ccccccccccccccccccccccc}
 & & 0 & & & & 0 & & & & 1 & & & & 1 & & & & 1 & & & & 0 \\
 & 0 & 0 & & & 1 & 0 & & & 0 & 0 & & & 1 & 1 & & & 0 & 1 & & & 1 & 1 \\
1 & 0 & 1 & \sim & 0 & 0 & 1 & \sim & 0 & 0 & 0 & , & 0 & 1 & 0 & \sim & 1 & 1 & 0 & \sim & 1 & 1 & 1\,. \\
0 & 1 & & & 0 & 0 & & & 0 & 1 & & & 1 & 0 & & & 1 & 1 & & & 1 & 0 & \\
0 & & & & 1 & & & & 1 & & & & 1 & & & & 0 & & & & 0 & &
\end{array}$$

Note that the diagonal form facilitates checking cyclic weights, which in this case are $2^0+2\cdot 2^2=2 \bmod 7$ and $2\cdot 2^0+3\cdot 2^1+2^2=5 \bmod 7$ respectively.

We next sketch the twisted version of t-back splicing. Again this preserves the cyclic weight, by Theorem 27.1.8. We start with a block B having a digit 1 in position $(i-1,j+t-1)$ and 0 in positions $(i-1,j+s)$ for $0\leq s<t-1$, as well as in position $(i,j-1)$. The block B' is formed from B by inserting digits 1 in positions $(i,j-1)$ and $(i-1,j+s)$ for $0\leq s<t-1$ and inserting 0 in position $(i-1,j+t-1)$. Then $Sq^{2^{j-2}}$ is applied to B' to reproduce B and a sum F of other blocks, giving an equivalence $B\sim F$. Unlike the case of 1-back

splicing, error blocks with ω-sequence higher that $\omega(B)$ can occur in F, just as for the standard action.

Example 27.2.7 The following equivalence is produced by 2-back splicing of B at $(2,1)$. This is achieved by applying Sq^1 to the block B'.

$$B = \begin{matrix} 0\,0\,1 \\ 0\,1\,0 \\ 1\,0\,0 \end{matrix} \sim \begin{matrix} 0\,1\,0 \\ 1\,0\,1 \\ 0\,0\,0 \end{matrix}, \quad B' = \begin{matrix} 0\,1\,0 \\ 1\,1\,0 \\ 1\,0\,0 \end{matrix}.$$

For the standard action of A_2, the α-sequence of a block counts the number of digits 1 in each row. The analogue for the twisted action counts the number of digits 1 in each diagonal.

Definition 27.2.8 For an n-block B let $\widetilde{\alpha}(B) = (\widetilde{\alpha}_1, \widetilde{\alpha}_2, \dots, \widetilde{\alpha}_n)$, where $\widetilde{\alpha}_i$ is the sum of the digits in the ith diagonal of B.

Thus $\mathrm{cw}(B) = \sum_{i=1}^n 2^{i-1}\widetilde{\alpha}_i \bmod 2^n - 1$. In Example 27.2.6, the $\widetilde{\alpha}$-sequence of the blocks is $(1,0,2)$ or $(2,3,1)$. It is easy to see that 1-back splicing of a block B produces blocks with $\widetilde{\alpha}$-sequence $\widetilde{\alpha}(B)$ and ω-sequence $\leq_l \omega(B)$.

27.3 The Kameko map for $\widetilde{\mathsf{P}}(n)$

We next introduce some elementary maps which relate the spaces $\widetilde{\mathsf{Q}}^{\omega}(n,j)$ for certain sequences ω and cyclic weights j which are not necessarily connected by rotation of the variables $t_1, \dots, t_n$ (Proposition 27.1.9). These maps also make sense for the standard action of A_2, but are not so useful in this case.

Proposition 27.3.1 (i) *Let L be a block in $\widetilde{\mathsf{P}}^{\gamma}(n,\ell)$, where* $\mathrm{size}(\gamma) = t$ *and sequences are left ordered. Then the prefixing map $R \mapsto L|R$ induces a map $\psi_L : \widetilde{\mathsf{Q}}^{\sigma}(n,r) \to \widetilde{\mathsf{Q}}^{\gamma|\sigma}(n, \ell + 2^t r)$ for all sequences σ and cyclic weights r.*

(ii) *Let R be a block in $\widetilde{\mathsf{P}}^{\sigma}(n,r)$, where sequences are right ordered. Then the postfixing map $L \mapsto L|R$ induces a map $\psi^R : \widetilde{\mathsf{Q}}^{\gamma}(n,\ell) \to \widetilde{\mathsf{Q}}^{\gamma|\sigma}(n, \ell + 2^t r)$ for all sequences γ of size t and all cyclic weights ℓ.*

Proof We prove (i), as the proof of (ii) is similar. Let $R = R' + E + \sum_{i>0} Sq^i(H_i)$ be a hit equation, where $R, R' \in \widetilde{\mathsf{P}}^{\sigma}(n,r)$ and $\omega(E) <_l \sigma$. Then $R \sim R'$, and $L|R = L|R' + L|E + \sum_{i>0} L|Sq^i(H_i)$. Applying the χ-trick to the last summation, we obtain $L|R \sim L|R' + L|E + \sum_{i>0} L'|H_i'$, where $\omega(L|E) <_l \gamma|\sigma$ because $\omega(E) <_l \sigma$, and $\omega(L'|H_i') <_l \gamma|\sigma$ because $\omega(L') <_l \gamma$. Hence $L|R \sim L|R'$, and so ψ_L is well defined. □

There are several variations on the basic theme of Proposition 27.3.1. We note that if $\omega = \gamma|\sigma$ is minimal then the maps ψ_L and ψ^R are well defined in either the left or the right order, because all blocks with ω-sequence lower than the minimal spike in left or right order are hit.

Proposition 27.3.2 *Let R be a block in $\widetilde{\mathsf{P}}^\sigma(n,r)$, where ω-sequences are left ordered. Suppose that for $L, L' \in \widetilde{\mathsf{P}}^\gamma(n,\ell)$ there is a restricted hit equation $L = L' + E + \sum_{i=1}^{2^t-1} Sq^i(H_i)$, where $t = len(\gamma)$ and $\omega(E) <_l \gamma$. Then $L|R \sim L'|R$ in $\widetilde{\mathsf{P}}^{\gamma|\sigma}(n, \ell + 2^t r)$.*

Proof In the notation of the preceding proof, we have $Sq^i(H_i)|C = Sq^i(H_i|R)$. Hence $L|R = L'|R + E|R + \sum_{i=1}^{2^t-1} Sq^i(H_i|R)$ and $L|R \sim L'|R$, as required. □

As a special case of the map ψ_L, we define the twisted analogue of the local up Kameko map υ.

Definition 27.3.3 Let $\sigma \in \mathsf{Seq}(n)$ and let $\omega = (n)|\sigma$. The twisted local up Kameko map $\widetilde{\upsilon} : \widetilde{\mathsf{Q}}^\sigma(n) \to \widetilde{\mathsf{Q}}^\omega(n)$ is defined by $\widetilde{\upsilon}(f) = t_1 \cdots t_n \rho^{-1}(f)^2$.

In terms of blocks, the map $\widetilde{\upsilon}$ applies the inverse of the rotation ρ and the map ψ_L where L is an $(n,1)$-block having all entries 1, as for the standard up Kameko map υ. The next result explains why the rotation ρ^{-1} appears in the formula for $\widetilde{\upsilon}$.

Proposition 27.3.4 *The twisted up Kameko map $\widetilde{\upsilon}$ is an $\mathbb{F}_2$-isomorphism which sends $\widetilde{\mathsf{Q}}^\sigma(n,j)$ to $\widetilde{\mathsf{Q}}^{(n)|\sigma}(n,j)$ for each j.*

Proof Let $f \in \widetilde{\mathsf{P}}(n,j)$ be a monomial. Then $\mathrm{cw}(f^2) = 2\mathrm{cw}(f)$ and $\mathrm{cw}(\rho^{-1}(f)) = \mathrm{cw}(f)/2$. Hence $\mathrm{cw}(\widetilde{\upsilon}(f)) = (2^n - 1) + 2(j/2) = j \bmod 2^n - 1$. The fact that $\widetilde{\upsilon}$ is an isomorphism follows from a modification of the arguments in Section 6.5 involving the twisted version of the inverse map κ. □

As in the case of the standard action of A_2, the up Kameko map $\widetilde{\upsilon}$ is not always well defined globally (see Example 6.3.5). On the other hand, the down Kameko map for the standard action $\kappa : \mathsf{P}^{n+2d}(n) \to \mathsf{P}^d(n)$ is defined globally by sending a block $F = L|R$ to R if L is the single column block with all entries 1, and to 0 otherwise. In the former case, it is easy to see that if F is hit then so is R. Hence κ induces a map $\kappa : \mathsf{Q}^{n+2d}(n) \to \mathsf{Q}^d(n)$. In the case of $\widetilde{\mathsf{P}}(n)$, the definition of $\widetilde{\kappa} : \widetilde{\mathsf{Q}}^{n+2d}(n) \to \widetilde{\mathsf{Q}}^d(n)$ is similar, except that R is rotated by ρ.

The maps ψ_L and ψ^R of Proposition 27.3.1 make sense for the standard action of A_2, but are rarely surjective. However, this is true for the twisted version in some interesting situations. This can already be seen in the

2-variable case. To determine the dimension of $\widetilde{\mathrm{Q}}^d(2,j)$, attention may be restricted to decreasing sequences $(2,\ldots,2,1,\ldots,1)$, as $\mathrm{Q}^\omega(2)=0$ if ω is not decreasing. Iterated use of the twisted Kameko isomorphism in its local form then reduces the problem to tail sequences $\omega=(1,\ldots,1)$ of size s. If $s=1$, there are two blocks A_1,A_2 and if $s=2$ there are four blocks B_0,B_0',B_1,B_2 where

$$A_1=\begin{matrix}1\\0\end{matrix},\ A_2=\begin{matrix}0\\1\end{matrix},\ B_0=\begin{matrix}1\,1\\0\,0\end{matrix},\ B_0'=\begin{matrix}0\,0\\1\,1\end{matrix},\ B_1=\begin{matrix}0\,1\\1\,0\end{matrix},\ B_2=\begin{matrix}1\,0\\0\,1\end{matrix}.$$

Here the suffix indicates the cyclic weight. We see that $B_0\sim B_0'$ by 1-back splicing of B_0 at $(2,1)$. Hence $\dim\widetilde{\mathrm{Q}}^{(1,1)}(2,j)=1$ for $j=0,1,2$. Of the eight blocks for $s=3$, those with an entry 1 in position $(2,3)$ are

$$C_0=\begin{matrix}0\,1\,0\\1\,0\,1\end{matrix},\ C_1=\begin{matrix}1\,0\,0\\0\,1\,1\end{matrix},\ C_2=\begin{matrix}0\,0\,0\\1\,1\,1\end{matrix},\ C_2'=\begin{matrix}1\,1\,0\\0\,0\,1\end{matrix}.$$

From the case $s=2$, $C_2\sim C_2'$. By 1-back splicing C_1 and C_2 at $(1,2)$ and 2-back splicing C_0 at $(1,1)$, we obtain the equivalent blocks

$$D_0=\begin{matrix}1\,0\,1\\0\,1\,0\end{matrix},\ D_1=\begin{matrix}1\,1\,1\\0\,0\,0\end{matrix},\ D_2=\begin{matrix}0\,1\,1\\1\,0\,0\end{matrix},$$

in which a digit 1 now occupies the position $(1,3)$. Thus we can arrange to have a digit 1 in position $(1,3)$ or in position $(2,3)$ in the blocks of a basis for $\widetilde{Q}^\omega(2)$ when $s\geq 3$.

Proposition 27.3.5 *Let $\omega=(1,\ldots,1)$ be a tail sequence of size $s\geq 2$. Write $\omega=\gamma|\sigma$ where* size$(\gamma)=2$ *and* size$(\sigma)=s-2$. *Then the blocks*

$$F_0=\begin{array}{c|c}1\,0&1\cdots1\\0\,1&0\cdots0\end{array},\ F_1=\begin{array}{c|c}1\,1&1\cdots1\\0\,0&0\cdots0\end{array},\ F_2=\begin{array}{c|c}0\,1&1\cdots1\\1\,0&0\cdots0\end{array}$$

form a basis for $\widetilde{\mathrm{Q}}^\omega(2)$, i.e. $F_i=B_i|R$ for $i=0,1,2$, where $R=\begin{matrix}1\cdots1\\0\cdots0\end{matrix}$ and $\omega(R)=\sigma$. More generally, for any tail block $R\in\widetilde{\mathrm{P}}^\sigma(2)$ the postfixing map ψ^R is an isomorphism $\widetilde{\mathrm{Q}}^\gamma(2,j)\cong\widetilde{\mathrm{Q}}^\omega(2,j+\mathrm{cw}(R))$. In particular, $\dim\widetilde{\mathrm{Q}}^\omega(2,j)=1$ for $j=0,1,2$.

Proof For $s=2$ the result is true by the preceding remarks, the block R being empty. The fact that ψ^R is surjective follows from Proposition 27.3.5. For $B\in\widetilde{\mathrm{P}}^\gamma(2,j)$, we have $\mathrm{cw}(\psi^R(B))=\mathrm{cw}(B)+4\mathrm{cw}(R)=j+\mathrm{cw}(R)\bmod 3$. From

Theorem 1.8.2, $\dim \mathsf{Q}^d(2) = 3$ when $d = 2^s - 1$ and $s > 1$. It follows that the blocks F_i are not hit, and $\dim \widetilde{\mathsf{Q}}^\omega(2,j) = 1$ for $j = 0, 1, 2$. □

In the standard action of A_2, the map ψ^R in Proposition 27.3.5 is not surjective, as its image does not contain the spike block in $\mathsf{P}^\omega(2)$ with first row zero.

In the case $n = 2$, we may assume by Theorem 2.5.5 that $\mu(d) \leq 2$, and write $d = 2^{t+s} + 2^t - 2$. By iteration of $\widetilde{\upsilon}$, $\dim \widetilde{\mathsf{Q}}^d(2,j)$ is independent of t. The results are summarized in the table below.

s	$j=0$	$j=1$	$j=2$
0	1	0	0
1	0	1	1
≥ 2	1	1	1

(27.4)

Since $\dim \widetilde{\mathsf{Q}}^d(2) = \sum_{j=0}^{2} \dim \widetilde{\mathsf{Q}}^d(2,j)$, we can check using Theorem 1.8.2 that $\dim \widetilde{\mathsf{Q}}^d(2) = \dim \mathsf{Q}^d(2)$ for all d. Since the summand $\widetilde{\mathsf{P}}(2,0)$ corresponds to $\mathsf{C}(2)$ invariants of $\mathsf{P}(2)$, we can also check the result for cyclic weight $j = 0$ using Proposition 15.4.3.

27.4 Twisted tails and heads

The aim of this section is to find normalized forms for the generators of $\widetilde{\mathsf{Q}}^\omega(n)$, where ω is a tail or a head sequence. Proofs will be given in the tail case. The results in the head case are formally similar with the roles of 0 and 1 interchanged. The splicing operations used in the proofs are also the same, but in the head case we work modulo blocks with lower ω-sequences, using either the left or the right order. As the case $n = 2$ is dealt with in Section 27.2, we assume $n \geq 3$.

We view blocks in their diagonal format, and recall that the cyclic weight of a block is calculated by assigning the power 2^{j-1} to each entry 1 in the jth diagonal. The following result is analogous to Proposition 6.7.2.

Proposition 27.4.1 *Let ω be a tail sequence. Suppose that B and B' are blocks in $\mathsf{P}^\omega(n)$ such that $\widetilde{\alpha}(B) = \widetilde{\alpha}(B')$. Then $B \sim B'$.*

Proof Using diagonal format, the tail blocks B and B' have the same $\widetilde{\alpha}$-sequence if and only if the columns of B are a permutation of the columns of B'. Since we can exchange two adjacent columns by 1-back splicing, there is a sequence of 1-back splicing operations which changes B into B'. Since B and B' are tail blocks, they are equivalent modulo hit elements. $\square$

Proposition 27.4.2 *Let B be a tail n-block for $n \geq 3$, with $\widetilde{\alpha}(B) = (\widetilde{\alpha}_1, \ldots, \widetilde{\alpha}_n)$, and suppose that either* (i) $\widetilde{\alpha}_r \geq 1$ *and* $\widetilde{\alpha}_s \geq 2$, *where* $s > r > 1$, *or* (ii) $\widetilde{\alpha}_s \geq 3$, *where* $s > 1$. *Then* $B \sim B'$ *where* $\widetilde{\alpha}(B') >_l \widetilde{\alpha}(B)$.

Proof By 1-back splicing, we may assume using Proposition 27.4.1 that B has two adjacent entries 1 at positions $(s,1), (s-1,2)$ and an entry 1 in position $(r-2,3)$, where $r = s$ in case (ii). We 2-back splice B at position $(r-1,1)$, which is possible since $r > 1$ by assumption. Only the first three columns of B are affected. The diagrams below illustrate two examples of B and the resulting block B' after splicing:

$$(1)\ B = \begin{array}{ccc} & & 0 \\ & 0 & 0 \\ 0 & 0 & 1 \\ 0 & 0 & 0 \\ 0 & 1 & \\ 1 & & \end{array}, \ B' = \begin{array}{ccc} & & 1 \\ & 0 & 0 \\ 0 & 1 & 0 \\ 1 & 0 & 0 \\ 0 & 0 & \\ 0 & & \end{array}; \quad (2)\ B = \begin{array}{ccc} & & 0 \\ & 0 & 0 \\ 0 & 0 & 1 \\ 0 & 1 & 0 \\ 1 & 0 & \\ 0 & & \end{array}, \ B' = \begin{array}{ccc} & & 0 \\ & 0 & 0 \\ 0 & 1 & 0 \\ 1 & 0 & 1 \\ 0 & 0 & \\ 0 & & \end{array}.$$

The $\widetilde{\alpha}$-sequence is raised from $(0,0,1,2)$ to $(1,2,0,0)$ in example (1), and from $(0,0,3,0)$ to $(0,2,0,1)$ in example (2). We see in particular that the $(r+1)$-diagonal of B gains two entries 1 in B' and no diagonal below position r loses any entries 1. Hence $\widetilde{\alpha}(B') >_l \widetilde{\alpha}(B)$. $\square$

We will show that to each cyclic weight j such that $\alpha(j) \leq c$, so that $\widetilde{\mathrm{P}}^\omega(n,j)$ is represented by (n,c)-blocks, there is a corresponding maximal $\widetilde{\alpha}$-sequence $\widetilde{\alpha}(j)$. The blocks B in $\widetilde{\mathrm{P}}^\omega(n,j)$ with $\widetilde{\alpha}(B) = \widetilde{\alpha}(j)$ are all equivalent modulo hits by 1-back splicing, and any of them may be taken as a normalized representative of $\widetilde{\mathrm{Q}}^\omega(n,j)$. We begin by defining the maximal $\widetilde{\alpha}$-sequences as those which cannot be raised in the left order using Proposition 27.4.2.

Definition 27.4.3 Let $\widetilde{\alpha} = (\widetilde{\alpha}_1, \ldots, \widetilde{\alpha}_n)$ be a sequence with $|\widetilde{\alpha}| \leq c$, and let i be the smallest integer > 1 such that $\widetilde{\alpha}_i \neq 0$. Then $\widetilde{\alpha}$ is **maximal** if $\alpha_i \leq 2$ and $\alpha_j \leq 1$ for all $j > i$.

Example 27.4.4 For $n=c=3$, the maximal $\widetilde{\alpha}$-sequences are $(3,0,0)$, $(2,1,0)$, $(1,2,0)$, $(2,0,1)$, $(1,1,1)$, $(0,2,1)$ and $(1,0,2)$, corresponding to cyclic weights $3,4,5,6,0,1$ and 2. When $n=4$ and $c=3$, the corresponding sequences with $\widetilde{\alpha}_4 = 0$ are maximal for cyclic weights $3 \leq j \leq 9$, while the $\widetilde{\alpha}$-sequences $(2,0,0,1)$, $(1,1,0,1)$, $(0,2,0,1)$, $(1,0,1,1)$, $(0,1,1,1)$, $(0,0,2,1)$ and $(1,0,0,2)$ are maximal for cyclic weights $10,11,12,13,14,1$ and 2. Since $\alpha(15)=4$, there is no maximal $\widetilde{\alpha}$-sequence for cyclic weight 0 when $c=3$. When $\alpha(j)=c$, $\widetilde{\alpha}(j)$ is simply the reverse binary expansion of j.

Proposition 27.4.5 *For $n,c \geq 1$, let j be an integer such that $c \leq j \leq 2^n+c-2$ and $\alpha(j) \leq c$. Then there is a unique maximal $\widetilde{\alpha}$-sequence $\widetilde{\alpha}(j) = (\widetilde{\alpha}_1,\ldots,\widetilde{\alpha}_n)$ such that $|\widetilde{\alpha}(j)| = c$ and $\sum_{i=1}^{n} 2^{i-1}\widetilde{\alpha}_i = j$.*

Proof By Proposition 2.4.4, the condition $\alpha(j) \leq c$ is equivalent to $\mu(j-c) \leq c$. By Theorem 5.4.2, there is a unique spike partition $(2^{i_m}-1,\ldots,2^{i_2}-1,2^{i_1}-1)$ of $j-c$ such that $i_1 \leq i_2 < i_3 < \ldots < i_m$, where $m = \mu(j-c)$. Hence $j = (c-m)2^0 + 2^{i_1} + 2^{i_2} + \cdots + 2^{i_m}$ is a binary partition of j. If $i_m \geq n$, then $j \geq (c-m) + 2^n \geq 2^n + c - 1$, contrary to hypothesis. Hence $i_m \leq n-1$, and we define $\widetilde{\alpha}(j)$ as the corresponding 'reverse binary expansion of j' obtained by taking the coefficients of the 2-powers $2^0,\ldots,2^{n-1}$. Since the parts of a minimal spike partition are distinct except possibly for the two smallest parts, $\widetilde{\alpha}(j)$ is a minimal $\widetilde{\alpha}$-sequence of cyclic weight j. The uniqueness of $\widetilde{\alpha}(j)$ follows from the uniqueness of the minimal spike partition of $j-c$. □

Proposition 27.4.6 *For $n,c \geq 1$, all tail blocks in $\widetilde{\mathsf{P}}^{\omega}(n)$ with the same cyclic weight are equivalent modulo hits. Thus* $\dim \widetilde{\mathsf{Q}}^{\omega}(n,j) = 1$ *if $\alpha(j) \leq c$, and 0 otherwise. In particular,* $\dim \widetilde{\mathsf{Q}}^{\omega}(n) = \sum_{k=1}^{c} \binom{n}{k}$, *and if $c \geq n$ then* $\dim \widetilde{\mathsf{Q}}^{\omega}(n) = 2^n - 1$.

Proof The first statement follows immediately from the previous results in this section, and the second from Proposition 27.1.11. The third statement follows by counting the number of cyclic weights j such that $\alpha(j) \leq c$, or from Theorem 6.7.6 and Proposition 27.2.2. □

We next describe normalized blocks for the equivalence classes of tail (n,c)-blocks.

Definition 27.4.7 Let $n,c \geq 1$ and let $\alpha(j) \leq c$. A **normalized block** $B \in \widetilde{\mathsf{P}}^{\omega}(n)$ is a block such that $\widetilde{\alpha}(B) = \widetilde{\alpha}(j)$ and, in diagonal format, if the entries 1 in columns c_1 and c_2, where $c_1 < c_2$, are in rows r_1 and r_2, then $r_1 > r_2$.

Example 27.4.8 The six blocks below are normalized tail blocks for $n=4$ and $c=3$ with cyclic weights $2,3,7,8,10,14$ respectively.

$$\begin{array}{ccc} & & 1\\ & 0 & 0\\ 0 & 0 & 0\\ 0 & 0 & 0\\ 0 & 1 & \\ 1 & & \end{array}, \quad \begin{array}{ccc} & & 1\\ & 1 & 0\\ 1 & 0 & 0\\ 0 & 0 & 0\\ 0 & 0 & \\ 0 & & \end{array}, \quad \begin{array}{ccc} & & 1\\ & 0 & 0\\ 0 & 1 & 0\\ 0 & 0 & 0\\ 1 & 0 & \\ 0 & & \end{array}, \quad \begin{array}{ccc} & & 0\\ & 0 & 1\\ 0 & 1 & 0\\ 0 & 0 & 0\\ 1 & 0 & \\ 0 & & \end{array}, \quad \begin{array}{ccc} & & 1\\ & 1 & 0\\ 0 & 0 & 0\\ 0 & 0 & 0\\ 0 & 0 & \\ 1 & & \end{array}, \quad \begin{array}{ccc} & & 0\\ & 0 & 1\\ 0 & 0 & 0\\ 0 & 1 & 0\\ 0 & 0 & \\ 1 & & \end{array}.$$

In the head case, on formally interchanging 0s and 1s the splicing arguments used in Propositions 27.4.1 and 27.4.2 go through modulo blocks with left or right lower ω-sequences. The following result corresponds to Proposition 27.4.6. Note that exchange of 0s and 1s replaces $(\widetilde{\alpha}_1,\ldots,\widetilde{\alpha}_n)$ by $(c-\widetilde{\alpha}_1,\ldots,c-\widetilde{\alpha}_n)$, and replaces a block of cyclic weight j with one of cyclic weight $-j$.

Proposition 27.4.9 *For $n,c\geq 1$, all head blocks in $\widetilde{\mathrm{P}}^{\omega}(n)$ with the same cyclic weight are equivalent modulo hits. Thus* $\dim\widetilde{\mathrm{Q}}^{\omega}(n,-j)=1$ *if $\alpha(j)\leq c$, and* 0 *otherwise. In particular,* $\dim\widetilde{\mathrm{Q}}^{\omega}(n)=\sum_{k=1}^{c}\binom{n}{k}$, *and if $c\geq n$ then* $\dim\widetilde{\mathrm{Q}}^{\omega}(n)=2^n-1$.

Proof The first statement follows immediately from the analogues of Propositions 27.4.1 and 27.4.2. Hence $\dim\widetilde{\mathrm{Q}}^{\omega}(n,-j)\leq 1$ if $\alpha(j)\leq c$, and 0 otherwise. By counting the number of cyclic weights j such that $\alpha(j)\leq c$, Theorem 6.8.7 and Proposition 27.2.2 then give the third statement, and the second statement follows. □

In particular, it follows from these results that a monomial $t_1^{a_1}\cdots t_n^{a_n}$ in $\widetilde{\mathrm{P}}(n)$ whose ω-sequence ω is a tail or a head sequence is not hit, and that any such monomial of cyclic weight j generates $\widetilde{\mathrm{Q}}^{\omega}(n,j)$, for either the left or the right order on ω-sequences. The concatenation maps ψ_L and ψ^R of Proposition 27.3.1 are therefore always isomorphisms (of 1-dimensional spaces) when $\omega=\gamma|\sigma$ is a tail or head sequence.

27.5 The twisted cohits $\widetilde{\mathrm{Q}}^{\omega}(3)$

In this section we determine $\dim\widetilde{\mathrm{Q}}^{\omega}(3,j)$ for $\omega\in\mathrm{Seq}(3)$, where $\widetilde{\mathrm{Q}}^{\omega}(3)$ is defined using the left order $\leq_{\ell}$. By Proposition 8.3.1, $\dim\widetilde{\mathrm{Q}}^{\omega}(3)=\dim\mathrm{Q}^{\omega}(3)=0$ if ω is not decreasing, and by Proposition 27.3.4 the twisted

Kameko isomorphism $\widetilde{v}$ deals with the cases where ω is decreasing and has an entry 3. Hence we restrict attention to sequences of the form $\omega = (2,\ldots,2,1,\ldots,1)$, with t 2s and s 1s. Using the rotation isomorphism ρ of Proposition 27.1.9, we may further restrict attention to cyclic weights $j = 0,1,3$.

The table below may be viewed as a refinement of Theorem 8.3.3. The regularity of the distribution of dimensions of the cyclic summands, even when not related by rotation, can be explained by the existence of homomorphisms obtained by prefixing and postfixing with suitable fixed blocks. In particular, there are isomorphisms for the corresponding $\widetilde{\mathsf{Q}}^{\omega}(3,j)$ horizontally from the rightmost column of the chart and vertically from the lowest row. This is also true starting at $(t,s) = (2,2)$ and true horizontally starting at $(t,s) = (2,1)$. In suitable situations, prefixing and postfixing by fixed blocks give maps between cohit spaces for the standard action of A_2 on P(n), but these are rarely surjective.

Theorem 27.5.1 *In the following table, the three numbers in the* (t,s)*-box give* $\dim \widetilde{\mathsf{Q}}^{\omega}(3,j)$ *in the order* $j = 0,1,3$.

$t\backslash s$	0	1	2	≥ 3
0	1 0 0	0 1 0	0 1 1	1 1 1
1	0 0 1	2 1 1	3 2 2	2 2 2
2	0 1 1	2 2 2	3 3 3	3 3 3
≥ 3	1 1 1	2 2 2	3 3 3	3 3 3

Proof The cases $s = 0$ and $t = 0$ have been dealt with in Section 27.4. We deal with the remaining four cases where $s,t \leq 2$ in Propositions 27.5.2 to 27.5.5. The first column of each diagram gives the cyclic weight j, the next column lists relations on the blocks of cyclic weight j and the third column selects a spanning set for $\widetilde{\mathsf{Q}}^{\omega}(3,j)$. The number c_j of elements in the spanning set is listed in the fourth column, giving an upper bound for $\dim \widetilde{\mathsf{Q}}^{\omega}(3,j)$. In each case, we verify from the chart in Theorem 8.3.3 that $c_0 + 3c_1 + 3c_3 = \dim \mathsf{Q}^{\omega}(3)$. As explained in Section 27.1, this implies that $c_j = \dim \widetilde{\mathsf{Q}}^{\omega}(3,j)$, and hence the blocks in the third column give a monomial basis of $\widetilde{\mathsf{Q}}^{\omega}(3,j)$. The proof of the theorem is completed by the results on prefixing and postfixing maps given in Propositions 27.5.6 and 27.5.7 respectively. □

Proposition 27.5.2 *The diagram below gives details for* $\widetilde{Q}^{(2,1)}(3,j)$.

cw	relations	basis	dim
0	11 10 00 00 ∼ 11 + 10 10 00 11	10 00 11 , 10 00 11	2
1		01 10 10	1
3		00 11 10	1

Proof The relation is proved by 1-back splicing the first block at $(2,1)$. □

In this case it is easy to see that we have a maximum set of independent relations between the blocks. This is not always clear. In the cases that follow, most relations on the blocks are obtained by 1-back splicing. We shall only draw attention to situations where a more elaborate procedure is required.

Proposition 27.5.3 *The diagram below gives details for* $\widetilde{Q}^{(2,1,1)}(3,j)$.

cw	relations	generators	dim
0		110 100 001 100 , 001 , 110 001 110 100	3
1	101 100 000 110 100 ∼ 111 ∼ 101 + 001 010 000 110 100 100 110 000 ∼ 001 111 100	000 110 101 , 001 110 100	2
3	100 101 011 ∼ 000 100 110	101 010 000 , 100 110 101	2

Proof There are no relations on the three blocks of cyclic weight 0, and there is one relation on the three blocks of cyclic weight 3. Also $3+3\cdot 2+3\cdot 2 = 15 = \dim \widetilde{\mathsf{Q}}^{\omega}(3)$, and so the bounds are sharp. □

Proposition 27.5.4 *The diagram below gives details for* $\widetilde{\mathsf{Q}}^{(2,2,1)}(3,j)$.

cw	relations	generators	dim
0	110 010 101 101 ~ 110 + 010 010 101 110	010 101 110 , 010 101 110	2
1	111 001 010 ~ 110 100 110	001 110 110 , 100 110 011	2
3	110 100 110 111 000 ~ 110 ~ 111 + 100 111 011 000 010 010 110 101 ~ 111 110 000	110 111 111 , 100 000 010	2

Proof The equivalence for cyclic weight 0 is produced by 2-back splicing at $(3,1)$. Since $2+3\cdot 2+3\cdot 2 = 14 = \dim \widetilde{\mathsf{Q}}^{\omega}(3)$, the upper bounds are sharp. □

Proposition 27.5.5 *The diagrams below give details for* $\widetilde{\mathsf{Q}}^{(2,2,1,1)}(3,j)$.

cw	relations	generators	dim
0	1100 0001 0101 1110 ~ 1100 + 1110 0001 1110 1000 1110 1100 1110 1000 ~ 1110 + 0001 0101 0001 1100 1000 1110 0001 0101 ~ 0001 + 1100 1110 1100 1110	0101 1110 1000 1110 0001 1100 0001 1100 1110	3

cw	relations	generators	dim
	0100 0101 1000 1011	0101 1011	
	1111 ∼ 1100, 0110 ∼ 0100	1100, 0100	
	1000 1010 1101 1100	1010 1100	
1			3
	1100 0000 0100	0000	
	1100 ∼ 1101 + 1111	1101	
	0011 1110 1000	1110	

cw	relations	generators	dim
	0010 0000 0111 0100	0000 0100	
	1101 ∼ 1100, 1100 ∼ 1110	1100, 1110	
	1100 1111 1000 1001	1111 1001	
3			3
	1101 1100 0000 0100	1100	
	0010 ∼ 1001 + 1100 + 1110	1001	
	1100 0110 1111 1001	0110	

Proof We regard a block in $\widetilde{P}^{\omega}(3)$ as the concatenation of a block B in $\widetilde{P}^{(2,2,1)}(3)$ with a block in $\widetilde{P}^{(1)}(3)$. Since ω is minimal, we can manipulate B within its equivalence class independently of the fourth column. Hence we can choose B to be one of two generators of $\widetilde{Q}^{(2,2,1)}(3)$ in each cyclic weight. For each such choice of B, we adjoin an entry 1 in positions $(1,4)$, $(2,4)$ and $(3,4)$. This changes the cyclic weight by adding $1,2,4$ respectively to $\mathrm{cw}(B)$. Hence we obtain six blocks in $\widetilde{P}^{\omega}(3)$ for each choice of cyclic weight. For example, in cyclic weight 0 we adjoin 1 in positions $(1,4)$, $(2,4)$ and $(3,4)$ to two independent blocks in $\widetilde{P}^{(2,2,1)}(j)$ for $j = 6,5,3$. For cyclic weights 1 and 3, we do the same to blocks B of cyclic weights $0,6,4$ and $2,1,6$ respectively. The relevant blocks are found by rotating the rows of the blocks shown in the table of Proposition 27.5.4 and choosing two independent ones.

Most of these equivalences can be obtained by iterated 1-back splicing, but the bottom line of cyclic weight 3 deserves more discussion. Here we start with a 2-back splicing at $(2,2)$ to produce the relation

$$\begin{array}{ccccc} 1101 & & 1110 & & 1010 \\ 0010 & \sim & 0101 & + & 0110\,. \\ 1100 & & 1000 & & 1110 \end{array}$$

This introduces a block with ω-sequence $(2,2,3,0)$, which is higher in left order than $(2,2,1,1)$. The problem is resolved by 2-back splicing this block at $(2,1)$ to obtain the relation

$$\begin{array}{ccccc} 1010 & & 0100 & & 1100 \\ 0110 & \sim & 1110 & + & 1001\,. \\ 1110 & & 1001 & & 0110 \end{array}$$

Further 1-back splicing produces the stated relation. Since $3+3\cdot 3+3\cdot 3 = 21 = \dim \widetilde{\mathsf{Q}}^{\omega}(3)$, the upper bounds are sharp. □

Proposition 27.5.6 *Let $\omega = (2,\ldots,2,1,\ldots,1)$, with t 2s and s 1s, and let $\omega' = (2)|\omega$. Then the map $\psi_L : \widetilde{\mathsf{Q}}^{\omega}(3) \to \widetilde{\mathsf{Q}}^{\omega'}(3)$ induced by prefixing with the block $L = \begin{smallmatrix}1\\1\\0\end{smallmatrix}$ is an isomorphism.*

Proof From Section 8.3, $\dim \widetilde{\mathsf{Q}}^{\omega}(3) = \dim \mathsf{Q}^{\omega}(3)$ for $t \geq 2$ and $s \geq 1$. Therefore it is enough to show that ψ_L is a surjection. A block F in $\mathsf{P}^{\omega'}(3)$ is of the form $F = A|B$, where $\omega(A)$ is a head sequence of length $t+1$ and $\omega(B)$ is a tail sequence of length s. Since $s \geq 1$ and $t \geq 2$, $\omega' = \omega(F)$ is the ω-sequence of a minimal spike, and so we can change A and B in their equivalence classes without changing the equivalence class of F. If $t \geq 3$, then the result follows by arranging A to have first column L. So we now assume $t = 2$. If $\mathrm{cw}(A) \neq 3$, then A is again equivalent to a block with first column L. If $\mathrm{cw}(A) = 3$, then manipulations of A and B separately are not sufficient. It is enough to consider the case $s = 1$. There are three equivalent blocks

$$A_1 = \begin{array}{c} 011 \\ 111 \\ 100 \end{array},\ A_2 = \begin{array}{c} 001 \\ 110 \\ 111 \end{array},\ A_3 = \begin{array}{c} 111 \\ 010 \\ 101 \end{array}$$

of cyclic weight 3 in $\widetilde{\mathsf{P}}^{(2,2,2)}(3)$. There are 3 blocks $B_1 = \begin{smallmatrix}1\\0\\0\end{smallmatrix}$, $B_2 = \begin{smallmatrix}0\\1\\0\end{smallmatrix}$, $B_3 = \begin{smallmatrix}0\\0\\1\end{smallmatrix}$ in $\mathsf{P}^{(1)}(3)$. Hence $F = A|B$ is equivalent to a block of one of the following three forms:

$$F_1 = A_2|B_1 = \begin{array}{c|c} 0\,0\,1 & 1 \\ 1\,1\,0 & 0 \\ 1\,1\,1 & 0 \end{array}, \quad F_2 = A_1|B_2 = \begin{array}{c|c} 0\,1\,1 & 0 \\ 1\,1\,1 & 1 \\ 1\,0\,0 & 0 \end{array}, \quad F_3 = A_1|B_3 = \begin{array}{c|c} 0\,1\,1 & 0 \\ 1\,1\,1 & 0 \\ 1\,0\,0 & 1 \end{array}.$$

By 1-back splicing F_1 at $(2,3)$ and F_2 at $(3,3)$, and 3-back splicing F_3 at $(1,1)$, we obtain

$$F_1 \sim \begin{array}{c|c} 001 & 0 \\ 111 & 1 \\ 110 & 0 \end{array} + \begin{array}{c|c} 000 & 0 \\ 111 & 0 \\ 111 & 1 \end{array}, F_2 \sim \begin{array}{c|c} 011 & 1 \\ 110 & 0 \\ 101 & 0 \end{array} + \begin{array}{c|c} 010 & 0 \\ 111 & 1 \\ 101 & 1 \end{array}, F_3 \sim \begin{array}{c|c} 100 & 1 \\ 011 & 0 \\ 111 & 0 \end{array} + \begin{array}{c|c} 111 & 0 \\ 100 & 1 \\ 011 & 0 \end{array}.$$

Since the left subblocks of the new blocks have cyclic weight $\neq 3$, the first part of the argument completes the proof. □

Proposition 27.5.7 *Let $\omega = (2,\ldots,2,1,\ldots,1)$, with t 2s and s 1s, where $t \geq 1$ and $s \geq 2$, and let $\omega' = \omega|(1)$. Then the map $\psi^R : \widetilde{\mathsf{Q}}^{\omega}(3) \to \widetilde{\mathsf{Q}}^{\omega'}(3)$ induced by postfixing with $R = \begin{smallmatrix}1\\0\\0\end{smallmatrix}$ is surjective, and is an isomorphism if $t > 1$ or if $s > 2$.*

Proof From Section 8.3, $\dim \widetilde{\mathsf{Q}}^{\omega}(3) = \dim \mathsf{Q}^{\omega}(3)$ for $t \geq 1$ and $s \geq 2$, except in the case $t = 1$, $s = 2$. Therefore it is enough to show that ψ^R is a surjection. A block F in $\mathsf{P}^{\omega'}(3)$ is of the form $F = A|B$ where $\omega(A)$ is a head sequence of length t and $\omega(B)$ is a tail sequence of length $s+1$. Since $s \geq 2$ and $t \geq 1$, $\omega(F) = \omega'$ is the ω-sequence of a minimal spike, and so we can change A and B in their equivalence classes without changing the equivalence class of F. If $s \geq 3$, then the result follows by arranging B to have last column R. So we now assume $s = 2$. If $\mathrm{cw}(B) \neq 4$, then B is again equivalent to a block with last column R. If $\mathrm{cw}(B) = 4$, then manipulations of A and B separately are not sufficient. It is enough to consider the case $t = 2$. There are three equivalent blocks

$$B_1 = \begin{array}{c} 1\,0\,0 \\ 0\,0\,0 \\ 0\,1\,1 \end{array}, \quad B_2 = \begin{array}{c} 1\,1\,0 \\ 0\,0\,1 \\ 0\,0\,0 \end{array}, \quad B_3 = \begin{array}{c} 0\,0\,0 \\ 1\,0\,1 \\ 0\,1\,0 \end{array}$$

of cyclic weight 4 in $\widetilde{\mathsf{P}}^{(1,1,1)}(3)$. There are three blocks $A_1 = \begin{smallmatrix}1\\1\\0\end{smallmatrix}$, $A_2 = \begin{smallmatrix}1\\0\\1\end{smallmatrix}$ and $A_3 = \begin{smallmatrix}0\\1\\1\end{smallmatrix}$ in $\mathsf{P}^{(2)}(3)$. Hence $F = A|B$ is equivalent to one of the three blocks

$$F_1 = A_1|B_3 = \begin{array}{c|c} 1 & 0\,0\,0 \\ 1 & 1\,0\,1 \\ 0 & 0\,1\,0 \end{array}, F_2 = A_2|B_1 = \begin{array}{c|c} 1 & 1\,0\,0 \\ 0 & 0\,0\,0 \\ 1 & 0\,1\,1 \end{array}, F_3 = A_3|B_1 = \begin{array}{c|c} 0 & 1\,0\,0 \\ 1 & 0\,0\,0 \\ 1 & 0\,1\,1 \end{array}.$$

By 1-back splicing F_1 at $(3,1)$ and F_2 at $(2,1)$, and 2-back splicing F_3 at $(1,1)$, we obtain

$$F_1 \sim \begin{array}{c|c} 1 & 100 \\ 0 & 001 \\ 1 & 010 \end{array} + \begin{array}{c|c} 0 & 000 \\ 1 & 001 \\ 1 & 110 \end{array}, F_2 \sim \begin{array}{c|c} 1 & 000 \\ 1 & 100 \\ 0 & 011 \end{array} + \begin{array}{c|c} 0 & 000 \\ 1 & 000 \\ 1 & 111 \end{array},$$

$$F_3 \sim \begin{array}{c|c} 1 & 010 \\ 0 & 000 \\ 1 & 101 \end{array} + \begin{array}{c|c} 1 & 100 \\ 1 & 100 \\ 0 & 101 \end{array}.$$

Writing each new block in the same concatenated form, all the right subblocks except the last one have ω-sequence $(1,1,1)$ and cyclic weight $\neq 4$. The proof is then concluded in the first two cases from the first part of the argument. In the third case, we deal with the block with ω-sequence $(2,3,0,1)$ by using the equation

$$Sq^4 \left(\begin{array}{c|c} 1 & 1\,1\,0 \\ 1 & 1\,0\,0 \\ 0 & 1\,0\,0 \end{array} \right) = \begin{array}{c|c} 1 & 1\,0\,0 \\ 1 & 1\,0\,0 \\ 0 & 1\,0\,1 \end{array} + \begin{array}{c|c} 1 & 0\,0\,1 \\ 1 & 0\,0\,0 \\ 0 & 1\,1\,0 \end{array} + \begin{array}{c|c} 1 & 1\,0\,1 \\ 1 & 0\,1\,0 \\ 0 & 0\,0\,0 \end{array} + \begin{array}{c|c} 1 & 0\,1\,0 \\ 1 & 1\,1\,0 \\ 0 & 0\,1\,0 \end{array} + E$$

where E is a sum of blocks with $\omega_1 = 0$, which are therefore hit. The second and third blocks on the right have right subblocks with ω-sequence $(1,1,1)$ and cyclic weight $\neq 4$, while 1-back splicing at $(3,2)$ shows that the last block is equivalent modulo hits to

$$\begin{array}{c|c} 1 & 0\,0\,1 \\ 1 & 0\,0\,0 \\ 0 & 1\,1\,0 \end{array},$$

which has the same property. This completes the proof in the third case. □

27.6 Remarks

The cyclic decomposition of $\mathsf{P}(n)$ was introduced by Campbell and Selick [25], who showed that $\zeta \in \mathbb{F}_{2^n}$ can be chosen so that $\mathsf{P}(n)$ and $\widetilde{\mathsf{P}}(n)$ are isomorphic as A_2-modules (but not as A_2-algebras). Our treatment is partly based on Helen Weaver's Ph.D. thesis [224].

28
The cyclic splitting of $\mathsf{DP}(n)$

28.0 Introduction

In this chapter we introduce the twisted down action of A_2 on the divided polynomial algebra $\mathsf{DP}(n)$. As in Chapter 9, we treat this as a left action of the opposite algebra of A_2. In Section 28.1 we compute the twisted action of Sq_k on d-monomials. The twisted analogue $\widetilde{\mathsf{K}}(n)$ of the Steenrod kernel $\mathsf{K}(n)$ is a subalgebra of $\widetilde{\mathsf{DP}}(n)$, and is dual as a $\mathbb{F}_2$-space to the twisted cohit module $\widetilde{\mathsf{Q}}(n)$. We also define a twisted analogue $\widetilde{\mathsf{J}}(n)$ of the d-spike module $\mathsf{J}(n)$. The generators $\sigma_d(j)$ of the subring $\widetilde{\mathsf{J}}(n)$ are the sums of all d-monomials with cyclic weight j and degree d, where $0 \leq j \leq 2^n - 2$ and $d = 2^r - 1$ for some $r \geq 1$.

In Section 28.2 we first obtain a family of quadratic relations in $\widetilde{\mathsf{J}}(n)$ with coefficients in $\mathbb{F}_{2^n}$, and derive relations of the form $\sum \sigma_{d_1}(i)\sigma_{d_2}(j) = 0$ over $\mathbb{F}_2$, where $d_1 = 2^{s+t} - 1$, $d_2 = 2^t - 1$ for $s \geq 0$ and $t \geq 1$, and all the terms have the same cyclic weight. The simplest relation of this type is $\sigma_{d_1}(k)\sigma_{d_2}(0) = 0$. Proposition 28.2.6 gives a procedure which can be applied to one of these relations to obtain a further relation of the same type of twice the length. These relations allow us write all products $\sigma_{d_1}(i)\sigma_{d_2}(j)$ of generators of $\widetilde{\mathsf{J}}(n)$ as sums of the products for which $1 \leq j \leq 2^{n-1} - 1$.

In Section 28.3 we consider the dual $\widetilde{\kappa}_* : \widetilde{\mathsf{DP}}^d(n) \to \widetilde{\mathsf{DP}}^{2d+n}(n)$ of the twisted down Kameko map $\widetilde{\kappa}$, and show that $\widetilde{\kappa}_*$ maps $\widetilde{\mathsf{K}}^d(n)$ to $\widetilde{\mathsf{K}}^{2d+n}(n)$ for all $d \geq 0$, and gives an isomorphism between $\widetilde{\mathsf{J}}^d(n)$ and $\widetilde{\mathsf{J}}^{2d+n}(n)$ when $\mu(2d+n) = n$. We compute $\widetilde{\mathsf{K}}(n)$ for $n = 2$ in Section 28.4 and for $n = 3$ in Section 28.5.

28.1 The twisted A_2-module $\widetilde{\mathsf{DP}}(n)$

Recall from Chapter 27 that the A_2-module $\widetilde{\mathsf{P}}(n)$ is given by the twisted action of A_2 on the polynomial algebra $\mathbb{F}_2[t_1, \ldots, t_n]$ defined on the generators $t_1, \ldots, t_n$ by $\mathsf{Sq}(t_i) = t_i + t_{i-1}^2$, where the variables are indexed mod n.

Definition 28.1.1 We denote by $\widetilde{\mathsf{DP}}(n)$ the A_2-module dual to $\widetilde{\mathsf{P}}(n)$. Thus $\widetilde{\mathsf{DP}}(n)$ is the divided power algebra over $\mathbb{F}_2$ in variables $u_1,\ldots,u_n$, where the d-monomials $u_1^{(a_1)}\cdots u_n^{(a_n)}$ form the basis dual to the monomial basis $t_1^{a_1}\cdots t_n^{a_n}$ of $\widetilde{\mathsf{P}}(n)$, and the action of A_2 is defined by $\langle\theta(f),g\rangle = \langle f,\theta(g)\rangle$, where $f\in\widetilde{\mathsf{DP}}(n)$, $g\in\widetilde{\mathsf{P}}(n)$ and $\theta\in\mathsf{A}_2$.

As for $\mathsf{P}(n)$ and $\mathsf{DP}(n)$, the same n-block is used to represent the monomial $t_1^{a_1}\cdots t_n^{a_n}$ in $\widetilde{\mathsf{P}}(n)$ and the corresponding d-monomial $u_1^{(a_1)}\cdots u_n^{(a_n)}$ in $\widetilde{\mathsf{DP}}(n)$. As in Chapter 27, we regard the variables as indexed mod n, so that $u_n = u_0$, etc.

Proposition 28.1.2 *The action of* A_2 *on* $\widetilde{\mathsf{DP}}(n)$ *is given by*

$$Sq_k(u_i^{(j)}) = \begin{cases} u_i^{(j-2k)}u_{i+1}^{(k)}, & \text{if } 0\le k\le j/2,\\ 0, & \text{if } k>j/2\end{cases}$$

and the Cartan formula $Sq_k(uv) = \sum_{i+j=k} Sq_i(u)Sq_j(v)$ *for* $u,v\in\widetilde{\mathsf{DP}}(n)$.

Proof It is enough to show that

$$\langle Sq_k(u_i^{(j)}),g\rangle = \begin{cases} 1, & \text{if } g = t_i^{j-2k}t_{i+1}^k,\\ 0, & \text{if } g \text{ is any other monomial in } \widetilde{\mathsf{P}}^{j-k}(n).\end{cases}$$

By definition, $\langle Sq_k(u_i^{(j)}),g\rangle = \langle u_i^{(j)},Sq^k(g)\rangle$. Applying the total twisted squaring operation to a monomial in $\widetilde{\mathsf{P}}(n)$, we obtain

$$\mathsf{Sq}(t_1^{a_1}\cdots t_n^{a_n}) = (t_1+t_n^2)^{a_1}(t_2+t_1^2)^{a_2}\cdots(t_n+t_{n-1}^2)^{a_n}.$$

This polynomial contains the term t_i^j if and only if $a_r = 0$ for $r\ne i,i+1$ and $a_i+2a_{i+1} = j$. To locate t_i^j in $Sq^k(g)$, we then require $j = k+a_i+a_{i+1}$. Hence $a_{i+1} = k$ and $a_i = j-2k$, and so $g = t_i^{j-2k}t_{i+1}^k$.

The Cartan formula for $\widetilde{\mathsf{DP}}(n)$ follows from the Cartan formula for $\widetilde{\mathsf{P}}(n)$ by duality, as in Proposition 9.3.4. □

Definition 28.1.3 The **twisted Steenrod kernel** $\widetilde{\mathsf{K}}(n) = \sum_{d\ge 0}\widetilde{\mathsf{K}}^d(n)$, where $\widetilde{\mathsf{K}}^d(n)$ is the set of elements $u\in\widetilde{\mathsf{DP}}^d(n)$ such that $Sq_k(u) = 0$ for all $k>0$.

By the Cartan formula, $\widetilde{\mathsf{K}}(n)$ is a subalgebra of $\widetilde{\mathsf{DP}}(n)$. It is the dual of the twisted cohits $\widetilde{\mathsf{Q}}(n)$. Since the twisted and standard actions of A_2 coincide in the case $n=1$, $\widetilde{\mathsf{K}}(1)$ is spanned by the d-spikes $v_1^{(2^r-1)}$, $r\ge 1$.

Definition 28.1.4 The **cyclic weight** of a d-monomial $f = u_1^{(a_1)}\cdots u_n^{(a_n)}$ in $\widetilde{\mathsf{DP}}(n)$ is the cyclic weight of the dual monomial $t_1^{a_1}\cdots t_n^{a_n}$ in $\widetilde{\mathsf{P}}(n)$, i.e. $\mathrm{cw}(f) = \sum_{i=1}^n 2^{i-1}a_i \bmod 2^n-1$. For $0\le j\le 2^n-2$, we denote by $\widetilde{\mathsf{DP}}(n,j)$ the $\mathbb{F}_2$-subspace of $\widetilde{\mathsf{DP}}(n)$ spanned by all d-monomials f of cyclic weight j.

It follows from Proposition 28.1.2 that the operations Sq_k preserve cyclic weight.

Proposition 28.1.5 *The spaces* $\widetilde{\mathsf{DP}}(n,j)$ *are* A_2*-submodules of* $\widetilde{\mathsf{DP}}(n)$*. There is a direct sum splitting of* A_2*-modules*

$$\widetilde{\mathsf{DP}}(n) = \sum_{j=0}^{2^n-2} \widetilde{\mathsf{DP}}(n,j).$$

Proof This follows from Theorem 27.1.8 by duality. □

We regard the summands as indexed mod 2^n-1, so that $\widetilde{\mathsf{DP}}(n,2^n-1) = \widetilde{\mathsf{DP}}(n,0)$, etc. Thus the twisted Steenrod kernel $\widetilde{\mathsf{K}}(n)$ is the direct sum of 2^n-1 summands $\widetilde{\mathsf{K}}(n,j)$ dual to $\widetilde{\mathsf{Q}}(n,j)$. The down action of Steenrod squares increases the ω-sequence of a block, and gives rise to filtrations of $\widetilde{\mathsf{K}}(n)$ and $\widetilde{\mathsf{K}}(n,j)$ with associated graded spaces $\widetilde{\mathsf{K}}^\omega(n)$ and $\widetilde{\mathsf{K}}^\omega(n,j)$ dual to $\widetilde{\mathsf{Q}}^\omega(n)$ and $\widetilde{\mathsf{Q}}^\omega(n,j)$.

Recall that the rotation map ρ is the automorphism of $\widetilde{\mathsf{P}}(n)$ defined by $\rho(t_i) = t_{i+1}$, where the variables are indexed mod n. The next result provides isomorphisms between the cyclic summands of $\widetilde{\mathsf{DP}}(n)$ corresponding to those for $\widetilde{\mathsf{P}}(n)$. The proof follows by duality from Proposition 27.1.9.

Proposition 28.1.6 *Let* $\rho_* : \widetilde{\mathsf{DP}}(n) \to \widetilde{\mathsf{DP}}(n)$ *be the linear dual of* ρ*. Then* ρ_* *is the algebra automorphism of* $\widetilde{\mathsf{DP}}(n)$ *which maps* $u_i^{(r)}$ *to* $u_{i-1}^{(r)}$ *for* $r \geq 0$*. The map* ρ_* *commutes with the action of* A_2 *on* $\widetilde{\mathsf{DP}}(n)$ *and induces an* A_2*-module isomorphism of* $\widetilde{\mathsf{DP}}(n,2j)$ *with* $\widetilde{\mathsf{DP}}(n,j)$ *for all cyclic weights* j *mod* 2^n-1. □

We would like to have a twisted analogue $\widetilde{\mathsf{J}}(n)$ of the d-spike subalgebra $\mathsf{J}(n)$ of $\mathsf{K}(n)$. However, d-monomials in $u_1,\dots,u_n$ with exponents of the form 2^r-1 are not generally in $\widetilde{\mathsf{K}}(n)$. For example, when $n=2$ we have $Sq_1(u_1^{(3)}) = u_1u_2 = Sq_1(u_2^{(3)})$, so the d-polynomial $u_1^{(3)} + u_2^{(3)}$ is in $\widetilde{\mathsf{K}}^3(2)$, but its terms are not.

Definition 28.1.7 For $n \geq 1$ and $d \geq 0$, we denote the sum of all d-monomials of degree d by σ_d, so that $\sigma_d = \sum_{j=0}^{2^n-2} \sigma_d(j)$, where $\sigma_d(j)$ is the sum of all d-monomials of degree d and cyclic weight j.

Proposition 28.1.8 *Let* $r \geq 1$ *and let* $d = 2^r-1$*. Then* (i) $\sigma_d \in \widetilde{\mathsf{K}}(n)$*, and for all cyclic weights* j*,* (ii) $\sigma_d(j) \in \widetilde{\mathsf{K}}(n,j)$ *and* (iii) $\rho_*(\sigma_d(2j)) = \sigma_d(j)$.

Proof Since we have a direct sum splitting into the cyclic summands, (ii) follows from (i). By the d-binomial theorem (9.3), $\sigma_d = (u_1 + \cdots + u_n)^{(d)}$. We need to show that $Sq_k(\sigma_d) = 0$ for all $k > 0$. Equivalently, $\langle Sq_k(\sigma_d), g\rangle =$

$\langle \sigma_d, Sq^k(g)\rangle = 0$ for all monomials g in $\widetilde{\mathsf{P}}^{d-k}(n)$. By specializing the variables $t_1,\ldots,t_n$ in $\widetilde{\mathsf{P}}(n)$ to one variable t, we see that the polynomial $Sq^k(g) \in \widetilde{\mathsf{P}}^d(n)$ has an even number of terms, as t^d is not hit in the 1-variable case. Since all d-monomials of degree $d = 2^r - 1$ are terms of σ_d, $\langle \sigma_d, Sq^k(g)\rangle = 0$. Finally (iii) is clear from Proposition 28.1.6. □

Definition 28.1.9 The **twisted d-spike module** $\widetilde{\mathsf{J}}(n)$ is the subalgebra of $\widetilde{\mathsf{DP}}(n)$ generated by the elements $\sigma_{2^r-1}(j)$ for all $r \geq 1$ and all cyclic weights j mod $2^n - 1$.

Proposition 28.1.10 (i) *The twisted d-spike module* $\widetilde{\mathsf{J}}(n)$ *is a subalgebra of* $\widetilde{\mathsf{K}}(n)$, *and* (ii) *all products of length* $> n$ *of the generators* $\sigma_{2^r-1}(j)$ *of* $\widetilde{\mathsf{J}}(n)$ *are* 0.

Proof Statement (i) follows from Proposition 28.1.8 and the Cartan formula. For (ii), we observe that by the superposition rule for multiplication in $\mathsf{DP}(n)$ a product of $> n$ elements of odd degree is 0. □

The twisted d-spike module $\widetilde{\mathsf{J}}(n)$ splits as the direct sum $\widetilde{\mathsf{J}}(n) = \bigoplus_{j=0}^{2^n-2} \widetilde{\mathsf{J}}(n,j)$, where $\widetilde{\mathsf{J}}(n,j) \subseteq \widetilde{\mathsf{K}}(n,j)$ is the $\mathbb{F}_2$-subspace of $\widetilde{\mathsf{J}}(n)$ spanned by the elements of cyclic weight j. Multiplication in $\widetilde{\mathsf{DP}}(n)$ induces pairings $\widetilde{\mathsf{K}}^{d_1}(n,i) \otimes \widetilde{\mathsf{K}}^{d_2}(n,j) \to \widetilde{\mathsf{K}}^{d_1+d_2}(n,i+j)$ and $\widetilde{\mathsf{J}}^{d_1}(n,i) \otimes \widetilde{\mathsf{J}}^{d_2}(n,j) \to \widetilde{\mathsf{J}}^{d_1+d_2}(n,i+j)$. In particular, $\widetilde{\mathsf{J}}(n,0)$ and $\widetilde{\mathsf{K}}(n,0)$ are subalgebras of $\widetilde{\mathsf{K}}(n)$.

Example 28.1.11 The table below gives a basis for $\widetilde{\mathsf{J}}^d(2,j)$ in degrees $d \leq 7$. We shall prove in Section 28.4 that $\widetilde{\mathsf{J}}^d(2) = \widetilde{\mathsf{K}}^d(2)$ for all d.

$j \backslash d$	0	1	2	3	4	5	6	7
0	1	·	$\sigma_1(1)\sigma_1(2)$	$\sigma_3(0)$	·	·	$\sigma_3(1)\sigma_3(2)$	$\sigma_7(0)$
1	·	$\sigma_1(1)$	·	$\sigma_3(1)$	$\sigma_3(0)\sigma_1(1)$	·	·	$\sigma_7(1)$
2	·	$\sigma_1(2)$	·	$\sigma_3(2)$	$\sigma_3(0)\sigma_1(2)$	·	·	$\sigma_7(2)$
$\dim \widetilde{\mathsf{J}}^d(2)$	1	2	1	3	2	0	1	3

In terms of the generators u_1 and u_2, $\sigma_1(1) = u_1$, $\sigma_1(2) = u_2$, $\sigma_3(0) = u_1^{(3)} + u_2^{(3)}$, $\sigma_3(1) = u_1^{(2)}u_2$, $\sigma_3(2) = u_1u_2^{(2)}$, $\sigma_7(0) = u_1^{(5)}u_2^{(2)} + u_1^{(2)}u_2^{(5)}$, $\sigma_7(1) = u_1^{(7)} + u_1^{(4)}u_2^{(3)} + u_1u_2^{(6)}$ and $\sigma_7(2) = u_1^{(6)}u_2 + u_1^{(3)}u_2^{(4)} + u_2^{(7)}$.

Example 28.1.12 In the case $n = 3$, $\sigma_1(1) = u_1$, $\sigma_1(2) = u_2$, $\sigma_1(4) = u_3$ span $\widetilde{\mathsf{K}}^1(3)$, and the products $\sigma_1(1)\sigma_1(2)$, $\sigma_1(1)\sigma_1(4)$ and $\sigma_1(2)\sigma_1(4)$ of cyclic weights 3, 5 and 6 span $\widetilde{\mathsf{K}}^2(3)$. The expansion of $(u_1+u_2+u_3)^{(3)}$ also gives the elements $\sigma_3(0) = u_1u_2u_3$, $\sigma_3(1) = u_2^{(2)}u_3$, $\sigma_3(2) = u_1u_3^{(2)}$, $\sigma_3(3) = u_1^{(3)} + u_2u_3^{(2)}$, $\sigma_3(4) = u_1^{(2)}u_2$, $\sigma_3(5) = u_1u_2^{(2)} + u_3^{(3)}$ and $\sigma_3(6) = u_1^{(2)}u_3 + u_2^{(3)}$. These six elements have ω-sequence $\omega = (1,1)$, and so $\dim \widetilde{\mathsf{K}}^{\omega}(3,j) = 1$ for $1 \leq j \leq 6$. This is the dual result corresponding to the values of $\dim \widetilde{\mathsf{Q}}^{\omega}(3,j)$ in Table 27.5.1 for $t = 0$ and $s = 2$. In addition $\dim \widetilde{\mathsf{K}}^{(3)}(3,0) = 1$, giving $\dim \widetilde{\mathsf{K}}^3(3) = 7 = \dim \mathsf{K}^3(3)$.

It follows from Section 27.1 that the elements $v_1,\ldots,v_n$ in $\mathsf{DP}(n)$ are related to the elements $u_1,\ldots,u_n$ in $\widetilde{\mathsf{DP}}(n)$ by $\mathbf{v} = \mathbf{u}Z$ where $\mathbf{v} = (v_1,\ldots,v_n)$ and $\mathbf{u} = (u_1,\ldots,u_n)$ are row vectors and Z is the matrix (27.1). Thus for $1 \leq i \leq n$

$$v_i = \zeta^{i-1}u_1 + \zeta^{2(i-1)}u_2 + \cdots + \zeta^{2^{n-1}(i-1)}u_n, \tag{28.1}$$

where ζ generates $\mathbb{F}_{2^n}^* \cong \mathbb{Z}/(2^n - 1)$. In particular, $v_1 = u_1 + \cdots + u_n$. Since $\det(Z) = 1$ and since $u_1,\ldots,u_n$ generate an exterior algebra, $v_1 \cdots v_n = u_1 \cdots u_n$. For example, for $n = 2$ we have $v_1 = u_1 + u_2$, $v_2 = \zeta u_1 + \zeta^2 u_2$ where $\zeta^2 = 1 + \zeta \in \mathbb{F}_4$, and for $n = 3$, $v_1 = u_1 + u_2 + u_3$, $v_2 = \zeta u_1 + \zeta^2 u_2 + \zeta^4 u_3$, $v_3 = \zeta^2 u_1 + \zeta^4 u_2 + \zeta u_3$, where $\zeta^3 = 1 + \zeta \in \mathbb{F}_8$.

The next result shows that a choice of the generator ζ of $\mathbb{F}_{2^n}^*$ defines a bijection between the nonzero elements of the defining module $\mathsf{V}(n)$ for $\mathsf{GL}(n)$, regarded as $\mathsf{DP}^1(n)$, and the cyclic summands of $\widetilde{\mathsf{DP}}(n)$.

Proposition 28.1.13 *Let ζ be a generator of the cyclic group $\mathbb{F}_{2^n}^*$. Given a nonzero element $v = \sum_i c_i v_i$ of $\mathsf{DP}^1(n)$, where $c_i \in \mathbb{F}_2$, let $\zeta^a = \sum_{i=1}^n c_i \zeta^{i-1} \in \mathbb{F}_{2^n}^*$. Then* (i) *$v = \sum_{i=1}^n \zeta^{2^{i-1}a} u_i$, and* (ii) *$v^{(d)} = \sum_{j=0}^{2^n-2} \zeta^{ja}\sigma_d(j)$ for all $d \geq 0$.*

Proof (i) follows from (28.1) by linearity over $\mathbb{F}_2$. Hence

$$v^{(d)} = (\zeta^a u_1 + \zeta^{2a}u_2 + \cdots + \zeta^{2^{n-1}a}u_n)^{(d)}.$$

Expanding by the d-binomial theorem, the coefficient of a d-monomial $f = u_1^{(a_1)} \cdots u_n^{(a_n)}$ in $v^{(d)}$ is ζ^{aj} where $j = \sum_{i=1}^n 2^{i-1}a_i$ is the cyclic weight of f. Hence (ii) follows by collecting terms with the same cyclic weight. □

We write ℓ_a for the element of $\mathsf{V}(n)$ which corresponds to $a \in \mathbb{F}_{2^n}^*$, so that

$$\ell_a = \sum_{i=1}^n \zeta^{2^{i-1}a}u_i. \tag{28.2}$$

In particular, $\ell_0 = v_1$. equation (28.2) can be inverted as follows.

Proposition 28.1.14 *For $d \geq 0$ and $0 \leq j \leq 2^n - 2$, $\sigma_d(j) = \sum_{a=1}^{2^n-1} \zeta^{-ja}\ell_a^{(d)}$. In particular, $u_i = \sum_{a=0}^{2^n-2} \zeta^{-2^{i-1}a}\ell_a$ for $1 \leq i \leq n$.*

Proof This follows from Proposition 28.1.13 by observing that the matrices $A = (a_{i,j})$ and $B = (b_{i,j})$ in $\mathsf{M}(2^n - 1, \mathbb{F}_{2^n})$ with (i,j)th entries $a_{i,j} = \zeta^{ij}$ and $b_{i,j} = \zeta^{-ij}$ are inverses of each other. To see this, let $AB = C = (c_{i,j})$. Then $c_{i,j} = \sum_{k=1}^{2^n-1} a_{i,k}b_{k,j} = \sum_{k=1}^{2^n-1} \zeta^{(i-j)k}$. If $i = j$, then $\zeta^{i-j} = \zeta^0 = 1$ and so $c_{i,j} = 1$, while if $i \neq j$ then $\alpha = \zeta^{i-j} \neq 1$ in $\mathbb{F}_{2^n}^*$ and so $c_{i,j} = \sum_{k=0}^{2^n-2} \alpha^k = 0$. □

Proposition 28.1.15 *For $n \geq 1$ and $d \geq 0$, $\dim \mathsf{J}^d(n) = \dim \widetilde{\mathsf{J}}^d(n)$ as vector spaces over $\mathbb{F}_2$.*

Proof The twisted d-spike module $\widetilde{\mathsf{J}}(n)$ is the subalgebra of $\widetilde{\mathsf{DP}}(n)$ generated by the elements $\sigma_d(j)$ for $0 \leq j \leq 2^n - 2$ and $d = 2^r - 1$, $r \geq 1$. The d-spike module $\mathsf{J}(n)$ can alternatively be defined as the subalgebra of $\mathsf{DP}(n)$ generated by the elements $\ell_a^{(d)}$ for $a \in \mathbb{F}_{2^n}^*$ and the same values of d. The equations $\ell_a^{(d)} = \sum_{j=0}^{2^n-2} \zeta^{aj}\sigma_d(j)$ and $\sigma_d(j) = \sum_{a=1}^{2^n-1} \zeta^{-aj}\ell_a^{(d)}$ show that $\widetilde{\mathsf{J}}(n) \otimes_{\mathbb{F}_2} \mathbb{F}_{2^n} = \mathsf{J}(n) \otimes_{\mathbb{F}_2} \mathbb{F}_{2^n}$, and the result follows. □

Since the elements $\sigma_{2^k-1}(j)$ lie in different cyclic summands, they are linearly independent if they are nonzero. Since $\mathrm{cw}(u_i) = 2^{i-1}$ for $1 \leq i \leq n$, there is a d-monomial of cyclic weight j in $\widetilde{\mathsf{DP}}^d(n)$ if and only if $\alpha(j) \leq d$. Hence $\sigma_{2^k-1}(j) \neq 0$ if and only if $\alpha(j) \leq 2^k - 1$. Thus we have constructed $\sum_{s=1}^{d} \binom{n}{s}$ linearly independent elements of $\widetilde{\mathsf{J}}(n)$ in degree $d = 2^k - 1$. For $k \geq n$ this number is $2^n - 1$, and it follows that the generators $v^{(d)}$ corresponding to the nonzero elements $v = \sum_i c_i v_i$ of $\mathsf{V}(n)$ are linearly independent in $\mathsf{J}(n)$.

28.2 Relations in $\widetilde{\mathsf{J}}(n)$

In this section, we use the formulae of Section 28.1 to translate relations in $\mathsf{J}(n)$ into quadratic relations in $\widetilde{\mathsf{J}}(n)$. Since the generators of $\widetilde{\mathsf{J}}(n)$ are in degrees $2^s - 1$, these relations are in degrees d such that $\mu(d) = 2$. Propositions 28.2.1 and 28.2.2 give a preliminary formulation of the relations with coefficients in $\mathbb{F}_{2^n}$, from which relations over $\mathbb{F}_2$ can be extracted by comparing coefficients of ζ^i for $0 \leq i \leq n - 1$. There is no loss of generality in taking $a = 0$ in Propositions 28.2.1 and 28.2.2, since the relations as stated are recovered on multiplication by ζ^{ak}.

Proposition 28.2.1 *Let $d = d_1 + d_2$ where $d_1 = 2^{s+t} - 1$, $d_2 = 2^t - 1$ and $s \geq 0$, $t \geq 1$, and let $\zeta^a + \zeta^b + \zeta^c = 0$ in $\mathbb{F}_{2^n}$. Then in $\mathbb{F}_{2^n} \otimes_{\mathbb{F}_2} \widetilde{\mathsf{DP}}^d(n,k)$,*

$$\text{(i)} \sum \sigma_{d_1}(i)\sigma_{d_2}(j) = 0, \quad \text{(ii)} \sum \zeta^{ai}(\zeta^{bj} + \zeta^{cj})\sigma_{d_1}(i)\sigma_{d_2}(j) = 0,$$

for all cyclic weights k, where the sums are over all cyclic weights i and j such that $i+j=k \bmod 2^n-1$.

Proof For (i), we apply Proposition 28.1.13 to the relation $v^{(d_1)}v^{(d_2)}=0$, where $v \neq 0 \in \mathsf{V}(n)=\mathsf{DP}^1(n)$, and equate terms of cyclic weight k. The relation so obtained is independent of the choice of v.

For (ii), we use in the same way the relation $v^{(d_1)}(v+w)^{(d_2)}=v^{(d_1)}w^{(d_2)}$, where v and w are nonzero elements of $\mathsf{V}(n)$, $v \neq w$. Using Proposition 28.1.13(i), we may write $v=\sum_{i=1}^{n}\zeta^{2^{i-1}a}u_i$ and $w=\sum_{i=1}^{n}\zeta^{2^{i-1}b}u_i$, where $a \neq b \bmod 2^n-1$. Let $\zeta^a+\zeta^b=\zeta^c$, so that $\zeta^{2^{i-1}a}+\zeta^{2^{i-1}b}=\zeta^{2^{i-1}c}$ for $1 \leq i \leq n$, and hence $v+w=\sum_{i=1}^{n}\zeta^{2^{i-1}c}u_i$. By Proposition 28.1.13(ii), the relation $v^{(d_1)}(v+w)^{(d_2)}=v^{(d_1)}w^{(d_2)}$ becomes

$$\sum_{i=0}^{2^n-2}\zeta^{ai}\sigma_{d_1}(i)\sum_{j=0}^{2^n-2}\zeta^{cj}\sigma_{d_2}(j)=\sum_{i=0}^{2^n-2}\zeta^{ai}\sigma_{d_1}(i)\sum_{j=0}^{2^n-2}\zeta^{bj}\sigma_{d_2}(j),$$

from which (ii) follows on equating terms of cyclic weight k. □

In the case $s=1$, the same argument applied to the relation

$$(v+w)^{(2^{t+1}-1)}w^{(2^t-1)}=v^{(2^{t+1}-1)}w^{(2^t-1)}+w^{(2^{t+1}-1)}v^{(2^t-1)}$$

in $\mathsf{DP}(n)$ produces a further relation of the same type.

Proposition 28.2.2 *Let* $d=d_1+d_2$ *where* $d_1=2^{t+1}-1$, $d_2=2^t-1$ *and* $t \geq 1$, *and let* $\zeta^a+\zeta^b+\zeta^c=0$ *in* $\mathbb{F}_{2^n}$. *Then in* $\mathbb{F}_{2^n}\otimes_{\mathbb{F}_2}\widetilde{\mathsf{DP}}^d(n,k)$,

$$\sum_{i+j=k}(\zeta^{ci+bj}+\zeta^{ai+bj}+\zeta^{bi+aj})\sigma_{d_1}(i)\sigma_{d_2}(j)=0. \quad \square$$

Proposition 28.2.3 *Let* $d=d_1+d_2$ *where* $d_1=2^{s+t}-1$, $d_2=2^t-1$ *and* $s \geq 0$, $t \geq 1$. *Then* $\sigma_{d_1}(k)\sigma_{d_2}(0)=0$ *for all cyclic weights k.*

Proof Summing the relation 28.2.1(ii) over the $2^{n-1}-1$ distinct equations of the form $1+\zeta^b+\zeta^c=0$, we have $\sum_{i+j=k}(\zeta^j+\zeta^{2j}+\cdots+\zeta^{(2^n-2)j})\sigma_{d_1}(i)\sigma_{d_2}(j)=0$. Since $\sum_{a=1}^{2^n-2}\zeta^a=1$, the coefficient of $\sigma_{d_1}(i)\sigma_{d_2}(j)$ is 1 if $j \neq 0$, and is 0 if $j=0$. The result follows by comparing this with relation 28.2.1(i). □

We write a block F representing a d-monomial $f \in \widetilde{\mathsf{DP}}(n)$ as the concatenation $F=A|B$, where A is the first column of F. If $\deg f$ is odd, then $A \neq 0$. Since the generators $\sigma_d(n)$ of $\widetilde{\mathsf{J}}(n)$ are in odd degrees d, the superposition rule for products in $\widetilde{\mathsf{DP}}(n)$ shows that all products of length $> n$ in $\widetilde{\mathsf{J}}(n)$ are zero, and that products of length n in $\widetilde{\mathsf{J}}(n)$ can be calculated by ignoring all blocks F where $\omega_1(F)>1$, i.e. A has more than one entry 1.

For all $d \geq 0$ and all cyclic weights k, we can therefore expand $\sigma_{2d+1}(k)$ in the form

$$\sigma_{2d+1}(k) = \sum_{i=1}^{n} u_i \sigma'_{2d}(k - 2^{i-1}) + E \tag{28.3}$$

where E is a sum of blocks F with $\omega_1(F) > 1$, and $\sigma'_{2d}(2j)$ denotes the sum of all terms in $\sigma_{2d}(2j)$ with no odd exponents, i.e. the sum of d-monomials $u_1^{(2a_1)} \cdots u_n^{(2a_n)}$ obtained from the terms $u_1^{(a_1)} \cdots u_n^{(a_n)}$ of $\sigma_d(j)$ by doubling all exponents.

Example 28.2.4 For $n = 3$, $\sigma_7(3)$ is the sum of the five blocks

$$\begin{array}{c|c} 1 & 0\,0 \\ 0 & 0\,1 \\ 0 & 1\,0 \end{array}\,, \quad \begin{array}{c|c} 0 & 0\,1 \\ 1 & 1\,0 \\ 0 & 0\,0 \end{array}\,, \quad \begin{array}{c|c} 0 & 0\,0 \\ 0 & 1\,0 \\ 1 & 0\,1 \end{array}\,, \quad \begin{array}{c|c} 0 & 1\,1 \\ 0 & 0\,0 \\ 1 & 0\,0 \end{array}\,, \quad \begin{array}{c|c} 1 & 1 \\ 1 & 0 \\ 1 & 1 \end{array}\,,$$

as can be seen from the cyclic weight array

$$\begin{array}{cccccc} 1 & 2 & 4 & 1 & 2 & \cdots \\ 2 & 4 & 1 & 2 & 4 & \cdots \\ 4 & 1 & 2 & 4 & 1 & \cdots \end{array}\,.$$

Thus $\sigma_7(3) = u_1\sigma'_6(2) + u_2\sigma'_6(1) + u_3\sigma'_6(6) + E$, where E represents $e = u_1^{(3)} u_2 u_3^{(3)}$. Notice that e is a term of each of the products $u_1\sigma_6(2)$, $u_2\sigma_6(1)$ and $u_3\sigma_6(6)$.

The expansion (28.3) gives a recursive procedure for producing relations in $\widetilde{\mathrm{J}}(n)$, doubling the number of terms. The following example illustrates this.

Example 28.2.5 Let $n = 3$. By Proposition 28.2.3, we have the relation $\sigma_{15}(5)\sigma_7(0) = 0$. Expanding by (28.3), we obtain

$$\begin{aligned} \sigma_{15}(5) &= u_1\sigma'_{14}(4) + u_2\sigma'_{14}(3) + u_1\sigma'_{14}(1) + E_1 \\ \sigma_7(0) &= u_1\sigma'_6(6) + u_2\sigma'_6(5) + u_1\sigma'_6(3) + E_2 \end{aligned}$$

where $\omega_1(E_1) > 1$ and $\omega_1(E_2) > 1$. Equating the coefficients of u_1u_2, u_1u_3 and u_2u_3 to 0 and removing the first column from each block, we obtain the relations $\sigma_7(2)\sigma_3(6) = \sigma_7(5)\sigma_3(3)$ in $\widetilde{\mathrm{J}}^{10}(1)$, $\sigma_7(2)\sigma_3(5) = \sigma_7(4)\sigma_3(3)$ in $\widetilde{\mathrm{J}}^{10}(0)$ and $\sigma_7(5)\sigma_3(5) = \sigma_7(4)\sigma_3(6)$ in $\widetilde{\mathrm{J}}^{10}(3)$.

We allow negative integers as cyclic weights, with the usual meaning as integers mod $2^n - 1$.

Proposition 28.2.6 *Let* $d_1 = 2^{s+t} - 1 = 2d_1' + 1$, $d_2 = 2^t - 1 = 2d_2' + 1$ *where* $s \geq 0$, $t \geq 1$. *Given an m-term quadratic relation*

$$\sum_{\ell} \sigma_{d_1}(k+\ell)\sigma_{d_2}(-\ell) = 0 \tag{28.4}$$

and i, j *such that* $1 \leq i < j \leq n$, *there is a 2m-term quadratic relation*

$$\sum_{\ell} \sigma_{d_1'}(k+\ell-2^j)\sigma_{d_2'}(-\ell-2^i) = \sum_{\ell} \sigma_{d_1'}(k+\ell-2^i)\sigma_{d_2'}(-\ell-2^j) \tag{28.5}$$

where all the sums are over the same set of cyclic weights ℓ. *If* (28.4) *is true for all* $s \geq 0$, $t \geq 1$ *and all cyclic weights* k, *then so is* (28.5).

Proof By applying the inverse ρ_*^{-1} of the rotation map ρ_* to the given relation (28.4), we obtain an equivalent relation $\sum_\ell \sigma_{d_1}(2k+2\ell)\sigma_{d_2}(-2\ell) = 0$. Using (28.3), we expand both factors in each term. There is nothing to prove if $i = j$, so we may assume that $i \neq j$ and extract those d-monomials in which exactly the variables u_{i+2} and u_{j+2} have odd exponents. We then omit the first column from the corresponding blocks, to obtain the relation (28.5). □

By applying this result to the 1-term quadratic relations of Proposition 28.2.3, we obtain the following 2-term relations.

Proposition 28.2.7 *Let* $d_1 = 2^{s+t} - 1$, $d_2 = 2^t - 1$ *where* $s \geq 0$, $t \geq 1$, *and let* $d = d_1 + d_2$. *Then* $\sigma_{d_1}(k+1)\sigma_{d_2}(-1) = \sigma_{d_1}(k+2)\sigma_{d_2}(-2) = \ldots = \sigma_{d_1}(k+2^{n-1})\sigma_{d_2}(-2^{n-1})$ *for all cyclic weights* k. □

Example 28.2.8 In the case $n = 3$, these relations are $\sigma_{d_1}(k+1)\sigma_{d_2}(-1) = \sigma_{d_1}(k+2)\sigma_{d_2}(-2) = \sigma_{d_1}(k+4)\sigma_{d_2}(-4)$. Applying Proposition 28.2.6 to the first of these equations with $i = 1$, $j = 2$ gives the 4-term relation $\sigma_{d_1}(k+3)\sigma_{d_2}(-3) + \sigma_{d_1}(k+5)\sigma_{d_2}(-5) = \sigma_{d_1}(k+4)\sigma_{d_2}(-4) + \sigma_{d_1}(k+6)\sigma_{d_2}(-6)$. Rewriting the cyclic weights mod 7, we obtain $\sigma_{d_1}(k-1)\sigma_{d_2}(1) + \sigma_{d_1}(k-2)\sigma_{d_2}(2) + \sigma_{d_1}(k-4)\sigma_{d_2}(4) = \sigma_{d_1}(k-3)\sigma_{d_2}(3) = \sigma_{d_1}(k-5)\sigma_{d_2}(5) = \sigma_{d_1}(k-6)\sigma_{d_2}(6)$ and $\sigma_{d_1}(k)\sigma_{d_2}(0) = 0$. We shall see in Section 28.5 that all relations in $\widetilde{\mathrm{J}}(3)$ which hold for all $d_1 = 2^{s+t} - 1$ and $d_2 = 2^t - 1$, where $s \geq 0$ and $t \geq 1$, are consequences of these.

Example 28.2.9 For $n = 3$, let $d_1 = 2^{s+t+u} - 1$, $d_2 = 2^{2+t} - 1$, $d_3 = 2^s - 1$, where $s, t \geq 0$ and $u \geq 1$. We shall show that all products $f = \sigma_{d_1}(i)\sigma_{d_2}(j)\sigma_{d_3}(k)$ are linear combinations of those with $k = 1$ and $j = 1, 2, 3$.

By Proposition 28.2.3, $f = 0$ if $j = 0$ or $k = 0$. Applying Proposition 28.2.7 to the first and third factors and then to the second and third factors, we have $\sigma_{d_1}(i+3)\sigma_{d_2}(j)\sigma_{d_3}(3) = \sigma_{d_1}(i)\sigma_{d_2}(j)\sigma_{d_3}(6) = \sigma_{d_1}(i)\sigma_{d_2}(j+3)\sigma_{d_3}(3)$ for all i and j. This relation shifts j by 3, and so by iterating it we can shift j to 0 mod 7.

Hence $f = 0$ if $k = 3$, and using Proposition 28.2.7 again it follows that $f = 0$ if $k = 6$ or 5.

Applying the 4-term relation of Example 28.2.8, it follows that $\sigma_{d_1}(i)\sigma_{d_2}(j-4)\sigma_{d_3}(4) + \sigma_{d_1}(i)\sigma_{d_2}(j-2)\sigma_{d_3}(2) + \sigma_{d_1}(i)\sigma_{d_2}(j-1)\sigma_{d_3}(1) = \sigma_{d_1}(i)\sigma_{d_2}(j-3)\sigma_{d_3}(3) = 0$. Hence products with $k = 4$ are linear combinations of those with $k = 1$ or 2. Similarly $\sigma_{d_1}(i-4)\sigma_{d_2}(j)\sigma_{d_3}(4) + \sigma_{d_1}(i-2)\sigma_{d_2}(j)\sigma_{d_3}(2) + \sigma_{d_1}(i-1)\sigma_{d_2}(j)\sigma_{d_3}(1) = 0$. Thus $\sigma_{d_1}(i)\sigma_{d_2}(j+2)\sigma_{d_3}(2) + \sigma_{d_1}(i)\sigma_{d_2}(j+3)\sigma_{d_3}(1) = \sigma_{d_1}(i)\sigma_{d_2}(j)\sigma_{d_3}(4) = \sigma_{d_1}(i+2)\sigma_{d_2}(j)\sigma_{d_3}(2) + \sigma_{d_1}(i+3)\sigma_{d_2}(j)\sigma_{d_3}(1)$, which shows that for $k = 2$ we can shift j by 2 modulo terms with $k = 1$. By iterating this step we can shift j to 0, showing that all products f are linear combinations of those with $k = 1$. Finally we can reduce to $1 \leq j \leq 3$ by applying Example 28.2.8 to the first two factors,

We next obtain a set of quadratic relations which can be used to express all products $\sigma_{d_1}(i)\sigma_{d_2}(j)$ of generators of $\widetilde{\mathrm{J}}(n)$ as sums of the products for which $1 \leq j \leq 2^{n-1} - 1$. Recall that the mod 2 binomial coefficient $\binom{j}{i}$ is 1 if and only if $\mathrm{bin}(i) \subseteq \mathrm{bin}(j)$. Thus if $\alpha(j) = r$ then the sum in the following result has $2^r - 1$ terms, since $i = j$ is excluded, and can alternatively be written in the form $\sum_{i=0}^{j-1} \binom{j}{i} \sigma_{d_1}(k+i+2^{n-1})\sigma_{d_2}(-i-2^{n-1})$.

The following result allows us to write certain quadratic relations in cyclic weight k in the form of a mod 2 Pascal triangle (see Example 28.2.12). For $n \leq 3$, the relations given by Proposition 28.2.10 are already implied by those discussed in Example 28.2.8. The calculation of $\widetilde{\mathrm{J}}(n)$ in these cases is completed in Sections 28.4 and 28.5.

Proposition 28.2.10 *Let $d_1 = 2^{s+t} - 1$, $d_2 = 2^t - 1$ where $s \geq 0$, $t \geq 1$, and let k be a cyclic weight for $\widetilde{\mathrm{DP}}(n)$. The products of the generators of $\widetilde{\mathrm{J}}(n,k)$ satisfy the quadratic relations*

$$\sigma_{d_1}(k+j)\sigma_{d_2}(-j) = \sum_{bin(i) \subset bin(j)} \sigma_{d_1}(k+i+2^m)\sigma_{d_2}(-i-2^m)$$

for $0 \leq m \leq n-1$. In particular, this sum is independent of m.

Example 28.2.11 The proof is by induction on $\alpha(j)$. We illustrate it by carrying out the steps for $\alpha(j) \leq 3$, writing the relation in short notation $[-j] = \sum_i [-i - 2^m]$. In this notation, Proposition 28.2.6 can be stated as $\sum_\ell [-\ell] = 0$ implies $\sum_\ell [-\ell - 2^k] = \sum_\ell [-\ell - 2^j]$ for all k and j.

The case $\alpha(j) = 0$ occurs only when $j = 0$ and the relation of Proposition 28.2.3 is written as $[0] = 0$ in short notation. When $\alpha(j) = 1$, $j = 2^a$ and we obtain the relation $[-2^a] = [-2^m]$ of Proposition 28.2.7.

When $\alpha(j) = 2$, $j = 2^a + 2^b$ and applying Proposition 28.2.6 again, we obtain $[-j]+[-2^m-2^b]=[-2^a-2^m]+[-2^m-2^m]$. The term $[-2^m-2^m]=[-2^{m+1}]=[-2^m]$ by the case $\alpha(j)=1$, giving $[-j]=[-2^a-2^m]+[-2^b-2^m]+[-2^m]$ as required.

When $\alpha(j) = 3$, $j = 2^a + 2^b + 2^c$, so Proposition 28.2.6 gives the 8-term relation $[-j]+[-2^a-2^b-2^m]=[-2^a-2^m-2^c]+[-2^a-2^m-2^m]+[-2^b-2^m-2^c]+[-2^b-2^m-2^m]+[-2^m-2^c]+[-2^m-2^m]=0$. By the case $\alpha(j)=2$, the three terms involving -2^m-2^m can be replaced by $[-2^a-2^b]$, and using the case $\alpha(j)=2$ again this can be replaced by the corresponding three terms involving -2^m, as required.

Proof of Proposition 28.2.10 We argue by induction on $\alpha(j)$, where $0 \leq j \leq 2^n-2$. Proposition 28.2.3 is the base case $\alpha(j)=0$, which occurs when $j=0$. The case $\alpha(j)=1$ occurs when $j=2^a$ for $0 \leq a \leq n-1$, and this is Proposition 28.2.7.

For the inductive step, let $\alpha(j)=r>1$ and let $\text{bin}(j)=\{2^{a_1},\ldots,2^{a_r}\}$. Then $j=j'+2^{a_r}$, where $j'=2^{a_1}+\cdots+2^{a_{r-1}}$. Then $\alpha(j')=r-1$ and using the induction hypothesis we may assume that the result is true for j'. Thus $\sigma_{d_1}(k+j')\sigma_{d_2}(-j')$ is the sum of the $2^{r-1}-1$ products $\sigma_{d_1}(k+i'+2^m)\sigma_{d_2}(-i'-2^m)$ where $\text{bin}(i') \subset \text{bin}(j') = \{a_1,\ldots,a_{r-1}\}$. We index these terms by the cyclic weight $-i'-2^m$ of the factor in degree d_2.

We apply Proposition 28.2.6 to this relation, taking $k=a_r$ and $j=m$, where $m \neq a_r$. The term $\sigma_{d_1}(k+j')\sigma_{d_2}(-j')$ gives the terms $\sigma_{d_1'}(k+j)\sigma_{d_2'}(-j)$ and $\sigma_{d_1'}(k+j'+2^m)\sigma_{d_2'}(-j'-2^m)$. The first of these is the term on the left of the required relation, while the second is the term on the right of the required relation which corresponds to $i=j'$.

In the same way, each term on the right of the j' relation gives rise to two terms in a new relation, the term indexed by $-i'-2^m$ giving terms indexed by the cyclic weights $-i'-2^{a_r}-2^m$ and $-i'-2^m-2^m=-i'-2^{m+1} \bmod 2^n-1$ of their degree d_2' factors. The first term is the term in the required formula for $\sigma_{d_1}(k+j)\sigma_{d_2}(-j)$ corresponding to $i=i'+2^{a_r}$.

Using the inductive hypothesis with m replaced by $m+1$, the sum of the second terms is

$$\sum_{\text{bin}(i')\subset\text{bin}(j')} \sigma_{d_1}(k+i'+2^{m+1})\sigma_{d_2}(-i'-2^{m+1}) = \sigma_{d_1}(k+j')\sigma_{d_2}(j').$$

Using the inductive hypothesis again, this is

$$\sum_{\text{bin}(i')\subset\text{bin}(j')} \sigma_{d_1}(k+i'+2^m)\sigma_{d_2}(-i'-2^m).$$

Thus the terms involving 2^{m+1} can be replaced by the corresponding terms involving 2^m, as required to complete the inductive step. □

By choosing $m = n-1$ in Proposition 28.2.10, we can express the products $\sigma_{d_1}(k+j)\sigma_{d_2}(-j)$ for $0 \le j \le 2^{n-1}-1$ as sums of the corresponding products for $2^{n-1} \le j \le 2^n - 2$. We illustrate this in the case $n = 4$.

Example 28.2.12 Let $n=4, j=7$ and $m=n-1=3$. Then Proposition 28.2.10 gives $\sigma_{d_1}(k+7)\sigma_{d_2}(-7) = \sum_{i=0}^{6} \sigma_{d_1}(k+i+8)\sigma_{d_2}(-i-8)$. We tabulate this and the similar relations for $1 \le j \le 6$ are expressed by writing the (k,j)th entry in the table as 1 if the term $\sigma_{d_1}(k+j)\sigma_{d_2}(-j)$ appears in the jth relation, where $1 \le j \le 15$. Missing entries in the table are 0s.

$k \backslash j$	1	2	3	4	5	6	7
8	1	1	1	1	1	1	1
9		0	1	0	1	0	1
10			1	0	0	1	1
11				0	0	0	1
12					1	1	1
13						0	1
14							1
15							

28.3 The Kameko map for $\widetilde{\mathsf{DP}}(n)$

Recall from Section 6.5 that the down Kameko map $\kappa : \mathsf{P}^{2d+n}(n) \to \mathsf{P}^d(n)$ is defined by $\kappa(x_1^{2a_1+1} \cdots x_n^{2a_n+1}) = x_1^{a_1} \cdots x_n^{a_n}$ and $\kappa(f) = 0$ if f is a monomial with at least one even exponent.

Proposition 28.3.1 *The map $\kappa \otimes 1 : \mathsf{P}^{2d+n}(n) \otimes_{\mathbb{F}_2} \mathbb{F}_{2^n} \to \mathsf{P}^d(n) \otimes_{\mathbb{F}_2} \mathbb{F}_{2^n}$ restricts to the map $\widetilde{\kappa} : \widetilde{\mathsf{P}}^{2d+n}(n) \to \widetilde{\mathsf{P}}^d(n)$ defined by $\widetilde{\kappa}(t_1^{2a_1+1} \cdots t_n^{2a_n+1}) = t_2^{a_1} \cdots t_n^{a_{n-1}} t_1^{a_n}$, and $\kappa(f) = 0$ if f is a monomial in $\widetilde{\mathsf{P}}^{2d+n}(n)$ with at least one even exponent.*

Proof The variables $t_1, \ldots, t_n$ are given in terms of the generators $x_1, \ldots, x_n$ of $\mathsf{P}(n)$ by (27.2), which can be written as $\mathbf{t} = Z\mathbf{x}$, where $\mathbf{t}$ and $\mathbf{x}$ are the column vectors of the t- and x-variables and Z is the $n \times n$ matrix (27.1) with (i,j)th entry $\zeta^{(j-1)\cdot 2^{i-1}}$, for $1 \le i,j \le n$. Since $\det Z = 1$, the product $t_1 \cdots t_n$ is a homogeneous polynomial of degree n in $x_1,\ldots,x_n$ in which the coefficient of $\mathsf{c}(n) = x_1 \cdots x_n$ is 1.

A monomial f in $\widetilde{\mathsf{P}}^{2d+n}(n)$ with r odd exponents can be written in the form $f = t_{i_1} \cdots t_{i_r} g^2$, where g is a monomial in $t_1,\ldots,t_n$. To evaluate $\widetilde{\kappa}(f)$ we substitute for each variable t_i in terms of $x_1,\ldots,x_n$ and select the terms with all exponents odd. Since g^2 is a polynomial in $x_1^2,\ldots,x_n^2$, there are no such terms unless $r = n$. Hence $\widetilde{\kappa}(f) = 0$ unless all exponents in f are odd.

Thus let $f = t_1^{2a_1+1} \cdots t_n^{2a_n+1} = (t_1^2)^{a_1} \cdots (t_n^2)^{a_n} t_1 \cdots t_n$ where $a_1,\ldots,a_n \geq 0$ and $(t_i^2)^{a_i} = (x_1^2 + \zeta^{2^i} x_2^2 + \cdots + \zeta^{(n-1)\cdot 2^i} x_n^2)^{a_i}$. As observed above, we may replace the product $t_1 \cdots t_n$ by $\mathsf{c}(n)$ when selecting the terms in $x_1,\ldots,x_n$ with n odd exponents. Hence $\widetilde{\kappa}(f) = \widetilde{\kappa}(\prod_{i=1}^n (x_1^2 + \zeta^{2^i} x_2^2 + \cdots + \zeta^{(n-1)\cdot 2^i} x_n^2)^{a_i} \mathsf{c}(n)) = \prod_{i=1}^n (x_1 + \zeta^{2^i} x_2 + \cdots + \zeta^{(n-1)\cdot 2^i} x_n)^{a_i}$. Since $t_{i+1} = x_1 + \zeta^{2^i} x_2 + \cdots + \zeta^{(n-1)\cdot 2^i} x_n$, this is $t_2^{a_1} \cdots t_n^{a_{n-1}} t_1^{a_n}$. □

The map $\widetilde{\kappa}$ sends hit polynomials to hit polynomials for all $n \geq 1$ and $d \geq 0$, and preserves the cyclic weight. We define the up Kameko map $\widetilde{\kappa}_*$ on $\widetilde{\mathsf{DP}}(n)$ as the dual of the map $\widetilde{\kappa}$ on $\widetilde{\mathsf{P}}(n)$.

Proposition 28.3.2 *The up Kameko map* $\widetilde{\kappa}_* : \widetilde{\mathsf{DP}}^d(n) \to \widetilde{\mathsf{DP}}^{2d+n}(n)$ *acts on d-monomials by* $\widetilde{\kappa}_*(u_1^{(s_1)} \cdots u_n^{(s_n)}) = u_n^{(2s_1+1)} u_1^{(2s_2+1)} \cdots u_{n-1}^{(2s_n+1)}$. *It restricts to a map* $\widetilde{\kappa}_* : \widetilde{\mathsf{K}}^d(n,j) \to \widetilde{\mathsf{K}}^{2d+n}(n,j)$ *for each cyclic weight* j *mod* $2^n - 1$*, which is injective for all* n, d *and* j *and is an isomorphism when* $\mu(2d+n) = n$.

Proof The first statement follows from Proposition 28.3.1, since the d-monomials $u_1^{(a_1)} \cdots u_n^{(a_n)}$ form the basis of $\widetilde{\mathsf{DP}}(n)$ dual to the basis of monomials $t_1^{a_1} \cdots t_n^{a_n}$ of $\widetilde{\mathsf{P}}(n)$. It follows that $\widetilde{\kappa}_*$ is injective and that it preserves the cyclic weight. The formulae $Sq_{2k}\widetilde{\kappa}_*(f) = \widetilde{\kappa}_* Sq_k f$ and $Sq_{2k+1}\widetilde{\kappa}_*(f) = 0$ hold for all $f \in \widetilde{\mathsf{DP}}(n)$ and $k \geq 0$. These can be proved by direct calculation using Proposition 28.1.2 or by using the corresponding formulae for the map κ_* on $\mathsf{DP}(n)$. It follows that $\widetilde{\kappa}_*$ maps $\widetilde{\mathsf{K}}^d(n)$ to $\widetilde{\mathsf{K}}^{2d+n}(n)$. Since $\dim_{\mathbb{F}_2} \widetilde{\mathsf{K}}^d(n) = \dim_{\mathbb{F}_2} \mathsf{K}^d(n)$, the last statement follows from the corresponding result for $\mathsf{DP}(n)$. □

Proposition 28.3.3 *For* $1 \leq r \leq n$

$$\widetilde{\kappa}_*(\sigma_{d_1}(j_1) \cdots \sigma_{d_r}(j_r)) = \sigma_{2d_1}(j_1) \cdots \sigma_{2d_r}(j_r) u_1 \cdots u_n.$$

Proof Since the product of f with $u_1 \cdots u_n$ is 0 if f is a d-monomial in $\widetilde{\mathsf{DP}}(n)$ with an odd exponent, the product $\sigma_{2d_1}(j_1) \cdots \sigma_{2d_r}(j_r) u_1 \cdots u_n$ is unchanged if we replace each factor $\sigma_{2d_i}(j_i)$, $1 \leq i \leq r$, by the sum of the d-monomials of degree $2d_i$ and cyclic weight j_i with all exponents even. But this sum is obtained from $\sigma_{d_i}(j_i)$ by doubling exponents and rotating the variables using ρ_*. □

The next result shows that $\widetilde{\kappa}_*$ maps the generators $\sigma_{2^r-1}(j)$ of $\widetilde{\mathrm{J}}(n)$ to elements of $\widetilde{\mathrm{J}}(n)$. We shall see in Proposition 28.3.7 that $\widetilde{\kappa}_*$ is an isomorphism from $\widetilde{\mathrm{J}}^d(n)$ to $\widetilde{\mathrm{J}}^{2d+n}(n)$ when $\mu(2d+n) = n$.

Proposition 28.3.4 *For $1 \leq i \leq n$, $d \geq 0$ and all cyclic weights j mod $2^n - 1$,*

$$\widetilde{\kappa}_*(\sigma_d(j)) = \sigma_{2d+1}(j+2^{i-1})u_1 \cdots \widehat{u_i} \cdots u_n,$$

where the notation $\widehat{u_i}$ means that the factor u_i is omitted.

Proof By Proposition 28.3.3, $\widetilde{\kappa}_*(\sigma_d(j)) = \sigma_{2d}(j)u_1 \cdots u_n$. In this product, we can replace $\sigma_{2d}(j)u_i$ by $\sigma_{2d+1}(j+2^{i-1})$ since $f \in \sigma_{2d}(j)$ is a monomial with all exponents even if and only if $fu_i \in \sigma_{2d+1}(j+2^{i-1})$ is a monomial whose only odd exponent is that of u_i. □

Proposition 28.3.5 *When $d_1, \ldots, d_n$ are odd,*

$$\widetilde{\kappa}_*(\sigma_{d_1}(j_1) \cdots \sigma_{d_n}(j_n)) = \sigma_{2d_1+1}(j_1) \cdots \sigma_{2d_n+1}(j_n).$$

Proof Since $d_1, \ldots, d_n$ are odd, $\sigma_{d_1}(j_1) \cdots \sigma_{d_n}(j_n)$ is a sum of d-monomials represented by n-blocks $B = A|B'$ whose first column A consists entirely of 1s, each of which is contributed by a different factor $\sigma_{d_i}(j_i)$ for $1 \leq i \leq n$.

The effect of $\widetilde{\kappa}_*$ on B is to shift the rows upwards cyclically and prefix a further column of 1s. Thus each digit 1 in the block B is moved to the next position along its diagonal (see Definition 27.2.3). Since a full column of 1s is unchanged by the cyclic shift, we can regard $\widetilde{\kappa}_*(B)$ as a cyclic shift of B' followed by prefixing a column of 1s.

We first show how to write $\widetilde{\kappa}_*(B)$ as a term in $\sigma_{2d_1+1}(j_1) \cdots \sigma_{2d_n+1}(j_n)$. To do this, we match up the 1s in the first two columns of $\widetilde{\kappa}_*(B)$ with the same cyclic weight. These paired 1s then contribute the cyclic weights $1, 2, \ldots, 2^{n-1}$. The factor in $\sigma_{2d_i+1}(j_i)$ is made up of the cyclic shift of the contribution to B' of the factor in $\sigma_{d_i}(j_i)$, together with the paired 1s in the first two columns which give the same cyclic weight as the corresponding 1 in the first column of B.

For example, the block

$$B = \begin{array}{c|c} 1 & 1\,1\,1 \\ 1 & 1\,0\,1 \\ 1 & 0\,1\,0 \end{array}$$

appears as a term in $\sigma_7(5)\sigma_{19}(2)\sigma_5(1)$ using the factorization

$$B = \begin{array}{c|ccc} 1 & 1 & 1 & 1 \\ 1 & 1 & 0 & 1 \\ 1 & 0 & 1 & 0 \end{array} = \begin{array}{c|ccc} 1 & 1 & 0 & 0 \\ 0 & 0 & 0 & 0 \\ 0 & 0 & 1 & 0 \end{array} \cdot \begin{array}{c|ccc} 0 & 0 & 0 & 1 \\ 1 & 1 & 0 & 1 \\ 0 & 0 & 0 & 0 \end{array} \cdot \begin{array}{c|ccc} 0 & 0 & 1 & 0 \\ 0 & 0 & 0 & 0 \\ 1 & 0 & 0 & 0 \end{array}.$$

The corresponding factorization of $\widetilde{\kappa}_*(B)$ in $\sigma_{15}(5)\sigma_{39}(2)\sigma_{11}(1)$ is

$$\widetilde{\kappa}_*(B) = \begin{array}{cc|ccc} 1 & 1 & 1 & 0 & 1 \\ 1 & 1 & 0 & 1 & 0 \\ 1 & 1 & 1 & 1 & 1 \end{array} = \begin{array}{cc|ccc} 0 & 0 & 0 & 0 & 0 \\ 0 & 1 & 0 & 1 & 0 \\ 1 & 0 & 1 & 0 & 0 \end{array} \cdot \begin{array}{cc|ccc} 1 & 0 & 1 & 0 & 1 \\ 0 & 0 & 0 & 0 & 0 \\ 0 & 1 & 0 & 0 & 1 \end{array} \cdot \begin{array}{cc|ccc} 0 & 1 & 0 & 0 & 0 \\ 1 & 0 & 0 & 0 & 0 \\ 0 & 0 & 0 & 1 & 0 \end{array}.$$

This construction defines an injective map from d-monomials in $\sigma_{d_1}(j_1)\cdots\sigma_{d_n}(j_n)$ to d-monomials in $\sigma_{2d_1+1}(j_1)\cdots\sigma_{2d_n+1}(j_n)$. To complete the proof, we show that this is a surjection.

Thus let the block C represent a d-monomial in $\sigma_{2d_1+1}(j_1)\cdots\sigma_{2d_n+1}(j_n)$ where $d_1,\ldots,d_n$ are odd. Then the first two columns of C must consist entirely of 1s, and each factor $\sigma_{2d_i+1}(j_i)$, $1 \le i \le n$, must contribute one of these entries to each of the two columns. We may therefore associate to C the permutation $\pi \in \Sigma(n)$ defined by $\pi(k) = j$ if the entry of cyclic weight 2^k in the first column of C and the entry of cyclic weight 2^j in the second column of C arise from the same factor $\sigma_{2d_i+1}(j_i)$.

If π is not the identity permutation, then we may pair this factorization of C with a different factorization obtained by matching the entry of cyclic weight 2^j in the first column of C and the entry of cyclic weight 2^k in the second column of C, where $\pi(k) = j$ for each j. This gives a second term of the product $\sigma_{2d_1+1}(j_1)\cdots\sigma_{2d_n+1}(j_n)$ which cancels the original one.

The example below shows two cancelling factorizations of the block $C = \widetilde{\kappa}_*(B)$ above as a term of $\sigma_{15}(0)\sigma_{39}(6)\sigma_{11}(2)$.

$$\begin{array}{cc|ccc} 1 & 1 & 1 & 0 & 1 \\ 1 & 1 & 0 & 1 & 0 \\ 1 & 1 & 1 & 1 & 1 \end{array} = \begin{array}{cc|ccc} 1 & 1 & 0 & 0 & 0 \\ 0 & 0 & 0 & 1 & 0 \\ 0 & 0 & 1 & 0 & 0 \end{array} \cdot \begin{array}{cc|ccc} 0 & 0 & 1 & 0 & 1 \\ 1 & 1 & 0 & 0 & 0 \\ 0 & 0 & 0 & 0 & 1 \end{array} \cdot \begin{array}{cc|ccc} 0 & 0 & 0 & 0 & 0 \\ 0 & 0 & 0 & 0 & 0 \\ 1 & 1 & 0 & 1 & 0 \end{array},$$

$$= \begin{array}{cc|ccc} 0 & 0 & 0 & 0 & 0 \\ 1 & 0 & 0 & 1 & 0 \\ 0 & 1 & 1 & 0 & 0 \end{array} \cdot \begin{array}{cc|ccc} 0 & 1 & 1 & 0 & 1 \\ 0 & 0 & 0 & 0 & 0 \\ 1 & 0 & 0 & 0 & 1 \end{array} \cdot \begin{array}{cc|ccc} 1 & 0 & 0 & 0 & 0 \\ 0 & 1 & 0 & 0 & 0 \\ 0 & 0 & 0 & 1 & 0 \end{array}.$$

For the remaining factorizations of C, π is the identity. These are precisely the factorizations which arise as above from factorizations of blocks B. Hence $C = \widetilde{\kappa}_*(B)$, where B represents a d-monomial in $\sigma_{d_1}(j_1)\cdots\sigma_{d_n}(j_n)$. □

The following formula is useful in calculating $\widetilde{\kappa}_*$ on products of length $< n$ of odd degree elements $\sigma_d(j)$.

Proposition 28.3.6 *Let* $d = d_1 + \cdots + d_n$ *where* $\mu(2d+n) = n$. *Then*

$$\widetilde{\upsilon}_*(\sigma_{2d_1+1}(j_1)\cdots\sigma_{2d_n+1}(j_n)) = \begin{vmatrix} \sigma_{d_1}(j_1-1) & \cdots & \sigma_{d_1}(j_1-2^{n-1}) \\ \vdots & & \vdots \\ \sigma_{d_n}(j_n-1) & \cdots & \sigma_{d_n}(j_n-2^{n-1}) \end{vmatrix}.$$

Proof By expanding the sum in a similar way to (28.3), we have

$$\sigma_{2d+1}(j) = u_1\sigma_{2d}(j-1) + \cdots + u_n\sigma_{2d}(j-2^{n-1}) + E$$

where $\omega_1(E) > 1$. Since $u_1,\ldots,u_n$ generate an exterior subalgebra of $\widetilde{\mathsf{DP}}(n)$,

$$\begin{aligned}\sigma_{2d_1+1}(j_1)\cdots\sigma_{2d_n+1}(j_n) &= u_1\cdots u_n \begin{vmatrix} \sigma_{2d_1}(j_1-1) & \cdots & \sigma_{2d_1}(j_1-2^{n-1}) \\ \vdots & & \vdots \\ \sigma_{2d_n}(j_n-1) & \cdots & \sigma_{2d_n}(j_n-2^{n-1}) \end{vmatrix} \\ &= \widetilde{\kappa}_* \begin{vmatrix} \sigma_{d_1}(j_1-1) & \cdots & \sigma_{d_1}(j_1-2^{n-1}) \\ \vdots & & \vdots \\ \sigma_{d_n}(j_n-1) & \cdots & \sigma_{d_n}(j_n-2^{n-1}) \end{vmatrix},\end{aligned}$$

and the result follows, since $\widetilde{\kappa}_*$ is injective. □

Proposition 28.3.7 *For degrees d such that* $\mu(2d+n) = n$, *the cyclic Kameko map* $\widetilde{\kappa}_*$ *maps* $\widetilde{\mathsf{J}}^d(n)$ *isomorphically on to* $\widetilde{\mathsf{J}}^{2d+n}(n)$.

Proof Since $\mu(2d+n) = n$, there are no products of the generators $\sigma_{2^r-1}(j)$ of $\widetilde{\mathsf{J}}(n)$ of length $< n$ in degree $2d+n$, and all products of length $> n$ are 0. Hence $\widetilde{\mathsf{J}}^{2d+n}(n)$ is spanned by products of length exactly n in these generators, and the proof of Proposition 28.3.6 shows that all such products are in the image of $\widetilde{\kappa}_*$. The result follows, since $\widetilde{\kappa}_*$ is injective. □

Remark 28.3.8 It seems to be difficult to prove Proposition 28.3.7 by means of an explicit formula for evaluating $\widetilde{\kappa}_*$ on arbitrary products of the generators $\sigma_{2^r-1}(j)$ of $\widetilde{\mathsf{J}}(n)$, so as to generalize Propositions 28.3.4 and 28.3.5.

28.4 Calculation of $\widetilde{\mathsf{K}}(2)$

It follows from Chapter 9 that $\widetilde{\mathsf{K}}(2) = \widetilde{\mathsf{J}}(2)$ and from Theorem 2.5.5 that $\widetilde{\mathsf{K}}^d(2) = 0$ when $\mu(d) > 2$. Let $d = d_1 + d_2$, where $d_1 = 2^{s+t} - 1$, $d_2 = 2^t - 1$ and $s, t \geq 0$. Table 27.4 gives $\dim \widetilde{\mathsf{K}}^d(2,j) = \dim \widetilde{\mathsf{Q}}^d(2,j)$, which is 0 or 1 in all cases.

Theorem 28.4.1 *The tables below give the nonzero elements of* $\widetilde{\mathsf{K}}^d(2,j)$ *as products in* $\widetilde{\mathsf{J}}(2)$. *Note that* $\sigma_1(1) = u_1$ *and* $\sigma_1(2) = u_2$.

$t = 0$

$s \backslash j$	0	1	2
0	1	·	·
1	·	u_1	u_2
≥ 2	$\sigma_d(0)$	$\sigma_d(1)$	$\sigma_d(2)$

$t > 0$

$s \backslash j$	0	1	2
0	$\sigma_{d_1}(2)\sigma_{d_2}(1)$	·	·
1	·	$\sigma_{d_1}(0)\sigma_{d_2}(1)$	$\sigma_{d_1}(1)\sigma_{d_2}(1)$
≥ 2	$\sigma_{d_1}(2)\sigma_{d_2}(1)$	$\sigma_{d_1}(0)\sigma_{d_2}(1)$	$\sigma_{d_1}(1)\sigma_{d_2}(1)$

Proof The generators of $\widetilde{\mathsf{J}}(2)$ are the nonzero elements $\sigma_d(j)$ for $d = 2^r - 1$ and $j = 0, 1, 2$, and the first table simply lists these. Since $n = 2$, all products of length ≥ 3 are 0.

For $t \geq 1$, the quadratic relations given by Propositions 28.2.3 and 28.2.7 are $\sigma_{d_1}(j)\sigma_{d_2}(0) = 0$ and $\sigma_{d_1}(j-1)\sigma_{d_2}(1) = \sigma_{d_1}(j-2)\sigma_{d_2}(2)$ for $j = 0, 1, 2$. In the case $s \geq 2$, there are no further relations. In the case $s = 1$, there is an additional relation $\sigma_{d_1}(2)\sigma_{d_2}(1) = 0$ in $\widetilde{\mathsf{K}}(2,0)$. When $s = 0$, $d_1 = d_2$ and by commutativity we obtain $\sigma_{d_1}(0)\sigma_{d_1}(1) = 0$ and $\sigma_{d_1}(0)\sigma_{d_1}(2) = 0$. □

When $s = 0$, $t \geq 1$, $d_1 = d_2 = 2^t - 1$, the unique nonzero element of $\widetilde{\mathrm{K}}^d(2)$ is $u_1^{(d_2)}u_2^{(d_2)}$. The superposition rule shows that every term in a product given in the table for $t \geq 1$ must have ω-sequence $(2,\ldots,2,1,\ldots,1)$, with t 2s and s 1s. In particular, $\sigma_{d_2}(2)\sigma_{d_2}(1) = u_1^{(d_2)}u_2^{(d_2)}$, but verifying this by direct calculation is equivalent to showing that the number of d-monomials in $\sigma_{d_2}(1)$ is odd.

More generally, by starting with $t = 0$ and iterating $\widetilde{\kappa}_*$ we can express all the elements of $\widetilde{\mathrm{K}}^d(2,j)$ given by Proposition 28.4.1 as sums of d-monomials. The general case can be expressed as follows.

Proposition 28.4.2 *Let $d_1 = 2^{s+t} - 1$ and $d_2 = 2^t - 1$ where $s \geq 0$ and $t \geq 1$. Then $\sigma_{d_1}(j-1)\sigma_{d_2}(1)$ is the sum of all d-monomials in $\widetilde{\mathrm{DP}}^\omega(2,j)$, where $\omega = (2,\ldots,2,1,\ldots,1)$, with t 2s and s 1s.* □

Example 28.4.3 Let $d = 16$, so that $s = 4$ and $t = 1$, and $j = 1$, corresponding to the entry $\sigma_{15}(0)\sigma_1(1)$ in the table above. Since $\sigma_1(1) = u_1$ and $\sigma_{15}(0) = \sum_{a+b=5} u_1^{(3a)}u_2^{(3b)}$, this product is $u_1u_2^{(15)} + u_1^{(7)}u_2^{(9)} + u_1^{(13)}u_2^{(3)} = u_1u_2f$ where $f = u_2^{(14)} + u_1^{(6)}u_2^{(8)} + u_1^{(12)}u_2^{(2)}$.

Comparing Example 28.1.11, we see that f is obtained from $\sigma_7(2)$ by doubling exponents. In terms of blocks, this corresponds to shifting the blocks in $\sigma_7(2)$ one column to the right, thus doubling the cyclic weight.

28.5 Calculation of $\widetilde{\mathrm{K}}(3)$

In this section we give an additive basis for $\widetilde{\mathrm{K}}(3) = \sum_{j=0}^{6} \widetilde{\mathrm{K}}(3,j)$. We give details only for $j = 0,1,3$, as the results for other cyclic weights follow by rotating the variables using the map ρ_*.

We have $\widetilde{\mathrm{K}}^d(3) = \widetilde{\mathrm{J}}^d(3)$ except in the degrees $d = (2^{u+3} - 1) + (2^{u+1} - 1) + (2^u - 1)$, $u \geq 0$, when $\widetilde{\mathrm{J}}^d(3)$ has codimension 1 in $\widetilde{\mathrm{K}}^d(3)$. The missing generators are given by $u_1^{(4)}u_2^{(3)}u_3$ and its images under iteration of $\widetilde{\kappa}_*$. The generators of $\widetilde{\mathrm{J}}(3)$ are the nonzero elements $\sigma_d(j)$ for $d = 2^r - 1$ and $0 \leq j \leq 6$. All products of length ≥ 4 in these generators are 0.

In all cases, we prove only upper bounds for $\widetilde{\mathrm{J}}^d(3)$. It follows from Proposition 28.1.15 that these bounds are exact.

Theorem 28.5.1 *For $\omega = (2,\ldots,2,1,\ldots,1)$ with t 2s and s 1s, the table below gives bases of $\widetilde{\mathrm{K}}^\omega(3,j)$ as products of the generators $\sigma_d(j)$ of $\widetilde{\mathrm{J}}(3)$. In the tables for $j = 1$ and $j = 3$, we use the short notation (i,j) to represent the product $\sigma_{d_1}(i)\sigma_{d_2}(j)$. These products are chosen so as to minimize $j \geq 0$. Note that $\sigma_1(1) = u_1$ and $\sigma_1(2) = u_2$.*

Basis for $\widetilde{\mathsf{K}}^{\omega}(3,0)$, $u=0$

$t\backslash s$	0	1	2	≥ 3
0	1	·	·	$\sigma_d(0)$
1	·	$\sigma_3(6)u_1$ $\sigma_3(5)u_2$	$\sigma_7(6)u_1$ $\sigma_7(5)u_2$ $u_1^{(4)}u_2^{(3)}u_3$	$\sigma_{d_1}(6)u_1$ $\sigma_{d_1}(5)u_2$
2	·	$\sigma_7(6)\sigma_3(1)$ $\sigma_7(5)\sigma_3(2)$	$\sigma_{15}(6)\sigma_3(1)$ $\sigma_{15}(5)\sigma_3(2)$ $\sigma_{15}(4)\sigma_3(3)$	$\sigma_{d_1}(6)\sigma_{d_2}(1)$ $\sigma_{d_1}(5)\sigma_{d_2}(2)$ $\sigma_{d_1}(4)\sigma_{d_2}(3)$
≥ 3	$\sigma_{d_1}(6)\sigma_{d_2}(1)$	$\sigma_{d_1}(6)\sigma_{d_2}(1)$ $\sigma_{d_1}(5)\sigma_{d_2}(2)$	$\sigma_{d_1}(6)\sigma_{d_2}(1)$ $\sigma_{d_1}(5)\sigma_{d_2}(2)$ $\sigma_{d_1}(4)\sigma_{d_2}(3)$	$\sigma_{d_1}(6)\sigma_{d_2}(1)$ $\sigma_{d_1}(5)\sigma_{d_2}(2)$ $\sigma_{d_1}(4)\sigma_{d_2}(3)$

Basis for $\widetilde{\mathsf{K}}^{\omega}(3,1)$, $u=0$

$t\backslash s$	0	1	≥ 2
0	·	u_1	$\sigma_d(1)$
1	·	$(6,2)$	$(0,1)$ $(6,2)$
≥ 2	$(6,2)$	$(0,1)$ $(6,2)$	$(0,1)$ $(6,2)$ $(5,3)$

Basis for $\widetilde{\mathsf{K}}^{\omega}(3,3)$, $u=0$

$t\backslash s$	0	1	≥ 2
0	·	·	$\sigma_d(3)$
1	u_1u_2	$(1,2)$	$(2,1)$ $(1,2)$
≥ 2	$(2,1)$	$(1,2)$ $(0,3)$	$(2,1)$ $(1,2)$ $(0,3)$

Proof As in the case $n=2$, when $\mu(d)\leq 2$ we write $d=d_1+d_2$ where $d_1=2^{s+t}-1$, $d_2=2^t-1$ and $s,t\geq 0$. Degrees with $\mu(d)=1$ appear in the $t=0$ row of the tables. In these degrees, there is one generator $\sigma_d(j)$ for each cyclic weight j, and the only relations are $\sigma_1(j)=0$ for $j=0,3,5,6$ in degree 1.

When $\mu(d)=2$, so that $t\geq 1$, the products $\sigma_{d_1}(i)\sigma_{d_2}(j)$, $i+j=k \bmod 7$ span $\widetilde{\mathsf{J}}^d(3,k)$ for each cyclic weight k, since products of even length >2 are 0. These elements satisfy the four linearly independent relations $\sigma_{d_1}(k)\sigma_{d_2}(0)=0$ and $\sigma_{d_1}(k-1)\sigma_{d_2}(1)+\sigma_{d_1}(k-2)\sigma_{d_2}(2)+\sigma_{d_1}(k-4)\sigma_{d_2}(4)=\sigma_{d_1}(k-3)\sigma_{d_2}(3)=$

$\sigma_{d_1}(k-6)\sigma_{d_2}(6) = \sigma_{d_1}(k-5)\sigma_{d_2}(5)$ of Example 28.2.8. When $s,t \geq 2$ there are no further relations, and we obtain the basis given in the tables for cyclic weights $k = 0, 1$ and 3.

For $s = 0$, $d_1 = d_2 = 2^t - 1$ and there are additional relations $\sigma_{d_2}(i)\sigma_{d_2}(j) = \sigma_{d_2}(j)\sigma_{d_2}(i)$. In particular, $\sigma_{d_2}(0)\sigma_{d_2}(k) = \sigma_{d_2}(k)\sigma_{d_2}(0) = 0$ for each cyclic weight k. In cyclic weight 0, it follows that all products $\sigma_{d_2}(-j)\sigma_{d_2}(j), j \neq 0$ mod 7 are equal. In cyclic weight 1, we obtain $\sigma_{d_1}(6)\sigma_{d_1}(2) = \sigma_{d_1}(2)\sigma_{d_1}(6) = \sigma_{d_1}(3)\sigma_{d_1}(5) = \sigma_{d_1}(5)\sigma_{d_1}(3)$ and $\sigma_{d_1}(4)\sigma_{d_1}(4) = 0$. In cyclic weight 3, we obtain $\sigma_{d_1}(4)\sigma_{d_1}(6) = \sigma_{d_1}(6)\sigma_{d_1}(4) = \sigma_{d_1}(5)\sigma_{d_1}(5) = 0$ and $\sigma_{d_1}(2)\sigma_{d_1}(1) = \sigma_{d_1}(1)\sigma_{d_1}(2)$.

For $t \geq 3$, the remaining element is nonzero for all cyclic weights, but for $t < 3$ it is easy to check that this is also 0 for the cyclic weights shown in the tables.

For $s = 1$, the relation of Proposition 28.2.2 holds in each cyclic weight. Choosing ζ such that $1 + \zeta + \zeta^3 = 0$, we obtain the relation

$$\sum_{i+j=k} (\zeta^{3i+j} + \zeta^j + \zeta^i)\sigma_{d_1}(i)\sigma_{d_2}(j) = 0.$$

In cyclic weight 0, this reduces to $(1+\zeta)\sigma_{d_1}(1)\sigma_{d_2}(6) + (1+\zeta^2)\sigma_{d_1}(2)\sigma_{d_2}(5) + (1 + \zeta + \zeta^2)\sigma_{d_1}(4)\sigma_{d_2}(3) = 0$, giving a new relation $\sigma_{d_1}(1)\sigma_{d_2}(6) + \sigma_{d_1}(2)\sigma_{d_2}(5) + \sigma_{d_1}(4)\sigma_{d_2}(3) = 0$ on equating the coefficient of 1 to zero. Combining this with the relations which hold for all s and t, it follows that $\sigma_{d_1}(6)\sigma_{d_2}(1) + \sigma_{d_1}(5)\sigma_{d_2}(2) + \sigma_{d_1}(3)\sigma_{d_2}(4) = \sigma_{d_1}(1)\sigma_{d_2}(6) = \sigma_{d_1}(2)\sigma_{d_2}(5) = \sigma_{d_1}(4)\sigma_{d_2}(3) = 0$.

Similar calculations lead to the relations $\sigma_{d_1}(6)\sigma_{d_2}(2) = \sigma_{d_1}(4)\sigma_{d_2}(4)$ and $\sigma_{d_1}(0)\sigma_{d_2}(1) = \sigma_{d_1}(2)\sigma_{d_2}(6) = \sigma_{d_1}(3)\sigma_{d_2}(5) = \sigma_{d_1}(5)\sigma_{d_2}(3)$ in cyclic weight 1, and the relations $\sigma_{d_1}(2)\sigma_{d_2}(1) = 0$, $\sigma_{d_1}(1)\sigma_{d_2}(2) + \sigma_{d_1}(6)\sigma_{d_2}(4) = \sigma_{d_1}(0)\sigma_{d_2}(3) = \sigma_{d_1}(4)\sigma_{d_2}(6) = \sigma_{d_1}(5)\sigma_{d_2}(5)$ in cyclic weight 3.

This leads to the results tabulated for $s = 1$, $t \geq 2$. For $t = 1$, $d = 4$ and so $\sigma_3(i)\sigma_1(j) = 0$ unless $j = 1,2$ or 4. This does not add a new relation in cyclic weight 0, but in cyclic weight 1 it follows that $\sigma_3(5)\sigma_1(3) = 0$, and in cyclic weight 3 it follows that $\sigma_3(1)\sigma_1(2) = \sigma_3(6)\sigma_1(4)$. For $t = 0$, $d = 1$, and no comment is needed. For $s \geq 2$ and $t = 1$, there are additional relations $\sigma_{d_1}(i)\sigma_{d_2}(j) = 0$ for $j \neq 1,2,4$. It follows that $\sigma_{d_1}(k-1)\sigma_{d_2}(1) + \sigma_{d_1}(k-2)\sigma_{d_2}(2) + \sigma_{d_1}(k-4)\sigma_{d_2}(4) = 0$, giving the results tabulated. □

Theorem 28.5.2 *For $\omega = (3,\ldots,3,2,\ldots,2,1,\ldots,1)$ with $u > 0$ 3s, t 2s and s 1s, we tabulate basis elements of $\widetilde{\mathrm{K}}^\omega(3,j)$ as products in $\widetilde{\mathrm{J}}(3)$, using the short notation (i,j,k) for $\sigma_{d_1}(i)\sigma_{d_2}(j)\sigma_{d_3}(k)$.*

Basis for $\widetilde{\mathsf{K}}^\omega(3,0)$, $u > 0$

$t\backslash s$	0	1	2	≥ 3
0	$(4,2,1)$	$\cdot$	$\cdot$	$(4,2,1)$
1	$\cdot$	$(5,1,1)$ $(3,3,1)$	$(5,1,1)$ $(3,3,1)$ $\widetilde{\upsilon}_*^u(u_1^{(4)}u_2^{(3)}u_3)$	$(5,1,1)$ $(3,3,1)$
2	$\cdot$	$(5,1,1)$ $(3,3,1)$	$(5,1,1)$ $(4,2,1)$ $(3,3,1)$	$(5,1,1)$ $(4,2,1)$ $(3,3,1)$
≥ 3	$(4,2,1)$	$(5,1,1)$ $(3,3,1)$	$(5,1,1)$ $(4,2,1)$ $(3,3,1)$	$(5,1,1)$ $(4,2,1)$ $(3,3,1)$

Basis for $\widetilde{\mathsf{K}}^\omega(3,1)$, $u > 0$

$t\backslash s$	0	1	≥ 2
0	$\cdot$	$(5,2,1)$	$(5,2,1)$
1	$\cdot$	$(6,1,1)$	$(6,1,1)$ $(4,3,1)$
≥ 2	$(5,2,1)$	$(6,1,1)$ $(5,2,1)$	$(6,1,1)$ $(5,2,1)$ $(4,3,1)$

Basis for $\widetilde{\mathsf{K}}^\omega(3,3)$, $u > 0$

$t\backslash s$	0	1	≥ 2
0	$\cdot$	$\cdot$	$(0,2,1)$
1	$(6,3,1)$	$(1,1,1)$	$(1,1,1)$ $(6,3,1)$
≥ 2	$(6,3,1)$	$(1,1,1)$ $(0,2,1)$	$(1,1,1)$ $(0,2,1)$ $(6,3,1)$

Proof The elements listed in the tables for $u > 0$ are calculated using the Kameko map $\widetilde{\upsilon}_* = \widetilde{\kappa}_*^{-1}$ and Proposition 28.3.6. Proposition 28.3.5 shows that the table is independent of u for $u > 0$. For example, the entry for $s \geq 2$, $t \geq 2$, $u = 1$ in $\widetilde{\mathsf{K}}(3,k)$ is calculated from the entry for $s \geq 2$, $t \geq 2$, $u = 0$ in $\widetilde{\mathsf{K}}(3,k)$ as follows.

$$\widetilde{\upsilon}_*(\sigma_{31}(k-2)\sigma_7(1)\sigma_1(1)) = \begin{vmatrix} \sigma_{15}(k-3) & \sigma_{15}(k-4) & \sigma_{15}(k-6) \\ \sigma_3(0) & \sigma_3(6) & \sigma_3(4) \\ 1 & 0 & 0 \end{vmatrix}$$

$$= \sigma_{15}(k-4)\sigma_3(4) + \sigma_{15}(k-6)\sigma_3(6).$$

Similarly we have $\widetilde{\upsilon}_*(\sigma_{31}(k-3)\sigma_7(2)\sigma_1(1)) = \sigma_{15}(k-5)\sigma_3(5)$ and $\widetilde{\upsilon}_*(\sigma_{31}(k-4)\sigma_7(3)\sigma_1(1)) = \sigma_{15}(k-6)\sigma_3(6) + \sigma_{15}(k-1)\sigma_3(1)$. Using the basis for the corresponding case with $u=0$, it follows that the elements listed give a basis for the case $u=1$. □

28.6 Remarks

Like Chapter 27, this chapter is partly based on [224], where linear independence of the generators of $\widetilde{\mathsf{K}}^d(3)$ is proved by a direct argument, rather than by duality.

29
The 4-variable hit problem, I

29.0 Introduction

In this chapter we deal with some special cases of the hit problem for 4 variables. A new phenomenon arises: whereas $\mathsf{Q}^{\omega}(n) = 0$ for all non-decreasing ω-sequences ω when $n \leq 3$ (see Chapter 1 and Proposition 8.3.1), this is no longer true for $n \geq 4$. Using the left order, $\mathsf{Q}^{\omega}(4) \neq 0$ when $\omega = (1,3)$, $(2,3)$ or $(2,3,2)$, and the dimension of $\mathsf{Q}^{\omega}(4)$ in these cases is 1, 4 and 4 respectively. The fact that $\mathsf{Q}^{(1,3)}(4) \neq 0$ was shown in Example 8.3.2, and the three cases are discussed in Section 29.1. We prove in Section 29.2 that when $n = 4$ there are no more exceptions besides these three and their iterates under the up Kameko map.

The other main topic of this chapter is the case $\omega = (2,\ldots,2)$ of the 4-variable hit problem, which is treated in Section 29.3. We show that $\dim \mathsf{Q}^{\omega}(4)$ is 6, 20 or 35 according as the length t of ω is 1, 2 or ≥ 3. The case $t = 3$ is exceptional, as, although the dimension of $\mathsf{Q}^{\omega}(4)$ is the same as for $t \geq 4$, it is not isomorphic to the partial flag module $\mathsf{FL}^2(4)$. The key result here is Proposition 29.3.4, which is an analogue of results obtained in Chapter 8 for blocks with 'large head' or 'long tail'. In Section 29.3 we show that the polynomial $x_1^6x_2^6x_3x_4 + x_1x_2x_3^6x_4^6$ is $\mathsf{GL}(4)$-invariant mod hits, and so generates a 1-dimensional submodule of $\mathsf{Q}^{(2,2,2)}(4)$.

29.1 $\mathsf{Q}^d(4)$ and $\mathsf{K}^d(4)$ for $d = 7$, 8 and 16

When ω is a decreasing sequence, $\mathsf{P}^{\omega}(n)$ contains a spike, and hence $\mathsf{Q}^{\omega}(n) \neq 0$. It follows from the results of Chapter 1 that the converse is true when $n = 1$ or 2, and we show in Proposition 8.3.1 that is also true when $n = 3$. In this section we discuss the three basic cases where this simplification fails for $n = 4$. These occur in degrees $d = 7, 8$ and 16, and by iteration under the local Kameko isomorphism each of them gives rise to an infinite sequence of similar examples, where non-decreasing sequences contribute in an essential way to the cohit module $\mathsf{Q}^d(n)$ for any choice of ordering on ω-sequences. These examples of course persist when $n > 4$, presumably along with many other cases of the same kind which appear as the number of variables increases.

Example 29.1.1 We show that $\dim \mathsf{Q}^7(4) = 35$. As in Example 5.3.5, the 2-dominance order on 4-bounded sequences of 2-degree 7 is shown below.

$$\begin{array}{ccc} & (3,2) & \\ \swarrow & & \searrow \\ (1,3) & & (3,0,1) \\ \searrow & & \swarrow \\ & (1,1,1) & \end{array} \tag{29.1}$$

First we consider $\mathsf{P}^{(3,2)}(4)$, which has dimension 24 and a basis of 12 spikes given by permuting the variables in $x_1^3x_2^3x_3$ and 12 monomials given by permuting the variables in $x_1^3x_2^2x_3x_4$.

The hit equation $Sq^1(x_1^3x_2x_3x_4) = x_1^4x_2x_3x_4 + x_1^3x_2x_3x_4(x_2 + x_3 + x_4)$ shows that $x_1^3x_2x_3x_4(x_2 + x_3 + x_4)$ is left and right reducible, so permutation of variables gives four hit polynomials. No other hit equation involves $\mathsf{P}^{(3,2)}(4)$, so $\dim \mathsf{Q}^{(3,2)}(4) = 12 + 8 = 20$.

Since no monomial in $\mathsf{P}^{(1,1,1)}(4)$ can involve all four variables, we can immediately reduce to the 3-variable case, and if three variables are involved then the relevant monomials are the permutations of $x_1^4x_2^2x_3$, which are all equivalent modulo hits by 1-back splicing. Since specialization to the 1-variable case implies that a monomial of degree 7 is not hit, $\dim \mathsf{Q}^{(1,1,1)}(\mathsf{Z}[3]) = 1$. Combining this with the results for $n = 1$ and 2 and using Proposition 1.4.7, we obtain $\dim \mathsf{Q}^{(1,1,1)}(4) = 4 \cdot 1 + 6 \cdot 1 + 4 \cdot 1 = 14$.

It remains to consider $\omega = (3,0,1)$ or $(1,3)$. Note that $(3,0,1) >_l (1,3)$ but $(1,3) >_r (3,0,1)$, so the definition of $\mathsf{Q}^\omega(4)$ in these cases depends on the choice of left or right order. Since $\mathsf{Q}^\omega(3) = 0$ in both cases, we need consider only $\mathsf{P}^\omega(\mathsf{Z}[4])$, which is spanned by the permutations of $x_1^4x_2x_3x_4$ and $x_1^2x_2^2x_3^2x_4$. The hit equations $Sq^1(x_1^2x_2^2x_3x_4) = x_1^2x_2^2(x_3^2x_4 + x_3x_4^2)$ and $Sq^2(x_1^2x_2x_3x_4) = x_1^4x_2x_3x_4 + x_1^2(x_2^2x_3^2x_4 + x_2^2x_3x_4^2 + x_2x_3^2x_4^2)$ and their permutations show that these eight monomials are equivalent modulo hits.

Since hit polynomials $f \in \mathsf{P}^7(4)$ are of the form $f = Sq^1(h_1) + Sq^2(h_2)$ and we need consider only $\mathsf{P}^\omega(\mathsf{Z}[4])$, there are no more hit equations which can be used to reduce these eight monomials. Hence $x_1^2x_2^2x_3^2x_4$ is not left reducible, and $x_1^4x_2x_3x_4$ is not right reducible. We conclude that $\dim \mathsf{Q}^{(1,3)}(4) = 1$ for $\leq_l$ and $\dim \mathsf{Q}^{(3,0,1)}(4) = 1$ for $\leq_r$, and so $\dim \mathsf{Q}^7(4) = 20+1+14 = 35$.

Example 29.1.2 We consider the dual of Example 29.1.1. The 2-dominance order on 4-bounded sequences of 2-degree 7 is shown in (29.1). First consider $\mathsf{DP}^{(3,2)}(4)$. This has dimension 24 and is spanned by 12 d-spikes, which are permutations of $v_1^{(3)}v_2^{(3)}v_3$, and 12 d-monomials which are permutations of $v_1^{(3)}v_2v_3v_4^{(2)}$. Pairs such as $v_1^{(3)}v_2v_3v_4^{(2)} + v_1^{(3)}v_2^{(2)}v_3v_4$ give 8 further linearly independent elements of $\mathsf{K}^{(3,2)}(4)$, which has dimension 20. By Theorem 20.5.1, this module is isomorphic to the irreducible $\mathbb{F}_2\mathsf{GL}(4)$-module $L(2,2,1)$. For the tail sequence $(1,1,1)$ we obtain $\mathsf{K}^{(1,1,1)}(4) \cong \mathsf{FL}_1(4)$, of dimension 14.

As for $\mathsf{Q}^7(4)$, $\mathsf{K}^7(4)$ has a further contribution of dimension 1 from $\omega = (1,3)$ or $\omega = (3,0,1)$ depending on the choice of left or right order. This is given by $f = \sigma_4(v_1^{(4)}v_2v_3v_4) + \sigma_4(v_1^{(2)}v_2^{(2)}v_3^{(2)}v_4) + \sigma_4(v_1^{(3)}v_2v_3v_4^{(2)})$, where σ_4 denotes symmetrization, so that f has $4+4+12=20$ terms. It suffices to check that $Sq_1(f) = 0$ and $Sq_2(f) = 0$. The $\leq_l$-minimal ω-sequence of the terms of f is $(1,3)$, so $f \in \mathsf{K}^{(1,3)}(4)$ for the left order, and similarly $f \in \mathsf{K}^{(3,0,1)}(4)$ for the right order. Thus $\dim \mathsf{K}^7(4) = 20+1+14 = 35$.

The d-polynomial f is the sum of all d-monomials in $\mathsf{DP}^7(Z[4])$. Hence $f = \sum_v v^{(7)}$, where the sum is over all $v \in \mathsf{DP}^1(4)$, and so $f \in \mathsf{J}^7(4)$. Thus the Crabb–Hubbuck map $\phi^{(1,1,1)}$ (Definition 23.1.2) maps the summand $\mathsf{FL}^1(4)$ of $\mathsf{FL}(4)$ isomorphically on to the submodule of $\mathsf{J}^7(4)$ spanned by the elements $v^{(7)}$, $v \neq 0 \in \mathsf{DP}^1(4)$. Hence $\mathsf{J}^7(4) = \mathsf{K}^7(4) \cong \mathsf{FL}^1(4) \oplus L(2,2,1)$.

Example 29.1.3 Omitting sequences starting with 0, the 2-dominance order on 4-bounded sequences of 2-degree 8 is shown below.

$$\begin{array}{ccc} & (4,2) & \\ \swarrow & & \searrow \\ (2,3) & & (4,0,1) \\ \searrow & & \swarrow \\ & (2,1,1) & \end{array} \tag{29.2}$$

Since $\mathsf{P}^{(4,2)}(4)$ has dimension 6 and a basis of spikes, $\dim \mathsf{Q}^{(4,2)}(4) = 6$. Since $\dim \mathsf{Q}^{(2,1,1)}(2) = 3$ and $\dim \mathsf{Q}^{(2,1,1)}(3) = 15 = 3 \cdot 3 + 6$, 2-variable monomials contribute $3 \cdot 6 = 18$ and 3-variable monomials contribute $6 \cdot 4 = 24$ to $\dim \mathsf{Q}^{(2,1,1)}(4)$. There are 12 4-variable monomials, which are permutations of $x_1x_2x_3^2x_4^4$. These reduce to 3 linearly independent monomials $x_1x_2x_3^2x_4^4$,

$x_1x_2^2x_3x_4^4$, $x_1x_2^2x_3^4x_4$ mod hits by 1-back splicing, and so $\dim \mathsf{Q}^{(2,1,1)}(4) = 18 + 24 + 3 = 45$.

For the left order, $\dim \mathsf{Q}^{(4,0,1)}(4) = \dim \mathsf{Q}^{(0,1)}(4) = 0$ using the local Kameko map. Since $\mathsf{Q}^{(2,3)}(3) = 0$, we need only consider the 12 blocks in $\mathsf{P}^{(2,3)}(4)$ which are row permutations of $x_1^3x_2x_3^2x_4^2$. Permutations of rows 2, 3, 4 give equivalent blocks mod $\mathsf{P}^{(2,1,1)}(4)$, and it follows that $x_1^3x_2x_3^2x_4^2 \sim x_1^5x_2x_3x_4$ mod $\mathsf{P}^{(2,1,1)}(4)$. However, this raises B in the left order. No further reduction is possible, and we obtain $\dim \mathsf{Q}^{(2,3)}(4) = 4$, with a spanning set given by cyclic row permutations of B. From the equivalence above, it is clear that $\mathsf{Q}^{(2,3)}(4) \cong L(1)$. We conclude that $\dim \mathsf{Q}^8(4) = 6 + 4 + 45 = 55$. For the right order, $\dim \mathsf{Q}^{(2,3)}(4) = 0$ and $\dim \mathsf{Q}^{(4,0,1)}(4) = 4$. Note that the result $\dim \mathsf{Q}^{(4,0,1)}(4) = 4$ shows that Proposition 6.5.4 fails for the right order.

Example 29.1.4 Omitting sequences starting with 0, the 2-dominance order on 4-bounded sequences of 2-degree 8 is shown in (29.2). As for $\mathsf{Q}^{(2,1,1)}(4)$, using the results for $n \leq 3$ we need only show that $\dim \mathsf{K}^{(2,1,1)}(Z[4]) = 3$. Let $f_7 = \sum_v v^{(7)} \in \mathsf{J}^7(3)$, where the sum is over all $v \in \mathsf{DP}^1(3)$. Then $f_7v_4 \in \mathsf{J}^8(4)$ is the sum of 6 terms $\sigma_3(v_1^{(4)}v_2^{(2)}v_3)v_4$ in $\mathsf{DP}^{(2,1,1)}(4)$ and terms with higher ω-sequences.

The four cyclic permutations of f_7v_4 are linearly independent in $\mathsf{K}^8(4)$, but their sum is $g = \sigma_4(v_1v_2^{(2)}v_3^{(2)}v_4^{(3)}) + \sigma_4(v_1v_2v_3v_4^{(5)})$, and all the terms of g have ω-sequence $(2,3)$ or $(4,0,1)$, and so the four cyclic permutations of f_7v_4 span only a 3-dimensional subspace of $\mathsf{K}^{(2,1,1)}(4)$. Hence $\dim \mathsf{K}^{(2,1,1)}(4) = 45$. It was shown in Section 24.4 that $\dim \mathsf{J}^{(2,1,1)}(4) \geq 41$ since $\mathsf{FL}_{1,2} \downarrow L(2,1,1) = 35$, with composition factors $L(2,1)$, $L(0)$ and $L(2,1,1)$, and $\mathsf{FL}_2 \downarrow L(1,1) \cong L(1,1)$ has dimension 6. We show in Section 30.3 that $\mathsf{K}^{(2,1,1)}(4)/\mathsf{J}^{(2,1,1)}(4) \cong L(1,1,1)$ has dimension 4.

Let $h = v_1v_2v_3v_4^{(5)} + v_1v_2^{(2)}v_3^{(2)}v_4^{(3)} + v_1^{(2)}v_2v_3^{(2)}v_4^{(3)} + v_1^{(2)}v_2^{(2)}v_3v_4^{(3)}$. Then $Sq_1(h) = 0$ and $Sq_2(h) = 0$, so $h \in \mathsf{K}^8(4)$. The d-polynomial g above is the sum of the cyclic row permutations of h, which span a subspace of $\mathsf{K}^8(4)$ of dimension 4. For the left order this is $\mathsf{K}^{(2,3)}(4)$, and for the right order it is $\mathsf{K}^{(4,0,1)}(4)$. We observe that $h \in \mathsf{J}^8(4)$, since the product of the d-polynomial $f = \sum_v v^{(7)}$ of Example 29.1.2 by v_4 is the sum of h and three d-spikes.

The calculation of $\mathsf{K}^8(4)$ is completed by including the six d-spikes which span $\mathsf{P}^{(4,2)}(4)$. We conclude that $\dim \mathsf{K}^8(4) = 45 + 4 + 6 = 55$.

Example 29.1.5 We show that $\dim \mathsf{Q}^{16}(4) = 73$. For the left order, we show that $\mathsf{Q}^{16}(4)$ has filtration quotients $\mathsf{Q}^\omega(4)$ for $\omega = (4,4,1)$, $(4,2,2)$, $(2,3,2)$ and $(2,1,1,1)$, of dimension 4, 20, 4 and 45 respectively. The 2-dominance order on ω-sequences for $\mathsf{P}^{16}(4)$ with $\omega_1 > 0$ is shown below.

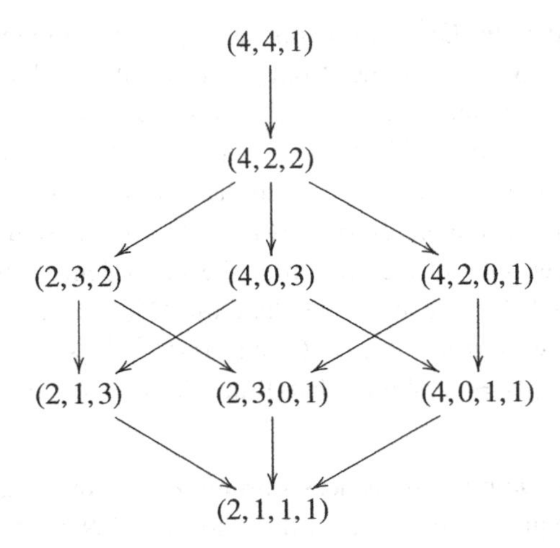

The cases $(4,4,1)$ and $(4,2,2)$ reduce to (1) and $(2,2)$ by the local Kameko isomorphism, and $\omega = (2,2)$ is treated in Section 29.3. Similarly $\mathsf{Q}^{\omega}(4) = 0$ for the left order for $\omega = (4,0,3)$, $(4,2,0,1)$ and $(4,0,1,1)$. Since $\dim \mathsf{Q}^{(2,1,1,1)}(2) = 3$ and $\dim \mathsf{Q}^{(2,1,1,1)}(3) = 14 = 3 \cdot 3 + 5$, monomials not involving all four variables contribute a subspace of dimension $6 \cdot 3 + 4 \cdot 5 = 38$ to $\mathsf{Q}^{(2,1,1,1)}(4)$.

By multiple 1-back splicing, all blocks in $\mathsf{P}^{(2,1,1,1)}(Z[4])$ can be expressed as formal sums of the 12 semi-standard blocks (Definition 7.3.1) shown below.

$A_1 = \begin{array}{l} 1\,1 \\ 0\,0\,1 \\ 0\,0\,0\,1 \\ 1 \end{array}$	$B_1 = \begin{array}{l} 1 \\ 0\,1\,1 \\ 0\,0\,0\,1 \\ 1 \end{array}$	$C_1 = \begin{array}{l} 1 \\ 0\,1 \\ 0\,0\,1\,1 \\ 1 \end{array}$	$D_1 = \begin{array}{l} 1 \\ 0\,1 \\ 1 \\ 0\,0\,1\,1 \end{array}$
$A_2 = \begin{array}{l} 1\,1 \\ 0\,0\,1 \\ 1 \\ 0\,0\,0\,1 \end{array}$	$B_2 = \begin{array}{l} 1 \\ 0\,1\,1 \\ 1 \\ 0\,0\,0\,1 \end{array}$	$C_2 = \begin{array}{l} 1 \\ 1 \\ 0\,1\,1 \\ 0\,0\,0\,1 \end{array}$	$D_2 = \begin{array}{l} 1 \\ 1 \\ 0\,1 \\ 0\,0\,1\,1 \end{array}$
$A_3 = \begin{array}{l} 1\,1 \\ 1 \\ 0\,0\,1 \\ 0\,0\,0\,1 \end{array}$	$B_3 = \begin{array}{l} 1 \\ 1\,1 \\ 0\,0\,1 \\ 0\,0\,0\,1 \end{array}$	$C_3 = \begin{array}{l} 1 \\ 0\,1 \\ 1\,0\,1 \\ 0\,0\,0\,1 \end{array}$	$D_3 = \begin{array}{l} 1 \\ 0\,1 \\ 0\,0\,1 \\ 1\,0\,0\,1 \end{array}$

By splicing in columns 2,3,4 as for $\mathsf{Q}^7(2)$, we obtain $B_1 \sim C_1$, $B_2 \sim D_1$ and $C_2 \sim D_2$. There are further linearly independent relations

$$(1)\ A_1 + A_2 \sim B_1 + B_2, \quad (2)\ A_1 + A_3 \sim B_1 + C_2,$$

and so $\dim \mathsf{Q}^{(2,1,1,1)}(Z[4]) \leq 7$. Relation (2) is obtained by switching rows 2 and 3 in relation (1) and reducing to semi-standard form. To prove (1), we first show

$$X = \begin{array}{l} 1\,1 \\ 1\,1 \\ 0\,0\,0\,1 \\ 0\,1 \end{array} \sim \begin{array}{l} 1\,1 \\ 1\,1 \\ 0\,1 \\ 0\,0\,0\,1 \end{array} = Y. \tag{29.3}$$

By splicing at $(3,3)$,

$$X \sim \begin{array}{l} 1\,0\,1 \\ 1\,0\,1 \\ 0\,0\,1 \\ 0\,1 \end{array} + \begin{array}{l} 1\,0\,1 \\ 1\,1 \\ 0\,0\,1 \\ 0\,0\,1 \end{array} + \begin{array}{l} 1\,1 \\ 1\,0\,1 \\ 0\,0\,1 \\ 0\,0\,1 \end{array}.$$

The second and third blocks are equivalent modulo hits by 1-back splicing, and the first block is equivalent to Y by the χ-trick $fSq^6(g) \sim Xq^6(f)g$, where $f = x_1x_2x_4^2$ and $g = x_1^2x_2^2x_3^2$.

The proof of (1) is completed by splicing A_1 at position $(2,1)$, to obtain $A_1 \sim X + A_1'$ where $A_1' = \begin{array}{l} 0\,0\,1 \\ 1\,1 \\ 0\,0\,0\,1 \\ 1 \end{array}$. Further splicing operations reduce A_1' to the sum $B_1 + B_3$ of semi-standard blocks, so $A_1 \sim X + B_1 + B_3$. Switching rows 3 and 4, the same argument gives $A_2 \sim Y + B_2 + B_3$. Adding these relations and using (29.3) gives (1).

This completes the proof that $\dim \mathsf{Q}^{(2,1,1,1)}(4) \leq 45$. Since $\dim \mathsf{J}^{(2,1,1,1)}(4) = 45$ by the results of Section 24.4, we conclude that $\dim \mathsf{Q}^{(2,1,1,1)}(4) = 45$.

The lower bound $\dim \mathsf{Q}^{(2,1,1,1)}(4) \geq 45$ can alternatively be established by showing that the 7 blocks A_1, A_2, A_3, B_1, B_3, C_3, D_3 are linearly independent modulo hits. We specialize by setting $x_2 = x_1$, and recall from Section 8.2 that $\mathsf{Q}^{(2,1,1,1)}(3)$ has dimension 14, with a basis given by special blocks (see Definition 8.2.2).

On specialization, A_1 and B_1 map to the same special block $\begin{array}{l} 1\,1\,1 \\ 0\,0\,0\,1 \\ 1 \end{array}$, while A_2, C_3, D_3 map to different special blocks and A_3, B_3 map to hit elements. Hence A_1 and B_1 have the same coefficient in any linear relation in $\mathsf{Q}^{(2,1,1,1)}(4)$, and A_2, C_3, D_3 have coefficient 0. On setting $x_4 = x_1$, A_1 and B_1 specialize to hit elements, while A_3 and B_3 map to different special blocks, so A_3 and B_3 also cannot appear in a relation. It remains to show that $A_1 + B_1$ is not hit. This follows on setting $x_4 = x_3 = x_2$, since B_1 specializes to a spike and A_1 does not.

It remains to consider $\omega = (2,3,0,1)$, $(2,1,3)$ and $(2,3,2)$. Blocks with $\omega = (2,3,0,1)$ or $(2,1,3)$ are left reducible to blocks B with $\omega(B) = (2,1,1,1)$ by 1-back splicing, together with multiple splicing in column 1 for blocks which are row permutations of $\begin{smallmatrix}1&0&0&1\\1&1&0&0\\0&1&0&0\\0&1&0&0\end{smallmatrix}$. Hence $\mathsf{Q}^{(2,3,0,1)}(4) = 0$ and $\mathsf{Q}^{(2,1,3)}(4) = 0$.

Finally we show that $\dim \mathsf{Q}^{(2,3,2)}(4) \le 4$. All blocks with $\omega = (2,3,2)$ are row permutations of one of the 6 blocks

$$A = \begin{matrix}0\,1\,0\\0\,1\,1\\1\,0\,1\\1\,1\,0\end{matrix},\; B = \begin{matrix}1\,1\,0\\1\,1\,0\\0\,1\,1\\0\,0\,1\end{matrix},\; C = \begin{matrix}1\,0\,0\\0\,1\,1\\0\,1\,1\\1\,1\,0\end{matrix},\; D = \begin{matrix}1\,1\,1\\1\,1\,0\\0\,0\,1\\0\,1\,0\end{matrix},\; E = \begin{matrix}1\,1\,1\\0\,1\,1\\1\,0\,0\\0\,1\,0\end{matrix},\; F = \begin{matrix}1\,1\,1\\1\,0\,1\\0\,1\,0\\0\,1\,0\end{matrix}.$$

Block D is reducible to $D_1 + D_2$ where $\omega(D_1) = (2,1,1,1)$ and $\omega(D_2) = (2,1,3)$ by 1-back splicing at $(4,2)$, and hence to a sum of blocks in $\mathsf{P}^{(2,1,1,1)}(4)$. Blocks E and F are reducible to blocks with $\omega = (2,1,1,1)$ by multiple 1-back splicing in column 1. Block B is reducible to the sum of two blocks with $\omega = (2,1,3)$ by 1-back splicing at $(3,1)$, and hence to a sum of blocks in $\mathsf{P}^{(2,1,1,1)}(4)$. By 1-back splicing at $(3,1)$, block C is reducible to the sum of block A and a block with $\omega = (2,1,3)$. Thus $\mathsf{Q}^{(2,3,2)}(4)$ is spanned by the row permutations of block A.

By 1-back splicing at $(2,1)$, block A is reducible to the sum of A' and a block with $\omega = (2,1,3)$, where A' is obtained by switching rows 2 and 3 of A. By 1-back splicing at $(3,2)$, block A is reducible to $A'' + B' + G$ where A'' is obtained by switching rows 3 and 4 of A, B' is a row permutation of B and so is reducible and $\omega(G) = (2,3,0,1)$. Thus all row permutations of A with the same α-sequence are equivalent modulo left reducible blocks, and it follows that $\mathsf{Q}^{(2,3,2)}(4)$ is spanned by the 4 cyclic row permutations of block A. We show in Example 29.1.6 that $\dim \mathsf{Q}^{(2,3,2)}(4) = 4$.

Example 29.1.6 For the left order, $\mathsf{K}^{16}(4)$ has filtration quotients $\mathsf{K}^{(4,4,1)}(4)$, $\mathsf{K}^{(4,2,2)}(4)$, $\mathsf{K}^{(2,3,2)}(4)$ and $\mathsf{K}^{(2,1,1,1)}(4)$ of dimension 4, 20, 4 and 45 respectively. We prove that $\dim \mathsf{K}^{(2,3,2)}(4) \ge 4$, as the rest of the statement follows by duality from Example 29.1.5.

We construct 4 linearly independent 16-term d-polynomials in $\mathsf{K}^{(2,3,2)}(4)$. Let $f_{16} = \sum_{i=1}^{5} g_i$ where $g_1 = v_1^{(2)}\sigma_3(v_2^{(6)}v_3^{(5)}v_4^{(3)})$, $g_2 = v_1\sigma_3(v_2^{(6)}v_3^{(6)}v_4^{(3)})$, $g_3 = v_1\sigma_3(v_2^{(9)}v_3^{(3)}v_4^{(3)})$, $g_4 = v_1v_2^{(5)}v_3^{(5)}v_4^{(5)}$ and $g_5 = v_1(v_2^{(3)}v_3^{(5)}v_4^{(7)} + v_2^{(5)}v_3^{(7)}v_4^{(3)} + v_2^{(7)}v_3^{(3)}v_4^{(5)})$.

Then $Sq_1(g_1) = Sq_1(g_2) = v_1\sigma_3(v_2^{(6)}v_3^{(5)}v_4^{(3)})$, $Sq_1(g_3) = Sq_1(g_4) = Sq_1(g_5) = 0$, so that $Sq_1(f_{16}) = 0$; $Sq_2(g_1) = 0$, $Sq_2(g_2) = Sq_2(g_4) = v_1\sigma_3(v_2^{(5)}v_3^{(5)}v_4^{(3)})$, and $Sq_2(g_3) = Sq_2(g_5) = v_1\sigma_3(v_2^{(3)}v_3^{(3)}v_4^{(7)})$, so that $Sq_2(f_{16}) = 0$; and $Sq_4(g_1) = Sq_4(g_2) = Sq_2(g_5) = 0$ and $Sq_4(g_4) = Sq_4(g_3) = v_1\sigma_3(v_2^{(5)}v_3^{(3)}v_4^{(3)})$, so that $Sq_4(f_{16}) = 0$. (We can ignore Sq_8.) It follows that $f_{16} \in \mathsf{K}^{16}(4)$, and by cyclic permutation of the rows we obtain 4 linearly independent elements of $\mathsf{K}^{16}(4)$.

Since $(2,3,2) <_l (4,0,3) <_l (4,2,0,1) <_l (4,2,2)$, these span a subspace of dimension 4 in $\mathsf{K}^{(2,3,2)}(4)$ for the left order. Hence $\dim \mathsf{K}^{(2,3,2)}(4) \geq 4$. As $\dim \mathsf{Q}^{(2,3,2)}(4) \leq 4$ by Example 29.1.5, and $\dim \mathsf{K}^{(2,3,2)}(4) = \dim \mathsf{Q}^{(2,3,2)}(4)$, it follows that $\dim \mathsf{K}^{(2,3,2)}(4) = 4$. For the right order, $(4,2,0,1) <_r (4,0,3) <_r (2,3,2) <_r (4,2,2)$, and so these elements would be assigned to $\mathsf{K}^{(4,2,0,1)}(4)$ for the right order.

Since $\mathsf{K}^{(4,2,2)}(4) \cong \mathsf{K}^{(2,2)}(4)$ for the left order by the local Kameko isomorphism, the case $\omega = (4,2,2)$ reduces to showing that $\dim \mathsf{K}^{(2,2)}(4) = 20$. Since $\dim \mathsf{K}^{(2,2)}(2) = 1$ and $\dim \mathsf{K}^{(2,2)}(3) = 6 = 3\cdot 1 + 3$, this reduces to showing that $\dim \mathsf{K}^{(2,2)}(Z[4]) = 20 - 6\cdot 1 - 4\cdot 3 = 2$. Let $u = v_1v_2v_3^{(2)}v_4^{(2)} + v_1^{(2)}v_2^{(2)}v_3v_4$. Then $Sq_1(u) = v_1v_2v_3v_4(v_1+v_2+v_3+v_4)$ and $Sq_2(u) = 0$, so the elements obtained by adding any two of the three distinct row permutations of u are in $\mathsf{K}(4)$.

It has been shown in Section 24.4 that $\dim \mathsf{J}^{(2,1,1,1)}(4) \geq 45$. Since $\mathsf{Q}^{(2,1,1,1)}(4)$ has dimension 45 by Example 29.1.5, $\mathsf{K}^{(2,1,1,1)}(4) = \mathsf{J}^{(2,1,1,1)}(4)$ has dimension 45. A basis for the 7-dimensional subspace dual to $\mathsf{Q}^{(2,1,1,1)}(Z[4])$ is given by the terms in $\mathsf{DP}^{16}(Z[4])$ of the four permutations of $(v_1+v_2+v_3)^{(15)}v_4$ and any three of the four permutations of $((v_1+v_2+v_3+v_4)^{(15)} + (v_1+v_2+v_3)^{(15)})v_4$.

29.2 Non-decreasing ω-sequences

Recall from Proposition 8.3.1 that $\mathsf{Q}^\omega(3) = 0$ if ω is non-decreasing. This is no longer the case in the 4-variable hit problem, but we can list the sequences ω for which it is not true, using the left order. There are three basic cases, $\omega = (1,3)$, $(2,3)$ and $(2,3,2)$, in degrees 7, 8 and 16, for which $\mathsf{Q}^\omega(4) \neq 0$. Examples 29.1.1, 29.1.3 and 29.1.5 show that $\dim \mathsf{Q}^{(1,3)}(4) = 1$, $\dim \mathsf{Q}^{(2,3)}(4) = 4$ and $\dim \mathsf{Q}^{(2,3,2)}(4) = 4$. All other cases are given by iteration of the Kameko isomorphism $\kappa : \mathsf{Q}^\omega(4) \to \mathsf{Q}^{(4,\omega)}(4)$, using Proposition 6.5.4.

Theorem 29.2.1 *If $\mathsf{Q}^\omega(4) \neq 0$ for the left order where ω is a 4-bounded non-decreasing sequence, then $\omega = (4,\ldots,4,\omega')$, where $\omega' = (1,3)$, $(2,3)$ or $(2,3,2)$.*

Proof Let $B = (b_{i,j})$ be a 4-block which is not left reducible. If $\omega(B) = \omega$ is non-decreasing, then $\omega_k < \omega_{k+1}$ for some k, and we write $B = A|F|C$, where F consists of columns k and $k+1$ of B. We may assume that all entries of ω are positive, since 1-back splicing into a zero column of B will reduce B in left order. We prove

(i) $\omega(F) = (1,3)$ or $(2,3)$;
(ii) all entries of A are 1;
(iii) if $\omega(F) = (1,3)$, then $C = 0$, the zero block;
(iv) if $\omega(F) = (2,3)$ and $C \neq 0$, then C has only one column, with two 1s.

The proof of (i) follows that of the 3-variable case, Proposition 8.3.1. Since $\omega_k < \omega_{k+1}$, there is a position (i,k) such that $b_{i,k} = 0$ and $b_{i,k+1} = 1$. If $\omega_{k+1} = 4$, then 1-back splicing at (i,k) reduces B in the left order. Thus we may assume $\omega_{k+1} < 4$. As row permutations do not affect the problem, we can reduce to the cases $F = F_j$, $1 \leq j \leq 6$, where

$$F_1 = \begin{matrix} 1\,0 \\ 1\,1 \\ 0\,1 \\ 0\,1 \end{matrix}, \quad F_2 = \begin{matrix} 1\,0 \\ 0\,1 \\ 0\,1 \\ 0\,1 \end{matrix}, \quad F_3 = \begin{matrix} 1\,1 \\ 1\,1 \\ 0\,1 \\ 0\,0 \end{matrix}, \quad F_4 = \begin{matrix} 1\,1 \\ 0\,1 \\ 0\,1 \\ 0\,0 \end{matrix}, \quad F_5 = \begin{matrix} 1\,1 \\ 0\,0 \\ 0\,1 \\ 0\,0 \end{matrix}, \quad F_6 = \begin{matrix} 1\,0 \\ 0\,1 \\ 0\,1 \\ 0\,0 \end{matrix}.$$

Let $B_j = AF_jC$. Clearly 1-back splicing of B_j at $(3,k)$ (corresponding to $(3,1)$ of F_j) reduces B_j in cases $j = 3,4,5$. To reduce B_6, we apply the χ-trick 2.5.3 on columns $k, k+1$ of B_6. This only involves the use of $Xq^{2^{k-1}}$ on column k of B_6 (effectively applying Sq^2 to the first column of F_6), which gives 0 and cannot affect C, while the action of Steenrod squares on A gives blocks A' with $\omega(A') \leq_l \omega(A)$. Hence B_6 is reduced. This proves (i).

To prove (ii), we separate the case B_1 and B_2 and argue by contradiction in each case. First we consider B_2. If A has an entry 0 in row i, then by iterated 1-back splicing we may assume that the entry in this row in the last column of A is 0. By 1-back splicing in column k, we may also assume that B_2 has a 1 at position (i,k). For example, when $i = 1$ columns $k-1, k, k+1$ of B_2 have the form

$$\begin{matrix} 0\,1\,0 \\ *\,0\,1 \\ *\,0\,1 \\ *\,0\,1 \end{matrix},$$

where the stars represent 0s or 1s. Then 1-back splicing at $(i,k-1)$ produces an equivalence of B_2 with a sum of blocks of type B_4 up to permutation of rows.

Hence B_2 is left reducible, contrary to hypothesis. Hence A has no 0 entries, and so B_1 is the iterated Kameko image of the base case $F_2|C$.

The case of B_1 is similar, but requires a preliminary move. By multiple 1-back splicing F_1 at $(3,1)$ and $(4,1)$, $F_1 \sim G' + G''$ where

$$G' = \begin{matrix} 100 \\ 101 \\ 100 \\ 100 \end{matrix}, \quad G'' = \begin{matrix} 010 \\ 001 \\ 100 \\ 100 \end{matrix}.$$

Let F', F'' denote the first two columns of these blocks. We note that $\omega(F') >_l \omega(F_1)$ and $\omega(F'') <_l \omega(F_1)$. Hence multiple 1-back splicing of B_1 at $(3,k)$ and $(4,k)$ produces an equivalence $B_1 \sim B'$, where $B' = A|F'|C'$ for some block C'. Although $\omega(B') >_l \omega(B_1)$, the significant property of B' is that all entries in column k are 1.

If A has an entry 0 in row i, then we may assume, as in the case of B_2, that the entry in this row in the last column of A is 0. For example, when $i = 3$ columns $k-1, k, k+1$ of B' have the form

$$\begin{matrix} *10 \\ *10 \\ 010 \\ *10 \end{matrix}.$$

Whatever the stars represent, 1-back splicing B' at $(i, k-1)$ produces blocks B'_1 such that $\omega(B'_1) <_l \omega(B_1)$. Since $B_1 \sim B'$, B_1 is left reducible, contrary to hypothesis. Hence A has no entries 0, and so B_1 is the iterated Kameko image of the base case $F_1|C$.

(iii) By (ii) and Proposition 6.5.4, we may assume that $B_2 = F_2|C$. If $C \neq 0$, then by 1-back splicing C we may assume that $C_{(i,1)} = 1$ for some i. By 1-back splicing F_2, we may also assume that $F_{(i,1)} = 1$ and $F_{(i,2)} = 0$. Then 1-back splicing at $(i,2)$ produces a sum of three blocks whose first two columns when $i = 3$ are

$$\begin{matrix} 01 \\ 01 \\ 11 \\ 00 \end{matrix}, \quad \begin{matrix} 01 \\ 00 \\ 11 \\ 01 \end{matrix}, \quad \begin{matrix} 00 \\ 01 \\ 11 \\ 01 \end{matrix}.$$

Finally, 1-back splicing in the first column reduces B_2.

(iv) By (ii) and Proposition 6.5.4, we may assume that $B_1 = F_1|C$ is not left reducible. If $C \neq 0$, then as in (iii) we may assume that $\omega_3(B_1) \neq 0$, and by 1-back splicing at (1,2) we may assume that $\omega_3(B_1) \neq 4$. We separate the cases $\omega_3(B_1) = 1$, $\omega_3(B_1) = 3$ and $\omega_3(B_1) = 2$.

Case 1: $\omega_3(B_1) = 1$. Let $B_1 = G|H|E$ where G is the first column of B_1 and H is the next two columns, and let $g, h \in \mathsf{P}(4)$ be the corresponding monomials. Then $\deg h = 5$ and since $\mu(5) = 3 > \deg g = 2$, Proposition 2.5.3 shows that the monomial gh^2 corresponding to the block $G|H$ is hit. More generally, let g be the monomial corresponding to the block $B' = G|O|E$, where O is a 2-column zero block. Then again $Xq^5(g) = 0$, because Xq^5 cannot affect any column of a block beyond the third, and the third column of B' is zero. Hence $B_1 = G|H|E$ is hit, and in particular it is reducible, contrary to hypothesis.

Case 2: $\omega_3(B_1) = 3$. With notation as in Case 1, the same argument shows that $G|H$ is hit, because $\deg h = 9$ and $\mu(9) = 3 > \deg g = 2$. However $Xq^9(g)$ can affect the fourth column of B_1, and so the argument of Case 1 cannot immediately be extended to B_1. The following alternative argument shows that B_1 is reducible if C has more than one column.

Suppose that there is a 1 in the fourth column of B_1. Since there are ≥ 9 entries 1 in the first four columns of B_1, at least one row has three 1s. By 1-back splicing, these may be moved into first three columns of B_1. Then the other three rows of this 3-column block have ≥ 4 entries 1, so there must be a row with at least two 1s. After further 1-back splicing, the first two columns of B_1 form a row permutation of the reducible block F_3, and so B is reducible, contrary to hypothesis. Hence C does not have more than one column. However, if C has just one column then B_1 is reducible (in fact hit) by the initial argument.

Case 3: $\omega_3(B_1) = 2$. The χ-trick argument with notation as in Case 1 fails, as $\deg h = 7$ and $\mu(7) = 1$. In fact, Example 29.1.5 shows that $G|H$ is not always reducible. If $G|H$ has a row with three 1s, then the argument of Case 2 shows that B_1 is reducible. If C has a second column with at least two 1s, then we can reduce to this case by 1-back splicing. If the second column of C has just a single 1, and no row of B_1 has three 1s in the first four columns, then the first four columns of B_1 must be the block

$$D = \begin{matrix} 1001 \\ 1100 \\ 0110 \\ 0110 \end{matrix}$$

up to permutation of rows and 1-back splicing. Since the last three columns of D form a block of degree 11 and $\mu(11) = 3$, the χ-trick shows as in Case 1 that D is hit. The argument of Case 1 used to extend the result to B_1 works again here, because Xq^{11} cannot affect columns after the fourth, and the fourth column of the block corresponding to g in Proposition 2.5.1 is zero. □

Theorem 29.2.1 reduces the 4-variable hit problem to the case where ω is decreasing. By Proposition 6.5.4 we can take $\omega = (3,\dots,3,2,\dots,2,1,\dots,1)$, where there are s 1s, t 2s and u 3s. Using Theorems 8.1.13 and 8.2.10, we find that $\mathsf{Q}^\omega(4) \cong \mathsf{FL}(4)$, of dimension 315, in the cases $s \geq 2$, $t \geq 3$, $u \geq 4$ and $s \geq 4$, $t \geq 3$, $u \geq 2$ respectively. In all cases with 'large head', i.e. $u \geq 4$, the dimension of $\mathsf{Q}^\omega(4)$ is determined by Theorem 8.1.11, and a basis given in terms of 'special blocks'. Likewise in cases with 'long tail', i.e. $s \geq 4$, the dimension of $\mathsf{Q}^\omega(4)$ is determined by Theorem 8.2.7, although we must be careful to check that the relevant 3-variable result can be achieved by restricted hit equations. The tail sequences $(1,\dots,1)$ and the head sequences $(3,\dots,3)$ have been treated in Sections 6.7, 6.8 and 6.9.

29.3 A basis for $\mathsf{Q}^\omega(4)$, $\omega = (2,\dots,2)$

In this section we determine $\mathsf{Q}^\omega(4)$, where ω is a sequence of length $t \geq 1$ with all entries equal to 2. We note that ω is the minimum decreasing sequence in its degree $d = 2(2^t - 1)$. The main result is as follows.

Theorem 29.3.1 *Let* $\omega = (2,\dots,2)$ *have length* t. *Then* $\dim \mathsf{Q}^\omega(4) = 6$, 20 *or* 35 *according as* $t = 1$, $t = 2$ *or* $t \geq 3$.

For $t = 1$ we have $\mathsf{Q}^\omega(4) = \mathsf{P}^\omega(4)$, and a basis in the case $t = 2$ is given in Examples 7.3.2 and 7.3.6. By Theorem 24.3.10, $\dim \mathsf{J}^\omega(4) = 35$ for $t \geq 4$, and it follows that $\dim \mathsf{Q}^\omega(4) = \dim \mathsf{K}^\omega(4) \geq 35$ in these cases. In the case $t = 3$, we prove in Proposition 29.3.2 that 35 is an upper bound and in Proposition 29.3.3 that it is a lower bound for $\dim \mathsf{Q}^\omega(4)$. The main tool required to obtain the upper bound in the case $t \geq 4$ is Proposition 29.3.4, and the argument is completed by Proposition 29.3.5.

The general idea for constructing a spanning set $\mathcal{T}$ of $\mathsf{Q}^\omega(4)$ is to consider the possible last rows of a block B, with the entries 1 on the right. We then find bases for the spaces of complementary blocks $\widehat{B}$ in the 3-variable case. In the situations where the entries of the last row are all 0s or all 1s, we obtain a linearly independent subset $\mathcal{S}$ of $\mathcal{T}$ which cannot appear in any linear relation on $\mathcal{T}$, but the proof that the complement of $\mathcal{S}$ in $\mathcal{T}$ is also linearly independent requires various specialization arguments.

Proposition 29.3.2 *For* $\omega = (2,2,2)$, $\dim \mathrm{Q}^{\omega}(4) \leq 35$.

Proof The possible last rows of a block B in $\mathrm{Q}^{\omega}(4)$ with entries 1 on the right are $R_0 = (0,0,0)$, $R_1 = (0,0,1)$, $R_2 = (0,1,1)$ and $R_3 = (1,1,1)$. We write $\mathcal{T}_i$ for a set of blocks with last row R_i for $0 \leq i \leq 3$. In the case of R_0, the complementary block $\widehat{B}$ is a head block in $\mathrm{Q}^{(2,2,2)}(3)$. From Section 8.1 we can find a basis $\widehat{\mathcal{T}_0}$ of $\mathrm{Q}^{(2,2,2)}(3)$ consisting of 7 blocks. Similarly the complementary block $\widehat{B}$ of R_3 is a tail block in $\mathrm{Q}^{(1,1,1)}(3)$, and from Section 8.2 we again have a basis $\widehat{\mathcal{T}_3}$ of 7 blocks. By Proposition 7.2.8 the set of 14 blocks $\mathcal{T}_0 \cup \mathcal{T}_3$ is linearly independent, and no member of this set can appear in a linear relation on a set of blocks obtained by adding blocks with last row R_1 or R_2 to it.

For a block B in $\mathrm{Q}^{(2,2,2)}(4)$ with last row R_2, the complementary block $\widehat{B}$ is in $\mathrm{Q}^{(2,1,1)}(3)$, which has dimension 15 by Proposition 8.3.5. We shall choose a suitable basis $\widehat{\mathcal{T}_2}$ for it. Similarly, by Proposition 8.3.7, for a block B in $\mathrm{Q}^{(2,2,2)}(4)$ with last row R_1, the complementary block $\widehat{B}$ is in the 14-dimensional vector space $\mathrm{Q}^{(2,2,1)}(3)$, for which we choose a basis $\widehat{\mathcal{T}_1}$. Potentially this gives a spanning set $\mathcal{T}_0 \cup \mathcal{T}_3 \cup \mathcal{T}_2 \cup \mathcal{T}_1$ with 43 blocks for $\mathrm{Q}^{(2,2,2)}(4)$. However, we show that $\mathcal{T}_2$ is the union of two subsets $\mathcal{A}$ and $\mathcal{B}$, with 7 and 8 elements respectively, such that each member of $\mathcal{B}$ is equivalent to a combination of members of $\mathcal{A}$ and $\mathcal{T}_1$. This cuts down the total spanning set to 35 blocks.

We now explain how the various sets of blocks are chosen and the relations between them. The crucial choice is the basis $\widehat{\mathcal{T}_2}$ for $\mathrm{Q}^{(2,1,1)}(3)$. It is convenient to list the blocks of $\widehat{\mathcal{T}_2}$ according to their α-sequences, with entries in increasing order, and to indicate the row permutations of a given block needed to make a total of 15. The possible sequences $\alpha(B)$ are permutations of $(0,1,3)$, $(0,2,2)$ and $(1,1,2)$.

The set $\widehat{\mathcal{T}_2}$ is made up as follows. For $\alpha(B)$ a permutation of $(0,1,3)$, we take the six row permutations of the block

$$\widehat{F_1} = \begin{matrix} 0\,0\,0 \\ 1\,0\,0 \\ 1\,1\,1 \end{matrix}.$$

These are spikes in $\mathrm{Q}^{(2,1,1)}(3)$ and are unavoidable. For $\alpha(B)$ a permutation of $(0,2,2)$, we take the three cyclic row permutations of the block

$$\widehat{F_2} = \begin{matrix} 0\,0\,0 \\ 1\,0\,1 \\ 1\,1\,0 \end{matrix}.$$

Switching the last two rows gives an equivalent block by 1-back splicing. For $\alpha(B)$ a permutation of $(1,1,2)$, there are two types, the cyclic row permutations of the blocks

$$\widehat{F}_3 = \begin{matrix} 1\,0\,0 \\ 1\,0\,0 \\ 0\,1\,1 \end{matrix}\ ,\ \widehat{F}_4 = \begin{matrix} 1\,0\,0 \\ 0\,0\,1 \\ 1\,1\,0 \end{matrix}\ .$$

This makes 15 elements in $\widehat{\mathcal{T}}_2$, and it can be verified from Section 8.3 that $\widehat{\mathcal{T}}_2$ is a basis of $Q^{(2,1,1)}(3)$. The set $\widehat{\mathcal{A}}$ consists of the cyclic row permutations of $\widehat{F}_2$ and of $\widehat{F}_4$ together with $\widehat{F}_3$, while the set $\widehat{\mathcal{B}}$ consists of the six row permutations of $\widehat{F}_1$ and the other two cyclic permutations of $\widehat{F}_3$.

We show that the eight blocks in $\mathcal{B}$ are linearly dependent on the blocks in $\mathcal{A} \cup \mathcal{T}_1$. By symmetry it is enough to deal with the blocks $\widehat{F}_1$ and $\widehat{F}_3$. For $\widehat{F}_1$ we replace the last row R_2 to regain F_1, and then move the first column to the third column to obtain the block

$$\begin{matrix} 0\,0\,0 \\ 0\,0\,1 \\ 1\,1\,1 \\ 1\,1\,0 \end{matrix}\ .$$

Splicing at $(2,1)$ shows that F_1 is equivalent to the block

$$\begin{matrix} 0\,0\,0 \\ 1\,1\,0 \\ 1\,1\,1 \\ 0\,0\,1 \end{matrix}$$

in $\mathcal{T}_1$. Restoring the last row R_2 to the block $\widehat{F}_3$, we obtain F_3, and, by moving the first column to the third column, we obtain the block

$$F = \begin{matrix} 0\,0\,1 \\ 0\,0\,1 \\ 1\,1\,0 \\ 1\,1\,0 \end{matrix}\ .$$

Splicing at $(1,1)$ shows that

$$F \sim \begin{matrix} 1\,1\,0 \\ 0\,0\,1 \\ 1\,1\,0 \\ 0\,0\,1 \end{matrix} + \begin{matrix} 1\,1\,0 \\ 0\,0\,1 \\ 0\,0\,1 \\ 1\,1\,0 \end{matrix}.$$

Moving the third column of the second block back to the first column then shows that the permutations of $\widehat{F}_3$ all give rise to the same element F_3 in $\mathrm{Q}^{\omega}(4)$, modulo the blocks in $\mathcal{T}_1$. This completes the proof. □

Proposition 29.3.3 *For* $\omega = (2,2,2)$, $\dim \mathrm{Q}^{\omega}(4) \geq 35$.

Proof We show that the spanning set for $\mathrm{Q}^{\omega}(4)$ constructed in Proposition 29.3.2 is linearly independent modulo hit polynomials. Recall that this set $\mathcal{S} = \mathcal{T}_0 \cup \mathcal{T}_3 \cup \mathcal{T}_1 \cup \mathcal{A}$ with 7, 7, 14 and 7 elements respectively. By the results for head and tail blocks in $\mathrm{Q}^{(2,2,2)}(3)$ and $\mathrm{Q}^{(1,1,1)}(3)$, the sets $\mathcal{T}_0$ and $\mathcal{T}_3$ are linearly independent in $\mathrm{Q}^{\omega}(4)$. Since blocks in $\mathcal{T}_0$ do not involve x_4 and all other blocks in $\mathcal{S}$ do, blocks in $\mathcal{T}_0$ cannot appear in any linear relation on $\mathcal{S}$ in $\mathrm{Q}^{\omega}(4)$. By Proposition 7.2.8 the same is true for blocks in $\mathcal{T}_3$. We choose the 14 blocks in $\mathcal{T}_1$ as in Proposition 8.3.7. Let

$$C_1 = \begin{matrix} 1\,1\,1 \\ 1\,1\,0 \\ 0\,0\,0 \\ 0\,0\,1 \end{matrix}, \quad C_2 = \begin{matrix} 1\,1\,1 \\ 1\,0\,0 \\ 0\,1\,0 \\ 0\,0\,1 \end{matrix}, \quad C_3 = \begin{matrix} 1\,1\,0 \\ 1\,0\,1 \\ 0\,1\,0 \\ 0\,0\,1 \end{matrix}.$$

Thus we have six blocks obtained from C_1 by permutation of the first three rows, three blocks obtained from C_2 by cyclic permutation of the first three rows, and five blocks obtained from C_3 by excluding any one permutation of the first three rows. As above, the 7 blocks in $\mathcal{A}$ have last row R_2 followed by cyclic permutations of $\widehat{F}_2$ or $\widehat{F}_4$ or the block $\widehat{F}_3$.

We next show that there is no linear relation between the 21 blocks $\mathcal{T}_1 \cup \mathcal{A}$ in $\mathrm{Q}^{\omega}(4)$. There are 3 blocks

$$\begin{matrix} 1\,1\,1 \\ 1\,1\,0 \\ 0\,0\,0 \\ 0\,0\,1 \end{matrix}, \quad \begin{matrix} 1\,1\,0 \\ 1\,1\,1 \\ 0\,0\,0 \\ 0\,0\,1 \end{matrix}, \quad \begin{matrix} 1\,0\,1 \\ 1\,1\,0 \\ 0\,0\,0 \\ 0\,1\,1 \end{matrix}$$

which do not involve x_3. Using the results for head blocks in $Q^{(2,2,2)}(3)$, it follows that these blocks are not involved in any linear relation on $\mathcal{T}_1 \cup \mathcal{A}$ in $Q^{\omega}(4)$. We deal similarly with the 3 blocks which do not involve x_1, and the 3 blocks which do not involve x_2. Of the remaining 12 blocks, 3 have a row R_3 and each can be eliminated in turn from a possible relation using Proposition 7.2.8. We denote the remaining 9 blocks

$$\begin{matrix}110\\101\\010\\001\end{matrix},\quad \begin{matrix}110\\010\\101\\001\end{matrix},\quad \begin{matrix}101\\110\\010\\001\end{matrix},\quad \begin{matrix}101\\010\\110\\001\end{matrix},\quad \begin{matrix}010\\110\\101\\001\end{matrix},\quad \begin{matrix}100\\001\\110\\011\end{matrix},\quad \begin{matrix}110\\100\\001\\011\end{matrix},\quad \begin{matrix}001\\110\\100\\011\end{matrix},\quad \begin{matrix}100\\100\\011\\011\end{matrix},$$

in the order shown, by B_i, $1 \le i \le 9$, and specialize the variables, as follows. Setting $x_1 = x_2$ in a possible relation $\sum_{i=1}^{9} b_i B_i = 0$ in $Q^{\omega}(4)$, where $b_i \in \mathbb{F}_2$, we obtain the relation

$$b_4 \begin{matrix}111\\110\\001\end{matrix} + b_6 \begin{matrix}101\\110\\011\end{matrix} + b_8 \begin{matrix}111\\100\\011\end{matrix} = 0$$

in $Q^{\omega}(3)$. The results on head blocks show that $b_4 = b_8$ and $b_6 = 0$. Similar arguments using the specializations $x_1 = x_3, x_1 = x_4, x_2 = x_3, x_2 = x_4$ and $x_3 = x_4$ lead to the equations $b_2 = b_4 = b_5 = b_6 = b_7 = b_8 = 0$ and $b_1 = b_3 = b_9$. Thus it remains to prove that $B_1 + B_3 + B_9$ is not hit.

By 1-back splicing B_3 at $(1,2)$, we have

$$C = \begin{matrix}110\\110\\001\\001\end{matrix} \sim \begin{matrix}110\\101\\010\\001\end{matrix} + \begin{matrix}101\\110\\010\\001\end{matrix} = B_1 + B_3.$$

Hence it is equivalent to prove that $C + B_9$ is not hit. By Proposition 7.2.6, B_9 is equivalent mod hits to the row reversal of C.

Let $C_{i,j}$ denote the row permutation of C with rows i and j equal to 0 0 1, so that $C = C_{3,4}$ and $B_9 \sim C_{1,2}$. The hit equation

$$Sq^1 \left(\begin{matrix}001\\110\\110\\110\end{matrix} + \begin{matrix}110\\001\\110\\110\end{matrix} \right) = C_{1,3} + C_{1,4} + C_{2,3} + C_{2,4}$$

shows that it is equivalent to prove that the sum $X = \sum_{i,j} C_{i,j}$ of all 6 row permutations of C is not hit. To complete the proof, we argue as in Proposition 8.3.5. We have a hit equation $Sq^2\begin{pmatrix} 1\,1 \\ 1\,1 \\ 1\,1 \\ 1\,1 \end{pmatrix} = X + Y$ where

$$Y = \begin{matrix} 1\,0\,1 \\ 1\,1 \\ 1\,1 \\ 1\,1 \end{matrix} + \begin{matrix} 1\,1 \\ 1\,0\,1 \\ 1\,1 \\ 1\,1 \end{matrix} + \begin{matrix} 1\,1 \\ 1\,1 \\ 1\,0\,1 \\ 1\,1 \end{matrix} + \begin{matrix} 1\,1 \\ 1\,1 \\ 1\,1 \\ 1\,0\,1 \end{matrix}.$$

Since none of the blocks in Y occurs in any other hit equation, Y is not hit, and since $X \sim Y$ we conclude that X is not hit. □

The following result is special to the 4-variable case.

Proposition 29.3.4 *Let* $\omega = (2,\ldots,2)$ *have length* $t \geq 4$. *Then* $\mathrm{Q}^{\omega}(4)$ *is spanned by blocks whose last row has* 0, 2 *or* t *digits* 1.

Proof We assume as induction hypothesis that the result is true for $4 \leq t < s$ for some s. Let $B \in \mathrm{P}^{\omega}(4)$ where $\omega = (2,\ldots,2)$ has length $t = s$, and let $\alpha(B)$ denote the number of digits 1 in the last row of B. We may assume that $0 < \alpha(B) < s$ and that all digits 1 in the last row of B are contiguous on the left. If $\alpha(B) < s-1$, we apply the inductive hypothesis to the block B' obtained from B by excluding the last column. Then $B' \sim_l B''$, where B'' is a sum of blocks with $\alpha = 2$, and so replacing B' by B'' in B produces blocks with $\alpha = 2$.

The error terms arising from the χ-trick involved in this process affect only the last column of B, and so they produce blocks with ω-sequence lower than ω. Hence these are hit, and so the equivalence class of B in $\mathrm{Q}^{\omega}(4)$ is unaltered. On the other hand, if $\alpha(B) = s-1$, then we first apply a similar argument to the block formed from the last four columns of B to reduce the problem to the previous case. This completes the inductive step.

It remains to prove the result in the initial case $t = 4$. We may assume that the last row R of B is either $(0,1,1,1)$ or $(0,0,0,1)$, and we wish to show that B can be replaced by an equivalent sum of blocks F with $\alpha(F) = 2$.

Case 1: $R = (0,0,0,1)$. Then $\widehat{B}$ is a 3-block and $\omega(\widehat{B}) = (2,2,2,1)$. Since the total number of entries 1 in $\widehat{B}$ is 7, there must be a row with at least three entries 1. By 1-back splicing we may assume, up to permutation of rows in $\widehat{B}$, that B has one of two forms

$$\begin{matrix} 1 & 1 & 1 & * \\ 1 & 0 & 0 & * \\ 0 & 1 & 1 & * \\ 0 & 0 & 0 & 1 \end{matrix}, \qquad \begin{matrix} 1 & 1 & 1 & * \\ 1 & 1 & 1 & * \\ 0 & 0 & 0 & * \\ 0 & 0 & 0 & 1 \end{matrix}$$

where two of the stars are 0s and the other is a 1. In all cases, splicing at $(4,2)$ produces an equivalent sum of blocks F with $\alpha(F) = 2$, as required.

Case 2: $R = (0,1,1,1)$. Then $\widehat{B}$ is a 3-block and $\omega(\widehat{B}) = (2,1,1,1)$ with a total of five entries 1. If there is a row of $\widehat{B}$ with three entries 1, then by 1-back splicing and row permutation of $\widehat{B}$ we may assume that B has the form

$$\begin{matrix} 1 & 1 & 1 & * \\ 1 & 0 & 0 & * \\ 0 & 0 & 0 & * \\ 0 & 1 & 1 & 1 \end{matrix}$$

where two of the stars are 0s and the other is a 1. Switching columns 1 and 3 gives the block

$$\begin{matrix} 1 & 1 & 1 & * \\ 0 & 0 & 1 & * \\ 0 & 0 & 0 & * \\ 1 & 1 & 0 & 1 \end{matrix}.$$

Then splicing at $(2,1)$ produces an equivalent sum of blocks F with $\alpha(F) = 2$. If $\widehat{B}$ does not contain a row with three entries 1, then it must contain a row with a single 1 and two rows with two entries 1. Following the usual routine, we may assume that B has the form

$$\begin{matrix} 1 & 0 & 0 & 0 \\ 1 & 1 & 0 & 0 \\ 0 & 0 & 1 & 1 \\ 0 & 1 & 1 & 1 \end{matrix}.$$

By permuting the columns cyclically, we obtain the equivalent block

$$\begin{matrix} 0\,0\,0\,1 \\ 1\,0\,0\,1 \\ 0\,1\,1\,0 \\ 1\,1\,1\,0 \end{matrix}.$$

Then splicing at $(2,2)$ produces the sum

$$\begin{matrix} 0\,0\,0\,1 \\ 1\,1\,1\,0 \\ 0\,0\,0\,1 \\ 1\,1\,1\,0 \end{matrix} + \begin{matrix} 0\,0\,0\,1 \\ 1\,1\,1\,0 \\ 0\,1\,1\,0 \\ 1\,0\,0\,1 \end{matrix}.$$

The first block is dealt with above, because its second row has three entries 1. The second block already has the desired form, with two entries 1 in the last row. □

The following result completes the proof of Theorem 29.3.1 in the cases $t \geq 4$.

Proposition 29.3.5 *For* $\omega = (2,\ldots,2)$ *of length* ≥ 4, $\dim \mathrm{Q}^{\omega}(4) \leq 35$.

Proof By Proposition 29.3.4 we may assume that $\mathrm{Q}^{\omega}(4)$ is spanned by blocks B such that $\alpha(B) = 0, t$ or 2 and the digits 1 are contiguous on the right. In the first case, we let the subblock $\widehat{B}$ run over a basis of $\mathrm{Q}^{\gamma}(3)$ where γ is a head sequence of length ≥ 4, and in the second case over a basis of $\mathrm{Q}^{\tau}(3)$ where τ is a tail sequence of length ≥ 4. This accounts for $7+7=14$ blocks. In the third case, $\omega' = \omega(\widehat{B}) = (2,\ldots,2,1,1)$, with at least two 2s. Hence $\dim \mathrm{Q}^{\omega'}(3) = 21$. By the χ-trick applied to the splitting of B into the last row and $\widehat{B}$, any effects on the last row of the operations used to reduce $\widehat{B}$ to a basis of $\mathrm{Q}^{\omega'}(3)$ produce blocks with ω-sequence $<_r \omega$, which are therefore hit. This makes a total of $14+21=35$ blocks which span $\mathrm{Q}^{\omega}(4)$. □

We conclude by showing that $\mathsf{P}^{(2,2,2)}(4)$ contains a polynomial which is invariant mod hits. In Section 30.3 we shall prove that this is dual to $\mathsf{K}^{14}(4)/\mathsf{J}^{14}(4)$.

Proposition 29.3.6 *The polynomial* $f = x_1^6x_2^6x_3x_4 + x_1x_2x_3^6x_4^6 \in \mathsf{P}^{(2,2,2)}(4)$ *represented by the sum of blocks*

$$\begin{array}{c} 0\,1\,1 \\ 0\,1\,1 \\ 1\,0\,0 \\ 1\,0\,0 \end{array} + \begin{array}{c} 1\,0\,0 \\ 1\,0\,0 \\ 0\,1\,1 \\ 0\,1\,1 \end{array}$$

generates a 1*-dimensional submodule of* $\mathsf{Q}^{(2,2,2)}(4)$.

Proof Since f is invariant under exchanges of rows 1 and 2, and of rows 3 and 4, we need only check that it is invariant mod hits under the exchange of rows 2 and 3, and under the standard transvection U. This follows from column permutation of the blocks (Proposition 7.2.6) and the first column splicings

$$\begin{array}{c} 1\,1\,0 \\ 1\,1\,0 \\ 0\,0\,1 \\ 0\,0\,1 \end{array} + \begin{array}{c} 1\,1\,0 \\ 0\,0\,1 \\ 1\,1\,0 \\ 0\,0\,1 \end{array} \sim \begin{array}{c} 0\,0\,1 \\ 1\,1\,0 \\ 1\,1\,0 \\ 0\,0\,1 \end{array} \sim \begin{array}{c} 0\,0\,1 \\ 1\,1\,0 \\ 0\,0\,1 \\ 1\,1\,0 \end{array} + \begin{array}{c} 0\,0\,1 \\ 0\,0\,1 \\ 1\,1\,0 \\ 1\,1\,0 \end{array}.$$

To show that f is not hit, we use the equivalent polynomial $f' = x_1^3x_2^3x_3^4x_4^4 + x_1^4x_2^4x_3^3x_4^3$. We have

$$\begin{array}{c} 1\,1\,0 \\ 1\,1\,0 \\ 0\,0\,1 \\ 0\,0\,1 \end{array} \xrightarrow{Sq^2} \begin{array}{c} 1\,0\,1 \\ 1\,1\,0 \\ 0\,0\,1 \\ 0\,0\,1 \end{array} + \begin{array}{c} 1\,1\,0 \\ 1\,0\,1 \\ 0\,0\,1 \\ 0\,0\,1 \end{array} + \begin{array}{c} 0\,0\,1 \\ 0\,0\,1 \\ 0\,0\,1 \\ 0\,0\,1 \end{array} \xrightarrow{Sq^2} \begin{array}{c} 0\,1\,1 \\ 0\,0\,1 \\ 0\,0\,1 \\ 0\,0\,1 \end{array} + \begin{array}{c} 0\,0\,1 \\ 0\,1\,1 \\ 0\,0\,1 \\ 0\,0\,1 \end{array} \xrightarrow{Sq^2} \begin{array}{c} 0\,0\,0\,1 \\ 0\,0\,1 \\ 0\,0\,1 \\ 0\,0\,1 \end{array} + \begin{array}{c} 0\,0\,1 \\ 0\,0\,0\,1 \\ 0\,0\,1 \\ 0\,0\,1 \end{array}$$

and so $(Sq^2)^3(f') = \sigma_4(x_1^8x_2^4x_3^4x_4^4)$. Assume that f' is hit. Then $f' = Sq^1(g_1) + Sq^2(g_2) + Sq^4(g_4)$ where $g_i \in \mathsf{P}^{14-i}(4)$ for $i = 1,2,4$. Applying $(Sq^2)^3$ to this equation, we obtain $Sq^9Sq^1(g_4) = \sigma_4(x_1^8x_2^4x_3^4x_4^4)$ since $(Sq^2)^3Sq^1 = 0$, $(Sq^2)^4 = 0$ and $(Sq^2)^3Sq^4 = Sq^9Sq^1$. However the following direct calculation shows that there is no polynomial $g_4 \in \mathsf{P}^{10}(4)$ such that $Sq^9Sq^1(g_4) = \sigma_4(x_1^8x_2^4x_3^4x_4^4)$.

Working up to permutations of the variables, we need to consider monomials in $\mathsf{P}^{10}(4)$ with positive exponents a,b,c,d. Since $\mathrm{Ker}(Sq^1)$ contains $x_1^4x_2^2x_3^2x_4^2$, $x_1^6x_2^2x_3x_4 + x_1^5x_2^2x_3^2x_4 + x_1^5x_2^2x_3x_4^2$, $x_1^4x_2^4x_3x_4 + x_1^3x_2^4x_3^2x_4 + x_1^3x_2^4x_3x_4^2$ and $x_1^3x_2^3x_3^2x_4^2 + x_1^4x_2^3x_3x_4^2 + x_1^3x_2^4x_3x_4^2$, we may exclude the cases $(a,b,c,d) = (4,2,2,2)$, $(6,2,1,1)$, $(4,4,1,1)$ and $(3,3,2,2)$.

The cases which remain are $(a,b,c,d) = (7,1,1,1)$, $(5,3,1,1)$, $(5,2,2,1)$, $(4,3,2,1)$ and $(3,3,3,1)$. Since only $Sq^9Sq^1(x_1^7x_2x_3x_4)$ contains $x_1^{16}x_2^2x_3x_4$ as a term, we can exclude the case $(7,1,1,1)$. Similarly we can exclude $(5,3,1,1)$ by using $x_1^{12}x_2^6x_3x_4$ and $(3,3,3,1)$ by using $x_1^8x_2^6x_3^5x_4$. Then $Sq^9Sq^1(x_1^5x_2^2x_3^2x_4) = Sq^9(x_1^6x_2^2x_3^2x_4 + x_1^5x_2^2x_3^2x_4^2)$ and the Cartan formula shows that no term with an exponent 8 occurs. Thus all terms in g_4 must have exponents $(4,3,2,1)$.

Finally $Sq^9Sq^1(x_1^4x_2^3x_3^2x_4) = Sq^9(x_1^4x_2^4x_3^2x_4 + x_1^4x_2^3x_3^2x_4^2)$. Using the relations $Sq^8(x_1^4x_2^4x_3^2x_4) = x_1^8x_2^8x_3^2x_4$ and $Sq^8(x_1^4x_2^3x_3^2x_4^2) = x_1^8x_2^5x_3^4x_4^2 + x_1^8x_2^5x_3^2x_4^4 + x_1^8x_2^3x_3^4x_4^4$, we obtain

$$Sq^9Sq^1(x_1^4x_2^3x_3^2x_4) = x_1^8x_2^8x_3^2x_4^2 + x_1^8x_2^6x_3^4x_4^2 + x_1^8x_2^6x_3^2x_4^4 + x_1^8x_2^4x_3^4x_4^4.$$

The terms in $Sq^9Sq^1(g_4)$ with exponents $(8,6,4,2)$ can only be removed by taking $g_4 = \sigma_4(x_1^4x_2^3x_3^2x_4)$. But in this case $Sq^9Sq^1(g_4) = 0$. □

29.4 Remarks

Many of the results of this chapter, for example Theorem 29.2.1, were stated without proof in [108]. This includes tabulations of $\dim \mathrm{Q}^d(4)$ and of $\dim \mathrm{Q}^\omega(4)$ where $d = \deg_2 \omega$ and $\omega = (3,\dots,3,2,\dots,2,1,\dots,1)$ with s 1s, t 2s and u 3s for all $s,t,u \geq 0$. A detailed solution of the whole 4-variable case of the hit problem has been given by Sum [204], who gives a monomial basis for $\mathrm{Q}^d(4)$ for each $d \geq 0$. This basis is chosen so as to be maximal for the linear ordering of monomials in $\mathrm{P}^d(4)$ obtained by extending the left order on ω-sequences ω with $\deg_2 \omega = d$, using the left lexicographic order on exponent sequences of monomials in each $\mathrm{P}^\omega(4)$.

Repka and Selick [173, p.287] show that $x_1^4x_2^4x_3^3x_4^3 + x_1^3x_2^3x_3^4x_4^4 \in \mathrm{Q}^{14}(4)$ is dual to $\mathrm{K}^{14}(4)/\mathrm{J}^{14}(4)$. This polynomial is equivalent mod hits to the polynomial f of Section 29.3 by Proposition 7.2.6. They also give a d-polynomial with 36 terms which generates $\mathrm{K}^{14}(4)/\mathrm{J}^{14}(4)$. A similar d-polynomial with 38 terms is given by Lê Minh Hà [71, p.102].

30

The 4-variable hit problem, II

30.0 Introduction

This chapter continues the study of the hit problem in the 4-variable case. Following the method and calculations of Kameko, Nguyen Sum has given a monomial basis for $\mathsf{Q}^d(4)$ for all $d \geq 0$. For the sake of completeness, we include in the tables in Section 30.4 a number of cases where we have proved only that the stated value of $\dim \mathsf{Q}^\omega(4)$ is a lower bound. These results are quoted from the work of Kameko and Sum, and are indicated by asterisks in the tables.

In Section 30.1 we turn to the upper bound problem for $\dim \mathsf{Q}^\omega(4)$, where ω is a decreasing sequence with u 3s, t 2s and s 1s. As before, we refer to the u, t, s subblocks of a block B representing a monomial in $\mathsf{P}^\omega(4)$ as the 'head', 'body' and 'tail' of B. This section complements the work of Sections 8.1 and 8.2 by obtaining a 'large body' theorem for the case $n = 4$ and $t \geq 4$, so reducing the hit problem for $\mathsf{P}(4)$ to a finite number of special cases.

Section 24.4 provides a method for calculating the submodule $\mathsf{J}^\omega(4)$ of $\mathsf{K}^\omega(4)$, and as in the 3-variable case we usually find that $\mathsf{J}^\omega(4) = \mathsf{K}^\omega(4)$. However, when $n = 4$ there are a substantial number of cases where the quotient $\mathsf{K}^\omega(4)/\mathsf{J}^\omega(4) \neq 0$.

In Section 30.2 we show that $\mathsf{K}^\omega(4)/\mathsf{J}^\omega(4)$ is dual to the 'spike-free' submodule $\mathsf{SF}^\omega(n)$ of $\mathsf{Q}^\omega(n)$ represented by 'strongly spike-free' polynomials f such that $f \cdot c$ contains no spikes for all $c \in \mathbb{F}_2\mathsf{GL}(n)$. The basic example of a strongly spike-free polynomial is Singer's polynomial $s_8 = x^6yz + xy^6z + xyz^6 \in \mathsf{P}^8(3)$. Dually, the d-polynomial f of Proposition 10.6.3 generates $\mathsf{K}^8(3)/\mathsf{J}^8(3)$.

In Section 30.3 we discuss cases where $\mathsf{SF}^\omega(4) \neq 0$. In some cases (in degrees 23, 39, 41 and their Kameko iterates) we do not give proofs, but rely on the result of computer calculations using MAPLE. The cases where $\mathsf{SF}^\omega(4)$ contains a nonzero $\mathsf{GL}(4)$-invariant deserve particular attention, as these polynomials are mapped by the Singer transfer to elements of filtration 4 in the E_2 term of the Adams spectral sequence. As we do not discuss the Singer transfer, these results are included for reference only, and are stated in Table (30.2) without proof. In Section 30.3, we use the abbreviated notation $[a_1,a_2,a_3,a_4]$ for the monomial $x_1^{a_1}x_2^{a_2}x_3^{a_3}x_4^{a_4} \in \mathsf{P}(4)$, and $\sigma_4([a_1,a_2,a_3,a_4])$ for its symmetrization.

30.1 The large body case $t \geq 4$

Let $n = 4$, $\omega = (3,\ldots,3,2,\ldots,2,1,\ldots,1)$ with u 3s, t 2s and s 1s, and let $d = \deg_2(\omega)$. We consider the case $t \geq 4$, so that blocks representing monomials in $\mathsf{P}^\omega(4)$ have a 'large body'. The lower bounds given by $\dim \mathsf{J}^\omega(4)$ when $t \geq 4$ are tabulated below. These are proved in Chapters 23, 24 and 29.

$u \backslash s$	0	1	2	≥ 3
0	35	70	105	105
1	70	140	210	210
≥ 2	105	210	315	315

(30.1)

Theorem 30.1.1 *For $t \geq 4$, $\dim \mathsf{Q}^\omega(4)$ is given by Table* (30.1).

Proof In the case $s \geq 4$, the 'long tail' theorem, Proposition 8.2.9, implies that the lower bounds are exact, i.e. $\mathsf{J}^\omega(4) = \mathsf{K}^\omega(4)$. The same is true if $u \geq 4$ by the 'large head' theorem 8.1.11, and if $u = 3$ by Theorem 8.4.6. The case $u = s = 0$ has already been treated in Theorem 29.3.1. The remaining 11 cases in (30.1) are considered separately as Cases 1 to 11 below.

The sequence $\omega = \omega^{\min}(d)$ when $s > 0$, and when $s = 0$ the sequences ρ such that $\rho \leq_l \omega$ and $\rho \leq_r \omega$ coincide. Hence, by combining Proposition 7.2.6 with the results of Sections 6.7, 6.8 and 29.3, we can write any block in $\mathsf{P}^\omega(4)$ as a sum of blocks with last row $R = R_3|R_2|R_1$ mod hits, where R_3, R_2, R_1 are sequences of the form $0\cdots0\,1\cdots1$ where the number of 1s is restricted to lie in the sets $\{0,u-1,u\}$, $\{0,2,t\}$ and $\{0,1,s\}$ respectively.

We refer to these as 'canonical forms' for the fourth row of a 4-block B, and encode them using the letters a, b, c to indicate a section consisting entirely of 1s, a section containing 0s followed by 1s, and a section consisting entirely of 0s respectively. A head b section of length ≥ 2 has exactly one entry 0, a body b section of length ≥ 4 has exactly two entries 1 and a tail b section of length ≥ 2 has exactly one entry 1. We call this three-letter code the **(last row) type** of B. We write $B = \begin{array}{c} \widehat{B} \\ R \end{array} = \begin{array}{c|c|c} \widehat{B}_3 & \widehat{B}_2 & \widehat{B}_1 \\ R_3 & R_2 & R_1 \end{array}$, where $\widehat{B}$ is a 3-block and R is the fourth row of B, and in Cases 3 to 6 below we use arrays $\begin{array}{c|c|c} b_3 & b_2 & b_1 \\ \hline r_3 & r_2 & r_1 \end{array}$, where $b_i = \omega(\widehat{B}_i)$ and $r_i = \omega(R_i)$ for $i = 1,2,3$, to illustrate last row types.

In the first part of the proof, we describe a 3-stage reduction procedure which produces a spanning set for $\mathrm{Q}^{\omega}(4)$ consisting of blocks B with last row R in canonical form and $\widehat{B}$ a basis element for $\mathrm{Q}^{\widehat{\omega}}(3)$, where $\widehat{\omega} = \omega(\widehat{B}) = \omega - \omega(R)$.

Stage 1. Using 1-back splicing operations, we left justify R, so as to collect all 1s on the left. If $\alpha(R) = u+t+s$, $u+t$, u or 0, then all the resulting blocks are of last row type *aaa*, *aac*, *acc* or *ccc* respectively. If $u+t < \alpha(R) < u+t+s$, then R_3 and R_2 are of type a, and by the tail theorem we can replace B by a block of last row type *aab*. If $u < \alpha(R) < u+t$, then R_3 is of type a, R_1 is of type c and we can reduce R_2 to type b by the body theorem. Finally if $0 < \alpha(R) < u$, then R_2 and R_1 are of type c and we can reduce R_3 to type b by the head theorem. Thus there is a spanning set of blocks for $\mathrm{Q}^{\omega}(4)$ in which the last row R of each block B has one of the 7 last row types

aaa, aab, aac, abc, acc, bcc, ccc.

Stage 2. At the second stage of the reduction, we use operations Sq^k to reduce $\widehat{B}$ to 'standard form', i.e. a linear combination of a spanning set for $\mathrm{Q}^{\widehat{\omega}}(3)$. By applying the χ-trick to the horizontal splitting $B = \begin{array}{c} \widehat{B} \\ R \end{array}$, this can be achieved at the cost of introducing further blocks in $\mathrm{P}^{\omega}(4)$ which result from the action of the operations $\chi(Sq^k)$ on R. Squaring operations which act on R_1 produce blocks with ω-sequences $<_r \omega$, which are hit since $\omega = \omega^{\min}(d)$. Operations Sq^{2^j} acting on R_3 replace some of the 1s in R_3 by a single 1 in R_2, and operations Sq^{2^j} acting on R_2 replace some of the 1s in R_2 by a single 1 in R_1.

After each such operation, we can use the body and tail theorems to restore the a,b,c pattern. We refer to this process informally as a 'flow' from one last row type to another. Since each operation Sq^{2^j} acting on a section of R increases the number of 1s in the section to its right only by 1, flows do not produce a body section R_2 of type a when $t \geq 4$, or a tail section R_1 of type a when $s \geq 3$, as in these cases at least one entry 0 remains in R_2 or R_1 respectively.

Stage 3. A further manoeuvre involves only the body and tail, and gives a further reduction of the last row types $*ba$ when $s = 1$ or 2 and $*$ denotes a, b

or c. We show first that these blocks can be omitted from a spanning set when $s = 2$.

Consider a block with ω-sequence $(2,2,2,2,1,1)$ and last row $0\ 0\ 1\ 1\ |1\ 1$, i.e. of type ba. In Stage 1 of the reduction, this is replaced by 1-back splicing with a sum of blocks with $R = 1\ 1\ 1\ 1\ |0\ 0$. Squaring operations on these will give blocks with last row $1\ 1\ 1\ 0\ |1\ 0$, $1\ 1\ 0\ 0|1\ 0$, $1\ 0\ 0\ 0|1\ 0$ or $0\ 0\ 0\ 0|1\ 0$. Using the body and tail theorems, we can replace these by blocks with last row $0\ 0\ 1\ 1|0\ 1$ or $0\ 0\ 0\ 0|0\ 1$. Thus blocks with $t \geq 4$, $s = 2$ and type $*ba$ can be replaced by blocks of type $*bb$, $*cb$ where $* = a, b$ or c.

The same argument applies when $u > 0$ or $t > 4$, as squaring operations acting on the head will produce blocks with ω-sequence right lower than ω at some position in the head, and so also right lower than any term produced by a splicing operation on the body and tail.

In the case $s = 1$, blocks of type $*ba$ can be replaced by blocks of the same type, but with the b part of the body sequence having only one entry 1 instead of two. To see this, consider a block with ω-sequence $(2,2,2,2,1)$ and last row $0\ 0\ 1\ 1\ |1$, i.e. of type ba.

Stage 1 of the reduction, by 1-back splicing, replaces this with a sum of blocks with last row $1\ 1\ 1\ 0\ |0$. By the body theorem, we can replace these with blocks with last row $R = 0\ 0\ 1\ 1\ |0$. Squaring operations on these give blocks with last row $0\ 0\ 1\ 0\ |1$ or $0\ 0\ 0\ 0\ |1$, and in the first case we permute columns to obtain last row $0\ 0\ 0\ 1\ |1$. This is still of type ba, but there is only one entry 1 in the body section b. As in the case $s = 2$, the argument applies when $u > 0$ or $t > 4$.

To complete the proof of Theorem 30.1.1, we consider separately the 11 cases of (30.1) where $t \geq 4$ and $(u,s) \neq (0,0)$.

Case 1: $u \geq 2$, $t \geq 4$ and $s \geq 3$. Since the flow types $*a*$ and $**a$ are not needed when $s \geq 3$, we require only the 8 flow types

abb, *acb*, *bbb*, *bbc*, *bcb*, *cbb*, *cbc*, *ccb*.

As $\dim \mathrm{Q}^{\widehat{\omega}}(3) = 21$ in each case, this gives a spanning set with $7 \cdot 21 + 8 \cdot 21 = 315$ blocks.

Case 2: $u = 2$, $t \geq 4$ and $s = 2$. Since $s = 2$, the tail section can be changed to type a by a squaring operation. Thus we have to consider the 6 flow types

aca, *bca*, *cca*, *aba*, *bba*, *cba*

in addition to the 7 initial types and the 8 flow types of Case 1. We use the argument of Stage 3 to omit the cases $*ba$. Of the remaining 11 flow types, we still have $\dim \mathrm{Q}^{\widehat{\omega}}(3) = 21$ in the 5 cases $*bb$ and $*bc$, but $\dim \mathrm{Q}^{\widehat{\omega}}(3) = 14$ in the 3 cases $*cb$, and $\dim \mathrm{Q}^{\widehat{\omega}}(3) = 7$ in the 3 cases $*ca$. Thus the spanning set obtained in this way consists of $7 \cdot 21 + 5 \cdot 21 + 3 \cdot 14 + 3 \cdot 7 = 315$ blocks.

Case 3: $u = 2, t \geq 4, s = 1$. As the last row type aab does not occur when $s = 1$, the first step reduces R to one of the 6 forms

$$aaa,\ aac,\ abc,\ acc,\ bcc,\ ccc.$$

The second step is to sort the subblocks $\widehat{B}$ into standard form. The χ-trick produces new last rows of the further 8 types

$$aba,\ aca,\ bba,\ bbc,\ cbc,\ bca,\ cba,\ cca.$$

We thus have a spanning set consisting of the 6 types

$$\begin{array}{c|c|c} 2\,2 & 1\,1\,1\,1 & 0 \\ \hline 1\,1 & 1\,1\,1\,1 & 1 \end{array}, \quad \begin{array}{c|c|c} 2\,2 & 1\,1\,1\,1 & 1 \\ \hline 1\,1 & 1\,1\,1\,1 & 0 \end{array}, \quad \begin{array}{c|c|c} 2\,2 & 2\,2\,1\,1 & 1 \\ \hline 1\,1 & 0\,0\,1\,1 & 0 \end{array},$$

$$\begin{array}{c|c|c} 2\,2 & 2\,2\,2\,2 & 1 \\ \hline 1\,1 & 0\,0\,0\,0 & 0 \end{array}, \quad \begin{array}{c|c|c} 3\,2 & 2\,2\,2\,2 & 1 \\ \hline 0\,1 & 0\,0\,0\,0 & 0 \end{array}, \quad \begin{array}{c|c|c} 3\,3 & 2\,2\,2\,2 & 1 \\ \hline 0\,0 & 0\,0\,0\,0 & 0 \end{array},$$

and the 8 types

$$\begin{array}{c|c|c} 2\,2 & 2\,2\,2\,1 & 0 \\ \hline 1\,1 & 0\,0\,0\,1 & 1 \end{array}, \quad \begin{array}{c|c|c} 2\,2 & 2\,2\,2\,2 & 0 \\ \hline 1\,1 & 0\,0\,0\,0 & 1 \end{array}, \quad \begin{array}{c|c|c} 3\,2 & 2\,2\,2\,1 & 0 \\ \hline 0\,1 & 0\,0\,0\,1 & 1 \end{array}, \quad \begin{array}{c|c|c} 3\,2 & 2\,2\,1\,1 & 1 \\ \hline 0\,1 & 0\,0\,1\,1 & 0 \end{array},$$

$$\begin{array}{c|c|c} 3\,3 & 2\,2\,1\,1 & 1 \\ \hline 0\,0 & 0\,0\,1\,1 & 0 \end{array}, \quad \begin{array}{c|c|c} 3\,2 & 2\,2\,2\,2 & 0 \\ \hline 0\,1 & 0\,0\,0\,0 & 1 \end{array}, \quad \begin{array}{c|c|c} 3\,3 & 2\,2\,2\,1 & 0 \\ \hline 0\,0 & 0\,0\,0\,1 & 1 \end{array}, \quad \begin{array}{c|c|c} 3\,3 & 2\,2\,2\,2 & 0 \\ \hline 0\,0 & 0\,0\,0\,0 & 1 \end{array},$$

where $\widehat{B}$ is in standard form. Here we have used the manoeuvre of Stage 3 above to take R_2 as $0\ 0\ 0\ 1$ in the 3 cases $*ba$. These 14 cases give an upper bound of $5 \cdot 21 + 6 \cdot 14 + 3 \cdot 7 = 210$ for $\dim \mathrm{Q}^{\omega}(4)$.

Case 4: $u = 1, t \geq 4, s = 3$. As the type bcc does not occur when $u = 1$, the first step reduces R to one of the 6 forms

$$aaa,\ aab,\ aac,\ abc,\ acc,\ ccc.$$

The second step is to sort the subblocks $\widehat{B}$ into standard form. The χ-trick produces blocks with last rows of the form $1|0\ 0\ 1\ 1|1\ 1$ and the further 7 last row types

$$abb,\ acb,\ aca,\ cbb,\ cbc,\ ccb,\ cca.$$

We thus have a spanning set consisting of the following 6 types, where $\widehat{B}$ is in standard form:

$$\begin{array}{c|c|c} 2 & 1\,1\,1\,1 & 0\,0\,0 \\ \hline 1 & 1\,1\,1\,1 & 1\,1\,1 \end{array}, \quad \begin{array}{c|c|c} 2 & 1\,1\,1\,1 & 1\,1\,0 \\ \hline 1 & 1\,1\,1\,1 & 0\,0\,1 \end{array}, \quad \begin{array}{c|c|c} 2 & 1\,1\,1\,1 & 1\,1\,1 \\ \hline 1 & 1\,1\,1\,1 & 0\,0\,0 \end{array},$$

$$\begin{array}{c|c|c} 2 & 2\,2\,1\,1 & 1\,1\,1 \\ \hline 1 & 0\,0\,1\,1 & 0\,0\,0 \end{array}, \quad \begin{array}{c|c|c} 2 & 2\,2\,2\,2 & 1\,1\,1 \\ \hline 1 & 0\,0\,0\,0 & 0\,0\,0 \end{array}, \quad \begin{array}{c|c|c} 3 & 2\,2\,2\,2 & 1\,1\,1 \\ \hline 0 & 0\,0\,0\,0 & 0\,0\,0 \end{array},$$

together with the following 5 types, where $\widehat{B}$ is again in standard form:

$$\begin{array}{c|c|c} 2 & 2\,2\,1\,1 & 1\,1\,0 \\ \hline 1 & 0\,0\,1\,1 & 0\,0\,1 \end{array}, \quad \begin{array}{c|c|c} 2 & 2\,2\,2\,2 & 1\,1\,0 \\ \hline 1 & 0\,0\,0\,0 & 0\,0\,1 \end{array}, \quad \begin{array}{c|c|c} 3 & 2\,2\,1\,1 & 1\,1\,0 \\ \hline 0 & 0\,0\,1\,1 & 0\,0\,1 \end{array},$$

$$\begin{array}{c|c|c} 3 & 2\,2\,2\,2 & 1\,1\,0 \\ \hline 0 & 0\,0\,0\,0 & 0\,0\,1 \end{array}, \quad \begin{array}{c|c|c} 3 & 2\,2\,1\,1 & 1\,1\,1 \\ \hline 0 & 0\,0\,1\,1 & 0\,0\,0 \end{array}.$$

These 11 cases produce an upper bound of $6 \cdot 21 + 5 \cdot 14 = 210$ for $\dim \mathrm{Q}^{\omega}(4)$.

Case 5: $u = 1, t \geq 4, s = 2$. The last row type bcc does not occur when $u = 1$, so the first step reduces the last row to one of the 6 forms

$$aaa,\ aab,\ aac,\ abc,\ acc,\ ccc.$$

The second step is to sort the subblocks $\widehat{B}$ into standard form. The χ-trick produces new last rows of the form $1|0\,0\,1\,1|1\,1$ and the further 7 types

$$abb,\ acb,\ aca,\ cbb,\ cbc,\ ccb,\ cca.$$

We thus have a spanning set consisting of the following 6 types, where $\widehat{B}$ is in standard form:

$$\begin{array}{c|c|c} 2 & 1\,1\,1\,1 & 0\,0 \\ \hline 1 & 1\,1\,1\,1 & 1\,1 \end{array}, \quad \begin{array}{c|c|c} 2 & 1\,1\,1\,1 & 1\,0 \\ \hline 1 & 1\,1\,1\,1 & 0\,1 \end{array}, \quad \begin{array}{c|c|c} 2 & 1\,1\,1\,1 & 1\,1 \\ \hline 1 & 1\,1\,1\,1 & 0\,0 \end{array},$$

$$\begin{array}{c|c|c} 2 & 2\,2\,1\,1 & 1\,1 \\ \hline 1 & 0\,0\,1\,1 & 0\,0 \end{array}, \quad \begin{array}{c|c|c} 2 & 2\,2\,2\,2 & 1\,1 \\ \hline 1 & 0\,0\,0\,0 & 0\,0 \end{array}, \quad \begin{array}{c|c|c} 3 & 2\,2\,2\,2 & 1\,1 \\ \hline 0 & 0\,0\,0\,0 & 0\,0 \end{array},$$

together with the following 7 types, where $\widehat{B}$ is again in standard form:

$$\begin{array}{c|c|c} 2 & 2\,2\,1\,1 & 1\,0 \\ \hline 1 & 0\,0\,1\,1 & 0\,1 \end{array}, \quad \begin{array}{c|c|c} 2 & 2\,2\,2\,2 & 1\,0 \\ \hline 1 & 0\,0\,0\,0 & 0\,1 \end{array}, \quad \begin{array}{c|c|c} 2 & 2\,2\,2\,2 & 0\,0 \\ \hline 1 & 0\,0\,0\,0 & 1\,1 \end{array}, \quad \begin{array}{c|c|c} 3 & 2\,2\,1\,1 & 1\,0 \\ \hline 0 & 0\,0\,1\,1 & 0\,1 \end{array},$$

$$\begin{array}{c|c|c} 3 & 2\,2\,1\,1 & 1\,1 \\ \hline 0 & 0\,0\,1\,1 & 0\,0 \end{array}, \quad \begin{array}{c|c|c} 3 & 2\,2\,2\,2 & 1\,0 \\ \hline 0 & 0\,0\,0\,0 & 0\,1 \end{array}, \quad \begin{array}{c|c|c} 3 & 2\,2\,2\,2 & 0\,0 \\ \hline 0 & 0\,0\,0\,0 & 1\,1 \end{array}.$$

Since $s = 2$, the argument of Stage 3 shows that the flow types *aba* and *cba* are not required. These 13 cases give an upper bound of $6 \cdot 21 + 5 \cdot 14 + 2 \cdot 7 = 210$ for $\dim \mathrm{Q}^{\omega}(4)$.

Case 6: $u = 1, t \geq 4, s = 1.$ Since the types *aab* and *bcc* do not occur, Stage 1 reduces R to one of the 5 last row types

$$aaa,\ aac,\ abc,\ acc,\ ccc.$$

The second step is to sort the subblocks $\widehat{B}$ into standard form. The χ-trick produces new last rows of the further 5 last row types

$$aba,\ aca,\ cba,\ cbc,\ cca.$$

We thus have a spanning set consisting of the following 5 types, where $\widehat{B}$ is in standard form:

$$\begin{array}{c|c|c} 2 & 1\,1\,1\,1 & 0 \\ \hline 1 & 1\,1\,1\,1 & 1 \end{array}, \quad \begin{array}{c|c|c} 2 & 1\,1\,1\,1 & 1 \\ \hline 1 & 1\,1\,1\,1 & 0 \end{array}, \quad \begin{array}{c|c|c} 2 & 2\,2\,1\,1 & 1 \\ \hline 1 & 0\,0\,1\,1 & 0 \end{array},$$

$$\begin{array}{c|c|c} 2 & 2\,2\,2\,2 & 1 \\ \hline 1 & 0\,0\,0\,0 & 0 \end{array}, \quad \begin{array}{c|c|c} 3 & 2\,2\,2\,2 & 1 \\ \hline 0 & 0\,0\,0\,0 & 0 \end{array},$$

together with the following 5 types, where $\widehat{B}$ is again in standard form:

$$\begin{array}{c|c|c} 2 & 2\,2\,2\,1 & 0 \\ \hline 1 & 0\,0\,0\,1 & 1 \end{array}, \quad \begin{array}{c|c|c} 2 & 2\,2\,2\,2 & 0 \\ \hline 1 & 0\,0\,0\,0 & 1 \end{array}, \quad \begin{array}{c|c|c} 3 & 2\,2\,2\,1 & 0 \\ \hline 0 & 0\,0\,0\,1 & 1 \end{array},$$

$$\begin{array}{c|c|c} 3 & 2\,2\,1\,1 & 1 \\ \hline 0 & 0\,0\,1\,1 & 0 \end{array}, \quad \begin{array}{c|c|c} 3 & 2\,2\,2\,2 & 0 \\ \hline 0 & 0\,0\,0\,0 & 1 \end{array}.$$

Here we have used the manoeuvre of Stage 3 above to take R_2 as 0 0 0 1 in the cases *aba* and *cba*. These 10 cases give the upper bound $2 \cdot 21 + 6 \cdot 14 + 2 \cdot 7 = 140$ for $\dim \mathrm{Q}^{\omega}(4)$.

Case 7: $u=2, t\geq 4, s=0$. In this case $\omega = (3,3,2,2,2,2)$ is not the minimal decreasing sequence, which is $(3,1,1,1,1,1,1)$, but Proposition 7.2.6 can still be used to permute columns, because there are no right error terms which are not also left lower.

If there is no 0 in R, then it has type aa. Otherwise we can reduce the last row of the body to b or c, giving a spanning set made up of the 7 types

$$aa,\ ab,\ bb,\ cb,\ ac,\ bc,\ cc.$$

Reducing $\widehat{B}$ to standard form in all cases gives no further last row types to consider, so we obtain the upper bound $4\cdot 21+3\cdot 7=105$ for $\dim \mathrm{Q}^{\omega}(4)$.

Case 8: $u=1, t\geq 4, s=0$. This case is similar to Case 7, but as the head type b does not arise there is a spanning set consisting of blocks of the 5 last row types

$$aa,\ ab,\ cb,\ ac,\ cc.$$

As before, reducing $\widehat{B}$ to standard form in all cases gives no further last rows to be considered, giving the upper bound $2\cdot 21+2\cdot 7+14=70$ for $\dim \mathrm{Q}^{\omega}(4)$.

Case 9: $u=0, t\geq 4, s=3$. Since $u=0$, there is a spanning set consisting of blocks of the 5 last row types

$$aa,\ ab,\ ac,\ bc,\ cc.$$

Since $s=3$ we do not need the flow types ba and ca but only bb and cb, giving the upper bound $4\cdot 21+3\cdot 7=105$ for $\dim \mathrm{Q}^{\omega}(4)$.

Case 10: $u=0, t\geq 4, s=2$. Since $u=0$, there is a spanning set consisting of blocks of the 5 last row types

$$aa,\ ab,\ ac,\ bc,\ cc.$$

We also require the 3 flow types bb, ca and cb, but since $s=2$ we can use the argument of Stage 3 to omit the flow type ba. Thus we obtain the upper bound $3\cdot 21+4\cdot 7+14=105$ for $\dim \mathrm{Q}^{\omega}(4)$.

Case 11: $u=0, t\geq 4, s=1$. In this case the tail type b does not arise, and so we obtain an initial spanning set of blocks of the 4 last row types

$$aa,\ ac,\ bc,\ cc.$$

In the case bc, reduction of $\widehat{B}$ to standard form introduces further blocks of types ba and ca. In the case ba, R_2 can be taken in the form 0 0 0 1 by Stage 3, giving the upper bound $21+2\cdot 14+3\cdot 7=70$ for $\dim \mathrm{Q}^{\omega}(4)$.

This completes the proof of Theorem 30.1.1. □

30.2 Strongly spike-free polynomials

Definition 30.2.1 A polynomial $f \in \mathsf{P}^d(n)$ is **strongly spike-free** if no polynomial in the $\mathbb{F}_2\mathsf{M}(n)$-submodule generated by f has a spike as a term.

It follows from the definition that a hit polynomial is strongly spike-free. The classic example of a strongly spike-free polynomial which is not hit is Singer's polynomial $s_8 = x^6yz + xy^6z + xyz^6 \in \mathsf{P}^8(3)$ (see Remark 8.3.6). Since s_8 is not hit but is invariant mod hit elements, it is strongly spike-free.

The set $\mathsf{SSF}^d(n)$ of strongly spike-free polynomials is a $\mathbb{F}_2\mathsf{M}(n)$-submodule of $\mathsf{P}^d(n)$, and we define the **spike-free submodule** of $\mathsf{Q}^d(n)$ to be $\mathsf{SF}^d(n) = \mathsf{SSF}^d(n)/\mathsf{H}^d(n)$.

Proposition 30.2.2 *The* $\mathbb{F}_2\mathsf{M}(n)$*-submodule* $\mathsf{SF}^d(n)$ *of* $\mathsf{Q}^d(n)$ *is the transpose dual of the quotient* $\mathsf{K}^d(n)/\mathsf{J}^d(n)$ *of* $\mathsf{K}^d(n)$.

Proof Let $f \in \mathsf{P}^d(n)$ be strongly spike-free. Then $\langle s_* \cdot \sigma, f\rangle = \langle s_*, f \cdot \sigma\rangle = 0$ for all $\sigma \in \mathbb{F}_2\mathsf{M}(n)$ and all d-spikes $s_* \in \mathsf{DP}^d(n)$. Since every element of $\mathsf{J}^d(n)$ is a sum of elements of the form $s_* \cdot \sigma$, f is orthogonal to $\mathsf{J}^d(n)$. Hence $\mathsf{SF}^d(n) \subseteq (\mathsf{DP}^d(n)/\mathsf{J}^d(n))^{\mathrm{tr}}$. This argument can be reversed to prove $(\mathsf{DP}^d(n)/\mathsf{J}^d(n))^{\mathrm{tr}} \subseteq \mathsf{SSF}^d(n)$. As $\mathsf{Q}^d(n) = \mathsf{K}^d(n)^{\mathrm{tr}}$, it follows from the definitions that $\mathsf{SF}^d(n) = (\mathsf{K}^d(n)/\mathsf{J}^d(n))^{\mathrm{tr}}$. □

The filtration of $\mathsf{P}^d(n)$ by the left order on ω-sequences induces a corresponding filtration on the submodule $\mathsf{SSF}^d(n)$, as follows.

Definition 30.2.3 We define $\mathsf{SSF}^\omega(n) = \mathsf{SSF}^d(n) \cap \mathsf{P}^{\leq_\ell \omega}(n)$, the submodule of $\mathsf{SSF}^d(n)$ spanned by polynomials whose terms with highest ω-sequence in the left order are in $\mathsf{P}^\omega(n)$. We also define $\mathsf{SF}^\omega(n) = \mathsf{SSF}^\omega(n)/(\mathsf{SSF}^\omega(n) \cap \mathsf{H}^d(n))$.

Thus the submodule $\mathsf{SF}^d(n)$ of $\mathsf{Q}^d(n)$ has a filtration by the modules $\mathsf{SF}^\omega(n)$. In general, $\mathsf{SF}^\omega(n)$ is not a submodule of $\mathsf{Q}^\omega(n)$, but a submodule of $\mathsf{Q}^\omega(n)$ isomorphic to $\mathsf{SF}^\omega(n)$ is obtained by taking the terms with ω-sequence ω in elements of $\mathsf{SF}^\omega(n)$.

By considering d-spikes $s_* \in \mathsf{DP}^\omega(n)$, Proposition 30.2.2 can be localized.

Proposition 30.2.4 *For any decreasing sequence* $\omega \in \mathsf{Dec}(n)$, *the* $\mathbb{F}_2\mathsf{M}(n)$*-submodule* $\mathsf{SF}^\omega(n)$ *of* $\mathsf{Q}^\omega(n)$ *is dual to the quotient* $\mathsf{K}^\omega(n)/\mathsf{J}^\omega(n)$ *of* $\mathsf{K}^\omega(n)$. □

It follows from Propositions 6.5.4 and 9.5.4 that the local Kameko maps preserve spike-free submodules. We next relate strongly spike-free polynomials to the duplication map δ.

Proposition 30.2.5 *Let $\omega = (n-1, \omega')$ be a decreasing sequence, let $\delta : \mathsf{Q}^{\omega'}(n) \to \mathsf{Q}^{\omega}(n)$ be the duplication map, and let $\mathsf{SF}^{\omega'}(n) \subseteq \mathsf{Q}^{\omega'}(n)$ be the $\mathbb{F}_2\mathsf{GL}(n)$-submodule spanned by strongly spike-free polynomials. Then* $\mathrm{Ker}(\delta) \subseteq \mathsf{SF}^{\omega'}(n)$.

Proof Let $\delta^* : \mathsf{K}^{\omega}(n) \to \mathsf{K}^{\omega'}(n)$ be the linear dual of δ, let s' be a spike in $\mathsf{Q}^{\omega'}(n)$ and let $s = \delta(s')$ be its duplication in $\mathsf{Q}^{\omega}(n)$. Let s_* and s'_* be the d-spikes corresponding to s and s'. Then for any monomial $g \in \mathsf{Q}^{\omega'}(n)$, $\langle g, \delta^*(s_*)\rangle = \langle \delta(g), s_*\rangle$ is 1 if $g = s'$ and is 0 otherwise. Hence $\delta^*(s_*) = s'_*$.

Since δ^* is a $\mathbb{F}_2\mathsf{GL}(n)$-module map and $\mathsf{J}^{\omega}(n)$, $\mathsf{J}^{\omega'}(n)$ are the modules generated by s_* and s'_* respectively, δ^* maps $\mathsf{J}^{\omega}(n)$ to $\mathsf{J}^{\omega'}(n)$. Moreover, the restriction of δ^* to $\mathsf{J}^{\omega}(n)$ is the map ε_* of Section 24.1 for the case where the first two columns are equal, i.e. δ^* acts on $\mathsf{J}^{\omega}(n)$ by deleting the first column of such blocks, and sends all other blocks to 0.

Now let $k \in \mathrm{Ker}(\delta)$ and let $f' \in \mathsf{J}^{\omega'}(n)$. By Proposition 24.1.1, ε_* is surjective, and so $f' = \delta^*(f)$ where $f \in \mathsf{J}^{\omega}(n)$. Hence $\langle k, f'\rangle = \langle k, \delta^*(f)\rangle = \langle \delta(k), f\rangle = 0$, so f' is in the submodule of $\mathsf{Q}^{\omega'}(n)$ orthogonal to $\mathsf{J}^{\omega'}(n)$. But by Proposition 30.2.4 this submodule is $\mathsf{SF}^{\omega'}(n)$. □

Example 30.2.6 By Remark 8.3.6 and the hit relation (8.3), the polynomial $x_1^6x_2x_3 + x_1x_2^6x_3 + x_1x_2x_3^6$ is in the kernel of the duplication map $\delta : \mathsf{Q}^{\omega'}(3) \to \mathsf{Q}^{\omega}(3)$, where $\omega = (2,2,1,1)$. Similarly, $s_9 = x_1^6x_2x_3x_4 + x_1x_2^6x_3x_4 + x_1x_2x_3^6x_4 + x_1x_2x_3x_4^6$ is in the kernel of the duplication map $\delta : \mathsf{Q}^{\omega'}(3) \to \mathsf{Q}^{\omega}(3)$, where $\omega = (3,3,1,1)$.

To see that $\delta(s_9)$ is hit, we use a similar argument. By 1-back splicing in the second column, we obtain the symmetrization $\sigma_4(f)$ of $f = x_1^{10}x_2^5x_3^3x_4^3$. Since the exponents 5 and 10 can be exchanged by further 1-back splicing, $\sigma_4(f)$ is hit.

To see that s_9 is not hit, we compute $(Sq^2)^3(s_9) = \sigma_4(x_1^8x_2^4x_3^2x_4)$. Assume that s_9 is hit. Then $s_9 = Sq^1(g_1) + Sq^2(g_2) + Sq^4(g_4)$, where $g_i \in \mathsf{P}^{9-i}(4)$ for $i = 1,2,4$. Applying $(Sq^2)^3$ to this equation, we obtain $Sq^9Sq^1(g_4) = \sigma_4(x_1^8x_2^4x_3^2x_4)$ since $(Sq^2)^3Sq^1 = 0$, $(Sq^2)^4 = 0$ and $(Sq^2)^3Sq^4 = Sq^9Sq^1$. However Sq^9Sq^1 has excess 8 and $\deg g_4 = 5$, so $Sq^9Sq^1(g_4) = 0$. Hence s_9 is not hit.

In the above examples, $\mathrm{Ker}(\delta) = \mathsf{SF}^{\omega}(n)$. In Section 30.3 we show that this is false in general for $n = 4$. For example if $\omega = (3,3,3,3,2,1,1)$ then δ is an isomorphism, but $\dim \mathsf{SF}^{\omega}(4) = 15$. As another example, the iterates $s_{19}, s_{41}, \ldots$ of s_8 under the up Kameko map for $n = 3$ are all strongly spike-free, but each of them is mapped to the next by δ, and none of them are hit.

30.3 The spike-free modules $\mathsf{SF}^{\omega}(4)$

It follows from the results of Chapter 10 that the spike-free module $\mathsf{SF}^{\omega}(3) = 0$ except when $\omega = (3,\ldots,3,2,1,1)$, when $\dim \mathsf{SF}^{\omega}(3) = 1$. In the same way, each base case below gives rise to an infinite sequence of cases by iteration of the up Kameko map υ for $n = 4$. Further, each exceptional case for $n = 3$ gives rise to a base case with $\omega = (3,\ldots,3,2,1,1)$ for $n = 4$, where $\mathsf{SF}^{\omega}(4)$ has a 4-dimensional submodule obtained by taking the variables three at a time. A second sequence of base cases with $\omega = (3,2,2,1,\ldots,1)$ for $n = 4$ arises by combining the 3-variable case $(2,1,1)$ with a spike in the fourth variable.

We tabulate cases where $\mathsf{SF}^{\omega}(4) \neq 0$ for $d = \deg_2 \omega \leq 41$. Table (30.2) includes all the base cases, i.e. those where $\omega_1 < 4$, in the sense that the sequences $d = 8, 19, 41, \ldots$ and $d = 15, 23, 39, \ldots$ stabilize for $u \geq 2$ and $s \geq 2$ respectively.

The results of Section 24.4 give lower bounds for $\dim \mathsf{J}^{\omega}(4)$. By combining these with the lower bounds for $\dim \mathsf{SF}^{\omega}(4)$ obtained in this section, we obtain lower bounds for $\dim \mathsf{Q}^{\omega}(4)$. In some cases, the $\mathbb{F}_2\mathsf{GL}(4)$-composition factors of $\mathsf{SF}^{\omega}(4)$ are quoted without proof from computer work using MAPLE.

The last column of the table is quoted from the sources cited in Section 30.5, and gives the image of the 1-dimensional quotient $L(0)$ of $\mathsf{K}^{\omega}(4)/\mathsf{J}^{\omega}(4)$ under the dual algebraic transfer map $\phi_4^* : \mathbb{F}_2 \otimes_{\mathsf{GL}(4)} \mathsf{K}^d(4) \to \mathrm{Ext}_{\mathsf{A}_2}^{4,4+d}(\mathbb{F}_2, \mathbb{F}_2)$.

d	ω	$\dim \mathsf{SF}^{\omega}$	$\dim \mathsf{Q}^{\omega}$	SF^{ω}	$\mathrm{Im}(\phi_4^*)$
8	$(2,1,1)$	4	45	$L(1,1,1)$	$\cdot$
9	$(3,1,1)$	1	36	$L(0)$	h_1c_0
14	$(2,2,2)$	1	35	$L(0)$	d_0
15	$(3,2,2)$	4	60	$L(1)$	$\cdot$
16	$(2,3,2)$	4	4	$L(1)$	$\cdot$
17	$(3,1,1,1)$	1	46	$L(0)$	e_0
18	$(2,2,1,1)$	1	91	$L(0)$	f_0
19	$(3,2,1,1)$	20	140	M_{20}	$\cdot$
20	$(4,2,1,1)$	4	45	$L(1,1,1)$	$\cdot$
21	$(3,3,1,1)$	4	80	$L(1,1,1)$	$\cdot$
22	$(4,3,1,1)$	1	36	$L(0)$	h_2c_1
23	$(3,2,2,1)$	15	155	M_{15}	h_4c_0
31	$(3,2,2,2)$	4	74	$L(1,1,1)$	$\cdot$
32	$(4,2,2,2)$	1	35	$L(0)$	d_1
33	$(3,3,2,2)$	1	91	$L(0)$	p_0
34	$(4,3,2,2)$	4	60	$L(1)$	$\cdot$
38	$(4,3,1,1,1)$	1	46	$L(0)$	e_1
39	$(3,2,2,1,1)$	15	225	M_{15}	h_5c_0
40	$(4,2,2,1,1)$	1	91	$L(0)$	f_1
41	$(3,3,2,1,1)$	15	225	M_{15}	h_0c_2

(30.2)

In Table (30.2), M_{20} represents some module of dimension 20 with composition factors $L(1)$, $L(1,1)$, $L(1,1)$ and $L(1,1,1)]$ and M_{15} represents some module of dimension 15, such as $\mathsf{FL}^1(4)$ or $\mathsf{FL}^3(4)$, with composition factors $L(0)$, $L(1)$, $L(1,1)$ and $L(1,1,1)$.

In the rest of this section, we comment on the 14 cases in Table (30.2) where $\omega_1 < 4$. As the standard generators of $\mathsf{GL}(4)$, we use the matrices S_1, S_2, S_3 representing the switches of variables $x_1 \leftrightarrow x_2$, $x_2 \leftrightarrow x_3$ and $x_3 \leftrightarrow x_4$ respectively, and the standard transvection U which maps x_1 to $x_1 + x_2$ and fixes x_2, x_3 and x_4. In this section we abbreviate monomials $x_1^{a_1}x_2^{a_2}x_3^{a_3}x_4^{a_4}$ in $\mathsf{P}(4)$ by listing their exponents in the form $[a_1,a_2,a_3,a_4]$.

$\underline{d = 8,\ \omega = (2,1,1).}$ Singer's polynomial

$$s_8 = \begin{array}{l} 0\,1\,1 \\ 1 \\ 1 \end{array} + \begin{array}{l} 1 \\ 0\,1\,1 \\ 1 \end{array} + \begin{array}{l} 1 \\ 1 \\ 0\,1\,1 \end{array}$$

represents a $\mathsf{GL}(3)$-invariant in $\mathsf{Q}^8(3)$. The 4 polynomials $b_i \in \mathsf{P}^8(4)$ obtained by inserting a zero ith row into s_8 are linearly independent in $\mathsf{Q}^\omega(4)$. By using the (ordered) basis (b_1,b_2,b_3,b_4) and checking the generators of $\mathsf{GL}(4)$, we see that this matrix representation of $\mathbb{F}_2\mathsf{M}(4)$ is given by $A \mapsto (A^{\mathrm{tr}})^{-1}$. Hence $\mathsf{SF}^\omega(4)$ and $\mathsf{K}^\omega(4)/\mathsf{J}^\omega(4)$ are isomorphic to $L(1,1,1)$. The d-polynomial

$$f_8 = \begin{array}{l} 0\,1\,1 \\ 1 \\ 1 \end{array} + \begin{array}{l} 1\,0\,1 \\ 0\,1 \\ 1 \end{array} + \begin{array}{l} 1\,1 \\ 0\,0\,1 \\ 1 \end{array} + \begin{array}{l} 1\,1 \\ 1\,1 \\ 0\,1 \end{array}$$

was shown in Chapter 9 to generate $\mathsf{K}^8(3)/\mathsf{J}^8(3)$. Hence we can construct a basis for $\mathsf{K}^\omega(4)/\mathsf{J}^\omega(4)$ in the same way, by inserting a zero row into f_8.

$\underline{d = 9,\ \omega = (3,1,1).}$ Since the polynomial

$$s_9 = \begin{array}{l} 0\,1\,1 \\ 1 \\ 1 \\ 1 \end{array} + \begin{array}{l} 1 \\ 0\,1\,1 \\ 1 \\ 1 \end{array} + \begin{array}{l} 1 \\ 1 \\ 0\,1\,1 \\ 1 \end{array} + \begin{array}{l} 1 \\ 1 \\ 1 \\ 0\,1\,1 \end{array}$$

is symmetric and $s_9 \cdot U \sim s_9$, s_9 represents a $\mathsf{GL}(4)$-invariant in $\mathsf{Q}^\omega(4)$. Hence s_9 is strongly spike-free. We have seen in Example 30.2.6 that $\delta(s_9) = 0$, where δ is the duplication map.

To prove that s_9 is not hit, we argue as in Proposition 29.3.6. Assume that s_9 is hit. Then $s_9 = Sq^1(g_1) + Sq^2(g_2) + Sq^4(g_4)$, where $g_i \in \mathsf{P}^{9-i}(4)$ for $i =$

1,2,4. Since $Sq^2([6,1,1,1]) = [8,1,1,1]+[6,2,2,1]+[6,2,1,2]+[6,1,2,2]$ and the last three terms are in the kernel of $Sq^2Sq^2 = Sq^3Sq^1$, $(Sq^2)^3(s_9) = \sigma_4([8,4,2,1])$, where σ_4 denotes symmetrization in $\mathsf{P}(4)$.

Thus, by applying $(Sq^2)^3 = Sq^5Sq^1$ to the hit equation and using the Adem relation $Sq^5Sq^5 = Sq^9Sq^1$, we obtain $Sq^9Sq^1(g_4) = \sigma_4([8,4,2,1])$. Since Sq^9Sq^1 has excess 8 and $\deg g_4 = 5$, $Sq^9Sq^1(g_4) = 0$, giving a contradiction.

Since $f_8 \in \mathsf{K}^8(3)$, it follows that the d-polynomial

$$f_8v_4 = \begin{array}{c} 0\,1\,1 \\ 1 \\ 1 \\ 1 \end{array} + \begin{array}{c} 1\,0\,1 \\ 0\,1 \\ 1 \\ 1 \end{array} + \begin{array}{c} 1\,1 \\ 0\,0\,1 \\ 1 \\ 1 \end{array} + \begin{array}{c} 1\,1 \\ 1\,1 \\ 0\,1 \\ 1 \end{array}$$

is in $\mathsf{K}^9(4)$. Since $\mathsf{J}^9(4)$ is the submodule of $\mathsf{DP}^9(4)$ orthogonal to $\mathsf{SSF}^9(4)$ by Proposition 30.2.2 and $\langle f_8v_4, s_9\rangle \neq 0$, it follows that $f_8v_4 \notin \mathsf{J}^9(4)$.

Since $\omega = \omega^{\min}(9)$, Proposition 7.3.5 gives $\dim \mathsf{K}^\omega(4) = \dim \mathsf{Q}^\omega(4) \leq 36$. Since $\dim \mathsf{J}^\omega(4) = 35$, it follows that $\mathsf{K}^\omega(4)/\mathsf{J}^\omega(4)$ has dimension 1, generated by f_8v_4. Computer calculations confirm that f_8v_4 generates $\mathsf{K}^9(4)$, which has dimension 46 since $\dim \mathsf{K}^{(3,3)}(4) = 10$.

$d = 14$, $\omega = (2,2,2)$. We have shown that $\mathsf{J}^\omega(4) \cong \mathsf{FL}_2(4)$ has dimension 34. From Section 29.3, $\dim \mathsf{Q}^\omega(4) = 35$. By Proposition 29.3.6 $s_{14} = [6,6,1,1]+[1,1,6,6]$ is invariant mod hits. Hence $\mathsf{SF}^\omega(4)$ has dimension 1 and is generated by s_{14}.

Generators of $\mathsf{K}^\omega(4)/\mathsf{J}^\omega(4)$ for $\omega = (2,2,2)$, with over 30 terms, can be found in [71] and [173]. Since the Kronecker product of these d-polynomials with s_{14} is 1, this gives an alternative proof that s_{14} is not hit.

$d = 15$, $\omega = (3,2,2)$. Since s_8 is not hit, its product with a spike in x_4

$$s_8x_4^7 = \begin{array}{c} 0\,1\,1 \\ 1\,0\,0 \\ 1\,0\,0 \\ 1\,1\,1 \end{array} + \begin{array}{c} 1\,0\,0 \\ 0\,1\,1 \\ 1\,0\,0 \\ 1\,1\,1 \end{array} + \begin{array}{c} 1\,0\,0 \\ 1\,0\,0 \\ 0\,1\,1 \\ 1\,1\,1 \end{array}$$

is not hit. More generally, the polynomials obtained from $s_8x_4^7$ by permuting the variables are linearly independent mod hits, as can be seen by equating terms in x_i^7, $1 \leq i \leq 4$, in a possible hit equation.

The polynomial $s_8x_4^7$ is not strongly spike-free, since identification of all four variables to x gives x^{15}. By permuting the variables and adding a tail block, we obtain a strongly spike-free polynomial

$$s_{15} = \begin{array}{c} 1\,1\,1 \\ 0\,1\,1 \\ 1\,0\,0 \\ 1\,0\,0 \end{array} + \begin{array}{c} 1\,1\,1 \\ 1\,0\,0 \\ 0\,1\,1 \\ 1\,0\,0 \end{array} + \begin{array}{c} 1\,1\,1 \\ 1\,0\,0 \\ 1\,0\,0 \\ 0\,1\,1 \end{array} + \begin{array}{c} 1\,0\,0\,0 \\ 0\,1\,0\,0 \\ 0\,0\,1\,0 \\ 0\,0\,0\,1 \end{array}$$

which generates a submodule of dimension 4 of $\mathsf{SF}^{\omega}(4)$. With (b_1, b_2, b_3, b_4) as basis, where $b_1 = s_{15}$, $b_2 = b_1 \cdot S_1$, $b_3 = b_2 \cdot S_2$ and $b_4 = b_3 \cdot S_3$, the standard generators S_1, S_2, S_3 and U of $\mathsf{GL}(4)$ are represented by the matrix itself, and it follows that s_{15} generates a submodule of $\mathsf{SF}^{\omega}(4)$ isomorphic to $L(1)$.

By using 1-back splicing operations, we see that the image under duplication

$$\delta(s_8 x_4^7) = \begin{array}{c} 0\,0\,1\,1 \\ 1\,1\,0\,0 \\ 1\,1\,0\,0 \\ 1\,1\,1\,1 \end{array} + \begin{array}{c} 1\,1\,0\,0 \\ 0\,0\,1\,1 \\ 1\,1\,0\,0 \\ 1\,1\,1\,1 \end{array} + \begin{array}{c} 1\,1\,0\,0 \\ 1\,1\,0\,0 \\ 0\,0\,1\,1 \\ 1\,1\,1\,1 \end{array}$$

of $s_8 x_4^7$ is hit modulo elements with ω-sequence $<_l (3,3,2,2)$, and so $\mathsf{SF}^{\omega}(4) \subseteq \mathrm{Ker}(\delta)$. By Proposition 7.3.5, which applies since all ω-sequences $<_r \omega$ are also $<_l \omega$, $\dim \mathsf{Q}^{\omega}(4) \leq 60$. Since $\dim \mathsf{J}^{\omega}(4) = 56$, it follows from Proposition 30.2.4 that $\dim \mathsf{SF}^{\omega}(4) \leq 4$, and hence $\mathsf{SF}^{\omega}(4) \cong L(1)$. Since the d-polynomial

$$f_8 v_4^{(7)} = \begin{array}{l} 0\,1\,1 \\ 1 \\ 1 \\ 1\,1\,1 \end{array} + \begin{array}{l} 1\,0\,1 \\ 0\,1 \\ 1 \\ 1\,1\,1 \end{array} + \begin{array}{l} 1\,1 \\ 0\,0\,1 \\ 1 \\ 1\,1\,1 \end{array} + \begin{array}{l} 1\,1 \\ 1\,1 \\ 0\,1 \\ 1\,1\,1 \end{array}$$

is in $\mathsf{K}^{15}(4)$ and $\langle f_8 v_4^{(7)}, s_{15} \rangle = 1$, it represents a generator of $\mathsf{K}^{\omega}(4)/\mathsf{J}^{\omega}(4) \cong L(1)$.

<u>$d = 16$, $\omega = (2,3,2)$.</u> We show that the polynomial

$$s_{16} = \begin{array}{l} 0\,1 \\ 1\,1 \\ 1\,0\,1 \\ 0\,1\,1 \end{array} + \begin{array}{l} 0\,1 \\ 1\,0\,1 \\ 0\,1\,1 \\ 1\,1 \end{array} + \begin{array}{l} 0\,1 \\ 0\,1\,1 \\ 1\,1 \\ 1\,0\,1 \end{array}$$

is strongly spike-free, and that the cyclic permutations of s_{16} span a 4-dimensional subspace of $\mathsf{Q}^\omega(4)$ isomorphic to $L(1)$. Since $\langle f_{16}, s_{16}\rangle \neq 0$, where $f_{16} \in \mathsf{K}^{16}(4)$ is defined in Example 29.1.6, s_{16} is not hit. Since $\dim \mathsf{Q}^\omega(4) \leq 4$ by Example 29.1.6, $\mathsf{Q}^\omega(4) = \mathsf{SF}^\omega(4)$ and so $\mathsf{J}^\omega(4) = 0$ by duality.

To see that s_{16} is strongly spike-free, we note that $s_{16} \cdot (U - I_4) = [0,5,6,5] + [0,7,3,6] + [0,8,5,3]$ and $[7,3,6] \sim [8,3,5] + [7,4,5] \sim [8,5,3] + [9,4,3] \sim [8,5,3] + [5,8,3] \sim [8,5,3] + [5,6,5]$. Also $s_{16} \sim s_{16} \cdot S_2$ using the hit relations $[2,3,5,6] + [2,3,6,5] \sim [2,4,5,5]$, $[2,4,5,5] + [2,5,4,5] + [2,5,5,4] \sim [1,4,6,5] + [1,4,5,6] + [1,6,4,5] + [1,5,4,6] + [1,6,5,4] + [1,5,6,4]$ and $[1,4,5,6] \sim [1,4,8,3] \sim [1,8,4,3] \sim [1,5,4,6]$.

$d = 17, \omega = (3,1,1,1)$. The proof that

$$s_{17} = \begin{matrix} 0\,1\,1\,1 \\ 1 \\ 1 \\ 1 \end{matrix} + \begin{matrix} 1 \\ 0\,1\,1\,1 \\ 1 \\ 1 \end{matrix} + \begin{matrix} 1 \\ 1 \\ 0\,1\,1\,1 \\ 1 \end{matrix} + \begin{matrix} 1 \\ 1 \\ 1 \\ 0\,1\,1\,1 \end{matrix}$$

represents a $\mathsf{GL}(4)$-invariant element of $\mathsf{Q}^{17}(4)$ follows, as in the case of s_9, by checking that $s_{17} \cdot U \sim s_{17}$. Since it is invariant mod hits and contains no spikes, s_{17} is strongly spike-free. It follows from a similar argument to that of Example 30.2.6 that $\delta(s_{17})$ is hit.

The proof that s_{17} is not hit is similar to that for s_9. Assume that s_{17} is hit. Then $s_{17} = Sq^1(g_1) + Sq^2(g_2) + Sq^4(g_4) + Sq^8(g_8)$, where $g_i \in \mathsf{P}^{17-i}(4)$ for $i = 1,2,4,8$. Since $Sq^2([14,1,1,1]) = [16,1,1,1] + [14,2,2,1] + [14,2,1,2] + [14,1,2,2]$ and the last three terms are in the kernel of $Sq^2Sq^2 = Sq^3Sq^1$, $(Sq^2)^3(s_{17}) = \sigma_4([16,4,2,1])$.

Thus by applying $Sq^2Sq^2Sq^2 = Sq^5Sq^1$ to the hit equation and using the Adem relations $Sq^5Sq^5 = Sq^9Sq^1$ and $Sq^5Sq^9 = Sq^{13}Sq^1$, we obtain $Sq^9Sq^1(g_4) + Sq^{13}Sq^1(g_8) = \sigma_4([16,4,2,1])$. Since $Sq^{13}Sq^1$ has excess 12 and $\deg g_8 = 9$, the term in g_8 is 0, and a direct calculation shows that there is no polynomial $g_4 \in \mathsf{P}^{13}(4)$ such that $Sq^9Sq^1(g_4) = \sigma_4([16,4,2,1])$, giving a contradiction.

In more detail, $Sq^9Sq^1(f) = 0$ if $f \in \mathsf{P}^{13}(4)$ is a monomial with three even exponents. Since $Sq^9Sq^1 = Sq^1Sq^8Sq^1$, the Cartan formula shows that $Sq^9Sq^1(f)$ cannot have a term $[16,4,2,1]$ unless $f = [7,4,1,1]$, $[7,3,2,1]$ or $[7,3,1,2]$. Now $Sq^1([7,4,1,1] + [7,3,2,1] + [7,3,1,2]) = Sq^1([8,3,1,1])$ and $Sq^1Sq^8Sq^1([8,3,1,1]) = 0$.

Thus $Sq^9Sq^1([7,4,1,1]) = Sq^9Sq^1([7,3,2,1]+[7,3,1,2])$, and we are reduced to considering permutations of $[7,3,2,1]$. But $Sq^9Sq^1(\sigma_4([7,3,2,1])) = \sigma_4([16,4,2,1])+$ left lower terms, such as $\sigma_4([12,8,2,1])$.

Since $\dim \mathsf{J}^\omega(4) = 45$, $\dim \mathsf{Q}^\omega(4) = \dim \mathsf{K}^\omega(4) \geq 46$. We refer to Sum [204] for a proof that $\dim \mathsf{Q}^\omega(4) = 46$. Lê Minh Hà [71] gives a 44-term generator f_{17}^* of $\mathsf{K}^{17}(4)/\mathsf{J}^{17}(4)$. Since $\langle f_{17}^*, s_{17}\rangle = 1$, this gives an alternative proof that s_{17} is not hit.

<u>$d = 18, \omega = (2,2,1,1)$.</u> The polynomial

$$s_{18} = \begin{array}{l} 1\,1 \\ 1\,1 \\ 0\,0\,1 \\ 0\,0\,0\,1 \end{array} + \begin{array}{l} 0\,0\,1 \\ 0\,0\,0\,1 \\ 1\,1 \\ 1\,1 \end{array} + \begin{array}{l} 1\,1 \\ 1\,1 \\ 0\,1\,1 \\ 0\,1\,1 \end{array} + \begin{array}{l} 0\,1\,1 \\ 0\,1\,1 \\ 1\,1 \\ 1\,1 \end{array}$$

represents a $\mathsf{GL}(4)$-invariant element of $\mathsf{Q}^{18}(4)$. It is clearly invariant under S_1 and S_3. For invariance under S_2, we note that $[3,3,4,8] \sim [3,3,8,4]$ by 1-back splicing and that $[3,3,4,8]+[3,4,3,8] \sim [4,3,3,8]$ by 2-back splicing. Thus by permuting variables we obtain $[3,3,4,8]+[4,8,3,3]+[3,4,3,8]+[4,3,8,3] \sim [4,3,3,8]+[8,4,3,3]+[8,3,4,3] \sim [4,3,3,8]+[8,3,3,4] \sim 0$.

We also have $[3,3,6,6] \sim [5,3,5,5]+[3,5,5,5]$ by the χ-trick applied to a horizontal splitting. It follows by permuting variables that $[3,3,6,6]+[6,6,3,3]+[3,6,3,6]+[6,3,6,3]$ is hit. Hence s_{18} is invariant mod hits under S_2. Finally, $s_{18}\cdot(U - I_4) = [1,5,6,6]+[4,8,3,3]+[2,10,3,3] \bmod \mathsf{P}^{<\omega}(4)$, and the χ-trick applied to a horizontal splitting gives $[1,5,6,6] \sim [2,6,5,5] \sim [4,8,3,3]+[2,10,3,3]$.

We note that s_{18} contains terms with ω-sequence $(2,4,2) >_l (2,2,1,1)$. However, the reductions $[3,3,6,6] \sim [4,3,5,6]+[3,4,5,6]$, $[3,4,5,6] \sim [3,2,5,8]+[5,2,5,6]$, $[5,2,5,6] \sim [6,1,6,5] \sim [8,1,6,3]+[6,1,8,3]$ and $[8,1,6,3] \sim [8,2,5,3]$ show that $s_{18} \sim k_{18} = [2,3,5,8]+[2,5,8,3]+[2,8,3,5]+[3,2,5,8]+[3,5,8,2]+[3,8,5,2]+[3,8,2,5]+[5,2,8,3]+[5,3,2,8]+[8,2,5,3]+[8,5,2,3]+[8,5,3,2]$.

We sketch a proof that s_{18} is not hit. This is similar to that given above for s_{17}, but a better argument is clearly desirable. We have $Sq^5Sq^1(s_{18}) = \sigma_4([8,6,6,4])+[4,8,8,4]+[8,4,4,8]$. To show that this is not $Sq^9Sq^1(g_4)$ for any $g_4 \in \mathsf{P}^{14}(4)$, we argue as follows.

Let $f \in \mathsf{P}^{14}(4)$ be a monomial with exponents in decreasing order. If f has all exponents even or some exponent ≥ 8, then $Sq^9Sq^1(f) = 0$. Monomials f with all exponents odd can be removed from g_4 by considering the maximal term in left order. By considering $Sq^1(m)$ for $m = [7,3,2,1]$, $[6,3,3,1]$ and

$[5,4,3,1]$ and noting that $Sq^5Sq^1([5,4,4,1]) = Sq^3Sq^3([5,4,4,1]) = 0$, we can remove permutations of $[7,3,2,2]$, $[6,3,3,2]$ and $[5,4,3,2]$. Of the remaining monomials f, $Sq^9Sq^1(f)$ contains a permutation of $[8,6,6,4]$ only for $f = [6,4,3,1]$ and a permutation of $[8,8,4,4]$ only for $f = [7,4,2,1]$, $[6,4,3,1]$ and $[4,4,3,3]$. Permutations of $[7,4,2,1]$ can be removed from g_4 by considering the left maximal term $[16,4,2,2]$.

Next let $g_4' = [1,4,3,6] + [1,6,3,4] + [1,6,4,3] + [3,1,4,6] + [3,1,6,4] + [3,4,1,6] + [3,4,6,1] + [3,6,1,4] + [3,6,4,1] + [4,1,3,6] + [4,6,1,3] + [4,6,3,1]$. Then $Sq^9Sq^1(g_4') = Sq^5Sq^1(s_{18}) + 12$ permutations of the monomials $[12,6,4,2]$ and $[10,8,4,2]$.

To complete the argument, we remove permutations of $[5,5,2,2]$ by using the relation $Sq^5Sq^1([6,6,1,1] + [5,5,2,2]) = 0$, and then consider the permutations of $[12,6,4,2]$ which occur in $Sq^9Sq^1(g_4')$, $Sq^9Sq^1([6,6,1,1])$ and $Sq^9Sq^1([6,5,2,1])$.

Since $\dim \mathsf{J}^{\omega}(4) = 90$, $\dim \mathsf{Q}^{\omega}(4) = \dim \mathsf{K}^{\omega}(4) \geq 91$. We refer to Sum [204] for a proof that $\dim \mathsf{Q}^{\omega}(4) = 91$, and hence that $\dim \mathsf{K}^{\omega}(4) = 91$. We do not know a generator of $\mathsf{K}^{\omega}(4)/\mathsf{J}^{\omega}(4)$.

<u>$d = 19$, $\omega = (3,2,1,1)$.</u> It follows from the 3-variable case that

$$s_{19} = \begin{matrix} 1\,0\,1\,1 \\ 1\,1 \\ 1\,1 \\ 0 \end{matrix} + \begin{matrix} 1\,1 \\ 1\,0\,1\,1 \\ 1\,1 \\ 0 \end{matrix} + \begin{matrix} 1\,1 \\ 1\,1 \\ 1\,0\,1\,1 \\ 0 \end{matrix}$$

and its permutations represent 4 linearly independent elements of $\mathsf{SF}^{\omega}(4)$. It follows by applying the transvection $U_{1,4}$ (i.e. $x_1 \mapsto x_1 + x_4$) to s_{19} that $t_{19} = [12,3,3,1] + [1,3,3,12] + [1,13,3,2] + [2,13,3,1] + [1,3,13,2] + [2,3,13,1]$ is in the submodule generated by s_{19}. We show below that t_{19} is not hit.

The polynomial t_{19} is the sum of two permutations of $g_{19} = [3,3,1,12] + [13,3,1,2] + [3,13,1,2]$. By 1-back splicing, it is easy to see that $t_{19} \cdot (U - I_4) = [9,7,1,2] + [9,7,2,1] + [5,11,1,2] + [5,11,2,1]$ is hit. Since t_{19} is invariant under S_2, it generates a submodule of $\mathsf{Q}^{\omega}(4)$ of dimension ≤ 6, with basis given by permutations of t_{19}. Thus s_{19} generates a submodule of dimension 10 of $\mathsf{SF}^{\omega}(4)$. It can be seen that the head of this module is isomorphic to $L(1,1,1)$, and the socle to $L(1,1)$.

The polynomial g_{19} generates a submodule of dimension 16 of $\mathsf{Q}^{\omega}(4)$, with a submodule generated by s_{19} and a quotient isomorphic to $L(1,1)$. To see this, note that g_{19} is invariant under S_2 and maps to s_{19} under the singular transformation $x_4 \mapsto x_1$. As for t_{19}, it follows that the head of the module generated by g_{19} is isomorphic to $L(1,1)$, with a basis given by permutations of g_{19}.

A final 4-dimensional composition factor of $\mathsf{SF}^\omega(4)$ is generated by $k_{19} = [3,12,1,3]+[3,1,12,3]+[3,4,1,11]+[3,1,4,11]+[1,14,1,3]+[1,1,14,3]+[1,6,1,11]+[1,1,6,11]$ and is isomorphic to $L(1)$, with basis given by permutations of h_{19}. To see this, we argue as follows.

As k_{19} is symmetric in x_2 and x_3, it suffices to prove that $f_1 = k_{19} \cdot (S_1 - I_4)$ and $f_2 = k_{19} \cdot (U - I_4)$ are in the submodule generated by g_{19}. We have $f_1 = [1,6,1,11]+[1,3,4,11]+[4,3,1,11]+[6,1,1,11]+[3,1,4,11]+[3,4,1,11]+[1,14,1,3]+[14,1,1,3]+[1,3,12,3]+[3,1,12,3]+[3,12,1,3]+[12,3,1,3]$. Using the 1-back splicing relations $[1,6,1,11]+[1,3,4,11]+[4,3,1,11] \sim [1,3,2,13]+[2,3,1,13]$, $[1,14,1,3] \sim [2,13,1,3]+[1,13,2,3]$, and the corresponding relations with x_1 and x_2 switched, f_1 is reduced to the sum of four permutations of g_{19}.

Also, $f_2 = [1,3,4,11]+[1,6,1,11]+[2,5,1,11]+[1,3,12,3]+[1,14,1,3]+[2,13,1,3] \bmod \mathsf{P}^{\leq\omega}(4)$, and the relations $[1,3,4,11] \sim [1,3,2,13]+[1,5,2,11]$, $[1,14,1,3] \sim [2,13,1,3]+[1,13,2,3]$ and $[2,5,1,11] \sim [1,5,2,11]+[1,6,1,11]$ reduce f_2 to a permutation of g_{19}. The proof that the head of the module generated by h_{19} is isomorphic to $L(1)$ follows by checking the matrix representation using a basis of permutations of h_{19}.

To prove that no nonzero element f in the 20-dimensional submodule M generated by k_{19} is hit, we show that $Sq^{17}Sq^5Sq^1$ maps M isomorphically to $\mathsf{P}^\rho(4)$, where $\rho = (0,1,2,2,1)$. The element $Sq^{17}Sq^5Sq^1 = Sq(7,3,1)$ is the top class of the subalgebra $\mathsf{A}_2(2)$ generated by Sq^1, Sq^2 and Sq^4. Hence a hit equation of the form $f = Sq^1(g_1) + Sq^2(g_2) + Sq^4(g_4) + Sq^8(g_8)$ implies that $Sq^{17}Sq^5Sq^1(f) = Sq^{17}Sq^5Sq^1Sq^8(g_8)$. The result follows, as $Sq^{17}Sq^5Sq^1Sq^8 = Sq^{25}Sq^5Sq^1$ has excess 19, and so it maps a polynomial g_8 of degree 11 to zero.

Thus we need to check that the images under $Sq^{17}Sq^5Sq^1$ in $\mathsf{P}^\rho(4)$ of the 20 basis elements described above are linearly independent. The 4 permutations of s_{19} can be disregarded, since each involves only three of the variables. For the remaining 16 basis elements, we check that the highest monomials in left order with $\rho = (0,1,2,2,1)$ in their images under $Sq^{17}Sq^5Sq^1$ are all different, so that no linear combination of them is in the kernel of $Sq^{17}Sq^5Sq^1$. We omit the details.

By Proposition 7.3.5, which applies since all ω-sequences $<_r \omega$ are also $<_l \omega$, $\dim \mathsf{Q}^\omega(4) \leq 140$. Since $\dim \mathsf{J}^\omega(4) \geq 120$ from Section 24.4, $\dim \mathsf{SF}^\omega(4) \leq 20$. It follows from the argument above that $M = \mathsf{SF}^\omega(4)$. We shall see below that for $\omega' = (3,3,2,1,1)$, $\dim \mathsf{SF}^{\omega'}(4) = 15$. Computer calculation confirms that $\delta(t_{19}) = 0$, where δ is the duplication map.

For the dual $\mathsf{K}^{\omega}(4)/\mathsf{J}^{\omega}(4)$, we have the Kameko iterate for $n = 3$ of f_8, i.e.

$$f_{19} = f_8^{(2)} v_1 v_2 v_3 = \begin{matrix} 1011 \\ 11 \\ 11 \end{matrix} + \begin{matrix} 1101 \\ 101 \\ 11 \end{matrix} + \begin{matrix} 111 \\ 1001 \\ 11 \end{matrix} + \begin{matrix} 111 \\ 111 \\ 101 \end{matrix},$$

and we can proceed, as in degree 8, to get 4 linearly independent elements of $\mathsf{K}^{19}(4)/\mathsf{J}^{19}(4)$. By computer calculation, the submodule of $\mathsf{K}^{19}(4)$ generated by f_{19} has codimension 6, and as $\mathsf{J}^{19}(4)$ has no quotient of dimension < 20, it follows that f_{19} generates a submodule of dimension 14 in $\mathsf{K}^{19}(4)/\mathsf{J}^{19}(4)$, with composition factors $L(1)$, $L(1,1)$ and $L(1,1,1)$. We do not know a generator of the quotient $\mathsf{K}^{19}(4)/\langle f_{19}\rangle \cong L(1,1)$.

$\underline{d = 21, \omega = (3,3,1,1).}$ We show below that the polynomial

$$s_{21} = \begin{matrix} 1 \\ 11 \\ 11 \\ 0111 \end{matrix} + \begin{matrix} 1 \\ 11 \\ 0111 \\ 11 \end{matrix} + \begin{matrix} 1 \\ 0111 \\ 11 \\ 11 \end{matrix} + \begin{matrix} 0011 \\ 11 \\ 11 \\ 11 \end{matrix}$$

is not hit. Iterated 1-back splicing gives the hit relation $[12,3,3,3] \sim [3,12,3,3]+[3,3,12,3]+[3,3,3,12]$, which shows that s_{21} is symmetric in the last 3 variables. The permutations of s_{21} span a 4-dimensional vector space V with basis (b_1,b_2,b_3,b_4) given by the cyclic permutations $b_1 = s_{21}$, $b_2 = b_1 \cdot S_1$, $b_3 = b_2 \cdot S_2$, $b_4 = b_3 \cdot S_3$. It is straightforward to verify by 1-back splicing arguments that V is invariant mod hits under the standard transvection U. More precisely, the matrix representation of $\mathsf{GL}(4)$ sends S_1, S_2 and S_3 to themselves, and U to its transpose. Hence V is a submodule of $\mathsf{SF}^{\omega}(4)$ and $V \cong L(1,1,1)$.

To prove that s_{21} is not hit, we argue in a similar way to the case of t_{19}. It suffices to show that $Sq^{17}Sq^5Sq^1(s_{21}) \neq 0$, since $Sq^{17}Sq^5Sq^1Sq^8 = Sq^{25}Sq^5Sq^1$ has excess 19, and so it vanishes on all polynomials of degree 13.

We show that the monomial $[24,12,6,2]$ is a term of $Sq^{17}Sq^5Sq^1(s_{21})$. By comparing the degree in each variable, we need consider only the term $[14,3,3,1]$ in s_{21}. By considering the α-count in each variable, we need consider only the term $[14,3,3,2]$ of $Sq^1([14,3,3,1])$. Hence we are reduced to showing that $[24,12,6]$ is a term in $Sq^{17}Sq^5([14,3,3])$. Once again, by considering the α-sequence, we can reduce to $Sq^{17}([14,5,6]+[14,6,5])$. Since $[24,12]$ is a term of $Sq^{16}([14,6])$ but not of $Sq^{17}([14,5])$, this completes the proof.

$\mathsf{K}^{21}(4)$ has a submodule $\mathsf{K}^{(3,3,3)}(4) = \mathsf{J}^{(3,3,3)}(4)$ of dimension 14 isomorphic to $\mathsf{FL}_3(4)$, with the quotient $\mathsf{K}^{\omega}(4)$. It was shown in Section 24.4 that $\mathsf{J}^{\omega}(4)$ has dimension ≥ 76, with a submodule isomorphic to $\mathsf{FL}_{1,3}(4)$. As the argument above shows that $\mathsf{SF}^{\omega}(4)$ has a $\mathbb{F}_2\mathsf{GL}(4)$-submodule of dimension 4, $\dim \mathsf{K}^{\omega}(4) = \dim \mathsf{Q}^{\omega}(4) \geq 80$. We refer to Sum [204] for a proof that this lower bound is exact.

$\underline{d = 23,\ \omega = (3,2,2,1).}$ The proof that

$$s_{23} = s_8 x_4^{15} = \begin{matrix} 011 \\ 100 \\ 100 \\ 1111 \end{matrix} + \begin{matrix} 100 \\ 011 \\ 100 \\ 1111 \end{matrix} + \begin{matrix} 100 \\ 100 \\ 011 \\ 1111 \end{matrix}$$

is not hit, is strongly spike-free and generates a submodule of $\mathsf{Q}^{\omega}(4)$ isomorphic to $L(1)$ follows the same pattern as for $s_8 x_4^7$ in degree 15. Using the basis $b_1 = s_{23}$, $b_2 = b_1 \cdot S_1$, $b_3 = b_2 \cdot S_2$, $b_4 = b_3 \cdot S_3$, a matrix $A \in \mathsf{GL}(4)$ is represented by A itself.

Let $g_{23} = [1,1,7,14] + [1,6,7,9] + [6,1,7,9]$. The singular transformation $x_3 \mapsto x_4$ sends g_{23} to s_{23}, so the 10-dimensional submodule of $\mathsf{Q}^{23}(4)$ has basis given by 6 permutations of g_{23} and 4 permutations of s_{23}. Clearly g_{23} is invariant under S_1. The following sequence of splices shows that g_{23} is invariant mod hits under S_3, so we need only 6 permutations of g_{23}. We have (1) $[1,6,7,9] + [1,6,9,7] \sim [1,8,7,7] \sim [1,4,11,7] + [1,4,7,11] \sim [1,2,13,7] +$ $1,2,7,13]$ and (2) $[1,1,7,14] \sim [2,1,7,13] + [1,2,7,13]$, and we conclude by applying $S_1 + I_4$ to relation (1) and applying $S_3 + I_4$ to relation (2).

A 14-dimensional submodule of $\mathsf{Q}^{\omega}(4)$ has basis given by these 10 polynomials, together with 4 permutations of $k_{23} = [1,3,5,14] + [1,3,14,5] + [3,3,5,12] + [3,3,12,5] + [3,7,5,8] + [3,7,8,5] + [1,7,3,12] + [1,7,12,3]$. To see this, we apply switches S_1, S_2, S_3 successively to k_{23}. The head of the module generated by k_{23} is isomorphic to $L(1,1,1)$. (These assertions are computer results.)

A 39-term representative for the generator of the $L(0)$ summand is given by $c_{23} = \sigma_4([1,1,6,15]) + [1,1,7,14] + [1,7,1,14] + [1,7,14,1] + [7,1,1,14] + [7,1,14,1] + [7,14,1,1] + [1,3,7,12] + [1,7,3,12] + [1,12,3,7] + [3,1,12,7] + [3,7,12,1] + [1,3,13,6] + [1,6,3,13] + [1,13,3,6] + [3,1,6,13] + [3,13,6,1] + [1,7,9,6] + [1,9,6,7] + [6,1,9,7] + [6,1,9,7] + [6,9,7,1] + [9,1,7,6] + [9,6,1,7] + [9,6,7,1] + [9,7,1,6] + [3,5,6,9] + [5,3,9,6]$.

To show that the set of 4 permutations of s_{23}, 6 of g_{23}, 4 of k_{23} and c_{23} is linearly independent mod hits, we argue in a similar way to the case of t_{19}. Since $Sq^{17}Sq^5Sq^1Sq^8 = Sq^{25}Sq^5Sq^1$ has excess 19, it vanishes on all polynomials of degree 15. Hence it suffices to show that $Sq^{17}Sq^5Sq^1(f) \neq 0$, where f is any nonzero linear combination of these 15 polynomials. Thus we apply $Sq^{17}Sq^5Sq^1$ and consider the terms with ω-sequence $<_l (0,1,1,3,1)$.

The first 14 polynomials map to linearly independent elements with $\omega = (0,1,1,1,2)$, while $Sq^{17}Sq^5Sq^1(c_{23})$ is the symmetrization $\sigma_4([32,8,4,2])$. It is not sufficient to take the left maximal monomial for the first 14 polynomials, since $[15,1,1,6]+[15,1,6,1]+[15,6,1,1]$ and $[7,9,1,6]+[7,9,6,1]+[7,14,1,1]$ give the same highest monomial $[24,16,4,2]$. One way to resolve this is first to eliminate $[15,1,1,6]+[15,1,6,1]+[15,6,1,1]$, as the only basis element giving the monomial $[18,16,8,4]$ on applying $Sq^{17}Sq^5Sq^1$, and then to use the left maximal monomial for the remaining 13 basis elements.

Computer calculation shows that $f_8 v_4^{(15)}$ generates $\mathsf{K}^{\omega}(4)$, which has dimension 155. The submodule $\mathsf{J}^{\omega}(4)$ has codimension 15. The arguments on $\mathsf{SF}^{\omega}(4)$ given above suggest that $\mathsf{K}^{\omega}(4)/\mathsf{J}^{\omega}(4) \cong \mathsf{FL}^1(4)$.

$\underline{d=31,\ \omega=(3,2,2,2).}$ We show that $\mathsf{SF}^{\omega}(4)$ has a $\mathbb{F}_2\mathsf{GL}(4)$-submodule V of dimension 4 isomorphic to $L(1,1,1)$, generated by $s_{31} = [8,7,7,9]+[12,3,3,13]$. It follows that $\dim \mathsf{K}^{\omega}(4) = \dim \mathsf{Q}^{\omega}(4) \geq 74$. We refer to Sum [204] for a proof that this lower bound is exact.

We show below that s_{31} is not hit. Let $b_1 = s_{31}$, $b_2 = b_1 \cdot S_1$, $b_3 = b_2 \cdot S_2$, $b_4 = b_3 \cdot S_3$. The hit relation $[3,3,13,12]+[3,13,3,12] \sim [7,7,9,8]+[7,9,7,8]$ shows that the permutations of s_{31} span a 4-dimensional vector space V with basis (b_1,b_2,b_3,b_4). Using 1-back splicing and the relation $[1,13,3,14]+[1,14,7,9]+[6,7,9,9]+[12,3,3,13] \sim 0$, we see that V is invariant mod hits under the standard transvection U. More precisely, these arguments show that the matrix representation of $\mathsf{GL}(4)$ sends S_1, S_2 and S_3 to themselves and U to its transpose. Hence V is a submodule of $\mathsf{SF}^{\omega}(4)$ and $V \cong L(1,1,1)$.

To show that s_{31} is not hit, we show $Sq^{17}Sq^5Sq^1(s_{31}) \neq 0$, which reduces the problem to evaluation of $Sq^{25}Sq^5Sq^1$ on monomials of degree 23. We omit the details.

$\underline{d=33,\ \omega=(3,3,2,2).}$ Computer calculation shows that the 12-term polynomial $t_{33} = [3,5,14,11]+[3,11,14,5]+[3,14,5,11]+[3,14,11,5]+[5,3,11,14]+[5,11,3,14]+[11,3,14,5]+[11,5,3,14]+[11,14,5,3]+[14,5,3,11]+[14,5,11,3]+[14,11,3,5]$ is invariant mod hits.

Since $Sq^{17}Sq^5Sq^1(t_{33}) \neq 0$, the proof that t_{33} is not hit is similar to that in degrees 23 and 31. It depends on evaluation of $Sq^{25}Sq^5Sq^1$ on polynomials of degree 25.

$\underline{d = 39, \omega = (3,2,2,1,1).}$ The proof that

$$s_{39} = \begin{matrix} 1\,1\,1\,1\,1 \\ 0\,1\,1 \\ 1\,0\,0 \\ 1\,0\,0 \end{matrix} + \begin{matrix} 1\,1\,1\,1\,1 \\ 1\,0\,0 \\ 0\,1\,1 \\ 1\,0\,0 \end{matrix} + \begin{matrix} 1\,1\,1\,1\,1 \\ 1\,0\,0 \\ 1\,0\,0 \\ 0\,1\,1 \end{matrix}$$

is not hit, is strongly spike-free and generates a submodule of $\mathsf{Q}^{\omega}(4)$ isomorphic to $L(1)$ follows the same pattern as for $s_8x_4^7$ in degree 15 and $s_8x_4^{15}$ in degree 23.

Computer calculation shows that 'tail duplication' (i.e. duplication of the last nonzero column of each block) gives an isomorphism from $\mathsf{SF}^{(3,2,2,1)}(4)$ to $\mathsf{SF}^{\omega}(4)$. In particular, a generator is given by tail duplication of c_{23}. Computer calculation also shows that $f_8v_4^{(31)}$ generates $\mathsf{K}^{39}(4)$, which has dimension 225.

$\underline{d = 41, \omega = (3,3,2,1,1).}$ The following results depend on computer calculation. The submodule of $\mathsf{Q}^{41}(4)$ generated by

$$s_{41} = \begin{matrix} 1\,1\,0\,1\,1 \\ 1\,1\,1 \\ 1\,1\,1 \\ 0 \end{matrix} + \begin{matrix} 1\,1\,1 \\ 1\,1\,0\,1\,1 \\ 1\,1\,1 \\ 0 \end{matrix} + \begin{matrix} 1\,1\,1 \\ 1\,1\,1 \\ 1\,1\,0\,1\,1 \\ 0 \end{matrix}$$

is isomorphic to $L(1,1,1)$, with the basis given by cyclic permutation of the variables. The duplicate $g_{41} = \delta(g_{19}) = [7,7,3,24] + [27,7,3,4] + [7,27,3,4]$ generates a 10-dimensional submodule of $\mathsf{Q}^{\omega}(4)$, with submodule generated by s_{41} and quotient $L(1,1)$. The duplicate $k_{41} = \delta(k_{19}) = [7,24,3,7] + [7,3,24,7] + [7,8,3,23] + [7,3,8,23] + [3,28,3,7] + [3,3,28,7] + [3,12,3,23] + [3,3,12,23]$ generates a submodule of $\mathsf{Q}^{\omega}(4)$ of dimension 14.

A generator of $\mathsf{SF}^{\omega}(4)$ is given by the 12-term polynomial $u_{41} = [3,3,5,30] + [3,3,30,5] + [3,7,5,26] + [4,7,11,19] + [7,4,11,19] + [7,3,5,26] + [3,7,25,6] + [7,3,25,6] + [7,7,8,19] + [7,7,19,8] + [5,6,11,19] + [19,11,5,6]$. The polynomial u_{41} is invariant mod hits under permutations of the variables, and generates the cokernel of the duplication map $\delta : \mathsf{SF}^{(3,2,1,1)}(4) \to \mathsf{SF}^{(3,3,2,1,1)}(4)$.

Adding 4 permutations of s_{41}, 6 of g_{41} and 4 of k_{41} to u_{41}, we obtain a 66-term spike-free polynomial w_{41} which is invariant mod hits. To prove that

w_{41} is not hit, we project on to $\mathsf{P}(3)$ by setting $x_1 = x_2$ to obtain $\upsilon^2(s_8) = [7,7,27]+[7,27,7]+[27,7,7] \in \mathsf{SF}^\omega(3)$, the second Kameko iterate of s_8.

For the Steenrod kernel $\mathsf{K}^\omega(4)$, we can proceed as in degrees 8 and 19 to get 4 linearly independent elements of $\mathsf{K}^\omega(4)/\mathsf{J}^\omega(4)$. Iterating the dual Kameko map for $\mathsf{P}(3)$ gives the d-polynomial

$$f_{41} = (v_1v_2v_3)^{(3)} f_8^{(4)} = \begin{matrix} 11011 \\ 111 \\ 111 \\ 0 \end{matrix} + \begin{matrix} 11101 \\ 1101 \\ 111 \\ 0 \end{matrix} + \begin{matrix} 1111 \\ 11001 \\ 111 \\ 0 \end{matrix} + \begin{matrix} 1111 \\ 1111 \\ 1101 \\ 0 \end{matrix}.$$

Computer calculation confirms that f_{41} generates $\mathsf{K}^\omega(4)$, of dimension 225, in agreement with Sum's result $\dim \mathsf{K}^{41}(4) = 225$.

30.4 $\mathsf{Q}^\omega(4)$ and $\mathsf{K}^\omega(4)$ for decreasing ω

In this section, we tabulate results on the dimension and module structure of $\mathsf{Q}^\omega(4)$ and $\mathsf{K}^\omega(4)$ for decreasing ω. We include cases (indicated by asterisks) where we have proved only that the stated dimensions are lower bounds, but results of Kameko and Sum establish these as exact.

Let $\omega = (3,\dots,3,2,\dots,2,1,\dots,1)$ with s 1s, t 2s and u 3s. There is one table for each nonempty subset $I(\omega) \subset \{1,2,3\}$, where $I(\omega)$ is the set of terms of ω, and two for the 1-dominant case $I(\omega) = \{1,2,3\}$, one for $u = 1$ and the other for $u \geq 2$.

Module descriptions are given for $\mathsf{K}^\omega(4)$, and should be replaced by their transpose duals to give the corresponding information for $\mathsf{Q}^\omega(4)$. For example, the case where $s = 0$ and $t,u > 0$, so that $I(\omega) = \{2,3\}$, is referred to as the 'head and body case'.

We use the partial flag modules $\mathsf{FL}^I = \mathsf{FL}^I(4)$ and $\mathsf{FL}_I = \mathsf{FL}_I(4)$ where these occur, and list composition factors in other cases. For $\mathsf{J}^\omega(4)$, these composition factors describe quotients of FL^I. Composition factors are indicated by $+$ and direct sums by $\oplus$.

Notation such as $\mathsf{FL}_2\backslash L(0)$, means that the submodule $\mathsf{J}^\omega(4) \cong \mathsf{FL}_2$ and the quotient $\mathsf{K}^\omega(4)/\mathsf{J}^\omega(4) \cong L(0)$, while notation such as $34\backslash 1 = 35$ for the dimension means that $\dim \mathsf{Q}^\omega(4) = \dim \mathsf{K}^\omega(4) = 35$, but $\dim \mathsf{J}^\omega(4) = 34$.

In the tables for the 1-dominant case, we give only the composition factors of the module $\mathsf{K}^\omega(4)/\mathsf{J}^\omega(4)$, or equivalently of $\mathsf{SF}^\omega(4)$. As in Table (30.2) M_{20} represents any module of dimension 20 with composition factors given by $L(1)]$, $L(1,1)$, $L(1,1)$ and $L(1,1,1)$, and M_{15} represents any module of dimension 15, such as $\mathsf{FL}^1(4)$ or $\mathsf{FL}^3(4)$, with composition factors $L(0)$, $L(1)$, $L(1,1)$ and $L(1,1,1)$.

Tail case: $s > 0, t = u = 0, I(\omega) = \{1\}$

s	1	2	3	≥ 4
K^ω	$L(1)$	$\mathsf{FL}_1 \downarrow L(1,1)$	FL_1	FL^1
$\dim \mathsf{K}^\omega$	4	10	14	15

Body case: $t > 0, s = u = 0, I(\omega) = \{2\}$

t	1	2	3	≥ 4
K^ω	$L(1,1)$	$\mathsf{FL}_2 \downarrow L(2,1,1)$	FL_2 $\backslash L(0)$	FL^2
$\dim \mathsf{J}^\omega \backslash \dim \mathsf{K}^\omega$	6	20	$34 \backslash 1 = 35$	35

Head case: $u > 0, s = t = 0, I(\omega) = \{3\}$

u	1	2	3	≥ 4
K^ω	$L(1,1,1)$	$\mathsf{FL}_3 \downarrow L(1,1)$	FL_3	FL^3
$\dim \mathsf{K}^\omega$	4	10	14	15

Body and tail case: $s, t > 0, u = 0, I(\omega) = \{1,2\}$

$t \backslash s$	1	2	3
1	$L(2,1)$ 20	$\mathsf{FL}_{1,2} \downarrow L(2,1,1) \oplus L(1,1)$ $\backslash L(1,1,1)$ $41 \backslash 4 = 45$	$\mathsf{FL}_{1,2} \downarrow L(2,1,1) \oplus \mathsf{FL}_2 \downarrow L(1)$ 45*
2	$\mathsf{FL}_{1,2}$ 56*	$\mathsf{FL}_{1,2} \oplus \mathsf{FL}_2$ $\backslash L(0)$ $90 \backslash 1 = 91$*	$\mathsf{FL}_{1,2} \oplus \mathsf{FL}_2$ 90*
≥ 3	$\mathsf{FL}_{1,2} \oplus \mathsf{FL}_1$ 70*	$\mathsf{FL}^{1,2}$ 105*	$\mathsf{FL}^{1,2}$ 105*

Head and tail case: $s, u > 0, t = 0, I(\omega) = \{1,3\}$

$u \setminus s$	1	2	3	≥ 4
1	$\Delta(2,1,1)$ 15	$\mathsf{FL}_{1,3} \downarrow L(2,2,1)$ $\setminus L(0)$ $35\setminus 1 = 36$	$\mathsf{FL}_{1,3} \downarrow L(2,2,1)$ $\oplus \mathsf{FL}_3 \downarrow L(1,1)$ $\setminus L(0)$ $45\setminus 1 = 46^*$	$\mathsf{FL}_{1,3} \downarrow L(2,2,1)$ $\oplus \mathsf{FL}_3 \downarrow L(1,1)$ 45
2	$\mathsf{FL}_{1,3} \downarrow L(2,1)$ 35*	$\mathsf{FL}_{1,3}$ $\setminus L(1,1,1)$ $76\setminus 4 = 80^*$	$\mathsf{FL}_{1,3} \oplus \mathsf{FL}_3$ 90*	$\mathsf{FL}_{1,3} \oplus \mathsf{FL}_3$ 90
≥ 3	$\mathsf{FL}_{1,3} \downarrow L(2,1)$ $\oplus \mathsf{FL}_1 \downarrow L(1,1)$ 45*	$\mathsf{FL}_{1,3} \oplus \mathsf{FL}_1$ 90*	$\mathsf{FL}^{1,3}$ 105*	$\mathsf{FL}^{1,3}$ 105

Head and body case: $t, u > 0, s = 0, I(\omega) = \{2,3\}$

$u \setminus t$	1	2	3	≥ 4
1	$L(2,2,1)$ 20	$\mathsf{FL}_{2,3}$ $\setminus L(1)$ $56\setminus 4 = 60$	$\mathsf{FL}_{2,3} \oplus \mathsf{FL}_3$ $\setminus L(1,1,1)$ $70\setminus 4 = 74^*$	$\mathsf{FL}_{2,3} \oplus \mathsf{FL}_3$ 70
2	$\mathsf{FL}_{2,3} \downarrow L(2,1,1) \oplus L(1,1)$ 41*	$\mathsf{FL}_{2,3} \oplus \mathsf{FL}_2$ $\setminus L(0)$ $90\setminus 1 = 91^*$	$\mathsf{FL}^{2,3}$ 105*	$\mathsf{FL}^{2,3}$ 105
≥ 3	$\mathsf{FL}_{2,3} \downarrow L(2,1,1)$ $\oplus \mathsf{FL}_2 \downarrow L(1,1,1)$ 45*	$\mathsf{FL}_{2,3} \oplus \mathsf{FL}_2$ 90*	$\mathsf{FL}^{2,3}$ 105*	$\mathsf{FL}^{2,3}$ 105

1-dominant case: $s,t > 0$, $u = 1$, $I(\omega) = \{1,2,3\}$

$t \backslash s$	1	2	≥ 3
1	$\mathsf{FL}_{1,2,3} = \mathsf{St}(4)$ 64	$\mathsf{FL}_{1,2,3} \oplus \mathsf{FL}_{2,3}$ $\backslash M_{20}$ $120 \backslash 20 = 140$	$\mathsf{FL}_{1,2,3} \oplus \mathsf{FL}_{2,3}$ 120*
2	$\mathsf{FL}_{1,2,3} \oplus \mathsf{FL}_{1,3}$ $\backslash M_{15}$ $140 \backslash 15 = 155$*	$\mathsf{FL}_{1,2,3} \oplus \mathsf{FL}_{1,3}$ $\oplus \mathsf{FL}_{2,3} \oplus \mathsf{FL}_{3}$ $\backslash M_{15}$ $210 \backslash 15 = 225$*	$\mathsf{FL}_{1,2,3} \oplus \mathsf{FL}_{1,3}$ $\oplus \mathsf{FL}_{2,3} \oplus \mathsf{FL}_{3}$ $\backslash M_{15}$ $210 \backslash 15 = 225$*
≥ 3	$\mathsf{FL}_{1,2,3} \oplus \mathsf{FL}_{1,3}$ 140*	$\mathsf{FL}_{1,2,3} \oplus \mathsf{FL}_{1,3}$ $\oplus \mathsf{FL}_{2,3} \oplus \mathsf{FL}_{3}$ 210*	$\mathsf{FL}_{1,2,3} \oplus \mathsf{FL}_{1,3}$ $\oplus \mathsf{FL}_{2,3} \oplus \mathsf{FL}_{3}$ 210*

1-dominant case: $s,t > 0$, $u \geq 2$, $I(\omega) = \{1,2,3\}$

$t \backslash s$	1	2	≥ 3
1	$\mathsf{FL}_{1,2,3} \oplus \mathsf{FL}_{1,2}$ 120*	$\mathsf{FL}_{1,2,3} \oplus \mathsf{FL}_{1,2}$ $\oplus \mathsf{FL}_{2,3} \oplus \mathsf{FL}_{2}$ $\backslash M_{15}$ $210 \backslash 15 = 225$*	$\mathsf{FL}_{1,2,3} \oplus \mathsf{FL}_{1,2}$ $\oplus \mathsf{FL}_{2,3} \oplus \mathsf{FL}_{2}$ 210*
≥ 2	$\mathsf{FL}_{1,2,3} \oplus \mathsf{FL}_{1,2}$ $\oplus \mathsf{FL}_{1,3} \oplus \mathsf{FL}_{1}$ 210*	$\mathsf{FL}(4)$ 315*	$\mathsf{FL}(4)$ 315*

30.5 Remarks

This chapter includes results announced by Masaki Kameko in [108] and proved by Nguyen Sum in [204], [207] and [209]. The method of Kameko and Sum aims to produce the basis for $\mathsf{Q}^{\omega}(n)$ which is minimal in the left order. The upper bound is achieved by finding a sufficient set of blocks B which are left reducible using restricted hit equations and applying Proposition 6.3.11 to reduce concatenated blocks $A|B$ in the left order. The lower bound on $\dim \mathsf{Q}^{\omega}(n)$ which we obtain in Chapter 24 by considering $\mathsf{J}^{\omega}(n)$ is obtained by Kameko and Sum by considering linear projections $\mathsf{P}(n) \to \mathsf{P}(n-1)$.

The notion of strongly spike-free polynomials (Section 30.2) is due to Judith Silverman and William Singer [185], who ask whether $\mathsf{SF}^d(n)$ can contain any monomials: i.e. if a monomial is strongly spike-free, is it necessarily hit?

The algebraic transfer map $\phi_n : \mathrm{Tor}^{\mathsf{A}_2}_{n,n+d}(\mathbb{F}_2,\mathbb{F}_2) \to \mathsf{Q}^d(n)^{\mathsf{GL}(n)}$, with values in the $\mathsf{GL}(n)$-invariant elements of the cohit module $\mathsf{Q}(n)$ was defined by W. M. Singer [186]. The dual map $\phi_n^* : \mathbb{F}_2 \otimes_{\mathsf{GL}(n)} \mathsf{K}^d(n) \to \mathrm{Ext}^{n,n+d}_{\mathsf{A}_2}(\mathbb{F}_2,\mathbb{F}_2)$, is defined on the $\mathsf{GL}(n)$-coinvariants in $\mathsf{K}(n)$, and takes values in the E_2 term of the classical Adams spectral sequence for the 2-primary components of the stable homotopy groups of spheres. The results on ϕ_4^* quoted in the table of Section 30.3 can be found in the work of Hung, Ha, Nam and Quynh [71], [82], [94], [155]. The polynomial $s_{18} = [3,3,4,8]+[3,3,8,4]+[3,3,6,6]+[6,6,3,3]$ was shown by Nam [155, p.1828] to represent a $\mathsf{GL}(4)$-invariant element of $\mathsf{Q}^{18}(4)$. Sum [204, (7.4.8.7)] gives an equivalent polynomial with 9 terms. No generator of $\mathsf{K}^{(2,2,1,1)}(4)/\mathsf{J}^{(2,2,1,1)}(4)$ seems to be known. Sum's work [204] contains much information about generators of $\mathsf{SF}^d(4)$. In particular, bases of $\mathsf{SF}^d(4)$ in degrees 15 and 31 appear in Remark 5.18 and in degree 21 in Remark 6.2.9.

In the cases where ω is strictly decreasing, the upper bound on $\dim \mathsf{Q}^\omega(n)$ given by the number of semi-standard blocks, Proposition 7.3.5, is exact by Theorem 20.5.2. In the cases $n \leq 4$ this is also true when ω is obtained by repeating the last part of a strictly decreasing sequence, i.e. in the cases $\omega = (1,1)$, $(2,2)$, $(3,3)$, $(2,1,1)$, $(3,1,1)$, $(3,2,2)$ and $(3,2,1,1)$. In the last four cases, it is the lower bound that is difficult to prove, since $\mathsf{SF}^\omega(4) \neq 0$. It would be interesting to know whether this pattern continues for $n > 4$.

Table (30.1) shows that the product formula $\dim \mathsf{Q}^\omega(4) = 35 \cdot \dim \mathsf{Q}^\rho(2) \cdot \dim \mathsf{Q}^\tau(2)$ holds when $t \geq 4$, where ρ and τ are sequences $(1,\ldots,1)$ of lengths u and s. This suggests that Theorems 8.1.11 and 8.2.7 may generalize to give a multiplicative formula for $\dim \mathsf{Q}^\omega(n)$ when ω is a decreasing sequence with any constant subsequence $(k,\ldots,k)$ of length $\geq n$, where $1 \leq k \leq n-1$. The case $\omega = (3,1,1,1)$, where $\dim \mathsf{Q}^\omega(4) = 46$, is an exception to the analogue of Proposition 8.4.5 for blocks with tail of length $n-1$.

Bibliography

[1] J. F. Adams, On the structure and applications of the Steenrod algebra, Comment. Math. Helv. **32** (1958), 180–214.

[2] J. F. Adams, J. Gunawardena and H. Miller, The Segal conjecture for elementary abelian 2-groups, Topology **24** (1985), 435–460.

[3] J. F. Adams and H. R. Margolis, Sub-Hopf algebras of the Steenrod algebra, Math. Proc. Cambridge Philos. Soc. **76** (1974), 45–52.

[4] J. Adem, The iteration of Steenrod squares in algebraic topology, Proc. Nat. Acad. Sci. U.S.A. **38** (1952), 720–726.

[5] J. Adem, The relations on Steenrod powers of cohomology classes, in Algebraic Geometry and Topology, a symposium in honour of S. Lefschetz, 191–238, Princeton Univ. Press, Princeton, NJ, 1957.

[6] J. L. Alperin and Rowen B. Bell, Groups and Representations, Graduate Texts in Mathematics 162, Springer-Verlag, New York, 1995.

[7] M. A. Alghamdi, M. C. Crabb and J. R. Hubbuck, Representations of the homology of BV and the Steenrod algebra I, Adams Memorial Symposium on Algebraic Topology vol. 2, London Math. Soc. Lecture Note Ser. **176**, Cambridge Univ. Press 1992, 217–234.

[8] D. J. Anick and F. P. Peterson, A_2-annihilated elements in $H_*(\Omega\Sigma(\mathbb{R}P^2))$, Proc. Amer. Math. Soc. **117** (1993), 243–250.

[9] D. Arnon, Monomial bases in the Steenrod algebra, J. Pure App. Algebra **96** (1994), 215–223.

[10] D. Arnon, Generalized Dickson invariants, Israel J. Maths **118** (2000), 183–205.

[11] M. F. Atiyah and F. Hirzebruch, Cohomologie-Operationen und charakteristische Klassen, Math. Z. **77** (1961), 149–187.

[12] Shaun V. Ault, Relations among the kernels and images of Steenrod squares acting on right $\mathcal{A}$-modules, J. Pure. Appl. Algebra **216**, (2012), no. 6, 1428–1437.

[13] Shaun Ault, Bott periodicity in the hit problem, Math. Proc. Camb. Phil. Soc. **156** (2014), no. 3, 545–554.

[14] Shaun V. Ault and William Singer, On the homology of elementary Abelian groups as modules over the Steenrod algebra, J. Pure App. Algebra 215 (2011), 2847–2852.

[15] M. G. Barratt and H. Miller, On the anti-automorphism of the Steenrod algebra, Contemp. Math. **12** (1981), 47–52.

[16] David R. Bausum, An expression for $\chi(Sq^m)$, Preprint, Minnesota University (1975).

[17] D. J. Benson, Representations and cohomology II: Cohomology of groups and modules, Cambridge Studies in Advanced Mathematics **31**, Cambridge University Press (1991).

[18] D. J. Benson and V. Franjou, Séries de compositions de modules instables et injectivité de la cohomologie du groupe $\mathbb{Z}/2$, Math. Zeit **208** (1991), 389–399.

[19] P. C. P. Bhatt, An interesting way to partition a number, Information Processing Letters **71** (1999), 141–148.

[20] Anders Björner and Francesco Brenti, Combinatorics of Coxeter Groups, Graduate Texts in Mathematics 231, Springer-Verlag, 2005.

[21] J. M. Boardman, Modular representations on the homology of powers of real projective spaces, Algebraic Topology, Oaxtepec 1991, Contemp. Math. **146** (1993), 49–70.

[22] Kenneth S. Brown, Buildings, Springer-Verlag, New York, 1989.

[23] Robert R. Bruner, Lê M Hà, and Nguyen H. V. Hung, On the algebraic transfer, Trans. Amer. Math. Soc. **357** (2005), 473–487.

[24] S. R. Bullett and I. G. Macdonald, On the Adem relations, Topology **21** (1982), 329–332.

[25] H. E. A. Campbell and P. S. Selick, Polynomial algebras over the Steenrod algebra, Comment. Math. Helv. **65** (1990), 171–180.

[26] David P. Carlisle, The modular representation theory of $\mathsf{GL}(n,p)$ and applications to topology, Ph.D. dissertation, University of Manchester, 1985.

[27] D. Carlisle, P. Eccles, S. Hilditch, N. Ray, L. Schwartz, G. Walker and R. Wood, Modular representations of $\mathsf{GL}(n,p)$, splitting $\Sigma(\mathbb{C}P^\infty \times \ldots \times \mathbb{C}P^\infty)$, and the β-family as framed hypersurfaces, Math. Zeit. **189** (1985), 239–261.

[28] D. P. Carlisle and N. J. Kuhn, Subalgebras of the Steenrod algebra and the action of matrices on truncated polynomial algebras, Journal of Algebra **121** (1989), 370–387.

[29] D. P. Carlisle and N. J. Kuhn, Smash products of summands of $B(\mathbb{Z}/p)^n_+$, Contemp. Math. **96** (1989), 87–102.

[30] David P. Carlisle and Grant Walker, Poincaré series for the occurrence of certain modular representations of $\mathsf{GL}(n,p)$ in the symmetric algebra, Proc. Roy. Soc. Edinburgh **113A** (1989), 27–41.

[31] D. P. Carlisle and R. M. W. Wood, The boundedness conjecture for the action of the Steenrod algebra on polynomials, Adams Memorial Symposium on Algebraic Topology, Vol. 2, London Math. Soc. Lecture Note Ser. **176**, Cambridge University Press, (1992), 203–216.

[32] D. P. Carlisle, G. Walker and R. M. W. Wood, The intersection of the admissible basis and the Milnor basis of the Steenrod algebra, J. Pure App. Algebra **128** (1998), 1–10.

[33] Séminaire Henri Cartan, **2** Espaces fibrés et homotopie (1949–50), **7** Algèbre d'Eilenberg-MacLane et homotopie (1954–55), **11** Invariant de Hopf et opérations cohomologiques secondaires (1958–59), available online at http://www.numdam.org

[34] H. Cartan, Une théorie axiomatique des carrés de Steenrod, C. R. Acad. Sci. Paris **230** (1950), 425–427.

[35] H. Cartan, Sur l'itération des opérations de Steenrod, Comment. Math. Helv. **29** (1955), 40–58.

[36] R. W. Carter, Representation theory of the 0-Hecke algebra, J. of Algebra **104** (1986), 89–103.

[37] R. W. Carter and G. Lusztig, Modular representations of finite groups of Lie type, Proc. London Math. Soc. (3) **32** (1976), 347–384.

[38] Chen Shengmin and Shen Xinyao, On the action of Steenrod powers on polynomial algebras, Proceedings of the Barcelona Conference on Algebraic Topology, Lecture Notes in Mathematics **1509**, Springer-Verlag (1991), 326–330.

[39] D. E. Cohen, On the Adem relations, Math. Proc. Camb. Phil. Soc. **57** (1961), 265–267.

[40] M. C. Crabb, M. D. Crossley and J. R. Hubbuck, K-theory and the anti-automorphism of the Steenrod algebra, Proc. Amer. Math. Soc. **124** (1996), 2275–2281.

[41] M. C. Crabb and J. R. Hubbuck, Representations of the homology of BV and the Steenrod algebra II, Algebraic Topology: new trends in localization and periodicity (Sant Feliu de Guixols, 1994) 143–154, Progr. Math. **136**, Birkhaüser, Basel, 1996.

[42] M. D. Crossley and J. R. Hubbuck, Not the Adem relations, Bol. Soc. Mat. Mexicana (2) **37** (1992), No. 1–2, 99–107.

[43] M. D. Crossley, $\mathcal{A}(p)$-annihilated elements of $H_*(\mathbb{C}P^\infty \times \mathbb{C}P^\infty)$, Math. Proc. Cambridge Philos. Soc. **120** (1996), 441–453.

[44] M. D. Crossley, H^*V is of bounded type over $\mathcal{A}(p)$, Group Representations: Cohomology, group actions, and topology (Seattle 1996), Proc. Sympos. Pure Math. **63**, Amer. Math. Soc. (1998), 183–190.

[45] M. D. Crossley, $\mathcal{A}(p)$ generators for H^*V and Singer's homological transfer, Math. Zeit. **230** (1999), No. 3, 401–411.

[46] M. D. Crossley, Monomial bases for $H^*(\mathbb{C}P^\infty \times \mathbb{C}P^\infty)$ over $\mathcal{A}(p)$, Trans. Amer. Math. Soc. **351** (1999), No. 1, 171–192.

[47] M. D. Crossley and Sarah Whitehouse, On conjugation invariants in the dual Steenrod algebra, Proc. Amer. Math. Soc. **128** (2000), 2809–2818.

[48] Charles W. Curtis and Irving Reiner, Representation theory of finite groups and associative algebras, Wiley, New York, 1962.

[49] D. M. Davis, The antiautomorphism of the Steenrod algebra, Proc. Amer. Math. Soc. **44** (1974), 235–236.

[50] D. M. Davis, Some quotients of the Steenrod algebra, Proc. Amer. Math. Soc. **83** (1981), 616–618.

[51] J. Dieudonné, A history of algebraic and differential topology 1900–1960, Birkhäuser, Basel, 1989.

[52] A. Dold, Über die Steenrodschen Kohomologieoperationen, Annals of Math. **73** (1961), 258–294.

[53] Stephen Donkin, On tilting modules for algebraic groups, Math. Zeitschrift **212** (1993), 39–60.

[54] Stephen Doty, Submodules of symmetric powers of the natural module for GL_n, Invariant Theory (Denton, TX 1986) 185–191, Contemp. Math. **88**, Amer. Math. Soc., Providence, RI, 1989.

[55] Stephen Doty and Grant Walker, The composition factors of $\mathbb{F}_p[x_1,x_2,x_3]$ as a $\mathsf{GL}(3,\mathbb{F}_p,$-module, J. of Algebra **147** (1992), 411–441.

[56] Stephen Doty and Grant Walker, Modular symmetric functions and irreducible modular representations of general linear groups, J. Pure App. Algebra **82** (1992), 1–26.

[57] Stephen Doty and Grant Walker, Truncated symmetric powers and modular representations of GL_n, Math. Proc. Cambridge Philos. Soc. **119** (1996), 231–242.

[58] Jeanne Duflot, Lots of Hopf algebras, J. Algebra **204** (1998), No. 1, 69–94.

[59] V. Franjou and L. Schwartz, Reduced unstable A-modules and the modular representation theory of the symmetric groups, Ann. Scient. Ec. Norm. Sup. **23** (1990), 593–624.

[60] W. Fulton, Young Tableaux, London Math. Soc. Stud. Texts **35**, Cambridge Univ. Press, 1997.

[61] A. M. Gallant, Excess and conjugation in the Steenrod algebra, Proc. Amer. Math. Soc. **76** (1979), 161–166.

[62] L. Geissinger, Hopf algebras of symmetric functions and class functions, Springer Lecture Notes in Mathematics **579** (1977), 168–181.

[63] V. Giambalvo, Nguyen H. V. Hung and F. P. Peterson, $H^*(\mathbb{R}P^\infty \times \cdots \times \mathbb{R}P^\infty)$ as a module over the Steenrod algebra, Hilton Symposium 1993, Montreal, CRM Proc. Lecture Notes **6**, Amer. Math. Soc. Providence RI (1994), 133–140.

[64] V. Giambalvo and H. R. Miller, More on the anti-automorphism of the Steenrod algebra, Algebr. Geom. Topol. **11** (2011), No. 5, 2579–2585.

[65] V. Giambalvo and F. P. Peterson, On the height of Sq^{2^n}, Contemp. Math. **181** (1995), 183–186.

[66] V. Giambalvo and F. P. Peterson, The annihilator ideal of the action of the Steenrod algebra on $H^*(\mathbb{R}P^\infty)$, Topology Appl. **65** (1995), 105–122.

[67] V. Giambalvo and F. P. Peterson, $\mathcal{A}$-generators for ideals in the Dickson algebra, J. Pure Appl. Algebra **158** (2001), 161–182.

[68] D. J. Glover, A study of certain modular representations, J. Algebra **51** (1978), No. 2, 425–475.

[69] M. Y. Goh, P. Hitczenko and Ali Shokoufandeh, s-partitions, Information Processing Letters **82** (2002), 327–329.

[70] Brayton I. Gray, Homotopy Theory, Academic Press, New York, 1975.

[71] Lê Minh Hà, Sub-Hopf algebras of the Steenrod algebra and the Singer transfer, Proceedings of the school and conference on algebraic topology, Hanoi 2004, Geom. Topol. Publ. Coventry, **11** (2007), 81–105.

[72] Nguyen Dang Ho Hai, Generators for the mod 2 cohomology of the Steinberg summand of Thom spectra over $B(\mathbb{Z}/2)^n$, J. Algebra **381** (2013), 164–175.

[73] G. H. Hardy and E. M. Wright, An Introduction to the Theory of Numbers, Clarendon Press, Oxford, 1979.

[74] J. C. Harris and N. J. Kuhn, Stable decomposition of classifying spaces of finite abelian p-groups, Math. Proc. Cambridge Philos. Soc. **103** (1988), 427–449.

[75] J. C. Harris, T. J. Hunter and R. J. Shank, Steenrod algebra module maps from $H^*(B(\mathbf{Z}/p)^n$ to $H^*(B(\mathbf{Z}/p)^s$, Proc. Amer. Math. Soc. **112** (1991), 245–257.

[76] T. J. Hewett, Modular invariant theory of parabolic subgroups of $\mathsf{GL}_n(\mathbf{F}_q)$ and the associated Steenrod modules, Duke Math. J. **82** (1996), 91–102.

[77] Florent Hivert and Nicolas M. Thiéry, The Hecke group algebra of a Coxeter group and its representation theory, J. Algebra **321**, No. 8 (2009), 2230–2258.

[78] Florent Hivert and Nicolas M. Thiéry, Deformation of symmetric functions and the rational Steenrod algebra, Invariant Theory in all Characteristics, CRM Proc. Lecture Notes **35**, Amer. Math. Soc, Providence, RI, 2004, 91–125.

[79] J. E. Humphreys, Modular Representations of Finite Groups of Lie Type, London Math. Soc. Lecture Note Ser. **326**, Cambridge Univ. Press, 2005.

[80] Nguyen H. V. Hung, The action of Steenrod squares on the modular invariants of linear groups, Proc. Amer. Math. Soc. **113** (1991), 1097–1104.

[81] Nguyen H. V. Hung, The action of the mod p Steenrod operations on the modular invariants of linear groups, Vietnam J. Math. **23** (1995), 39–56.

[82] Nguyen H. V. Hung, Spherical classes and the algebraic transfer, Trans. Amer. Math. Soc. **349** (1997), 3893–3910: Erratum, ibid. **355** (2003), 3841–3842.

[83] Nguyen H. V. Hung, The weak conjecture on spherical classes, Math. Z. **231** (1999), 727–743.

[84] Nguyen H. V. Hung, Spherical classes and the lambda algebra, Trans. Amer. Math. Soc. **353** (2001), 4447–4460.

[85] Nguyen H. V. Hung, On triviality of Dickson invariants in the homology of the Steenrod algebra, Math. Proc. Camb. Phil. Soc. **134** (2003), 103–113.

[86] Nguyen H. V. Hung, The cohomology of the Steenrod algebra and representations of the general linear groups, Trans. Amer. Math. Soc. **357** (2005), 4065–4089.

[87] Nguyen H. V. Hung, On A_2-generators for the cohomology of the symmetric and the alternating groups, Math Proc. Cambridge Philos. Soc. **139** (2005), 457–467.

[88] Nguyen H. V. Hung and Tran Dinh Luong, The smallest subgroup whose invariants are hit by the Steenrod algebra, Math. Proc. Cambridge Philos. Soc. **142** (2007), 63–71.

[89] Nguyen H. V. Hung and Pham Anh Minh, The action of the mod p Steenrod operations on the modular invariants of linear groups, Vietnam J. Math. **23** (1995), 39–56.

[90] Nguyen H. V. Hung and Tran Ngoc Nam, The hit problem for modular invariants of linear groups, J. Algebra **246** (2001), 367–384.

[91] Nguyen H. V. Hung and Tran Ngoc Nam, The hit problem for the Dickson algebra, Trans. Amer. Math. Soc. **353** (2001), 5029–5040.

[92] Nguyen H. V. Hung and F. P. Peterson, A_2-generators for the Dickson algebra, Trans. Amer. Math. Soc. **347** (1995), 4687–4728.

[93] Nguyen H. V. Hung and F. P. Peterson, Spherical classes and the Dickson algebra, Math. Proc. Cambridge Philos. Soc. **124** (1998), 253–264.

[94] Nguyen H. V. Hung and Võ T. N. Quynh, The image of Singer's fourth transfer, C. R. Acad. Sci. Paris, Ser I **347** (2009), 1415–1418.

[95] B. Huppert and N. Blackburn, Finite Groups II, Chapter VII, Springer-Verlag, Berlin, Heidelberg, 1982.

[96] Masateru Inoue, A_2-generators of the cohomology of the Steinberg summand $\mathsf{M}(n)$, Contemp. Math. **293** (2002), 125–139.

[97] Masateru Inoue, Generators of the cohomology of $\mathsf{M}(n)$ as a module over the odd primary Steenrod algebra, J. Lond. Math. Soc. **75**, No. 2 (2007), 317–329.

[98] G. D. James and A. Kerber, The representation theory of the symmetric group, Encyclopaedia of Mathematics, vol. **16**, Addison-Wesley, Reading, Mass., 1981.

[99] A. S. Janfada, The hit problem for symmetric polynomials over the Steenrod algebra, Ph.D. thesis, University of Manchester, 2000.

[100] A. S. Janfada, A criterion for a monomial in $\mathsf{P}(3)$ to be hit, Math. Proc. Cambridge Philos. Soc. **145** (2008), 587–599.

[101] A. S. Janfada, A note on the unstability conditions of the Steenrod squares on the polynomial algebra, J. Korean Math. Soc **46** (2009), No. 5, 907–918.

[102] A. S. Janfada, On a conjecture on the symmetric hit problem, Rend. Circ. Mat. Palermo, **60**, 2011, 403–408.

[103] A. S. Janfada, Criteria for a symmetrized monomial in $B(3)$ to be non-hit, Commun. Korean Math. Soc. **29** (2014), No. 3, 463–478.

[104] A. S. Janfada and R. M. W. Wood, The hit problem for symmetric polynomials over the Steenrod algebra, Math. Proc. Cambridge Philos. Soc. **133** (2002), 295–303.

[105] A. S. Janfada and R. M. W. Wood, Generating $H^*(BO(3),\mathbb{F}_2)$ as a module over the Steenrod algebra, Math. Proc. Camb. Phil. Soc. **134** (2003), 239–258.

[106] M. Kameko, Products of projective spaces as Steenrod modules, Ph.D. thesis, Johns Hopkins Univ., 1990.

[107] M. Kameko, Generators of the cohomology of BV_3, J. Math. Kyoto Univ. **38** (1998), 587–593.

[108] M. Kameko, Generators of the cohomology of BV_4, preprint, Toyama Univ., 2003.

[109] M. Kaneda, M. Shimada, M. Tezuka and N. Yagita, Representations of the Steenrod algebra, J. of Algebra **155** (1993), 435–454.

[110] Ismet Karaca, On the action of Steenrod operations on polynomial algebras, Turkish J. Math. **22** (1998), No. 2, 163–170.

[111] Ismet Karaca, Nilpotence relations in the mod p Steenrod algebra, J. Pure App. Algebra **171** (2002), No. 2–3, 257–264.

[112] C. Kassel, Quantum Groups, Graduate Texts in Mathematics **155**, Springer-Verlag, 1995.

[113] N. Kechagias, The Steenrod algebra action on generators of subgroups of $\mathsf{GL}(n,\mathbb{Z}/p\mathbb{Z})$, Proc. Amer. Math. Soc. **118** (1993), 943–952.

[114] D. Kraines, On excess in the Milnor basis, Bull. London Math. Soc. **3** (1971), 363–365.

[115] L. Kristensen, On a Cartan formula for secondary cohomology operations, Math. Scand. **16** (1965), 97–115.

[116] Nicholas J. Kuhn, The modular Hecke algebra and Steinberg representation of finite Chevalley groups, J. Algebra **91** (1984), 125–141.

[117] N. J. Kuhn, Generic representations of the finite general linear groups and the Steenrod algebra: I, Amer. J. Math. **116** (1994), 327–360; II, K-Theory **8** (1994), 395–428; III, K-theory **9** (1995), 273–303.

[118] N. J. Kuhn and S. A. Mitchell, The multiplicity of the Steinberg representation of $GL_n\mathbb{F}_q$ in the symmetric algebra, Proc. Amer. Math. Soc. **96** (1986), 1–6.
[119] J. Lannes and L. Schwartz, Sur la structure des $\mathcal{A}$-modules instables injectifs, Topology **28** (1989), 153–169.
[120] J. Lannes and S. Zarati, Sur les $\mathcal{U}$-injectifs, Ann. Scient. Ec. Norm. Sup. **19** (1986), 593–603.
[121] M. Latapy, Partitions of an integer into powers, in Discrete Mathematics and Theoretical Computer Science Proceedings, Paris, 2001, 215–228.
[122] Cristian Lenart, The combinatorics of Steenrod operations on the cohomology of Grassmannians, Adv. Math. **136** (1998), 251–283.
[123] Li Zaiqing, Product formulas for Steenrod operations, Proc. Edinburgh Math. Soc. **38** (1995), 207–232.
[124] Arunas Liulevicius, The factorization of cyclic reduced powers by secondary cohomology operations, Mem. Amer. Math. Soc. No. 42 (1962).
[125] Arunas Liulevicius, On characteristic classes, Lectures at the Nordic Summer School in Mathematics, Aarhus University, 1968.
[126] L. Lomonaco, A basis of admissible monomials for the universal Steenrod algebra, Ricer. Mat. **40** (1991), 137–147.
[127] L. Lomonaco, The iterated total squaring operation, Proc. Amer. Math. Soc. **115** (1992), 1149–1155.
[128] I. G. Macdonald, Symmetric Functions and Hall Polynomials (second edition), Oxford mathematical monographs, Clarendon Press, Oxford, 1995.
[129] Harvey Margolis, Spectra and the Steenrod algebra, North Holland Math Library, vol. 29, Elsevier, Amsterdam (1983).
[130] J. P. May, A general algebraic approach to Steenrod operations, The Steenrod Algebra and its Applications, Lecture Notes in Mathematics **168**, Springer-Verlag (1970), 153–231.
[131] Dagmar M. Meyer, Stripping and conjugation in the Steenrod algebra and its dual, Homology, Homotopy and Applications **2** (2000), 1–16.
[132] Dagmar M. Meyer, Hit polynomials and excess in the mod p Steenrod algebra, Proc. Edinburgh Math. Soc. (2) **44** (2001), 323–350.
[133] Dagmar M. Meyer and Judith H. Silverman, Corrigendum to 'Hit polynomials and conjugation in the dual Steenrod algebra', Math. Proc. Cambridge Philos. Soc. **129** (2000), 277–289.
[134] John Milnor, The Steenrod algebra and its dual, Annals of Math. **67** (1958), 150–171.
[135] J. Milnor and J. C. Moore, On the structure of Hopf algebras, Annals of Math. **81** (1965), 211–264.
[136] J. W. Milnor and J. D. Stasheff, Characteristic Classes, Princeton University Press, 1974.
[137] Pham Anh Minh and Ton That Tri, The first occurrence for the irreducible modules of the general linear groups in the polynomial algebra, Proc. Amer. Math. Soc. **128** (2000), 401–405.
[138] Pham Anh Minh and Grant Walker, Linking first occurrence polynomials over $\mathbb{F}_p$ by Steenrod operations, Algebr. Geom. Topol. **2** (2002), 563–590.
[139] S. A. Mitchell, Finite complexes with $A(n)$-free cohomology, Topology **24** (1985), 227–248.

[140] S. A. Mitchell, Splitting $B(\mathbb{Z}/p)^n$ and BT^n via modular representation theory, Math. Zeit. **189** (1985), 285–298.
[141] S. A. Mitchell and S. B. Priddy, Stable splittings derived from the Steinberg module, Topology **22** (1983), 285–298.
[142] K. Mizuno and Y. Saito, Note on the relations on Steenrod squares, Proc. Jap. Acad. **35** (1959), 557–564.
[143] K. G. Monks, Nilpotence in the Steenrod algebra, Bol. Soc. Mat. Mex. **37** (1992), 401–416.
[144] K. G. Monks, Polynomial modules over the Steenrod algebra and conjugation in the Milnor basis, Proc. Amer. Math. Soc. **122** (1994), 625–634.
[145] K. G. Monks, The nilpotence height of P^s_t, Proc. Amer. Math. Soc. **124** (1996), 1296–1303.
[146] K. G. Monks, Change of basis, monomial relations, and the P^s_t bases for the Steenrod algebra, J. Pure App. Algebra **125** (1998), 235–260.
[147] R. E. Mosher and M. C. Tangora, Cohomology operations and applications in homotopy theory, Harper and Row, New York, 1968.
[148] M. F. Mothebe, Generators of the polynomial algebra $\mathbb{F}_2[x_1,\dots,x_n]$ as a module over the Steenrod algebra, Communications in Algebra **30** (2002), 2213–2228.
[149] M. F. Mothebe, Dimensions of subspaces of the polynomial algebra $\mathbb{F}_2[x_1,\dots,x_n]$ generated by spikes, Far East J. Math. Sci. **28** (2008), 417–430.
[150] M. F. Mothebe, Admissible monomials and generating sets for the polynomial algebra as a module over the Steenrod algebra, Afr. Diaspora J. Math. **16** (2013), 18–27.
[151] M. F. Mothebe, Dimension result for the polynomial algebra $\mathbb{F}_2[x_1,\dots,x_n]$ as a module over the Steenrod algebra, Int. J. Math. Math. Sci. (2013) Art. ID 150704, 6pp., MR3144989.
[152] Huynh Mui, Dickson invariants and Milnor basis of the Steenrod algebra, Topology, theory and application, Coll. Math. Soc. Janos Bolyai **41**, North Holland (1985), 345–355.
[153] Huynh Mui, Modular invariant theory and cohomology algebras of symmetric groups, J. Fac. Sci. Univ. Tokyo Sec. 1A **22** (1975), 319–369.
[154] Tran Ngoc Nam, A_2-générateurs génériques pour l'algèbre polynomiale, Adv. Math. **186** (2004), 334–362.
[155] Tran Ngoc Nam, Transfert algébrique et action du groupe linéaire sur les puissances divisées modulo 2, Ann. Inst. Fourier (Grenoble) **58** (2008), 1785–1837.
[156] P. N. Norton, 0-Hecke algebras, J. Austral. Math. Soc. (Ser. A) **27** (1979), 337–357.
[157] John H. Palmieri and James J. Zhang, Commutators in the Steenrod algebra, New York J. Math. **19** (2013), 23–37.
[158] S. Papastavridis, A formula for the obstruction to transversality, Topology **11** (1972), 415–416.
[159] David J. Pengelley, Franklin P. Peterson and Frank Williams, A global structure theorem for the mod 2 Dickson algebras, and unstable cyclic modules over the Steenrod and Kudo-Araki-May algebras, Math. Proc. Cambridge Philos. Soc. **129** (2000), 263–275.

[160] D. J. Pengelley and F. Williams, Sheared algebra maps and operation bialgebras for mod 2 homology and cohomology, Trans. Amer. Math. Soc. **352** (2000), No. 4, 1453–1492.

[161] D. J. Pengelley and F. Williams, Global Structure of the mod 2 symmetric algebra $H^*(BO, \mathbb{F}_2)$ over the Steenrod algebra, Algebr. Geom. Topol. **3** (2003), 1119–1138.

[162] D. J. Pengelley and F. Williams, The global structure of odd-primary Dickson algebras as algebras over the Steenrod algebra, Math. Proc. Cambridge Philos. Soc. **136** (2004), No. 1, 67–73.

[163] D. J. Pengelley and F. Williams, Beyond the hit problem: minimal presentations of odd-primary Steenrod modules, with application to $\mathbb{C}P^\infty$ and BU, Homology, Homotopy and Applications, **9**, No. 2 (2007), 363–395.

[164] D. J. Pengelley and F. Williams, A new action of the Kudo-Araki-May algebra on the dual of the symmetric algebras, with applications to the hit problem, Algebraic and Geometric Topology **11** (2011), 1767–1780.

[165] D. J. Pengelley and F. Williams, The hit problem for $H^*(BU(2); \mathbb{F}_p)$, Algebraic and Geometric Topology **13** (2013), 2061–2085.

[166] D. J. Pengelley and F. Williams, Sparseness for the symmetric hit problem at all primes, Math. Proc. Cambridge Philos. Soc. **158** (2015), No. 2, 269–274.

[167] F. P. Peterson, Some formulas in the Steenrod algebra, Proc. Amer. Math. Soc. **45** (1974), 291–294.

[168] F. P. Peterson, Generators of $\mathbf{H}^*(RP^\infty \wedge RP^\infty)$ as a module over the Steenrod algebra, Abstracts Amer. Math. Soc. (1987), 833-55-89.

[169] F. P. Peterson, $\mathcal{A}$-generators for certain polynomial algebras, Math. Proc. Camb. Phil. Soc. **105** (1989), 311–312.

[170] Dang Vo Phuc and Nguyen Sum, On the generators of the polynomial algebra as a module over the Steenrod algebra, C. R. Acad. Sci. Paris, Ser. 1 **353** (2015), 1035–1040.

[171] Dang Vo Phuc and Nguyen Sum, On a minimal set of generators for the polynomial algebra of five variables as a module over the Steenrod algebra, Acta Math. Vietnam. **42** (2017), 149–162.

[172] Geoffrey M. L. Powell, Embedding the flag representation in divided powers, J. of Homotopy and Related Structures **4**(1) (2009), 317–330.

[173] J. Repka and P. Selick, On the subalgebra of $H_*((\mathbb{R}P^\infty)^n; \mathbb{F}_2)$ annihilated by Steenrod operations, J. Pure Appl. Algebra **127** (1998), 273–288.

[174] J. Riordan, Combinatorial Identities, John Wiley & Sons, New York, 1968.

[175] B. E. Sagan, The Symmetric Group, Graduate Texts in Mathematics **203**, Springer (2001).

[176] Robert Sandling, The lattice of column 2-regular partitions in the Steenrod algebra, MIMS EPrint 2011.101, University of Manchester 2011, http://www.manchester.ac.uk/mims/eprints

[177] L. Schwartz, Unstable modules over the Steenrod algebra and Sullivan's fixed point set conjecture, Chicago Lectures in Mathematics, University of Chicago Press, 1994.

[178] J. Segal, Notes on invariant rings of divided powers, CRM Proceedings and Lecture Notes **35**, Invariant Theory in All Characteristics, ed. H. E. A. Campbell and D. L. Wehlau, Amer. Math. Soc. 2004, 229–239.

[179] J.-P. Serre, Cohomologie modulo 2 des complexes d'Eilenberg-MacLane, Comment. Math. Helv. **27** (1953), 198–232.

[180] Judith H. Silverman, Conjugation and excess in the Steenrod algebra, Proc. Amer. Math. Soc. **119** (1993), 657–661.

[181] Judith H. Silverman, Multiplication and combinatorics in the Steenrod algebra, J. Pure Appl. Algebra **111** (1996), 303–323.

[182] Judith H. Silverman, Hit polynomials and the canonical antiautomorphism of the Steenrod algebra, Proc. Amer. Math. Soc. **123** (1995), 627–637.

[183] Judith H. Silverman, Stripping and conjugation in the Steenrod algebra, J. Pure Appl. Algebra **121** (1997), 95–106.

[184] Judith H. Silverman, Hit polynomials and conjugation in the dual Steenrod algebra, Math. Proc. Cambridge Philos. Soc. **123** (1998), 531–547.

[185] Judith H. Silverman and William M. Singer, On the action of Steenrod squares on polynomial algebras II, J. Pure App. Algebra **98** (1995), 95–103.

[186] William M. Singer, The transfer in homological algebra, Math. Z. **202** (1989), 493–523.

[187] William M. Singer, On the action of Steenrod squares on polynomial algebras, Proc. Amer. Math. Soc. **111** (1991), 577–583.

[188] William M. Singer, Rings of symmetric functions as modules over the Steenrod algebra, Algebr. Geom. Topol. **8** (2008), 541–562.

[189] Larry Smith and R. M. Switzer, Realizability and nonrealizability of Dickson algebras as cohomology rings, Proc. Amer. Math. Soc. **89** (1983), 303–313.

[190] Larry Smith, Polynomial Invariants of Finite Groups, A. K. Peters, Wellesley, Mass., 1995.

[191] Larry Smith, An algebraic introduction to the Steenrod algebra, in: Proceedings of the School and Conference in Algebraic Topology, Hanoi, 2004, Geometry and Topology Monographs **11** (2007), 327–348.

[192] R. P. Stanley, Enumerative Combinatorics, vol. 2, Cambridge Studies in Advanced Mathematics **62**, Cambridge University Press (1999).

[193] N. E. Steenrod, Products of cocycles and extensions of mappings, Ann. of Math. **48** (1947), 290–320.

[194] N. E. Steenrod, Reduced powers of cohomology classes, Ann. of Math. **56** (1952), 47–67.

[195] N. E. Steenrod, Homology groups of symmetric groups and reduced power operations, Proc. Nat. Acad. Sci. U.S.A. **39** (1953), 213–217.

[196] N. E. Steenrod and D. B. A. Epstein, Cohomology Operations, Annals of Math. Studies 50, Princeton University Press (1962).

[197] R. Steinberg, Prime power representations of finite general linear groups II, Can. J. Math. **9** (1957), 347–351.

[198] R. Steinberg, Representations of algebraic groups, Nagoya Math. J. **22** (1963), 33–56.

[199] R. Steinberg, On Dickson's theorem on invariants, J. Fac. Sci. Univ. Tokyo, Sect. 1A Math. **34** (1987), No. 3, 699–707.

[200] P. D. Straffin, Identities for conjugation in the Steenrod algebra, Proc. Amer. Math. Soc. **49** (1975), 253–255.

[201] Nguyen Sum, On the action of the Steenrod-Milnor operations on the modular invariants of linear groups, Japan J. Math. **18** (1992), 115–137.

[202] Nguyen Sum, On the action of the Steenrod algebra on the modular invariants of special linear group, Acta Math. Vietnam **18** (1993), 203–213.

[203] Nguyen Sum, Steenrod operations on the modular invariants, Kodai Math. J. **17** (1994), 585–595.

[204] Nguyen Sum, The hit problem for the polynomial algebra of four variables, Quy Nhon University, Vietnam, Preprint 2007, 240pp. Available online at http://arxiv.org/abs/1412.1709.

[205] Nguyen Sum, The negative answer to Kameko's conjecture on the hit problem, C. R. Acad. Sci. Paris, Ser I **348** (2010), 669–672.

[206] Nguyen Sum, The negative answer to Kameko's conjecture on the hit problem, Adv. Math. **225** (2010), 2365–2390.

[207] Nguyen Sum, On the hit problem for the polynomial algebra, C. R. Acad. Sci. Paris, Ser I **351** (2013), 565–568.

[208] Nguyen Sum, On the Peterson hit problem of five variables and its application to the fifth Singer transfer, East-West J. Math. **16** (2014), 47–62.

[209] Nguyen Sum, On the Peterson hit problem, Adv. Math. **274** (2015), 432–489.

[210] René Thom, Une théorie intrinsèque des puissances de Steenrod, Colloque de Topologie de Strasbourg, Publication of the Math. Inst. University of Strasbourg (1951).

[211] René Thom, Espaces fibrés en sphères et carrés de Steenrod, Ann. Sci. Ec. Norm. Sup. **69** (1952), 109–182.

[212] René Thom, Quelque propriétés globales des variétés différentiables, Comment. Math. Helv. **28** (1954), 17–86.

[213] Ton That Tri, The irreducible modular representations of parabolic subgroups of general linear groups, Communications in Algebra **26** (1998), 41–47.

[214] Ton That Tri, On a conjecture of Grant Walker for the first occurrence of irreducible modular representations of general linear groups, Comm. Algebra **27** (1999), No. 11, 5435–5438.

[215] Neset Deniz Turgay, An alternative approach to the Adem relations in the mod p Steenrod algebra, Turkish J. Math. **38** (2014), No. 5, 924–934.

[216] G. Walker and R. M. W. Wood, The nilpotence height of Sq^{2^n}, Proc. Amer. Math. Soc. **124** (1996), 1291–1295.

[217] G. Walker and R. M. W. Wood, The nilpotence height of P^{p^n}, Math. Proc. Cambridge Philos. Soc. **123** (1998), 85–93.

[218] G. Walker and R. M. W. Wood, Linking first occurrence polynomials over $\mathbb{F}_2$ by Steenrod operations, J. Algebra **246** (2001), 739–760.

[219] G. Walker and R. M. W. Wood, Young tableaux and the Steenrod algebra, Proceedings of the School and Conference in Algebraic Topology, Hanoi 2004, Geometry and Topology Monographs **11** (2007), 379–397.

[220] G. Walker and R. M. W. Wood, Weyl modules and the mod 2 Steenrod Algebra, J. Algebra **311** (2007), 840–858.

[221] G. Walker and R. M. W. Wood, Flag modules and the hit problem for the Steenrod algebra, Math. Proc. Cambridge Philos. Soc. **147** (2009), 143–171.

[222] C. T. C. Wall, Generators and relations for the Steenrod algebra, Annals of Math. **72** (1960), 429–444.

[223] William C. Waterhouse, Two generators for the general linear groups over finite fields, Linear and Multilinear Algebra **24**, No. 4 (1989), 227–230.

[224] Helen Weaver, Ph.D. thesis, University of Manchester, 2006.

[225] C. Wilkerson, A primer on the Dickson invariants, Proc. of the Northwestern Homotopy Theory Conference, Contemp. Math. **19** (1983), 421–434.

[226] W. J. Wong, Irreducible modular representations of finite Chevalley groups, J. Algebra **20** (1972), 355–367.

[227] R. M. W. Wood, Modular representations of $\mathsf{GL}(n,\mathbf{F}_p)$ and homotopy theory, Algebraic Topology, Göttingen, 1984, Lecture Notes in Mathematics **1172**, Springer-Verlag (1985), 188–203.

[228] R. M. W. Wood, Splitting $\Sigma(\mathbb{C}P^\infty \times \ldots \times \mathbb{C}P^\infty)$ and the action of Steenrod squares on the polynomial ring $\mathbf{F}_2[x_1,\ldots,x_n]$, Algebraic Topology Barcelona 1986, Lecture Notes in Mathematics **1298**, Springer-Verlag (1987), 237–255.

[229] R. M. W. Wood, Steenrod squares of Polynomials, Advances in homotopy theory, London Mathematical Society Lecture Notes 139, Cambridge University Press (1989), 173–177.

[230] R. M. W. Wood, Steenrod squares of polynomials and the Peterson conjecture, Math. Proc. Cambridge Philos. Soc. **105** (1989), 307–309.

[231] R. M. W. Wood, A note on bases and relations in the Steenrod algebra, Bull. London Math. Soc. **27** (1995), 380–386.

[232] R. M. W. Wood, Differential operators and the Steenrod algebra, Proc. London Math. Soc. **75** (1997), 194–220.

[233] R. M. W. Wood, Problems in the Steenrod algebra, Bull. London Math. Soc. **30** (1998), 194–220.

[234] R. M. W. Wood, Hit problems and the Steenrod algebra, Proceedings of the Summer School 'Interactions between Algebraic Topology and Invariant Theory', Ioannina University, Greece (2000), 65–103.

[235] R. M. W. Wood, Invariants of linear groups as modules over the Steenrod algebra, Ingo2003, Invariant Theory and its interactions with related fields, University of Göttingen (2003).

[236] R. M. W. Wood, The Peterson conjecture for algebras of invariants, Invariant Theory in all characteristics, CRM Proceedings and Lecture Notes **35**, Amer. Math. Soc., Providence R.I. (2004), 275–280.

[237] Wu Wen Tsün, Les *i*-carrés dans une variété grassmanniènne, C. R. Acad. Sci. Paris **230** (1950), 918–920.

[238] Wu Wen Tsün, Sur les puissances de Steenrod, Colloque de Topologie de Strasbourg, Publication of the Math. Inst. University of Strasbourg (1952).

[239] Hadi Zare, On the Bott periodicity, $\mathcal{A}$-annihilated classes in $H_*(QX)$, and the stable symmetric hit problem, submitted to Math. Proc. Cambridge Philos. Soc. 2015.

Index of Notation for Volume 2

Index for Volume 2

Index of Notation for Volume 1

Index for Volume 1